CONTEMPORARY SOVIET MATHEMATICS

Series Editor: Revaz Gamkrelidze, *Steklov Institute, Moscow, USSR*

COHOMOLOGY OF INFINITE-DIMENSIONAL LIE ALGEBRAS
D. B. Fuks

LINEAR DIFFERENTIAL EQUATIONS OF PRINCIPAL TYPE
Yu. V. Egorov

THEORY OF SOLITONS: The Inverse Scattering Method
S. Novikov, S. V. Manakov, L. P. Pitaevskii, and V. E. Zakharov

TOPICS IN MODERN MATHEMATICS: Petrovskii Seminar No. 5
Edited by O. A. Oleinik

Linear Differential Equations of Principal Type

Linear Differential Equations of Principal Type

Yu. V. Egorov
Moscow State University
Moscow, USSR

Translated from Russian by
Dang Prem Kumar

CONSULTANTS BUREAU • NEW YORK AND LONDON

Library of Congress Cataloging in Publication Data

Egorov, IU. V. (IUrii Vladimirovich)
Linear differential equations of principal type.

(Contemporary Soviet mathematics)
Translation of: Lineinye differentsial'nye uravneniia glavnogo tipa.
Bibliography: p.
1. Differential equations, Linear. I. Title. II. Series.
QA372.E3513 1986 515.3'54 86-25408
ISBN 0-306-10992-1

This translation is published under an agreement with VAAP, the Copyright Agency of the USSR

A Division of Plenum Publishing Corporation
233 Spring Street, New York, N.Y. 10013

Printed in the United States of America

CONTENTS

NOTATION

$\mathbb{R}^n$ is an n-dimensional Euclidean space of x points with coordinates $(x^1, \ldots, x^n)$. Here $|x|=[(x^1)^2+\ldots+(x^n)^2]^{1/2}$ and $x^\alpha=(x^1)^{\alpha_1}\ldots(x^n)^{\alpha_n}$ if $\alpha=(\alpha_1, \ldots, \alpha_n)$, $\alpha_j \in \mathbb{Z}_+$; $\mathbb{Z}_+$ is the set of non-negative integers. $dx=dx^1\ldots dx^n$ is the Lebesgue measure on $\mathbb{R}^n$.

$\mathbb{R}_n$ is the dual of $\mathbb{R}^n$; it consists of points ξ with coordinates $(\xi_1, \ldots, \xi_n)$. Here $|\xi|=[\xi_1^2+\ldots+\xi_n^2]^{1/2}$ and $\xi^\alpha=\xi_1^{\alpha_1}\ldots\xi_n^{\alpha_n}$ if $\alpha_j \in \mathbb{Z}_+$; $d\xi=d\xi_1\ldots d\xi_n$ is the Lebesgue measure on $\mathbb{R}_n$.

If $\alpha=(\alpha_1, \ldots, \alpha_n)\in\mathbb{Z}_+^n$, then $|\alpha|=\alpha_1+\ldots+\alpha_n$; $\alpha!=\alpha_1!\ldots\alpha_n!$; $D^\alpha=D_1^{\alpha_1}\ldots D_n^{\alpha_n}$, where $D=(D_1, \ldots, D_n)$. $\partial_j u=\partial_{x^j}u=\frac{\partial u}{\partial x^j}=iD_j u$. $f_{(\beta)}^{(\alpha)}(x, \xi)=\partial_1^{\beta_1}\ldots\partial_n^{\beta_n}(\partial_{\xi_1})^{\alpha_1}\ldots(\partial_{\xi_n})^{\alpha_n}f(x, \xi)$. $i^2=-1$, $\arg i=\pi/2$; Re z, Im z are the real and imaginary parts of the complex number z.

L(E, F) is the space of bounded linear operators from E into F, where E and F are linear topological spaces.

$C^m(\Omega)$, where $m\in\mathbb{Z}_+$ and Ω is a domain in $\mathbb{R}^n$, is the space of functions f continuous in Ω and having continuous derivatives $D^\alpha f$ for $|\alpha|\leqslant m$.

$C_0^m(\Omega)$ is the subspace of functions from $C^m(\Omega)$ equal to zero in a neighborhood of the boundary $b\Omega$ and in a neighborhood of infinity.

$L_p(\Omega)$ is the space of measurable functions in Ω for which the integral $\int_\Omega |f(x)|^p\,dx$ converges; $\|f\|_{L_p(\Omega)}=\left(\int_\Omega |f(x)|^p\,dx\right)^{1/p}$.

$L_p^{\mathrm{loc}}(\Omega)$ is the space of measurable functions in Ω having a finite integral $\int_K |f(x)|^p\,dx$ for an arbitrary compact set $K\subset\Omega$.

INTRODUCTION

1. The theory of linear partial differential equations has seen a remarkable development over the last thirty years. Until then, with the exception of some remarkable results obtained by S. Kovalevsky, D. Hilbert, J. Hörmander, I. G. Petrowsky, and S. L. Sobolev, one studied mainly second-order elliptic, parabolic, and hyperbolic equations in a two- or three-dimensional space.

Today, the theory, which has not lost contact with mathematical physics and its significance for mathematic modelling of real processes, can be used to study equations of any order in a space of any dimension. Unlike the classical theory which involves a large number of theorems and methods for studying different equations of mathematical physics, the present-day theory is typified by the creation and use of general methods for investigating wide classes of equations and the expectation of a complete description of the various classes of differential equations having some given property.

2. The general theory of linear partial differential equations is believed to have developed in three stages. The first stage may be related to the period of development of distribution theory and, closely related to it, the theory of differential equations with constant coefficients. Distribution theory, whose foundations were laid in S. L. Sobolev's works, developed intensively after the appearance of L. Schwartz's book[1] in 1951. Within the framework of this theory, the solutions of differential equations with constant coefficients can be reduced to the division of a Fourier distribution by a polynomial, that is, to a problem of the classical operational calculus. Some general problems such as solvability, the structure of solutions for a homogeneous equation, correctness classes of the Cauchy problem, etc., were thoroughly studied by B. Malgrange[1], I. M. Gelfand and G. E. Shilov,[1-3] F. Tréves,[1] L. Hörmander,[1] V. P. Palamodov,[1] and L. Ehrenpreis.[1]

The second stage is typified by the development of the theory of elliptic boundary problems and of the closely related theory of pseudo-differential operators. Elliptic differential operators are very similar to differential operators with constant coefficients and can be studied by the methods of perturbation theory. Until 1950 the theory of elliptic boundary problems was developed mainly for second-order equations. A review of the results concerning higher-order equations is available in C. Miranda.[1] Investigations of many-dimensional general elliptic problems were undertaken by Ya. B. Lopatinski[1] and Z. Ya. Shapiro,[1] who succeeded in finding operator conditions algebraic in nature and equivalent to the coercivity conditions of these problems in the general case. The extensive investigations of many scientists, which resulted in the growth of elliptic theory, are reviewed in detail in M. S. Agronovich (1965).[1] Since then, many important and interesting results have been obtained.

Investigations of the regularity of solutions of elliptic equations and the study of the index theorem, in particular, generated a need for research in singular integral operators and in the composition of such operators with differential equations. These are treated in the works of G. Giraud, S. G. Mikhlin, A. P. Calderón, and A. V. Zygmund. The application of singular integral operators made it possible, in particular, to simplify the Petrowsky theorem concerning Cauchy's problem for hyperbolic systems of equations (see S. Mizohata[1]), and also to obtain a very general theorem on the uniqueness of the solution to the Cauchy problem for a wide class of differential equations (see A. P. Calderón[1]). The singular integro-differential operators, that is, the composition of singular integral and differential operators, are discussed in M. S. Agranovich[1] and A. S. Dynin.[1] To investigate boundary problems for elliptic equations, M. I. Vishik and G. I. Eskin considered "convolution operators," that is, integral operators where the kernel is a distribution. A. Unterberger and J. Bokobza[1] have developed a theory for Calderón-Zygmund operators. All these theories are different variants of the theory of pseudo-differential operators, and many facts from the general calculus of such operators are listed in the papers mentioned above.

The work of J. J. Kohn and L. Nirenberg[1] on the algebra of such operators, the description and substantiation of their calculus, and the possibilities of applying such operators to the study of elliptic equations, has played an important part in the development of the theory of pseudo-differential operators. The name of this theory was suggested by K. Friedrichs and P. Lax,[1] although operators of this type had already appeared in H. Weyl's book in 1927. Nevertheless, the new calculus was distinctly presented and clearly described by J. J. Kohn and L. Nirenberg.[1] Thus, the considerable interest and intense development of the theory of pseudo-differential operators owe a great deal to this exposition.

Pseudo-differential operators can be conveniently determined using the Fourier transform $\tilde{u}(\xi) = \int u(x)e^{-ix\xi}dx$ so that

$$Pu(x) = (2\pi)^{-n}\int p(x,\xi)\tilde{u}(\xi)e^{ix\xi}\,d\xi\,,$$

where p is the symbol of the operator P belonging to the $S^m(\Omega)$ class, that is, a smooth function in $T^*\Omega$ satisfying inequalities of the form

$$|D_\xi^\alpha D_x^\beta p(x,\xi)| \leq C_{\alpha,\beta}(1+|\xi|)^{m-|\alpha|}$$

for any indices α and β. It is not hard to see that P is a differential operator provided p is a polynomial in ξ.

The theory of pseudo-differential operators makes it possible to form a minimal algebra of operators containing all of the differential operators and the operators inverse to elliptic differential ones.

Time has proved the vitality of the theory of pseudo-differential operators and their broad applications to the general theory of differential equations. A class of pseudo-differential equations may be defined, for instance, as a class of operators naturally arising from the class of differential operators under the action of an operator giving a solution to the Cauchy problem for hyperbolic equations. Most problems of the general theory can be conveniently studied by considering simultaneously the differential and pseudo-differential operators.

3. Finally, the last decade has witnessed the development and wide application of the theory of Fourier integral operators. In a local system of coordinates, such an operator has a form similar to the pseudo-differential operator

$$\Phi u(x) = (2\pi)^{-n}\int p(x,\xi)\tilde{u}(\xi)3^{iS(x,\xi)}d\xi,$$

where the symbol p has the same properties as in a pseudo-differential operator, and the phase function S is characterized by the non-singularity: $\det|S_{x\xi}| \geqslant C_0 > 0$. Earlier, a canonical operator theory was proposed by V. P. Maslov[1] which made it possible to solve several asymptotic problems, and in particular, to provide a solution to the Cauchy problem for hyperbolic equations in the large. The existence of such a solution was proved earlier in the works of S. L. Sobolov and I. G. Petrowsky. Although the Maslov canonical operators differ in appearance from the operator φ, their association with Fourier integral operators is beyond doubt (see V. E. Nazaikinsky, V. G. Oshmyan, B. Yu. Sternin, and V. E. Shatalov[1]). Very profound studies of Fourier integral operators were made by L. Hörmander.[12,13]

In Hörmander[12] a new important concept, that of the wave front set of a distribution, was introduced which permitted its singularities to be characterized not only in terms of the location in the base space but also in terms of its direction. As an illustration, we may indicate that a wave front for a plane-wave type distribution is always directed along the normal to the wave front. An analogous idea was earlier introduced by M. Sato for microdistributions.[1-2]

The concept of wave front set enables one to describe more accurately the singularities of distributions as a certain subset in the cotangent bundle. If the distribution u satisfies the differential or pseudo-differential equation Pu = f with a real-valued characteristic form and with a smooth right-hand side, the wave front of u, as proved by L. Hörmander, will be invariant with respect to the Hamiltonian flow of the principal symbol. This is one of the most important results of the modern theory of differential equations. Systematic application of Fourier integral operators and of the concept of wave front set made it possible to significantly develop the theory of boundary problems for hyperbolic equations, to obtain important results in the spectral theory of elliptic boundary problems, and to develop a theory of equations of principal type with a complex-valued characteristic form.

4. We note explicitly that due to space limitations, this book does not include such important topics as the theory of Maslov canonical operators, Hörmander's theory of canonical relations, the theory of boundary problems for hyperbolic equations, particularly the results of R. Melrose, G. Eskin, and M. Taylor, the recent results in spectral theory obtained by L. Hörmander, V. Ivrii, and R. Melrose, the theory of integral operators with a complex-valued phase function developed in the works of A. Melin and J. Sjostrand, V. P. Maslov, V. V. Kucherenko, A. S. Mishenko, B. Yu. Sternin, and V. E. Shatalov, the theory of operators with multiple characteristics, in particular the results of V. V. Grishin, P. R. Popivanov, A. Melin, L. Hörmander, O. A. Olenik, and E. V. Radkevich, and the theory of the propagation of analytic singularities developed in the articles of M. Sato, P. Shapiro, M. Kashivara, T. Kawai, and others. Note also that parabolic operators are not of principal type and therefore are not discussed in this book.

5. The class of differential and pseudo-differential operators of principal type, to which the book is dedicated, is next in simplicity to the classes of operators with constant coefficients and elliptic operators. This includes, in particular, elliptic and strictly hyperbolic operators, but is not confined to them. The basic characteristic of this class is that the properties of such operators are independent of the lower-order terms and determined only by the principal symbol.

The initial definition of an operator of principal type, given by L. Hörmander in 1955, is stated thus: An operator of principal type is an operator of strength equal to an operator having the same characteristic form. The operators P and Q are equally strong if there exists a constant $C > 0$ such that

$$c^{-1}\|Pu\|_{L_2} \leq \|Qu\|_{L_2} \leq C\|Pu\|_{L_2}, \qquad u \in C_0^\infty(\Omega).$$

For operators with constant coefficients, this definition is equivalent to the condition $dp^0(\xi) \neq 0$ when $\xi \neq 0$.

An analogous condition, that is, $d_\xi p^0(x, \xi) \neq 0$ when $\xi \neq 0$, was first used for operators with variable coefficients. The modern definition is broader and requires the trajectory of the Hamiltonian system corresponding to the principal symbol to fall outside the fiber of the cotangent bundle. Analytically it means that the forms $dp^0(x, \xi)$ and ξdx must not be proportional when $\xi \neq 0$. By virtue of Euler's identity, this condition is essential only for those points where $p^0(x, \xi) = 0$. Therefore, operators of principal type are operators having only real characteristics.

6. Chapter I deals with the foundations of distribution theory. A reader familiar with this theory could skip this chapter, which contains a summary of basic facts used in the further discussion, for instance the properties of Fourier transform, a description of the Sobolev spaces $H_s(\Omega) = W_2^s(\Omega)$, and definitions of distributions on an arbitrary smooth manifold. In the section devoted to differential operators with constant coefficients we discuss the construction, due to L. Hörmander, of the fundamental solution for an arbitrary operator, and consider the properties of elliptic operators and operators of principal type.

7. Chapter II gives the theory of pseudo-differential operators. The calculus of such operators is constructed on the basis of Lemma 2.1 which shows that the function

$$a(x, \xi) = (2\pi)^{-n} \iint A(x, y, \xi, \eta) e^{i(x-y, \eta-\xi)}\, dy\, d\eta$$

is of class $S^{m_1+m_2}$ if function A satisfies the condition

$$|D_\xi^\alpha D_\eta^{\alpha'} D_x^\beta D_y^{\beta'} A(x, y, \xi, \eta)| \leq C_{\alpha,\alpha',\beta,\beta'}(1 + |\xi|)^{m_1-|\alpha|}(1 + |\eta|)^{m_2-|\alpha|}.$$

The symbol a in this case can be expressed in an asymptotic series of the form

$$a(x, \xi) \sim \sum \frac{i^{|\alpha|}}{\alpha!} D_y^\alpha D_\eta^\alpha A(x, y, \xi, \eta)|_{y=x, \eta=\xi}.$$

Using this lemma one can immediately obtain formulas for the symbol of the composition of two operators, for the symbol of the adjoint operator, a theorem on the invariance of the principal symbol with respect to a change of variables, and other results.

A section is devoted to proving the invariance of the principal symbol with respect to homogeneous canonical transformations, and also to the invariance of microlocal inequalities with respect to general canonical transformations. In the past few years several generalized theories of pseudo-differential operators have appeared, chiefly in the works of R. Beals and L. Hörmander. Since it is not possible to describe such generalizations in detail, we only mention some of them in Section 6. One of them, required for further discussion, is treated in detail in Section 7 which also contains the proof of the Calderón-Vaillancourt theorem on the boundedness of operators with symbols of class

$$S_\rho^m(\Omega) = \{a \in C^\infty(T^*\Omega), |D_\xi^\alpha D_x^\beta a(x, \xi)| \leq C_{\alpha,\beta}(1 + |\xi|)^{m+\rho(|\beta|-|\alpha|)}\}$$

when $\rho < 1$.

8. In Chapter III we briefly describe the theory of boundary-value problems for a differential elliptic operator of order m. This chapter explains the construction of a parametrix for a problem satisfying the Lopatinsky condition. Another approach to such problems, which consists in reducing the problem to a pseudo-differential equation on a boundary manifold and to the investigation of this equation, is discussed in the last section of this chapter. This approach is developed in Y. B. Lopatinsky,[1] R. Seeley,[2-3] L. Hörmander,[6] and others.

9. Chapter IV is devoted to a particular boundary-value problem for a second-order elliptic equation, specifically to the Poincaré oblique derivative problem. Such a problem for a space having more than two dimensions is discussed in A. V. Bitsadze,[1] R. Borrelli,[1] L. Hörmander,[6] M. B. Malyutov,[1] V. G. Mazya and B. P. Paneyakh,[1] K. Taira,[1-4] and B. Winzell.[1] The discussion is based on the work of the author and V. A. Kondratyev.[1] The Poincaré problem is of interest, because after its reduction on the boundary we obtain a pseudo-differential equation of principal type to which the theory described in the following chapters is applicable. Section 7 of Chapter VIII contains a more precise discussion of the smoothness of the solution depending on the order of contact of the field, defined at the boundary of the domain, with this frontier. Another reason for interest in this problem is that it can be solved using a method which permits a simple geometric interpretation and which is also suitable for investigating general pseudo-differential operators of principal type.

10. In Chapter V a concept of the wave front set of a distribution is introduced. Particular attention is given to its functorial properties. Here we discuss the approach treated by V. Guillemin and D. Schaeffer and based on the concept that a distribution is a sum of plane waves. In this case a distribution wave front is a combination of wave fronts of plane waves determined naturally. This chapter also contains Hörmander's theorem on the propagation of wave fronts for solutions of equations of principal type with a real-valued principal symbol, and other appropriate theorems on solvability of such equations.

11. Solvability (even local) of the general linear differential equation $P(x, D)u = f(x)$ remains one of the central problems of the general theory of differential equations. As early as 1946, I. G. Petrowsky[4] remarked that even for the most simple nonanalytic equation we do not know, as a rule, whether such an equation has any solution. Therefore it is important to study this problem. The solvability problem has not been completely investigated, but the results obtained in the last few years are complete for differential operators with real characteristics.

In 1956 H. Lewy proved that there exists a function $f \in C^\infty(\mathbb{R}^3)$ for which the equation

$$\frac{\partial u}{\partial x^1} + i\frac{\partial u}{\partial x^2} + 2i(x^1 + ix^2)\frac{\partial u}{\partial x^3} = f(x^1, x^2, x^3)$$

has no solution in any domain $\Omega \supset \mathbb{R}^3$. As per Kovalevsky's theorem, such a solution always exists provided f is an analytic function. This is the reason why Lewy's example attracted the attention of many analysts and had a significant impact upon the further development of the theory of partial differential equations. Four year later, in 1960, L. Hörmander showed that any differential equation $Pu = f$ has no solution (even locally) if at any characteristic point $(x, \xi) \in T^*\Omega$ (i.e., at a point where $p^0(x, \xi) = 0$) of the equation the function

$$C^0(x, \xi) = 2\,\mathrm{Im} \sum_{j=1}^{n} \overline{\frac{\partial p^0(x, \xi)}{\partial \xi_j}} \frac{\partial p^0(x, \xi)}{\partial x^j}$$

assumes a value different from zero. Later he extended this theorem to pseudo-differential operators (see Hörmander[6]). Examples of a second-order differential equation with real-valued analytic coefficients which does not have solutions in any neighborhood of the origin can be found in Hörmander's *Linear Partial Differential Operators*.[3] In this case the set of functions f from $C^\infty(\Omega)$ for which the equation Pu = f cannot be solved in the distribution class is a secondary category set.

L. Nirenberg and F. Trèves[2] and the author[11] (simultaneously and independently, for slightly different conditions) have generalized the results obtained by L. Hörmander. They have shown that every equation Pu = f is insoluble if at a characteristic point the number j of the first operator $C_j = (adP^*)^j P$, having nonvanishing symbol, is odd and the imaginary part of this symbol is positive.

The proofs of these theorems are given in Chapter VI. There we consider a more general theorem permitting the symbols of all commutators C_j to go to zero. This theorem also generalizes the results of Yu. Egorov and P. R. Popivanov.[1]

12. Chapter VII contains the proof of a theorem on sufficient conditions for local solvability of equations of principal type. Apparently, the following condition, introduced in L. Nirenberg and F. Trèves,[2] is the most exact condition of this type.

Suppose $p^0(x, \xi)$ is the principal symbol of operator P(x, D), $a = \operatorname{Re} p^0$, $b = \operatorname{Im} p^0$ and the vectors $\operatorname{grad}_{x,\xi} b(x, \xi)$ and $(\xi, 0)$ are linearly independent. Let $x = x(t)$, $\xi = \xi(t)$ be an integral curve of the system of equations

$$\dot{x}(t) = \partial_\xi b(x(t), \xi(t)), \qquad \dot{\xi}(t) = -\partial_x b(x(t), \xi(t))$$

along which $b(x(t), \xi(t)) = 0$. Now, if $a(x(t_1), \xi(t_1)) \geqslant 0$, then $a(x(t), \xi(t)) \geqslant 0$ for all $t \geqslant t_1$. However, the sufficiency of this condition for local solvability has not been proved thus far. In L. Nirenberg and F. Trèves[2] it is proved that, for local solvability of differential equations with analytic coefficients, the condition (Ψ) suffices; this condition differs from condition ($\mathscr{P}$) in that the last phrase is replaced by the following: "The function $a(x(t), \xi(t))$ cannot change its sign." In R. Beals and C. Fefferman[1] it is proved that ($\mathscr{P}$) is a sufficient condition for the solvability of pseudo-differential equations with smooth symbol.

Note that when $\mathscr{P}$ is a differential operator, the conditions (Ψ) and ($\mathscr{P}$) are equivalent because $(x_0, -\xi^0)$ is also a characteristic point for every characteristic point (x_0, ξ^0), and if the function $a(x(t), \xi(t))$ changes sign at these points, then condition (Ψ) is always violated at any of these points.

Here we present a theorem where local solvability is proved with the satisfaction of condition (Ψ) and some other conditions.

13. The constructions given in Chapter VII are also suited to study of the propagation of singularities of solutions to equations of principal type. From the theorem on propagation of singularities one can easily deduce the smoothness theorem, the theorem on hypoellipticity conditions, the semi-global solvability theorem, etc. The main theorem of this chapter generalizes the earlier formulated theorem due to L. Hörmander and also the results of Duistermaat and Hörmander[1] and Hörmander.[15]

14. Chapter VIII is devoted to subelliptic operators first studied by L. Hörmander in 1966, who showed that if $C^0(x, \xi) > 0$ at every characteristic point, then there exists an estimate

$$\|u\|_s \leq C_K(\|Pu\|_{s-m+\delta} + \|u\|_{s-1}), \qquad u \in C_0^\infty(K),$$

where K is a compact in Ω, $\delta = \frac{1}{2}$, and $\|\cdot\|_s$ is a norm in the Sobolev space H_s.

Subelliptic operators are characterized by the same estimate when $0 \leqslant \delta < 1$. Different classes of such operators are described in Yu. Egorov.[1,3,8] In 1969, the author[4] stated an exact theorem giving necessary and sufficient algebraic conditions to characterize subelliptic operators. Complete proofs were published in Yu. Egorov[14,15] in 1975 (see also[18]). A simplified proof for sufficiency conditions of this theorem was proposed in L. Hörmander.[18] Our proof in Section 5 of Chapter VIII is based on this idea of L. Hörmander.

It may be mentioned that subelliptic operators inherit most properties of elliptic operators, that is, they are hypoelliptic, smoothness theorems hold for them, conjugate operators are solvable, etc.

We shall discuss only two examples of the application of this theorem: to the oblique derivative problem when the order of tangency of the given field with the boundary is finite; and to the $\bar{\partial}$-Neumann operator in the pseudo-convex domain of the space $\mathbb{C}$. The last problem was studied by J. J. Kohn[3] and P. Greiner.[1] We shall restrict our discussion to a two-dimensional space, as in this case the problem may be reduced to a scalar pseudo-differential equation on a (three-dimensional) boundary manifold.

15. Chapter IX discusses the Cauchy problem for equations of principal type. It contains a proof due to L. V. Ovsyannikov of the Cauchy-Kovalevsky theorem (see F. Trèves[2,11]), the parametrix of the Cauchy problem (in the small) for hyperbolic equations. Using the results of Chapters VI and VII we obtain Carleman estimates for sets of equations of principal type, which enable the uniqueness of the solution of the Cauchy problem to be proved. In particular, we succeed in generalizing somewhat the Calderon theorem and the results of K. T. Smith, F. Trèves, L. Hörmander, and L. Nirenberg.

16. Thus, the book may be conditionally divided into two parts. The first part (Chapters I-III and the first three sections of Chapter IX) is intended for nonspecialists and may be regarded as an introduction to the second, more specific, part. Chapters I-III can be easily understood by third year university students. In the second part, Chapters IV and VI are also on a fairly elementary level.

I am indebted to M. S. Agranovich and E. A. Kolesnikov for reading the manuscript and for their constructive criticism.

CHAPTER I. DISTRIBUTION THEORY

This chapter contains a summary of those basic facts about distributions and differential equations with constant coefficients used most frequently in the following chapters. Some basic facts about Sobolev spaces are given in Section 3.

§1. Definitions and Basic Properties of Distributions

1. The concept of a distribution, which is of utmost importance in modern mathematics and, particularly, in the theory of partial differential equations, initially appeared as a response to the need to generalize the concept of a solution of a differential equation. Many problems of mathematical physics demand such an expansion for simulation, since from the physical meaning of the problem it is often evident that the solution to the problem cannot have derivatives in the usual, classical sense and, frequently, cannot even be a continuous function.

The idea underlying the definition of a distribution is very simple. We no longer require the solution of the differential equation $P(x, D) \times u(x) = f(x)$, where $P(x, D) = \sum_{|\alpha| \leqslant m} a_\alpha(x) D^\alpha$, to have derivatives of order up to m inclusive in the domain Ω as is assumed in classical analysis. Instead we consider u to be a continuous functional over the space of functions $C_0^m(\Omega)$ having continuous derivatives of order m and vanishing in some neighborhood of the boundary of the domain Ω and in a neighborhood of infinity. Integral identity forms the basis of the definition as follows.

First we consider a smooth function u belonging to the $C^m(\Omega)$ class. Let us denote by f the result of applying the operator $P(x, D)$ to u. It is clear that if the coefficients are continuous then f will also be continuous in the domain Ω. The equality $P(x, D)\, u(x) = f(x)$ is equivalent to the equality

$$\int P(x, D)\, u(x)\, \varphi(x)\, dx = \int f(x)\, \varphi(x)\, dx, \tag{1.1}$$

if we assume that it is simultaneously valid for all the functions φ from $C_0^m(\Omega)$ [or $C(\Omega)$ or $C_0^\infty(\Omega)$].

For simplicity we assume the coefficients a_α of operator P to be infinitely differentiable functions in Ω. This enables us to perform integration by parts in Eq. (1.1) and to transfer the derivatives from function u to function φ. As a result we obtain an equality

$$\int u(x)\, {}^tP(x, D)\, \varphi(x)\, dx = \int f(x)\, \varphi(x)\, dx, \tag{1.2}$$

where there is no derivative of function u. Here

$${}^tP(x, D)\, \varphi(x) = \sum_{|\alpha| = 0}^{m} (-1)^{|\alpha|} D^\alpha [a_\alpha(x)\, \varphi(x)].$$

Equation (1.2) is equivalent to (1.1) when $u \in C^m(\Omega)$, and defines the generalized solution. It is not difficult to understand that this equation makes sense for functions u which are not necessarily continuous. It is sufficient to assume that $u \in L_1^{\mathrm{loc}}(\Omega)$. Thus we arrive at the following definition.

Definition 1.1. A function $u \in L_1^{\text{loc}}(\Omega)$ for which Eq. (1.2) is valid for all $\varphi \in C_0^m(\Omega)$ is called the *generalized solution* of the equation $P(x, D)\, u(x) = f(x)$, where $f \in L_1^{\text{loc}}(\Omega)$.

If, in particular, u is a generalized solution of equation $D^\alpha u = f$, then f is called the *generalized derivative* of function u.

From this definition we can immediately see that the generalized derivative $D^\alpha u$ is determined uniquely (if we neglect its values on a set of zero measure) and is independent of the order of differentiation.

1.2. Definition 1.1 suffices for most of the problems of mathematical physics. But it is convenient, from a mathematical viewpoint, to extend it, assuming that arbitrary functions from the space $C_0^\infty(\Omega)$ can be taken as φ functions and without confining the set of generalized solutions within the limits of the space $L_1^{\text{loc}}(\Omega)$.

The concept of a continuous linear functional on $C_0^\infty(\Omega)$ is not very elementary since it is based on the determination of the topology of this non-metrizable space (see, for example, K. Yosida[1]). However, this concept can be introduced using only the definition of sequence convergence in this space, and this suffices for applications.

Thus, we shall say that $\varphi_j \to \varphi$ in $C_0^\infty(\Omega)$ provided

1) there is a compact submanifold $K \subset \Omega$ such that $\varphi_j = 0$ in $\Omega \setminus K$;

2) for any α the sequence $D^\alpha \varphi_j$ converges to K uniformly in $D^\alpha \varphi$.

Let us denote by $\mathscr{D}(\Omega)$ the space $C_0^\infty(\Omega)$ equipped with such a topology. (This notation was introduced by L. Schwartz.) The completeness of the space $\mathscr{D}(\Omega)$ follows from well-known theorems of classical analysis.

Definition 1.2. A linear functional u continuous on the space $\mathscr{D}(\Omega)$ is called a *distribution*.

The set of all the distributions is denoted by $\mathscr{D}'(\Omega)$.

Definition 1.3. The distribution u is of *order* k on the domain $M \Subset \Omega$ if there exists a constant C such that

$$|u(\varphi)| \leqslant C \|\varphi\|_{C^k(M)} \quad \text{for} \quad \varphi \in C_0^\infty(M).$$

An example of a distribution of order m in the domain Ω is the functional

$$u(\varphi) = \int u\, {}^tP\varphi\, dx$$

in the left-hand side of Eq. (1.2), when the function u is locally integrable.

The distribution $\delta(\varphi) = \varphi(0)$ is called the *Dirac measure* and is of order 0. Analogously, the distribution $\delta^{(\alpha)}(\varphi) = (-1)^{|\alpha|}(D)^\alpha \varphi(0)$ is of order $|\alpha|$.

Proposition 1.1. A distribution $f \in \mathscr{D}'(\Omega)$ is of finite order m(K) in every compact subset K of the domain Ω.

Proof. Let us assume that this is not true and there is a compact subset $K \Subset \Omega$ and a sequence φ_k of functions from $C_0^\infty(K)$ which is such that

$$f(\varphi_k) \geqslant k \|\varphi_k\|_{C^k(K)}.$$

The sequence $\psi_k = \varphi_k \|\varphi_k\|^{-1}_{C^k(K)} k^{-1/2}$ tends to 0 in $\mathscr{D}(\Omega)$ when $k \to \infty$ since $\operatorname{supp} \psi_k \subset K$, and for $|\beta| \leqslant k$ the inequality

$$|D^\beta \psi_k(x)| \leqslant k^{-1/2}$$

is valid. Therefore $f(\psi_k) \to 0$ when $k \to \infty$. But, on the other hand,

$$f(\psi_k) = \frac{f(\varphi_k)}{\|\varphi_k\|_{C^k(K)} \sqrt{k}} \geq \sqrt{k},$$

such that $f(\psi_k) \to \infty$ when $k \to \infty$. This contradiction proves the theorem.

In the linear space $\mathscr{D}'(\Omega)$ we can introduce a weak topology in which the convergence of the sequence u_j of elements from $\mathscr{D}'(\Omega)$ to $u \in \mathscr{D}'(\Omega)$ means that $u_j(\varphi) \to u(\varphi)$ for any function φ from $C_0^\infty(\Omega)$. The completeness of the space $\mathscr{D}'(\Omega)$ follows from the Banach-Steinhaus theorem (see Yosida[1]).

It is not difficult to see that every function u in $L_1^{\text{loc}}(\Omega)$ defines a distribution $u(\varphi) = \langle u, \varphi \rangle \equiv \int u(x) \varphi(x)\, dx$. Therefore the space $L_1^{\text{loc}}(\Omega)$ is a subspace in $\mathscr{D}'(\Omega)$.

Definition 1.4. A distribution u is *equal to 0* in a subdomain ω if $u(\varphi) = 0$ for all functions $\varphi \in C_0^\infty(\omega)$.

Definition 1.5. The least closed set outside of which u = 0 is called the *support* of the distribution u and is denoted by supp u.

Furthermore, for the study of singularities of a distribution, the following definitions, analogous to 1.4 and 1.5, will be required.

Definition 1.6. A distribution u is called *infinitely differentiable* in ω, if in the space $C^\infty(\Omega)$ there exists a function v such that $u(\varphi) = \langle v, \varphi \rangle$ for all $\varphi \in C_0^\infty(\omega)$.

Definition 1.7. The least closed manifold outside of which $u \in C^\infty(\omega)$ is called the *singular support* of the distribution u and is denoted by sing supp u.

Let us now return to Definition 1.1 of the generalized solution and replace the space $L_1^{\text{loc}}(\Omega)$ by the space $\mathscr{D}'(\Omega)$. Then this definition will take the following form:

Definition 1.8. A distribution u for which $u({}^t P\varphi) = f(\varphi)$ is valid for all $\varphi \in C_0^\infty(\Omega)$ is called the *generalized solution* of the equation $P(x, D) u = f(x)$, where $f \in \mathscr{D}'(\Omega)$.

A large part of this book is dedicated to the study of necessary and sufficient conditions for the existence of a generalized solution, and to the investigation of its properties, particularly the structure of the singular support.

1.3. The above definition of distribution is not the only possible one. We may consider, for example, the space $C^\infty(\Omega)$ as the base space instead of $C_0^\infty(\Omega)$. Naturally, in this case the dual space of linear continuous functionals also varies.

The concept of convergence in the space $C^\infty(\Omega)$ is introduced as follows. We say that a sequence φ_j converges to φ in $C^\infty(\Omega)$, if for every α and any compact manifold $K \Subset \Omega$ the sequence $D^\alpha \varphi_j$ converges to $D^\alpha \varphi$ uniformly in K.

Schwartz denotes by $\mathscr{E}(\Omega)$ the space $C^\infty(\Omega)$ equipped with such a topology. The completeness of the space $\mathscr{E}(\Omega)$ follows from well-known theorems of classical analysis.

Definition 1.9. $\mathscr{E}'(\Omega)$ is the space of *linear continuous functionals* on the space $\mathscr{E}(\Omega)$.

It is clear that every element of the space $\mathscr{E}'(\Omega)$ belongs to $\mathscr{D}'(\Omega)$. The space $\mathscr{E}'(\Omega)$ is characterized by the following.

Theorem 1.1. The space $\mathscr{E}'(\Omega)$ consists of distributions with compact support.

Proof. It is evident that every function $\varphi \in \mathscr{D}(\Omega)$ belongs to $\mathscr{E}(\Omega)$ and if $\{\varphi_j\} \in \mathscr{D}(\Omega)$ and $\varphi_j \to 0$ in $\mathscr{D}(\Omega)$, then $\varphi_j \to 0$ in $\mathscr{E}(\Omega)$. Therefore, if $f \in \mathscr{E}'(\Omega)$, then there exists $g \in \mathscr{D}'(\Omega)$ for which $f(\varphi) = g(\varphi)$, where $\varphi \in \mathscr{D}(\Omega)$.

Let us show that the support of distribution g is bounded. Indeed, if this is the case, then there is a sequence of functions $\{\varphi_j\} \in \mathscr{D}(\Omega)$ such that $\varphi_j(x) = 0$ when $|x| \leqslant j$ and $g(\varphi_j) = 1$. But $\varphi_j \to 0$ in $\mathscr{E}(\Omega)$, and since the functional f is continuous, $f(\varphi_j) \to 0$ when $j \to \infty$. Hence it is clear that the equality $f(\varphi_j) = g(\varphi_j)$ cannot be true for all j and we arrive at a contradiction such that supp $g \subset K$, where K is a compact subset of points in Ω.

Let us now show that $f(\varphi) = g(\varphi)$ for all $\varphi \in \mathscr{E}(\Omega)$. In fact we suppose $h_j \in C_0^\infty(\Omega)$, $h_j = 1$ in a neighborhood of the compact K and $h_j \to 1$ in $\mathscr{E}(\Omega)$. Let $\varphi \in \mathscr{E}(\Omega)$. Then $h_j\varphi \to \varphi$ in $\mathscr{E}(\Omega)$. But $f(h_j\varphi) = g(h_j\varphi)$ and $g(\varphi) = g(h_j\varphi) + g((1 - h_j)\varphi) = g(h_j\varphi)$ for all j. Therefore $f(h_j\varphi) = g(\varphi)$ for all j. Passing to the limit, as $j \to \infty$ we obtain

$$f(\varphi) = g(\varphi).$$

This proves the theorem.

1.4. Schwartz's book describes another important space of distributions defined on $\mathbb{R}^n$, the space $\mathscr{S}'$. As mentioned earlier, this is a space dual to the base space $\mathscr{S}$ of smooth functions and is called the space of tempered distributions. More exact definitions are given below.

Definition 1.10. By $\mathscr{S}$ we denote the space of infinitely differentiable functions φ in $\mathbb{R}^n$ such that

$$\sup_x |x^\beta D^\alpha \varphi(x)| < \infty$$

for all multi-indices α and β.

We see that the functions in the space $\mathscr{S}$ tend to 0 as $x \to \infty$ even if they are multiplied by a polynomial; this is also true for their derivatives of any order. Clearly $C_0^\infty(\mathbb{R}^n) \subset \mathscr{S}$. Another example of a function belonging to $\mathscr{S}$ is $e^{-|x|^2}$. From the definition it is clear that $\mathscr{S}$ is invariant under differentiation and multiplication by polynomials.

Definition 1.11. The space of linear continuous functionals on $\mathscr{S}$ is called the space of *tempered distributions* and denoted by $\mathscr{S}'$.

It is not hard to see that a linear functional on $\mathscr{S}$ belongs to $\mathscr{S}'$ if and only if there exist indices α and k and a constant C such that the inequality

$$|u(\varphi)| \leqslant C \sup_x (1 + |x|)^k |D^\alpha \varphi(x)|, \qquad \varphi \in \mathscr{S},$$

is valid.

Since

$$\mathscr{D}(\mathbb{R}^n) \subset \mathscr{S} \subset \mathscr{E}(\mathbb{R}^n)$$

and the corresponding imbedding operators are continuous, we see that

$$\mathscr{D}'(\mathbb{R}^n) \supset \mathscr{S}' \supset \mathscr{E}'(\mathbb{R}^n),$$

and the corresponding imbedding operators are also continuous.

1.5. Here we show that the space of distributions is invariant under multiplication by any function h from $C^\infty(\Omega)$ and also under differentiation.

Definition 1.12. If $u \in \mathscr{D}'(\Omega)$ and $h \in C^\infty(\Omega)$, then the *product* hu is defined by the functional

$$(hu)(\varphi) = u(h\varphi), \qquad \varphi \in \mathscr{D}(\Omega).$$

Definition 1.13. If $u \in \mathscr{D}'(\Omega)$, then the *derivative* $D^\alpha u$ is defined by the functional

$$(D^\alpha u)(\varphi) = (-1)^{|\alpha|} u(D^\alpha \varphi), \quad \varphi \in \mathscr{D}(\Omega).$$

We note that if $h \in C^\infty(\Omega)$ and $u \in \mathscr{D}'(\Omega)$, then

$$D_k(hu) = (D_k h)u + h(D_k u). \tag{1.3}$$

In fact, using Definitions 1.12 and 1.13, we can obtain this equation from

$$-u(hD_k\varphi) = u((D_k h)\varphi) - u(D_k(h\varphi)), \qquad \varphi \in \mathscr{D}(\Omega).$$

It is convenient to make use of Leibniz's formula in the general form. If $P(\xi)$ is an arbitrary polynomial in n variables $\xi_1, \dots, \xi_n$, then

$$P(D)(hu) = \sum_\alpha \frac{1}{\alpha!} D^\alpha h P^{(\alpha)}(D)u, \quad h \in C^\infty(\Omega), \quad u \in \mathscr{D}'(\Omega), \tag{1.4}$$

where

$$P^{(\alpha)}(\xi) = \partial^{|\alpha|} P(\xi)/\partial\xi_1^{\alpha_1} \dots \partial\xi_n^{\alpha_n}, \text{ and } P(D), \quad P^{(\alpha)}(D)$$

are differential operators obtained from the polynomials $P(\xi)$ and $P^{(\alpha)}(\xi)$ upon replacing the variables ξ_j by D_j.

Indeed, from (1.3) it is evident that

$$P(D)(hu) = \sum_\alpha (D^\alpha h) Q_\alpha(D)u, \quad h \in C^\infty(\Omega), \quad u \in \mathscr{D}'(\Omega), \tag{1.5}$$

where $Q_\alpha(\xi)$ are some polynomials which can be determined by taking $h(x) = e^{i(x,\xi)}$ and u as the function $u(x) = e^{i(x,\eta)}$. Note that $P(D)e^{i(x,\xi)} = P(\xi)e^{i(x,\xi)}$. Therefore, from (1.5) it follows that

$$P(\xi + \eta) = \xi^\alpha Q_\alpha(\eta).$$

It thus follows from Taylor's formula that $Q_\alpha(\eta) = P^{(\alpha)}(\eta)/\alpha!$

In further discussion we shall come across a generalization of equality (1.4).

1.6. For a theory of distributions it is important that every distribution can be approximated by smooth functions. For this the mollifier technique proves to be very convenient.

Definition 1.14. A function $\omega \in C_0^\infty(\mathbb{R}^n)$ is called a *mollifier* kernel if $\omega(x) = 0$ when $|x| \geqslant 1$, $\omega(x) \geqslant 0$, and $\int \omega(x)\,dx = 1$.

A classical example of a mollifier kernel is the function

$$\omega(x) = \begin{cases} Ce^{\frac{1}{|x|^2-1}} & \text{when } |x| < 1, \\ 0 & \text{when } |x| \geqslant 1, \end{cases}$$

with a constant C equal to $\left(\int_{|x|<1} e^{\frac{1}{|x|^2-1}} dx \right)^{-1}$. Assume $\omega_h(x) = h^{-n}\omega(x/h)$.

Definition 1.15. A function $u_h(x) = u_y(\omega_h(x - y))$ is called the *mollifier of a distribution* $u \in \mathscr{D}'(\mathbb{R}^n)$.

Here u_y indicates that the functional is computed on $\omega_h(x - y)$ as a function of y; that is, x plays the role of parametrix.

Theorem 1.2. The function $u_h(x)$ is infinitely differentiable when $h > 0$. When $h \to +0$ the function u_h converges to u in $\mathscr{D}'(\mathbb{R}^n)$. If

$u \in L_p(\mathbb{R}^n)$, then $\|u_h - u\|_{L_p} \to 0$ when $h \to 0$ and $1 \geqslant p < \infty$.

If $u \in C(\mathbb{R}^n)$, then $u_h \to u$ uniformly in every bounded subdomain.

Proof. 1°. The continuity of the function $u_h(x)$ in x follows from the definition of a distribution: when $x \to x_0$ we have $\omega_h(y - x) \to \omega_h(y - x_0)$ in $\mathscr{D}$ and therefore

$$u_y(\omega_h(x-y)) \to u_y(\omega_h(x_0-y)).$$

To prove the differentiability of the function $u_h(y)$ we consider the ratio $[u_h(y + \delta z) - u_h(y)]/\delta$, where z is an arbitrary vector of unit length in $\mathbb{R}^n$. We have

$$\frac{u_h(y+\delta z)-u_h(y)}{\delta} = u_x\left(h^{-n}\left[\omega\left(\frac{x-y-\delta z}{h}\right) - \omega\left(\frac{x-y}{h}\right)\right]\Big/\delta\right).$$

But when $\delta \to 0$ and y is fixed

$$\left[\omega\left(\frac{x-y-\delta z}{h}\right) - \omega\left(\frac{x-y}{h}\right)\right]\Big/\delta \to -\frac{\partial \omega}{\partial z}\left(\frac{x-y}{h}\right)$$

in the topology of the space $\mathscr{D}(\mathbb{R}^n)$. Hence, when $\delta \to 0$

$$\frac{u_h(y+\delta z)-u_h(y)}{\delta} \to u_x\left(h^{-n}\frac{\partial \omega}{\partial z}\left(\frac{x-y}{h}\right)\right).$$

This implies that the function $u_h(y)$ is differentiable. Arguing in a similar fashion we can prove that $u_h \in C^\infty$.

2°. Suppose $u \in C(\mathbb{R}^n)$. Then

$$|u_h(x) - u(x)| = \left|\int \omega_h(x-y)[u(y) - u(x)]\,dy\right| \leqslant \max_{|x-y| \leqslant h} |u(y) - u(x)|.$$

If Ω is a bounded domain in $\mathbb{R}^n$, then u is uniformly continuous in a compact neighborhood. Therefore

$$\max_{x \in \bar{\Omega}} |u_h(x) - u(x)| \leqslant \max_{x \in \bar{\Omega},\ |x-y| \leqslant h} |u(y) - u(x)|,$$

and if $h \to 0$, then $\max_{x \in \bar{\Omega}} |u_h(x) - u(x)| \to 0$, that is, $u_h \to u$ uniformly in $\bar{\Omega}$.

3°. Let $u \in L_p$ and $p > 1$. It is immediately evident that $\|u_h\|_{L_p} \leqslant \|u\|_{L_p}$. In fact, applying Hölder's inequality, we see that

$$|u_h(x)|^p = \left|\int \omega_h(x-y)u(y)\,dy\right|^p = \left|\int \omega_h(x-y)^{1/p} u(y)\,\omega_h(x-y)^{1/p'}\,dy\right|^p \leqslant$$

$$\left(\int \omega_h(x-y)|u(y)|^p\,dy\right)\left(\int \omega_h(x-y)\,dy\right)^{p/p'} \leqslant \int \omega_h(x-y)|u(y)|^p\,dy.$$

Therefore

$$\|u_h\|^p_{L_p} = \int |u_h(x)|^p\,dx \leqslant \int |u(y)|^p\left(\int \omega_h(x-y)\,dx\right)dy = \int |u(y)|^p\,dy = \|u\|^p_{L_p}.$$

Similarly, when

$$\int |u_h(x)|\,dx \leqslant \int |u(y)|\left(\int \omega_h(x-y)\,dx\right)dy = \int |u(y)|\,dy.$$

Let now $\varepsilon > 0$. If $u \in L_p$ when $1 \leqslant p < \infty$, then a function u_1 continuous and equal to 0 for $|x| \geqslant R = R(\varepsilon)$ can be determined for which

$$\|u - u_1\|_{L_p} < \varepsilon/3.$$

By the triangle inequality

$$\|u-u_h\|_{L_p} \leqslant \|u-u_1\|_{L_p} + \|u_1-u_{1h}\|_{L_p} + \|u_{1h}-u_h\|_{L_p} \leqslant \frac{2\varepsilon}{3} + \|u_1-u_{1h}\|_{L_p}.$$

Since u_1 is equal to 0 for $|x| \geqslant R$ and is continuous, then from what has been proved earlier we have

$$\max_x |u_1(x) - u_{1h}(x)| < \frac{\varepsilon}{3V_{R+1}^{1/p}},$$

provided h is sufficiently small. Here V_R stands for the volume of a ball of radius R. Therefore

$$\|u_1-u_{1h}\|_{L_p} = \left(\int |u_1(x)-u_{1h}(x)|^p\,dx\right)^{1/p} < \frac{\varepsilon}{3}.$$

Thus, if h is sufficiently small, then the inequality $\|u - u_h\|_{L_p} < \varepsilon$ holds. This shows the convergence of u_h to u in L_p as $h \to 0$.

4°. Let now $u \in \mathscr{D}'(\mathbb{R}^n)$ and $\varphi \in \mathscr{D}(\mathbb{R}^n)$. Then

$$u_h(\varphi) = \int u_x(\omega_h(y-x))\,\varphi(y)\,dy = u_x\left(\int \omega_h(y-x)\,\varphi(y)\,dy\right) = u(\varphi_h).$$

From what has been proved it follows that $\varphi_h \to \varphi$ uniformly, $D^\alpha\varphi_h \to D^\alpha\varphi$ uniformly, and all of the functions φ_h vanish outside a compact neighborhood of the support of φ. Hence, $\varphi_h \to \varphi$ in $\mathscr{D}$. From the definition of the space $\mathscr{D}'$ it follows that

$$u_h(\varphi) = u(\varphi_h) \to u(\varphi) \quad \text{as} \quad h \to +0.$$

This implies that $u_h \to u$ in $\mathscr{D}'$

This completes the proof.

The following statement is given as an example of an application of the mollifier operation.

Theorem 1.3. If the support of the distribution u consists only of the point x = 0, then $u(x) = \sum_{|\alpha| \leqslant k} c_\alpha \delta^{(\alpha)}(x)$, where k is some nonnegative integer, $\delta(x)$ is the Dirac measure, and $c_\alpha \in \mathbb{C}$.

Proof. By Definition 1.1, the order k of the distribution u is finite. If $\varphi \in C_0^\infty(\Omega)$, where Ω is a neighborhood of the origin, we form the Taylor expansion of order k,

$$\varphi(x) = \sum_{|\alpha| \leqslant k} \frac{1}{\alpha!} (iD)^\alpha \varphi(0)\, x^\alpha g(x) + \psi(x),$$

where $g(x) \in C_0^\infty(\mathbb{R}^n)$ and $g(x) = 1$ in Ω. We shall prove that $u(\psi) = 0$. Hence we have

$$u(\varphi) = \sum_{|\alpha| \leqslant k} c_\alpha (iD)^\alpha \varphi(0),$$

where $c_\alpha = u(x^\alpha g)/\alpha!$. This proves the theorem.

We note that $D^\alpha\psi(0) = 0$ when $|\alpha| \leqslant k$. Let V_h be a ball of radius h centered at the origin.

Let

$$\chi_h(x) = \int_{V_{2h}} \omega_h(x-y)\,dy,$$

where ω is the mollifier kernel. Clearly $\chi_h \in C_0^\infty(\mathbb{R}^n)$ and $\chi_h(x) = 1$ when $|x| \leqslant h$, and $\chi_h(x) = 0$ when $|x| \geqslant 3h$. Further

$$u(\psi) = u(\psi\chi_h),$$

since $\psi(1 - \chi_h) = 0$ in the ball V_h. Hence

$$|u(\psi)| \leqslant C \sum_{|\alpha| \leqslant k} \sup |D^\alpha(\psi\chi_h)|.$$

But $|D^\alpha \chi_h(x)| \leqslant C_\alpha h^{-|\alpha|}$ and $|D^\alpha \psi(x)| \leqslant C'_\alpha h^{k+1-|\alpha|}$ when $|\alpha| \leqslant k$, and $|x| \leqslant 3h$. Computing the derivatives $D^\alpha(\psi\chi_h)$ using Leibniz's rule we obtain

$$|u_h(\psi)| \leqslant C'h$$

with the constant C' independent of h. Thus $u(\psi) = 0$, completing the proof.

§2. Fourier Transforms and the Convolution of a Distribution

2.1. If a function f is integrable over $\mathbb{R}^n$, its Fourier transform is defined by

$$\tilde{f}(\xi) = \int f(x) e^{-ix\xi} dx, \qquad (2.1)$$

where $x\xi = x^1\xi_1 + \ldots + x^n\xi_n$. If $\tilde{f} \in L_1$, one can express f in terms of $\tilde{f}$ by means of the Fourier inversion formula

$$f(x) = (2\pi)^{-n} \int \tilde{f}(\xi) e^{ix\xi} d\xi. \qquad (2.2)$$

Plancherel's theorem enables us to define the Fourier transform when $f \in L_2(\mathbb{R}^n)$:

$$\tilde{f}(\xi) = \lim_{R\to\infty} \int_{|x|<R} f(x) e^{-ix\xi} dx.$$

In doing so we find that $\tilde{f} \in L_2(\mathbb{R}^n)$ and

$$f(x) = \lim_{R\to\infty} (2\pi)^{-n} \int_{|\xi|<R} \tilde{f}(\xi) e^{ix\xi} d\xi.$$

The limits in the last two formulas are taken in $L_2(\mathbb{R}^n)$ and Parseval's equality holds:

$$\|f\|_{L_2(\mathbb{R}^n)} = (2\pi)^{-n/2} \|\tilde{f}\|_{L_2(\mathbb{R}_n)}. \qquad (2.3)$$

This is equivalent to

$$\int f(x)\overline{g(x)}\, dx = (2\pi)^{-n} \int \tilde{f}(\xi) \overline{\tilde{g}(\xi)}\, d\xi,$$

which is true for any pair of functions f, $g \in L_2(\mathbb{R}^n)$.

If $u \in L_1(\mathbb{R}^n)$ and $\varphi \in C_0^\infty(\mathbb{R}^n)$, we denote by $u * \varphi$ the function defined by

$$(u * \varphi)(x) = \int u(y) \varphi(x-y)\, dy.$$

In this case

$$\widetilde{(u * \varphi)}(\xi) = \tilde{u}(\xi)\tilde{\varphi}(\xi).$$

This formula also holds when $u \in L_2(\mathbb{R}^n)$, and $\varphi \in L_2(\mathbb{R}^n)$.

Note that the mollifier u_h is a convolution: $u_h = u * \omega_h$, where $\omega_h(x) = h^{-n}\omega(xh^{-1})$.

2.2. The above formulas hold, of course, for functions in the space $\mathcal{S}$. Also valid is

Theorem 2.1. If $f \in \mathcal{S}$, then $\tilde{f} \in \mathcal{S}$. Here the Fourier transform of the function $D^\alpha f$ is $\xi^\alpha \tilde{f}(\xi)$ and the Fourier transform of $x^\alpha f$ is $(-D_\xi)^\alpha \tilde{f}(\xi)$.

Proof. Differentiation of (2.1) with respect to ξ_j gives

$$D_{\xi_j}\tilde{f}(\xi) = \int (-x^j) f(x) e^{-ix\xi} dx.$$

and is legitimate since the integral obtained is uniformly convergent. Since the operation can be repeated, it follows that $f \in C^\infty$ and $D_\xi^\alpha \tilde{f}(\xi)$ is the Fourier transform of $(-x)^\alpha f(x)$.

On the other hand,

$$\xi^\beta \tilde{f}(\xi) = \int f(x)(-D_x)^\beta e^{-ix\xi}\,dx = \int D^\beta f(x)\cdot e^{-ix\xi}\,dx$$

and

$$D^\alpha[\xi^\beta \tilde{f}(\xi)] = \int (-x)^\alpha D_x^\beta f(x)\, e^{-ix\xi}\,dx. \qquad (2.4)$$

From the fact that $(-x)^\alpha D_x^\beta f \in L_1(\mathbb{R}^n)$, it follows that $D^\alpha[\xi^\beta \tilde{f}(\xi)]$ is defined and bounded for all values of α and β, so $\tilde{f} \in \mathcal{S}$. Also note that when $\alpha = 0$ we obtain that $\xi^\beta \tilde{f}(\xi)$ is the Fourier transform of $D^\beta f(x)$.

Corollary 2.1. The Fourier transform maps $\mathcal{S}$ continuously onto $\mathcal{S}$.

Proof. We know that the Fourier transform of a function $f \in \mathcal{S}$ is a function $\tilde{f} \in \mathcal{S}$. In this case $\tilde{f} \in L_1(\mathbb{R}_n)$ and the Fourier inversion formula (2.2) holds. Further, returning to formula (2.4), we can estimate the left-hand side:

$$\max_\xi |D^\alpha[\xi^\beta \tilde{f}(\xi)]| \leqslant \int (1+|x|)^{-n-1} \max_x |(1+|x|)^{n+1+|\alpha|} D_x^\beta f(x)|\,dx \leqslant$$

$$C_{\alpha,\beta}\int (1+|x|)^{-n-1}\,dx \leqslant C'_{\alpha,\beta}.$$

Therefore, the Fourier transformation maps the space $\mathcal{S}$ continuously onto $\mathcal{S}$ and this proves the corollary.

2.3. The convolution of a distribution $u \in \mathcal{D}'(\mathbb{R}^n)$ with a function $\varphi \in C_0^\infty(\mathbb{R}^n)$ is defined by

$$(u * \varphi)(x) = u_y(\varphi(x-y)).$$

Note that this formula can also be used for determining the convolution $u * \varphi$ when $u \in \mathcal{S}'$ and $\varphi \in \mathcal{S}$ or $u \in \mathcal{E}(\mathbb{R}^n)$ and $\varphi \in C^\infty(\mathbb{R}^n)$.

Theorem 2.2. If $u \in \mathcal{D}'(\mathbb{R}^n)$ and $\varphi \in C_0^\infty(\mathbb{R}^n)$ or $u \in \mathcal{E}'(\mathbb{R}^n)$ and $\varphi \in C^\infty(\mathbb{R}^n)$, then we have $u * \varphi \in C^\infty(\mathbb{R}^n)$ and $\operatorname{supp}(u * \varphi) \subset \operatorname{supp} u + \operatorname{supp} \varphi$. The derivatives D^α of the convolution for all values of α are given by

$$D^\alpha(u * \varphi) = D^\alpha u * \varphi = u * D^\alpha \varphi.$$

Proof. The smoothness of the convolution $u * \varphi$ can be proved in exactly the same manner as the smoothness of mollifiers in Theorem 1.4. From the definition of support we have $(u * \varphi)(x) = 0$, if $\operatorname{supp} u(y) \cap \operatorname{supp} \varphi(x-y) = \emptyset$, i.e., if there does not exist a point $y \in \operatorname{supp} u$ such that $x - y \in \operatorname{supp} \varphi$, or, in other words, $x \notin \operatorname{supp} u + \operatorname{supp} \varphi$. It remains to prove that

$$D_k(u * \varphi) = D_k u * \varphi = u * D_k \varphi.$$

Let e_k be a unit vector parallel to the x^k-axis and consider the difference quotient

$$\frac{(u*\varphi)(x+\delta e_k) - (u*\varphi)(x)}{\delta} = u_y\left(\frac{\varphi(x+\delta e_k - y) - \varphi(x-y)}{\delta}\right).$$

When $\delta \to 0$ and x is fixed, the difference quotient $[\varphi(x + \delta e_k - y) - \varphi(x - y]/\delta$ converges to $iD_k\varphi(x-y)$, in $C_0^\infty(\mathbb{R}^n)$, if $\varphi \in C_0^\infty(\mathbb{R}^n)$ or in $C^\infty(\mathbb{R}^n)$ if $\varphi \in C^\infty(\mathbb{R}^n)$. Hence we obtain

$$D_k(u * \varphi) = u * D_k \varphi.$$

From the definition of the derivative of a distribution it follows that

$$(u * D_k\varphi)(x) = u_y(D_{x^k}\varphi(x-y)) = -u_y(D_{y^k}\varphi(x-y)) = D_{y^k}u_y(\varphi(x-y)) = (D_k u * \varphi)(x),$$

and this completes the proof.

2.4. If a distribution u has compact support and $\varphi \in C_0^\infty(\mathbb{R}^n)$ then $u * \varphi \in C_0^\infty(\mathbb{R}^n)$ and $v * (u * \varphi)$ can be considered for $v \in \mathcal{D}'(\mathbb{R}^n)$. Similarly, if $v_1, \ldots, v_k \in \mathcal{D}'(\mathbb{R}^n)$ and all these distributions except one have compact support and if $\varphi \in C_0^\infty(\mathbb{R}^n)$, then the convolution

$$v_1 * v_2 * \ldots * v_k * \varphi \in C^\infty(\mathbb{R}^n).$$

is defined by Theorem 2.2.

In this case $D^\alpha(v_1 * \ldots * v_k * \varphi) = D^\alpha v_1 * \ldots * v_k * \varphi = \ldots = v_1 * \ldots * D^\alpha v_k * \varphi = v_1 * \ldots * v_k * D^\alpha \varphi$.

2.5. Definition 2.1. If $u \in \mathscr{S}'$, the *Fourier transform* ũ is defined by

$$\tilde{u}(\varphi) = u(\tilde{\varphi}), \qquad \varphi \in \mathscr{S}. \tag{2.5}$$

It follows from Theorem 2.1 that if $u \in \mathscr{S}'$, then $\tilde{u} \in \mathscr{S}'$.

Note that the Fourier transform of a distribution $u \in \mathscr{E}'$ may be defined as

$$\tilde{u}(\xi) = u_x(e^{-i(x,\xi)}).$$

In fact, if u is a function and $u(\varphi) = \int u(x)\varphi(x)\,dx$, then this definition agrees with (2.1). If $u \in \mathscr{E}'$, then we shall consider its mollifier u_h with an even kernel. By Theorem 1.2, $u_h \to u$ when $h \to 0$ in the topology $\mathscr{E}'$ and, therefore, in the topology $\mathscr{S}'$ also. From Definition 2.1 it follows that then $\tilde{u}_h \to \tilde{u} \in \mathscr{S}'$, and since $\tilde{u}_h = u_h(e^{-i(x,\xi)})$, we have

$$\tilde{u}(\xi) = u_x(e^{-i(x,\xi)}).$$

Example 2.1. If $u = \delta(x)$, then $\tilde{u}(\xi) = 1$.
If $u = \delta^{(\alpha)}(x)$, then $\tilde{u}(\xi) = (-i\xi)^\alpha$.

The following theorem plays a significant role in the general theory of differential equations.

Theorem 2.3. (Paley-Wiener). An entire analytic function $U(\zeta)$ is the Fourier-Laplace transform of a distribution u with support in the ball $S_A = \{x,\ x \in \mathbb{R}^n,\ |x| \leqslant A\}$ if and only if for some constants C and N we have

$$|U(\zeta)| \leqslant C(1 + |\zeta|)^N e^{A|\operatorname{Im}\zeta|}. \tag{2.6}$$

The distribution u coincides with a function in $C_0^\infty(S_A)$ if and only if its Fourier-Laplace transform $U(\zeta)$ is an entire analytic function and if for every integer N there exists a constant C_N such that

$$|U(\zeta)| \leqslant C_N(1 + |\zeta|)^{-N} e^{A|\operatorname{Im}\zeta|}. \tag{2.7}$$

Proof. To prove necessity we use the fact that if $u \in \mathscr{E}'$ and supp u $\subset S_A$, then there exist constants C and N such that

$$|u(\varphi)| \leqslant C \sum_{|\alpha| \leqslant N} \sup|D^\alpha \varphi|, \qquad \varphi \in C_0^\infty(S_A).$$

Let $h \in C_0^\infty(\mathbb{R})$ be equal to 1 when $|t| \leqslant \frac{1}{2}$ and 0 when $|t| \geqslant 1$. Then the function $\varphi_\zeta(x) = e^{-i(x,\zeta)} h(|\zeta|(|x| - A))$ is in C_0^∞ and agrees with $e^{-i(x,\zeta)}$ for $|x| \leqslant A + |2\zeta|^{-1}$. Therefore

$$|\tilde{u}(\zeta)| = |u(\varphi_\zeta)| \leqslant C \sum_{|\alpha| \leqslant N} \operatorname{supp}|D^\alpha \varphi_\zeta|, \tag{2.8}$$

but $|e^{-i(x,\zeta)}| \leqslant e^{A|\operatorname{Im}\zeta| + 1}$ when $x \in \operatorname{supp} \varphi_\zeta$ and the inequality (2.6) follows from (2.8). Also we see that $\tilde{u}(\zeta)$ is an entire analytic function.

The necessity of inequality (2.7) follows from (2.4) with $\alpha = 0$ and $f = u$.

b) The sufficiency of (2.7) is proved by shifting the integration in (2.2) into the complex domain, which gives

$$u(x) = (2\pi)^{-n} \int U(\xi) e^{ix\xi}\, d\xi = (2\pi)^{-n} \int U(\xi + i\eta) e^{i(x,\xi + i\eta)}\, d\xi.$$

Estimating the integral by means of (2.7) with N = n + 1, we obtain

$$|u(x)| \leqslant C_N e^{A|\eta| - (x,\eta)} (2\pi)^{-n} \int (1 + |\xi|)^{-n-1} d\xi.$$

If we choose $\eta = tx$ and let $t \to +\infty$, it then follows that $u(x) = 0$ if $|x| > A$. The fact that $u \in C^\infty$ follows readily from (2.2) if inequalities (2.2) are satisfied.

To prove the sufficiency of (2.6), we first note that $U \in \mathscr{S}'$; hence $U = \tilde{u}$ for some $u \in \mathscr{S}'$. The function $u_h = u * \omega_h$ is infinitely differentiable and $\tilde{u}_h = \tilde{u}\tilde{\omega}_h$. In this case, (2.6) with A replaced by A + h is satisfied for $\tilde{u}_h$. Hence it follows that when $h > 0$

$$\operatorname{supp} \tilde{u}_h \subset S_{A+h}.$$

Letting $h \to 0$, this implies that $\operatorname{supp} u \subset S_A$.

2.6. Now we consider the Fourier transform of the convolution $u * \varphi$, where $u \in \mathscr{D}'(\mathbb{R}^n)$ and $\varphi \in C_0^\infty(\mathbb{R}^n)$. We have

$$\widetilde{(u * \varphi)}(\psi(\xi)) = (u * \varphi)_x (\tilde{\psi}(x)) = u_y (\varphi(x-y)) (\tilde{\psi}(x)) =$$

$$u_y \left(\int \varphi(x-y) \tilde{\psi}(x)\, dx \right) = u_y \int e^{-i(y,\xi)} \tilde{\varphi}(\xi) \psi(\xi)\, d\xi = \tilde{u}(\tilde{\varphi}\psi) = \tilde{\varphi}(\xi) \tilde{u}(\psi),$$

i.e., $\widetilde{u * \varphi} = \tilde{u}\tilde{\varphi}$.

It is clear that the same arguments work when $u \in \mathscr{E}'(\mathbb{R}^n)$ and $\varphi \in C^\infty(\mathbb{R}^n)$. Hence, if $v_1, \ldots v_k \in \mathscr{D}'(\mathbb{R}^n)$ and all of these distributions but one have compact support, then, by Section 2.4 we have

$$\widetilde{(v_1 * \ldots * v_k * \varphi)}(\xi) = \tilde{v}_1(\xi) \ldots \tilde{v}_k(\xi) \tilde{\varphi}(\xi). \tag{2.9}$$

Note that the convolution is invariant under translation. In fact, let $z \in \mathbb{R}^n$ and $(\tau_z \varphi)(x) = \varphi(x - z)$. From the definition of a convolution it is evident that $u * (\tau_z \varphi) = \tau_z (u * \varphi)$. It may be shown that only the convolution has this property. This is proved as follows:

Theorem 2.4. If a linear operator A maps the space $C_0^\infty(\mathbb{R}^n)$ continuously into $C^\infty(\mathbb{R}^n)$ and $\tau_z A = A\tau_z$ for any $z \in \mathbb{R}^n$, then there exists a unique distribution u such that $A\varphi = u * \varphi$ and $\varphi \in C_0^\infty(\mathbb{R}^n)$.

Proof. By hypothesis the value of the function $A\varphi$ at $x = 0$ is a continuous linear functional of the function such that there exists a distribution u for which

$$A\varphi(0) = u(\varphi_1) = (u * \varphi)(0), \quad \varphi \in C_0^\infty(\mathbb{R}^n).$$

Replacing φ by $\tau_{-z}\varphi$ in this inequality, we obtain

$$A\varphi(z) = (u * \varphi)(z), \; \varphi \in C_0^\infty(\mathbb{R}^n), \; z \in \mathbb{R}^n,$$

and this proves the theorem.

We can now define the convolution $u * v$ of two distributions if one of them has compact support. In fact, the operator

$$A\colon \; \varphi \to u * (v * \varphi)$$

is linear, continuous, and commutes with τ_z for all $z \in \mathbb{R}^n$. Theorem 2.4 shows that there is a unique distribution w such that $A\varphi = w * \varphi$. Hence we have

$$w = u * v.$$

This enables us to determine the convolution of any number of distributions if all but one of them have compact support.

Example. If $u \in \mathscr{D}'(\mathbb{R}^n)$, then $u * \delta = u$.

Indeed, according to the definition of a convolution, the equalities

$$(u * \delta) * \varphi = u * (\delta * \varphi) = u * \varphi$$

are valid.

The commutativity of the convolution thus determined follows immediately from (2.9) for its Fourier transformation. Similarly

$$D^\alpha (u * v) = (D^\alpha u) * v = u * D^\alpha v,$$

since

$$\widetilde{D^\alpha (u * v)} = \xi^\alpha \tilde{u}(\xi)\, \tilde{v}(\xi) = (\widetilde{D^\alpha u})(\xi) \cdot \tilde{v}(\xi) = \tilde{u}(\xi)\, (\widetilde{D^\alpha v})(\xi). \tag{2.10}$$

§3. Sobolev Spaces

3.1. In this book distribution spaces, first introduced by S. L. Sobolev, play an important role.

If $s = 0$, then the space $H_0(\Omega)$ coincides with $L_2(\Omega)$. Let $s > 0$ be an integer. We define the Sobolev space $H_s(\Omega)$ to consist of the functions u belonging to $L_2(\Omega)$ which have generalized derivatives $D^\alpha u$ for $|\alpha| \leqslant s$ that also belong to $L_2(\Omega)$. The norm in this space is defined as

$$\| u \|_s = \Big(\int_\Omega \sum_{|\alpha| \leqslant s} |D^\alpha u|^2 \, dx \Big)^{1/2}.$$

It is easy to see that this is a Hilbert space with scalar product

$$(u, v)_s = \int_\Omega \sum_{|\alpha| \leqslant s} D^\alpha u(x) \cdot \overline{D^\alpha v(x)} \, dx.$$

Since the convergence of a sequence $\{u_k\}$ in the space $H_s(\Omega)$ and the boundedness of the norm $\|u_k\|_s$ follow from the convergence of a sequence $\{u_k\}$ in $L_2(\Omega)$, the completeness of such a space is implied by the following theorem:

Theorem 3.1. If the sequence $\{u_k\}$ of elements from $H_s(\Omega)$ weakly converges in $L_2(\Omega)$ to a function $u \in L_2(\Omega)$ and the norms $\|u_k\|_s$ are uniformly bounded by a constant M, then $u \in H_s(\Omega)$ and $\|u\|_s \leqslant M$.

Proof. By Definition 1.1

$$\int D^\alpha u_k \cdot \varphi \, dx = (-1)^{|\alpha|} \int u_k D^\alpha \varphi \, dx, \quad |\alpha| \leqslant s, \quad \varphi \in C_0^\infty(\Omega).$$

Since the right-hand side has a limit equal to $(-1)^{|\alpha|} \int u \cdot D^\alpha \varphi \, dx$, the sequence $\{D^\alpha u_k\}$ weakly converges to some function $v_\alpha \in L_2(\Omega)$ in $L_2(\Omega)$ and

$$\int v_\alpha \varphi \, dx = (-1)^{|\alpha|} \int u \cdot D^\alpha \varphi \, dx, \quad |\alpha| \leqslant s, \quad \varphi \in C_0^\infty(\Omega).$$

This equality shows that v_α are the generalized derivatives of the function u and hence $u \in H_s$. Since

$$\| u_k \|_s^2 = \sum_{|\alpha| \leqslant s} \| D^\alpha u_k \|_0^2 \leqslant M^2,$$

from the weak convergence of $D^\alpha u_k$ to v_α it follows that

$$\| u \|_s^2 = \sum_{|\alpha| \leqslant s} \| v_\alpha \|_0^2 \leqslant M^2,$$

and this proves the theorem.

3.2. The spaces $H_s(\Omega)$ with non-negative integers prove to be very useful, particularly in studying boundary problems for elliptic and hyperbolic equations. In this, as in many other cases, it is however necessary to consider H_s spaces with fractional or negative values of s. We shall first define spaces of functions $u: \mathbb{R}^n \to \mathbb{C}$ as subspaces of the distribution space $\mathcal{S}'(\mathbb{R}^n)$.

Definition 3.1. A distribution $u \in \mathcal{S}'$ belongs to the space H_s if the integral

$$\int |\tilde{u}(\xi)|^2 (1+|\xi|^2)^s \, d\xi$$

is finite. The value of this integral is equal to the square of the norm of u in H_s, which is denoted by $\|u\|_s$.

It is not hard to see that if s is a non-negative integer, then this definition is equivalent to the earlier one given for the case $\Omega = \mathbb{R}^n$.

Theorem 3.2. The space H_s may be defined also as the completion of the space $C_0^\infty(\mathbb{R}^n)$ in norm $\|\cdot\|_s$.

Proof. If a function u belongs to the completion, it follows that $u \in \mathcal{S}'$ and the norm $\|u\|_s$ is finite. Hence $u \in H_s$.

Conversely, if $u \in H_s$, then there exists a sequence u_k of elements of $\mathcal{S}$ for which $\|u_k - u\|_s \to 0$. It follows that the function $\tilde{u}(\xi)(1+|\xi|^2)^{s/2}$ belongs to $L_2(\mathbb{R}^n)$ and therefore may be approximated in $L_2(\mathbb{R}^n)$ by functions of the form $v_k(\xi)(1+|\xi|^2)^{s/2}$ in L_2, where $v_k \in C_0^\infty(\mathbb{R}^n)$. Hence $v_k \in \mathcal{S}$ and the inverse Fourier transforms u_k of these functions also belong to $\mathcal{S}$; thus

$$\|u_k - u\|_s \to 0 \quad \text{as} \quad k \to \infty.$$

It remains to prove that the space $C_0^\infty(\mathbb{R}^n)$ is dense in $\mathcal{S}$ for the norm $\|\cdot\|_s$. But if $v \in \mathcal{S}$ we shall choose a function $\varphi \in C_0^\infty(\mathbb{R}^n)$ which equals 1 when $|x| \leqslant 1$ and assume that $v_\varepsilon(x) = v(x)\varphi(\varepsilon x)$. It is clear that $v_\varepsilon \in C_0^\infty(\mathbb{R}^n)$ and

$$\|v_\varepsilon - v\|_s^2 = \|v(x)[1-\varphi(\varepsilon x)]\|_s^2 \to 0$$

as $\varepsilon \to 0$ because the difference $v_\varepsilon - v$ is different from 0 only when $|x| \geqslant \varepsilon^{-1}$ and tends to 0 even in the topology of the space $\mathcal{S}$.

3.3. It is well known that the space of continuous linear functionals in L_2 is isomorphic to the space L_2 (see, for example, A. N. Kolmogorov and S. V. Fomin[1] or K. Yosida[1]). Definition 3.1 shows that the space of continuous linear functionals on the space H_s is isomorphic to the space H_{-s} and this isomorphism (between H_s and L_2) is determined by the operator

$$A_s u = \tilde{u}(\xi)(1+|\xi|^2)^{s/2}.$$

Thus we have

Theorem 3.3. Every linear continuous functional on the space H_s can be represented in the form

$$l(u) = \int uv \, dx, \quad \text{where} \quad v \in H_{-s}(\Omega).$$

In this case $\|l\| = \|v\|_{-s}$.

3.4. Note that the inequality

$$\left|\int f\bar{g} \, dx\right| \leqslant \|f\|_s \|g\|_{-s} \tag{3.1}$$

for functions f and $g \in C_0^\infty(\mathbb{R}^n)$ is easily obtained by passing to Fourier transforms:

$$\left|\int f(x)\overline{g(x)} \, dx\right| = (2\pi)^{-n} \left|\int \tilde{f}(\xi)\overline{\tilde{g}(\xi)} \, d\xi\right| \leqslant$$

$$\left[\int |\tilde{f}(\xi)|^2 (1+|\xi|^2)^s d\xi \int |\tilde{g}(\xi)|^2 (1+|\xi|^2)^{-s} d\xi\right]^{1/2} = \|f\|_s \|g\|_{-s}.$$

Inequality (3.1) is valid, of course, for functions $f \in H_s$ and $g \in H_{-s}$. Given f and g in the domain Ω, inequality (3.1) holds when $f \in H_s(\Omega)$ and $g \in H_{-s}(\Omega)$.

3.5. We shall now consider the mollifier u_h of functions $u \in H_s$.

Theorem 3.4. If $u \in H_s$, then $\|u_h - u\|_s \to 0$ as $h \to 0$.

Proof. As noted earlier, $\tilde{u}_h(\xi) = \tilde{u}(\xi)\tilde{\omega}(h\xi)$. But the functions $\tilde{\omega}(h\xi)$ are uniformly bounded and $\tilde{\omega}(h\xi) \to \tilde{\omega}(0) = 1$ uniformly on every compact manifold when $h \to 0$. Hence

$$\|\tilde{u}(\xi)[\tilde{\omega}(h\xi) - 1](1 + |\xi|^2)^{s/2}\|_0 \to 0,$$

that is,

$$\|u_h - u\|_s \to 0$$

as $h \to 0$. This proves the theorem.

3.6. It is quite clear that a function belonging to H_s for large s must be smooth. This result was obtained by S. L. Sobolev (for integers $s > 0$) and L. N. Slobodsky (for the general case) in the following imbedding theorem indicating the relation of the space H_s to the space C^k of smooth functions.

Theorem 3.5. If $u \in H_s$ when $s > k + n/2$, where $k \geqslant 0$ is an integer, then u coincides almost everywhere with a function v of class C^k and

$$\|v\|_{C^k} \leqslant A\|u\|_s, \tag{3.2}$$

where the constant A is independent of u.

Proof. Let first $u \in C_0^\infty(\mathbb{R}^n)$. By (2.2)

$$u(x) = (2\pi)^{-n}\int \tilde{u}(\xi)\, e^{i(x,\xi)}\, d\xi$$

and therefore

$$D^\alpha u(x) = (2\pi)^{-n}\int \xi^\alpha \tilde{u}(\xi)\, e^{i(x,\xi)}\, d\xi.$$

By the Cauchy-Schwarz-Bunyakovsky inequality

$$|D^\alpha u(x)| \leqslant (2\pi)^{-n}\int |\tilde{u}(\xi)|(1 + |\xi|^2)^{s/2}|\xi|^{|\alpha|}(1 + |\xi|^2)^{-s/2}\, d\xi \leqslant$$
$$(2\pi)^{-n}\left(\int |\tilde{u}(\xi)|^2(1 + |\xi|^2)^s\, d\xi\right)^{1/2}\left(\int (1 + |\xi|^2)^{|\alpha| - s}\, d\xi\right)^{1/2}.$$

The inequality $|\alpha| - s < -n$ is satisfied when $|\alpha| \leqslant k$ and the last integral converges. Hence

$$|D^\alpha u(x)| \leqslant C\|u\|_s, \quad |\alpha| \leqslant k,$$

for all functions $u \in C_0^\infty(\mathbb{R}^n)$. Adding these inequalities over α, when $|\alpha| \leqslant k$, we obtain

$$\|u\|_{C^k} \leqslant A\|u\|_s.$$

By Theorem 3.2 there exists a sequence $\{u_m\}$ of functions from C_0^∞ which converges to a function u with respect to the norm $\|\cdot\|_s$. Substituting u_m in (3.2) we obtain

$$\|u_m - u_{m'}\|_{C^k} \leqslant A\|u_m - u_{m'}\|_s.$$

Hence, it follows that the sequence $\{u_m\}$ is a Cauchy sequence in C^k and by the completeness of this space has a limit $v \in C^k$. Since the equality

$$(u, \varphi) = \lim_{m\to\infty}(u_m, \varphi) = (v, \varphi)$$

is valid for $\varphi \in C_0^\infty(\mathbb{R}^n)$, the functions u and v coincide almost everywhere, and this proves the theorem.

3.7. On the other hand, the spaces H_s are found to contain all of the distributions with compact support.

Theorem 3.6. $\mathscr{E}'(\mathbb{R}^n) \subset \bigcup_s H_s$.

Proof. Let $u \in \mathscr{E}'(\mathbb{R}^n)$, supp $u \subset K$, and u be on the order of m such that

$$|u(\varphi)| \leqslant C \sup_{x\in K} \sum_{|\alpha|\leqslant m} |D^\alpha \varphi(x)|, \quad \varphi \in C^\infty(\mathbb{R}^n).$$

By inequality (3.1) we have

$$|u(\varphi)| \leq CA \|\varphi\|_s,$$

where $s = m + [n/2] + 1$. Thus u is a continuous linear functional on the space H_s and by Theorem 3.3 there exists an element v of the space H_{-s} for which

$$u(\varphi) = \int v\varphi \, dx.$$

Having identified the functional u with this element v, we obtain the desired inclusion. This completes the proof.

Remark 3.1. From the proof of Theorem 3.6 it is evident that the finiteness of order of the distribution u, but not the compactness of the support, is of significance in this theorem. If $u \in \mathscr{D}'(\mathbb{R}^n)$ and $\varphi \in C_0^\infty(\mathbb{R}^n)$, then there exists an $s \in \mathbb{R}$ for which $\varphi u \in H_s$. Therefore, every element of the space $\mathscr{D}'(\mathbb{R}^n)$ may be represented as a limit of distributions u_s when $s \to -\infty$ such that $u_s \in H_s$. An example of a distribution on a straight line

$$u(\varphi) = \sum_{j=1}^{\infty} D^j\varphi(j)$$

shows that $\mathscr{D}'(\mathbb{R}^n)$ is not contained in the union of the spaces H_s; that is, there exists a distribution not belonging to any space H_s.

3.8. We shall now define the space $H_s(\Omega)$ for an arbitrary domain $\Omega \subset R^n$. This is done for non-negative integers in Section 3.1.

Definition 3.2. A functional $u \in \mathscr{D}'(\Omega)$ belongs to the space $H_s(\Omega)$ if the considered functional has an extension U belonging to the space $H_s = H_s(\mathbb{R}^n)$ and coinciding with u in Ω. The infimum of the norms $\|u\|_s$ over all such U is taken as a norm $\|U\|_s$ in this space.

In most problems encountered in the theory of differential equations the space $\overset{\circ}{H}_s(\Omega)$ is of enormous significance. This space consists of functionals u for which the extension U to a functional from H_s may be obtained assuming $U = 0$ outside Ω.

Theorem 3.7. The completion of the space $C_0^\infty(\Omega)$ in the norm $\|\cdot\|_s$ is contained in $\overset{\circ}{H}_s(\Omega)$.

Proof. If a function u belongs to the completion of $C_0^\infty(\Omega)$ in the norm $\|\cdot\|_s$, then there exists a sequence $\{u_k\}$ of functions from $C_0^\infty(\Omega)$ which may be extended by zero to values outside Ω and converges in the norm $\|\cdot\|_s$. The limit v of this sequence belongs to H_s by Theorem 3.2. In this case, the function v equals 0 outside Ω, since

$$v(\varphi) = \lim_{k \to \infty} u_k(\varphi) = 0$$

for functions $\varphi \in C_0^\infty(\mathbb{R}^n)$ whose supports lie outside Ω.

On the other hand, if $\varphi \in C_0^\infty(\Omega)$, then

$$v(\varphi) = \lim u_k(\varphi) = u(\varphi),$$

and so $v = u$ in Ω. This proves the theorem.

Remark. If the boundary of a domain Ω is sufficiently smooth, then $\overset{\circ}{H}_s(\Omega)$ may be shown to coincide with the completion of $C_0^\infty(\Omega)$ in the norm $\|\cdot\|_s$. (See L. Hörmander[3] and L. P. Volevich and B. P. Paneyakh.[1])

For $s \geq 0$ we can define $H_s(\Omega)$ without using the extension as the closure of the space $C^\infty(\Omega)$ in norm $\|\cdot\|_{s,\Omega}$, which is defined as follows:

if $[s] = k$ and $s \neq k$, then

$$\|u\|_{s,\,\Omega}^2=\sum_{|\alpha|\leqslant k}\int_\Omega |D^\alpha u|^2\,dx+\sum_{|\alpha|=k}\int_\Omega\int_\Omega \frac{|D^\alpha u(x)-D^\alpha u(y)|^2}{|x-y|^{n+2(s-k)}}\,dx\,dy.$$

For more details see L. N. Slobodetsky.[1]

3.9. Let us now consider the restriction of an element $u \in H_s$ to the plane $x^n = 0$.

Theorem 3.8. If $u \in H_s$ and $s > \frac{1}{2}$, then the restriction γu to the plane $x^n = 0$ is defined and belongs to the space $H_{s-1/2}(\mathbb{R}^{n-1})$, and

$$\|u(x',\,0)\|_{s-1/2}\leqslant C\|u\|_s, \tag{3.3}$$

where the norm on the left-hand side is the norm in $H_{s-1/2}(\mathbb{R}^{n-1})$.

Proof. The theorem follows if we prove the inequality (3.3) for $u \in C_0^\infty(\mathbb{R}^n)$, since this is a dense set in H_s.

Let $v(x') = u(x', 0)$. Then

$$\tilde{v}(\xi')=\int \tilde{u}(\xi)\,d\xi_n$$

and hence, for $s > \frac{1}{2}$

$$|\tilde{v}(\xi')|^2\leqslant\int(1+|\xi|^2)^s|\tilde{u}(\xi)|^2\,d\xi_n\int(1+|\xi|^2)^{-s}\,d\xi_n=$$

$$c_0(1+|\xi'|^2)^{\frac{1}{2}-s}\int(1+|\xi|^2)^s|\tilde{u}(\xi)|^2\,d\xi_n.$$

Multiplying both sides by $(1 + |\xi'|^2)^{s-1/2}$ and integrating with respect to ξ', we obtain the desired inequality (3.2).

This completes the proof.

Note that the theorem does not hold if $s \leqslant \frac{1}{2}$ (see L. N. Slobodetsky[1]).

Corollary 3.2. If $u \in H_s$ and $s > k + \frac{1}{2}$, then the restriction $\gamma_j u$ of derivatives $D_n^j u$, for $j \leqslant k$, to the plane $x^n = 0$ is defined. In addition $D_n^j u(x',\,0)\in H_{s-j-1/2}$ and

$$\|D_n^j u(x',\,0)\|_{s-j-1/2}\leqslant C\|u\|_s,\qquad j=0,\,1,\,\ldots,\,k.$$

The proof is analogous to the one given above.

3.10. The converse of the statement is also valid.

Theorem 3.9. Let $\varphi_0, \varphi_1, \ldots, \varphi_k$ be defined in $\mathbb{R}^{n-1}$ and $\varphi_j\in H_{s-j-1/2}$. Let $s > k + \frac{1}{2}$. Then there exists a function $u \in H_s$ for which

$$D_n^j u(x',\,0)=\varphi_j(x'),\qquad j-0,\,1,\,\ldots,\,k.$$

Proof. Let $h\in C_0^\infty(\mathbb{R})$ and $h(t) = 1$ in a neighborhood of the point $t = 0$. Assume that

$$\tilde{u}(\xi',\,x^n)=\sum_{j=0}^{k}\frac{1}{j!}(ix^n)^j h(x^n(1+|\xi'|^2)^{1/2})\,\tilde{\varphi}_j(\xi').$$

It is clear that $\tilde{u}(\xi', 0) = \tilde{\varphi}_0(\xi')$ and $D_n^j\tilde{u}(\xi', 0) = \tilde{\varphi}_j(\xi')$ when $j = 1, 2, \ldots, k$. It remains to verify that $u \in H_s$. To do so we note that

$$\|u\|_s^2\leqslant\sum_{j=0}^{k}\frac{1}{j!^2}\int|\tilde{\varphi}_j(\xi')|^2(1+|\xi'|^2)^{-j-1}|h^{(j)}(\xi_n(1+|\xi'|^2)^{-1/2})|^2(1+|\xi|^2)^s\,d\xi,$$

since the Fourier transform with respect to x^n of the function $(ix^n)^j g(x^n)$ is equal to $\tilde{g}^{(j)}(\xi_n)$ and the Fourier transform with respect to x^n of the function $h(\rho x^n)$ equals $\rho^{-1}\tilde{h}(\xi_n\rho^{-1})$.

Thus, replacing ξ_n by $\xi_n(1 + |\xi'|^2)^{1/2}$ we obtain

$$\|\dot{u}\|_s^2 \leqslant \sum_{j=0}^{k} \frac{1}{j!^2} \int |\tilde{\varphi}_j(\xi')|^2 (1+|\xi'|^2)^{-\frac{2j+1}{2}+s} |h^{(j)}(\xi_n)|^2 (1+\xi_n^2)^s \, d\xi' \, d\xi_n \leqslant C \sum_{j=0}^{k} \|\varphi_j\|_{s-j-1/2}^2,$$

as the function $\tilde{h}(\xi_n)$ belongs to the space $\mathcal{S}(\mathbb{R})$.

This completes the proof.

3.11. In the study of the smoothness of solutions of differential equations it is convenient to use spaces in which a function belongs to H_s only locally.

<u>Definition 3.3.</u> A function u belongs to the space $H_s^{loc}(\Omega)$ if $\varphi u \in H_s$ for any function $\varphi \in C_0^\infty(\Omega)$.

It is clear that, for example, $C^k(\Omega) \subset H_k^{loc}(\Omega)$.

For further reference we note the following useful property of the space $H_s^{loc}(\Omega)$.

<u>Theorem 3.10.</u> If $u \in \mathcal{D}'(\Omega)$ and at every point $x_0 \in \Omega$ there exists a function $\varphi \in C_0^\infty(\Omega)$ such that $\varphi(x) \neq 0$ and $\varphi u \in H_s(\Omega)$, then $u \in H_s^{loc}(\Omega)$.

<u>Proof.</u> Let $\varphi \in C_0^\infty(\Omega)$. In view of the Borel-Lebesgue lemma we can find a finite number of functions $\varphi_1, \ldots, \varphi_N \in C_0^\infty(\Omega)$ such that $\varphi_j u \in H_s(\Omega)$ and $\Phi = \sum_{j=1}^{N} |\varphi_j|^2 > 0$ in the support of φ. Then it follows that $\psi = \varphi/\Phi$ is in $C_0^\infty(\Omega)$, and we obtain

$$\varphi u = \sum_{j=1}^{N} \psi |\varphi_j|^2 u = \sum_{j=1}^{N} (\psi \bar{\varphi}_j)(\varphi_j u) \in H_s(\Omega).$$

This implies that $u \in H_s^{loc}(\Omega)$.

3.12. In studying the smoothness of certain distributions we shall make use of auxiliary spaces $H_{s,\delta}$, where $s \in \mathbb{R}$, $\delta \in \mathbb{R}$, $\delta > 0$. A norm for this space is introduced as follows:

$$\|u\|_{s,\delta}^2 = \int (1+|\xi|^2)^s (1+|\delta\xi|^2)^{-1} |\tilde{u}(\xi)|^2 \, d\xi.$$

It is clear that if $\delta > 0$, then

$$C_1(\delta)\|u\|_{s-1} \leqslant \|u\|_{s,\delta} \leqslant C_2(\delta)\|u\|_{s-1},$$

where $C_1(\delta)$ and $C_2(\delta)$ are positive constants depending on δ but independent of u. However, when $\delta \to 0$

$$\|u\|_{s,\delta} \to \|u\|_s \text{ for } u \in H_s.$$

Hence, if there exists a constant C independent of δ for which $\|u\|_{s,\delta} \leqslant C$ when $0 < \delta \leqslant \delta_0$ and $u \in H_{s-1}$, then by Fatou's theorem (see F. Riesz and B. Sz.-Nagy,[1] p. 50) it follows that $u \in H_s$ and $\|u\|_s \leqslant C$.

3.13. <u>Theorem 3.11.</u> Let Ω be a bounded domain in $\mathbb{R}^n$ and s and t be real numbers, where $s < t$. Then the imbedding operator i: $\mathring{H}_t(\Omega) \to \mathring{H}_s(\Omega)$ is completely continuous.

<u>Proof.</u> Let $\{u_k\} \in \mathring{H}_t(\Omega)$ and $\|u_k\|_t \leqslant 1$. Now we need to prove that a subsequence converging in H_s can be selected from this sequence.

Let $h \in C_0^\infty(\mathbb{R}^n)$ and $h(x) = 1$ in a neighborhood $\bar{\Omega}$. Then $u_k = hu_k$ and hence

$$\tilde{u}_k(\xi) = \int h(x) u_k(x) e^{-i(x,\xi)} \, dx = (2\pi)^{-n} \int \tilde{u}_k(\eta) \tilde{h}(\xi - \eta) \, d\eta.$$

Since $1 + |\xi| \leqslant (1 + |\xi - \eta|)(1 + |\eta|)$, for all $s \in \mathbb{R}$ the inequality

$$(1+|\xi|)^s \leqslant (1+|\xi-\eta|)^{|s|}(1+|\eta|)^s$$

is valid and hence

$$(1+|\xi|)^s|\tilde{u}_k(\xi)| \leqslant (2\pi)^{-n}\|(1+|\eta|)^s\tilde{u}_k(\eta)\|_0\|(1+|\xi-\eta|)^{|s|}\tilde{h}(\xi-\eta)\|_0.$$

Applying the same estimate to the equality

$$D_{\xi_j}\tilde{u}_k(\xi) = (2\pi)^{-n}\int \tilde{u}_k(\eta)\,D_{\xi_j}\tilde{h}(\xi-\eta)\,d\eta,$$

we obtain

$$(1+|\xi|)^s|D_{\xi_j}\tilde{u}_k(\xi)| \leqslant (2\pi)^{-n}\|(1+|\eta|)^s\tilde{u}_k(\eta)\|_0\|(1+|\xi-\eta|)^{|s|}D_{\xi_j}\tilde{h}(\xi-\eta)\|_0.$$

These estimates show that the sequence $\tilde{u}_k$ is uniformly bounded and equicontinuous on every compact set. Thus, we find a subsequence $\tilde{u}_{kj}$ converging uniformly on all compact subsets in $\mathbb{R}_n$. Changing notation, we may assume that the sequence $\tilde{u}_k$ itself is uniformly convergent on compact subsets.

Let ε be an arbitrary positive number. Then there exists an R sufficiently large so that $(1+|\xi|^2)^{(s-t)/2}<\varepsilon$ when $|\xi| > R$.

Hence

$$\|u_k-u_l\|_s \leqslant \varepsilon\|u_k-u_l\|_t + \left(\int_{|\xi|\leqslant R}(1+|\xi|^2)^s|\tilde{u}_k(\xi)-\tilde{u}_l(\xi)|^2\,d\xi\right)^{1/2}.$$

The second term on the right-hand side tends to 0 as k, $l \to \infty$ and the first is always $\leqslant 2\varepsilon$. Hence the sequence u_k is a Cauchy sequence in $H_s(\Omega)$, which proves the theorem.

3.14. In applications the following property of the spaces H_s often proves useful.

<u>Theorem 3.12.</u> Let s and s' be real numbers; $s > s'$ and $s \geqslant -n/2$. Let ω be an open set in $\mathbb{R}^n$ and $u \in C_0^\infty(\omega)$. Then for every $\varepsilon > 0$ there exists a $\delta > 0$ such that if diam $\omega \leqslant \delta$, then $\|u\|_{s'} \leqslant \varepsilon\|u\|_s$.

<u>Proof.</u> Assume that this is not true. Then there exists a sequence $\{u_k\}$ of functions in $C_0^\infty(\omega)$ such that supp $u_k \subset \omega_k$, diam $\omega_k \leqslant k^{-1}$, $\|u_k\|_s = 1$, $\|u_k\|_{s'} \geqslant c_0 > 0$. It may be assumed without loss of generality that all ω_k are neighborhoods of the origin and $\omega_{k+1} \subset \omega_k$.

The sequence $\{u_k\}$ is compact in H_s', in view of Theorem 3.11, and we may therefore assume that it has a limit in this space. In this case $u \in H_s$, $\|u\|_s \leqslant 1$, $\|u\|_{s'} \geqslant c_0 > 0$ and supp $u = \{0\}$.

It thus follows from Theorem 1.3 that there exists a polynomial $P(\xi)$ of degree $m \geqslant 0$ such that $u(x) = P(D)\delta(x)$. But then $\tilde{u}(\xi) = P(\xi) \not\equiv 0$ and $P(\xi)(1 + |\xi|^2)^{s/2} \in L_2(R^n)$, which is possible only when $m + s < -n/2$. Hence, for $s \geqslant -n/2$ we reach a contradiction, and this proves the theorem.

<u>Remark.</u> The theorem is false if $s < -n/2$. Indeed, if $u_k = \omega_{1/k}$, where ω stands for the mollifier kernel, then supp $\omega_k \to \{0\}$ as $k \to \infty$. In this case, $\tilde{u}_k \to 1$ and $\|u_k\|_s^2 \to \int (1+|\xi|^2)^s d\xi < \infty$ for $s < -n/2$. This is why the constant C in the inequality $\|u_k\|_{s'} \leqslant C\|u_k\|_s$, where $s' < s < -n/2$, cannot vanish when $k \to \infty$.

3.15. Unlike Theorem 3.11, the following statement is valid for all real numbers s.

<u>Theorem 3.13.</u> Let s and s' be real numbers and $s > s'$. Then for $\varepsilon > 0$ we can find a constant $C = C(\varepsilon, s, s')$ such that the inequality

$$\|u\|_{s'} \leqslant \varepsilon\|u\|_s + C_\varepsilon\|u\|_{s'-1}$$

holds for all functions $u \in H_s(\mathbb{R}^n)$. Here $C_\varepsilon = c_0\varepsilon^{1/(s'-s)}$, c_0 being independent of ε.

<u>Proof.</u> Passing to Fourier transforms the desired inequality is obtained immediately from the inequality

$$\lambda^{s'} \leqslant \varepsilon\lambda^{s} + C_{\varepsilon}\lambda^{s'-1},$$

which is valid for all $\lambda \geqslant 1$, $\varepsilon > 0$.

§4. Differential Operators with Constant Coefficients

4.1. A complete theory of differential operators with constant coefficients was developed mainly in the 1950s on the basis of distribution theory.

The equation

$$P(D)u = f(x),$$

where

$$P(D) = \sum_{|\alpha| \leqslant m} a_\alpha D^\alpha, \quad a_\alpha \in \mathbb{C}, \quad f \in \mathscr{E}',$$

upon the application of a Fourier transformation changes to the equation

$$P(\xi)\tilde{u}(\xi) = \tilde{f}(\xi),$$

and the problem of solving it is reduced to determining the possibility of division by a polynomial of a class of Fourier transforms.

If the polynomial $P(\xi)$ is such that $|P(\xi)| \geqslant c_0$ when $\xi \in \mathbb{R}_n$, where $c_0 = \text{const} > 0$, then the formula

$$u(x) = (2\pi)^{-n} \int \frac{\tilde{f}(\xi)}{P(\xi)} e^{ix\xi}\, d\xi \tag{4.2}$$

gives the solution to equation (4.1).

In view of Theorem 3.6 the functional f is in the space H_s for some $s \in \mathbb{R}$. From (4.2) it follows that $u \in H_s$. Indeed,

$$\|u\|_s^2 = \int |\tilde{u}(\xi)|^2 (1 + |\xi|^2)^s\, d\xi = \int \left|\frac{\tilde{f}(\xi)}{P(\xi)}\right|^2 (1 + |\xi|^2)^s\, d\xi \leqslant c_0^{-2} \|f\|_s^2.$$

To study the solvability of equation (4.1) it is sufficient to consider a particular case, when $f(x) = \delta(x)$.

Definition 4.1. A distribution $E \in \mathscr{D}'(\mathbb{R}^n)$ is called a *fundamental solution* for the differential operator P(D) if

$$P(D)E(x) = \delta(x),$$

where δ is the Dirac measure.

In Section 2.6 we have shown that the equation $f = f * \delta$ is valid for any distribution $f \in \mathscr{D}'(\mathbb{R}^n)$.

Hence, if $f \in \mathscr{E}'(\mathbb{R}^n)$, then from (2.10) it follows that

$$f = f * P(D)E = P(D)(E * f),$$

that is, $u = E * f$ is a solution of equation (4.1).

4.2. Formula (4.2) will not make sense when the polynomial $P(\xi)$ has real zeros. Deforming the range of integration, this formula can, however, be transformed to represent an n-dimensional surface in $\mathbb{C}^n$, since $\tilde{f}$ is an entire analytic function. To do so, the integration surface is selected so that $P(\zeta) \neq 0$ on this surface.

Theorem 4.1. To every differential operator with constant coefficients there exists a fundamental solution.

Proof. 1°. If $n = 1$, then it is sufficient to find a $\tau \in \mathbb{R}$ such that $P(\xi + i\tau) \neq 0$ when $\xi \in \mathbb{R}$, and assume that

$$E(\varphi) = (2\pi)^{-n} \int \frac{\tilde{\varphi}(-\xi - i\tau)}{P(\xi + i\tau)} d\xi.$$

In fact

$$(P(D)E)(\varphi) = E(P(-D)\varphi) =$$

$$(2\pi)^{-n} \int \frac{\overline{P(-D)\varphi(-\xi - i\tau)}}{P(\xi + i\tau)} d\xi = (2\pi)^{-n} \int \frac{P(\xi + i\tau)\tilde{\varphi}(-\xi - i\tau)}{P(\xi + i\tau)} d\xi =$$

$$(2\pi)^{-n} \int \tilde{\varphi}(-\xi - i\tau) d\xi = (2\pi)^{-n} \int \tilde{\varphi}(\xi) d\xi = \varphi(0).$$

2°. For $n > 1$ we can always make the coefficient of η_1^m in the polynomial different from 0 by means of a rotation $\xi_j = \sum \alpha_j^k \eta_k$. This follows from the fact that this coefficient is equal to $P_m(\alpha_1^1, \alpha_2^1, \ldots, \alpha_n^1)$, where P_m denotes the homogeneous part of the polynomial P of degree m, which does not vanish identically.

Thus, we can assume that $D_1^m P = i^{-m} m!$ Fixing the value of the vector $\xi' = (\xi_2, \ldots, \xi_n)$ we arrive at a situation considered in 1°, and there exists a τ such that $|\tau| \leqslant m - 1$ and $|P(\xi_1 + i\tau, \xi')| > 1$, since $\min_{\xi_1} |\xi_1 + i\tau - \lambda| > 1$ for the roots λ of the equation $P(\lambda, \xi') = 0$. By continuity this inequality also holds for all points ξ' lying close to the fixed one.

Thus to every point ξ' there corresponds a τ and a neighborhood ω such that

$$|P(\xi_1 + i\tau, \xi')| > 1 \text{ when } \xi' \in \omega.$$

From the covering of the space $\mathbb{R}_{n-1}$ by neighborhoods ω we can select a locally finite covering by neighborhoods $\omega_1, \omega_2, \ldots$, to which correspond the values $\tau_1, \tau_2, \ldots$, $|\tau_j| \leqslant m + 1$. We shall now replace ω_2 by $\omega_2' = \bar{\omega}_2 \setminus \omega_1$ and ω_3 by $\omega_3' = \bar{\omega}_3 \setminus \omega_1 \cup \omega_2$ and so on. We obtain a partition of the space $\mathbb{R}_{n-1}$ into nonoverlapping regions. Now define H as the set of points $(\xi_1 + i\tau, \xi')$, where $\tau = \tau_j$ when $\xi' \in \omega_j'$. H is called a *Hörmander ladder*. Let

$$E(\varphi) = (2\pi)^{-n} \int_H \frac{\tilde{\varphi}(-\xi_1 - i\tau, \xi')}{P(\xi_1 + i\tau, \xi')} d\xi, \quad \varphi \in \mathscr{D}(\mathbb{R}^n).$$

Then the integral exists since $|P(\xi_1 + i\tau, \xi')| > 1$ and the function $\tilde{\varphi}(\xi_1 + i\tau, \xi')$ tends to 0 faster than any degree of $1/|\xi|$ as $|\xi| \to \infty$, uniformly in τ for $|\tau| \leqslant m + 1$. It is also clear that E is a distribution in $\mathscr{D}'(\mathbb{R}^n)$.

In this case

$$(P(D)E)(\varphi) = E(P(-D)\varphi) = (2\pi)^{-n} \int_H \tilde{\varphi}(-\xi_1 - i\tau, \xi') d\xi =$$

$$\sum_j (2\pi)^{-n} \int_{\xi' \in \omega_j} d\xi' \int_{-\infty}^{\infty} \tilde{\varphi}(\xi_1 - i\tau_j, \xi') d\xi_1.$$

According to Cauchy's theorem the internal integral taken in the plane $\operatorname{Im} \zeta_i = \tau_j$ along a line parallel to the real axis is equal to the integral along this axis. Therefore $(P(D)E)(\varphi) = (2\pi)^{-n} \int \tilde{\varphi}(\xi) d\xi = \varphi(0)$, that is $P(D)E(x) = \delta(x)$.

This completes the proof.

4.3. We shall now consider the important class of elliptic operators.

Definition 4.2. An operator P(D) is called *elliptic* of order m if its characteristic form $P_m(\xi)$ does not vanish when $\xi \in \mathbb{R}_n \setminus 0$.

The polynomial $P_m(\xi)$ is homogeneous. Therefore, from the absence of zeros on the sphere $|\xi| = 1$ it follows that

$$|P_m(\xi)| \geqslant c_0 |\xi|^m, \tag{4.3}$$

where

$$c_0 = \inf_{|\xi|=1} |P_m(\xi)| > 0.$$

Theorem 4.2. Let $P(D)$ be an elliptic operator of order m, Ω a domain in $\mathbb{R}^n$, $P(D)u = f$ in Ω, $u \in \mathscr{D}'(\Omega)$ and $f \in H_s(\Omega)$. Then $u \in H^{loc}_{s+m}(\Omega)$.

Let $h \in C_0^\infty(\Omega)$ and $h(x) = 1$ when $x \in \omega$ and let $\varphi \in C_0^\infty(\omega)$. Then there exists a constant $C = C(s)$ such that

$$\|\varphi u\|_{s+m} \leqslant C(\|Pu\|_s + \|hu\|_{s+m-1})$$

for all $u \in H^{loc}_{s+m}(\Omega)$.

Proof. Let $x_0 \in \Omega$, $\varphi \in C_0^\infty(\Omega)$ and $\varphi(x) = 1$ in a neighborhood ω of x_0. Then $\varphi u \in \mathscr{E}'(\Omega)$ and in view of Theorem 3.6 there exists $t \in \mathbb{R}$ for which $\varphi u \in H_t(\Omega)$. If $t < s + m$, then we can consider an equation satisfied by the function $u_1 = \varphi_1 u$, where $\varphi_1 \in C_0^\infty(\omega)$ and $\varphi_1 = 1$ in a neighborhood ω_1 of x_0. We have

$$P_m(D)\, u_1 = f_1, \tag{4.4}$$

where $f_1 \in H_{t_1}$ and $t_1 = \min(t - m + 1, s)$, since the operator $P - P_m$ is of order $m - 1$ and $(P - P_m)(\varphi_1 u) \in H_{t-m+1}$; $Qu = P\varphi_1 u - \varphi_1 Pu$ is also an operator of order $m - 1$; its coefficients are equal to 0 outside ω so that $Qu \in H_{t-m+1}$. Finally, $\varphi_1 Pu = \varphi_1 f \in H_s$, since $f \in H_s$.

After the application of a Fourier transformation to both sides of (4.4) we obtain $P_m(\xi)\tilde{u}_1(\xi) = \tilde{f}_1(\xi)$ such that

$$|\tilde{f}_1(\xi)| \geqslant c_0 |\xi|^m |\tilde{u}_1(\xi)|,$$

and therefore

$$(1 + |\xi|^2)^m |\tilde{u}_1(\xi)|^2 \leqslant C(|\tilde{f}_1(\xi)|^2 + |\tilde{u}_1(\xi)|^2).$$

Multiplying both sides by $(1 + |\xi|^2)^{t_1}(1 + |\delta\xi|^2)^{-1}$, where $\delta > 0$, and integrating with respect to $\xi \in \mathbb{R}_n$, we obtain

$$\int (1 + |\xi|^2)^{t_1+m}(1 + |\delta\xi|^2)^{-1} |\tilde{u}_1(\xi)|^2 d\xi \leqslant$$

$$C \int (1 + |\xi|^2)^{t_1}(1 + |\delta\xi|^2)^{-1}(|\tilde{f}_1(\xi)|^2 + |\tilde{u}_1(\xi)|^2)\, d\xi$$

or

$$\|u_1\|^2_{t_1+m,\,\delta} \leqslant C(\|f_1\|^2_{t_1,\,\delta} + \|u_1\|^2_{t_1,\,\delta}).$$

The right-hand side of this inequality is uniformly bounded in δ, since $u_1 \in H_t$, $t \geqslant t_1$, and $f \in H_t$. Hence the left-hand side of the inequality is also bounded uniformly in δ and, in accordance with Section 3.12, we obtain that $u_1 \in H_{t_1+m}$. If $t + 1 \geqslant s + m$, then $t_1 + m = \min(t + 1, s + m) = s + m$, and this proves the theorem by Theorem 3.10.

If $t + 1 < s + m$, then we can repeat the argument by introducing a function $\varphi_2 \in C_0^\infty(\omega_1)$ such that $\varphi_2 = 1$ in a neighborhood ω_2 of x. We now obtain that $u_2 = \varphi_2 u \in H_{t_2+m}$, where $t_2 = \min(t - m + 2, s)$. If $t + 2 \geqslant s + m$, then the proof is complete; for $t + 2 < s + m$ the argument is repeated.

To prove the second statement, note that

$$P(\varphi u) = \varphi Pu + \sum_{|\alpha|=1}^{m} \frac{1}{\alpha!} P^{(\alpha)}(D)(hu)\, D^\alpha \varphi.$$

Since, as noted above,

$$\|\varphi u\|_{s+m} \leqslant C(\|P(\varphi u)\|_s + \|\varphi u\|_{s+m-1}),$$

we obtain

$$\|\varphi u\|_{s+m} \leqslant C_1(\|Pu\|_s + \|hu\|_{s+m-1}).$$

and this proves the theorem.

Corollary 4.1. If $u \in D'(\Omega)$ and $Pu \in C_0^\infty(\Omega)$, then $u \in C^\infty(\Omega)$.

4.4. If an operator $P(D)$ of order m is elliptic and P_m is a real-valued function, then the so-called Gårding inequality

$$\operatorname{Re}\int Pu \cdot \bar{u}\, dx \geqslant c_0 \|u\|^2_{m/2} - C_1 \|u\|^2_{(m-1)/2} \tag{4.5}$$

holds for functions $u \in C_0^\infty(\Omega)$. Here c_0 is a constant similar to that in (4.3), and C_1 depends on the value of lower order coefficients of the operator P.

To prove (4.5) it is sufficient to pass to a Fourier transform $\tilde{u}$ of function u, which gives an inequality

$$\operatorname{Re}\int P(\xi)\,|\tilde{u}(\xi)|^2\, d\xi \geqslant c_0 \int (1+|\xi|^2)^{m/2} |\tilde{u}(\xi)|^2\, d\xi - C_1 \int (1+|\xi|^2)^{(m-1)/2} |\tilde{u}(\xi)|^2\, d\xi,$$

equivalent to the inequality

$$\operatorname{Re} P(\xi) \geqslant c_0 (1+|\xi|^2)^{m/2} - C_1 (1+|\xi|^2)^{(m-1)/2},$$

resulting from (4.3).

4.5. Let us consider the elliptic equation

$$P(D)\,u = f \tag{4.6}$$

in a bounded domain $\Omega \subset \mathbb{R}^n$ for $f \in H_s(\Omega)$ with arbitrary s.

If $\varphi \in \mathscr{E}'(\Omega)$ is a solution of the equation ${}^tP\varphi = 0$ and $\operatorname{supp} \varphi \subset \Omega$, then, by Theorem 4.2, $\varphi \in C_0^\infty(\Omega)$, since the operator ${}^tP(D) = \overline{P(D)}$ is obtained from P(D) upon replacing the coefficients by conjugate complex numbers; it is therefore also elliptic. We denote by $N(\Omega)$ the linear space of solutions of the equation ${}^tP\varphi = 0$ which belong to the space $C_0^\infty(\Omega)$.

The Gårding inequality (4.5) for the operator ${}^tP$ and the function $\varphi \in N(\Omega)$ shows that

$$c_0 \|\varphi\|^2_{m/2} \leqslant C_1 \|\varphi\|^2_{(m-1)/2}.$$

Thus, a set of functions belonging to $N(\Omega)$ and lying in the ball $\|\varphi\|_{(m-1)/2} \leqslant 1$ is compact by Theorem 3.11. From Kolmogorov's theorem it follows that the space $N(\Omega)$ must have finite dimension.

Multiplying both sides of (4.6) by $\varphi \in C_0^\infty(\Omega)$ and integrating with respect to Ω we obtain

$$\int u\, {}^tP\varphi\, dx = \int f\varphi\, dx. \tag{4.7}$$

From this it is evident that for (4.6) to be solvable the following equality must be valid:

$$\int f\varphi\, dx = 0, \quad \varphi \in N(\Omega). \tag{4.8}$$

We shall now show that this condition is sufficient for solvability.

Theorem 4.3. If a function $f \in H_s(\Omega)$ satisfies condition (4.8), then there exists a solution u of equation (4.6) of class $H_{s+m}(\Omega)$, and

$$\|u\|_{s+m} \leqslant C\|f\|_s$$

for some constant C independent of f.

Proof. To prove the theorem it is sufficient to prove that the inequality

$$\|\varphi\|_{-s} \leqslant C \|{}^tP\varphi\|_{-s-m} \tag{4.9}$$

is satisfied for all functions $\varphi \in C_0^\infty(\Omega)$ orthogonal to the space $N(\Omega)$. From this inequality, which by continuity is also valid for $\varphi \in H_{-s}$ orthogonal to $N(\Omega)$, by the Hahn-Banach theorem there exists an element

$u \in H_s(\Omega)$ satisfying (4.7) for such φ. Since this equality also holds for $\varphi \in N(\Omega)$, we find that it is valid for all $\varphi \in H_{-s}(\Omega)$; that is, u is a solution of equation (4.6) of class $H_s(\Omega)$ and

$$\|u\|_{s+m} \leqslant C\|f\|_s.$$

To prove inequality (4.9) note that

$$(1+|\xi|^2)^{-s} \leqslant C_1|\bar{P}(\xi)|^2(1+|\xi|^2)^{-s-m}+C_2(1+|\xi|^2)^{-s-1}.$$

Multiplying both sides by $|\tilde{\varphi}(\xi)|^2$ and integrating with respect to $\xi \in \mathbb{R}_n$, we obtain

$$\|\varphi\|_{-s}^2 \leqslant C_1\|{}^tP\varphi\|_{-s-m}^2+C_2\|\varphi\|_{-s-1}^2. \tag{4.10}$$

We now show that for functions φ orthogonal to $N(\Omega)$ the last term may be neglected so that (4.9) holds. Indeed, assume it to be false, so that there exists a sequence $\{\varphi_k\}$ of functions from $C_0^\infty(\Omega)$ for which $\|\varphi_k\|_{-s} = 1$, $\|{}^tP\varphi_k\|_{-s-m} \leqslant k^{-1}$, and $\varphi_k \perp N(\Omega)$. By Theorem 3.11 this sequence is compact in $H_{-s-1}(\Omega)$ and we can select a subsequence converging to φ. From (4.10), also valid for the difference $\varphi_k - \varphi_l$, it follows that this subsequence φ_{k_j} also converges in $H_{-s}(\Omega)$. Therefore, ${}^tP\varphi = 0$ and $\|\varphi\|_{-s} = 1$. But this contradicts the condition that the function φ is orthogonal to $N(\Omega)$. This proves inequality (4.9).

4.6. Equations of principal type, an important class of equations, were first introduced by L. Hörmander.

Definition 4.3. An operator $P(D)$ is called an *operator of principal type* if $|\text{grad } P_m(\xi)| \neq 0$ for $\xi \in \mathbb{R}_n \setminus 0$.

Since P_m is a homogeneous polynomial, the Euler identity holds; that is,

$$\sum_{j=1}^{n} \xi_j \frac{\partial P_m(\xi)}{\partial \xi_j} = mP_m(\xi).$$

Thus, if $\text{grad } P_m(\xi) = 0$, then $P_m(\xi) = 0$. This reveals that, in particular, elliptic operators are operators of principal type. Other examples of such operators are D_1 and $D_1 + iD_2$, operators which are not elliptic for $n > 2$.

We see that operators of principal type can have characteristic points; that is, the polynomial $P_m(\xi)$ can have real zeros.

Local solvability of equations of principal type with constant coefficients is proved, as above, using the Hahn-Banach theorem and the next theorem.

Theorem 4.4. If the operator $P(D)$ is of principal type, then there exists a sufficiently small neighborhood ω of the origin and a constant C such that for $\varphi \in C_0^\infty(\omega)$ the estimate

$$\|\varphi\|_{m-1} \leqslant C\varepsilon\|P(D)\varphi\|_0 \tag{4.11}$$

is valid. Here $\varepsilon = \text{diam } \omega$ and C is independent of ε and φ.

Proof. Note that the inequality

$$\sum_{k=1}^{n} |\partial P_m(\xi)/\partial \xi_k|^2 \geqslant c_0 > 0$$

is satisfied for $|\xi| = 1$. Therefore

$$\sum_{k=1}^{n} |\partial P_m(\xi)/\partial \xi_k|^2 \geqslant c_0|\xi|^{2(m-1)}$$

and for $m \geqslant 2$

$$\sum_{k=1}^{n} |\partial P(\xi)/\partial \xi_k|^2 \geqslant \frac{c_0}{2} |\xi|^{2(m-1)} - C_1 |\xi|^{2(m-2)}. \tag{4.12}$$

If m = 1, then the last inequality is valid for $C_1 = 0$. Also note that for the differential operator Q(D) with constant coefficients the equality

$$Q(D)\, x^k - x^k Q(D) = iQ^{(k)}(D)$$

is valid.

Let us consider the integral I_k of Im $2x^k P(D)\varphi \cdot \overline{P^{(k)}(D)\varphi}$ in $\mathbb{R}^n$. We have

$$I_k = -i \int x^k [P(D)\varphi \cdot \overline{P^{(k)}(D)\varphi} - \overline{P(D)\varphi} \cdot P^{(k)}(D)\varphi]\, dx =$$

$$-i \int [\overline{P^k(D)\, x^k P(\bar{D})} - \overline{P(\bar{D})}\, x^k P^{(k)}(D)]\, \varphi \cdot \bar{\varphi}\, dx =$$

$$\int [\overline{P^{(kk)}(\bar{D})}\, P(D) - P^{(k)}(D)\, \overline{P^k(\bar{D})}]\, \varphi \cdot \bar{\varphi}\, dx + O(\| P(D)\varphi \| \| x^k \varphi \|_{m-1}).$$

Hence

$$\int |P^{(k)}(D)\varphi|^2 dx = \int P(D)\varphi \cdot \overline{P^{(kk)}(D)\varphi}\, dx - I_k + O(\| P(D)\varphi \| \| x^k \varphi \|_{m-1}). \tag{4.13}$$

From (4.12) it easily follows that for $m \geqslant 2$

$$\|\varphi\|_{m-1}^2 = \int (1 + |\xi|^2)^{m-1} |\tilde{\varphi}(\xi)|^2 d\xi \leqslant C_2 \int \left(\sum_{k=1}^{n} |\partial P/\partial \xi_k|^2 + 1 \right) |\tilde{\varphi}(\xi)|^2 d\xi =$$

$$C_2 \sum_{k=1}^{n} \int |P^{(k)}(D)\varphi|^2 dx + C_2 \|\varphi\|_0^2.$$

Therefore

$$\|\varphi\|_{m-1} \leqslant C_2 \left(\sum_{k=1}^{n} \left| \int P(D)\varphi \cdot \overline{P^{(kk)}(D)\varphi}\, dx \right| + \sum_{k=1}^{n} |I_k| + \|\varphi\|_0^2 + \right.$$

$$\left. \sum_{k=1}^{n} \| P(D)\varphi \| \| x^k \varphi \|_{m-1} \right) \leqslant C_3 (\| P\varphi \| \|\varphi\|_{m-2} + \varepsilon \| P\varphi \| \|\varphi\|_{m-1} + \|\varphi\|_{m-2}^2),$$

where $\varepsilon = \max_{x \in \operatorname{supp} \varphi} |x^k|$. According to Friedrichs' inequality

$$\|\varphi\|_{m-2} \leqslant C_4 \varepsilon \|\varphi\|_{m-1}.$$

Hence

$$\|\varphi\|_{m-1}^2 \leqslant C_5 \varepsilon (\| P\varphi \| \|\varphi\|_{m-1} + \varepsilon \|\varphi\|_{m-1}^2).$$

If $2C_5 \varepsilon^2 < 1$, then it follows that

$$\|\varphi\|_{m-1} \leqslant 2C_5 \varepsilon \| P\varphi \|,$$

that is, inequality (4.11) holds.

If m = 1, then $P^{(kk)}(D) = 0$ and (4.13) takes the form

$$-I_k = \int |P^{(k)}(D)\varphi|^2 dx + O(\|P(D)\varphi\| \| x^k \varphi \|).$$

Hence

$$\|\varphi\|_0^2 \leqslant C_6 \sum_{k=1}^{n} \left| \int P^{(k)}(D)\varphi \right|^2 dx + O(\| P(D)\varphi \| \| x^k \varphi \|) \leqslant C_7 \varepsilon \| P(D)\varphi \| \|\varphi\|,$$

and we again come to (4.13).

This proves the theorem.

Corollary 4.2. If P(D) is an operator of principal type, then for every function f in H_{1-m} we can find a function $u \in L_2$ which satisfies the equation P(D)u = f in a sufficiently small neighborhood ω of the origin. In this case

$$\|u\|_0 \leqslant C\varepsilon \|f\|_{1-m}, \quad \varepsilon = \operatorname{diam} \omega.$$

Remark 4.1. It is not difficult to check that conversely from (4.11) it follows that the operator P is an operator of principal type.

4.7. We shall consider the Cauchy problem for a differential equation with constant coefficients in the following form: find a solution u of the equation $P(D)u = f$ for which supp u lies in the half-space $x^1 \geqslant 0$.

The more general problem, given a function φ, to find a solution u of the differential equation satisfying the condition $u - \varphi = O(|x^1|^m)$ as $x^1 \to 0$ (where m is the order of the operator P), reduces to the given one if we replace u by $u + \varphi$ and consider the problem for $x^1 \geqslant 0$ assuming $f = 0$ for $x^1 < 0$. Having solved this new problem and then the analogous one obtained after replacing x^1 by $-x^1$, we can find a solution to this formally more general problem.

4.8. We shall now show the uniqueness of the solution to the Cauchy problem.

Theorem 4.5. Let P(D) be an operator of order m where the coefficient of D_1^m is equal to 0. Then there exists a neighborhood ω of the origin and a function $u \in C^m(\omega)$ such that $P(D)u = 0$ and $u = 0$ when $x' > 0$, but $0 \in \operatorname{supp} u$.

Proof. Let $P = P_m + P_{m-1} + \ldots + P_0$, where P_j is homogeneous of degree j and $P_m(\xi) \not\equiv 0$. Let ξ be a vector such that $P_m(\xi) \neq 0$ and $e_1 = (1, 0, \ldots, 0)$. We shall study the solution of the equation

$$P(se_1 + t\xi) = 0 \tag{4.14}$$

for large s. To do so, we set $t = sz$ which reduces (4.14) to an equation of the form

$$P_m(e_1 + z\xi) + (1/s) P_{m-1}(e_1 + z\xi) + \ldots + (1/s)^m P_0(e_1 + z\xi) = 0.$$

When $s = \infty$ this equation has a root $z = 0$ since $P_m(e_1) = 0$ but $P_m(e_1 + z\xi) \not\equiv 0$. Such an equation has a solution $z = z(s)$ which is an analytic function of $(1/s)^{1/p}$ for some integer p (see A. I. Markushevich,[1] pp. 520-543). This means that (4.14) has a solution

$$t(s) = s \sum_{j=1}^{\infty} C_j (s^{-1/p})^j,$$

analytic for $|s^{1/p}| > M$, where M is a constant. Thus we have with a constant C

$$|t(s)| \leqslant C|s|^{1-1/p}, \quad |s| > (2M)^p.$$

Now choose a number ρ such that $1 - 1/p < \rho < 1$ and set

$$u(x) = \int_{i\tau-\infty}^{i\tau+\infty} e^{isx^1 + it(s)(x,\,\xi) - (s/i)^\rho} ds, \tag{4.15}$$

with $\tau > (2M)^p$. Here we define $(s/i)^\rho$ so that it is real and positive when s is on the positive imaginary axis, and we choose a fixed branch of this function for $\operatorname{Im} s > 0$.

The integral (4.15) converges when $x^1 \geqslant 0$ and is independent of τ, since

$$\operatorname{Re}(isx^1 + it(s)(x,\ \xi) - (s/i)^\rho) \leqslant -\tau x^1 + C|x||\xi||s|^{1-1/p} - |s|^\rho \cos\frac{\pi\rho}{2} \leqslant -\tau x^1 - c|s|^\rho,$$

where $C = \frac{1}{2}\cos(\pi\rho/2)$, $|x| \leqslant C_1$ and $|s|$ is sufficiently large. This estimate also shows that when x is in a compact set the integral (4.15) is uniformly convergent, even after an arbitrary number of differentiations with respect to x. Hence $u \in C^\infty$. Using (4.14) we conclude that $P(D)u = 0$.

We have shown that

$$|u(x)| \leqslant e^{-\tau x^1} \int_{-\infty}^{+\infty} e^{-c|\sigma|^\rho}\, d\sigma$$

for all $\tau \geqslant \tau_0$. Hence it follows when $\tau \to +\infty$ that $u(x) = 0$ if $x^1 > 0$.

To show that $0 \in \operatorname{supp} u$ we note that the function u depends only on x^1 when $(x, \xi) = 0$, where we use the notation

$$v(x^1) = \int_{i\tau-\infty}^{i\tau+\infty} e^{isx^1 - (s/i)^\rho}\, ds = \int_{-\infty}^{\infty} e^{i(\sigma+i\tau)x^1 - (\tau - i\sigma)^\rho}\, d\sigma,$$

such that

$$v(x^1)\, e^{\tau x^1} = \int_{-\infty}^{\infty} e^{ix^1\sigma - (\tau - i\sigma)^\rho}\, d\sigma.$$

By Parseval's formula we obtain

$$\int_{-\infty}^{\infty} |v(x^1)\, e^{\tau x^1}|^2\, dx^1 = 2\pi \int_{-\infty}^{\infty} |e^{-(\tau - i\sigma)^\rho}|^2\, d\sigma.$$

But we have already shown that $v(x^1) = 0$ when $x^1 > 0$. Now we have

$$|(\tau - i\sigma)^\rho| \leqslant (\tau + |\sigma|)^\rho \leqslant \tau^\rho + |\sigma|^\rho$$

since $\rho < 1$. Hence

$$\int_{-\infty}^{0} |v(x^1)\, e^{\tau x^1}|^2\, dx^1 \geqslant e^{-2\tau^\rho} \int_{-\infty}^{\infty} e^{-2|\sigma|^\rho}\, d\sigma. \tag{4.16}$$

But if v vanishes in $(-\varepsilon, 0)$ for some $\varepsilon > 0$, then the left-hand side must be $O(e^{-2\varepsilon\tau})$ when $\tau \to +\infty$, which contradicts (4.16). This completes the proof.

Corollary 4.3. If every solution of the equation $P(D)u = 0$ defined in some neighborhood of the origin is an analytic function, then P is an elliptic operator.

Proof. Assume P is not an elliptic operator. Then there exists a vector $\xi \neq 0$ for which $P_m(\xi) = 0$ and $P_m \not\equiv 0$. Assuming the vector ξ to be parallel to the x-axis, we can construct a solution of the equation $P(D)u = 0$ which satisfies Theorem 4.5 and is therefore nonanalytic.

4.9. I. G. Petrowsky has proved that the statement converse to Corollary 4.3 is valid.

Theorem 4.6. If P(D) is an elliptic operator, then every generalized solution u of the equation $P(D)u = 0$ in Ω is an analytic function.

Proof. By Corollary 4.1 the generalized solution belongs to $C^\infty(\Omega)$.

We shall show that u is an analytic function in a small neighborhood of an arbitrary point x_0 in Ω. Let a ball K_{2r} of radius 2r, centered at x_0, lie in Ω. Let $h \in C_0^\infty(K_{2r})$, where $h(x) = 1$ when $|x - x_0| \leqslant r_1$, $h(x) = 0$ when $|x - x_0| > r_1 + \varepsilon$, and $|D^\alpha h| \leqslant C\varepsilon^{-|\alpha|}$ when $|\alpha| \leqslant m$ for some $\varepsilon > 0$, where $r \leqslant r_1 \leqslant 2r - \varepsilon$.

The support of the function $f = P(hu)$ lies in the annulus $\{x: r_1 \leqslant |x - x_0| \leqslant r_1 + \varepsilon\}$ and $f = \sum_{|\alpha|=1}^{m} \frac{1}{\alpha!} P^{(\alpha)}(D)u \cdot D^\alpha h$. Let N be an integer, $N > m$ and $N > n/2 + 1$.

From Theorem 4.2 we have

$$\|hu\|_N \leqslant C(\|f\|_{N-m} + \|hu\|_{N-1}).$$

If r is sufficiently small, then $2C\|hu\|_{N-1} \leqslant \|hu\|_N$. Hence

$$\|hu\|_N \leqslant 2C\|f\|_{N-m} \leqslant C_1 \sum_{j=1}^{m} \varepsilon^{-j}\|u\|_{N-j,\, r_1+\varepsilon},$$

where $\|\cdot\|_{s,\rho}$ is a norm in the Sobolev space $H_s(K\rho)$. It is clear that $P(D) \times (D^\alpha u) = 0$ for all α, and therefore

$$\|hD^\alpha u\|_N \leqslant C_1 \sum_{j=1}^{m} \varepsilon^{-j}\|D^\alpha u\|_{N-j,\, r_1+\varepsilon}. \tag{4.17}$$

We shall now show that there exists a constant $A \geqslant 1$ such that for any $\varepsilon > 0$ and for all integers $j \geqslant 0$

$$\sum_{|\alpha|=j} \|D^\alpha u\|_{N,\, r_1} \leqslant \varepsilon^{-j} A^{j+1}, \tag{4.18}$$

where $j\varepsilon \leqslant 2r - r_1$ and the constant A is independent of ε and j.

This estimate is, of course, valid when $j = 0$ provided A is sufficiently large. Let us assume that it holds for $j < k$, where $k \geqslant 1$, and prove it for $j = k$. Let $h\varepsilon \leqslant 2r - r_1$. Then by (4.17) we have

$$\sum_{|\alpha|=k} \|D^\alpha u\|_{N,\, r_1} \leqslant \sum_{|\alpha|=k} \|hD^\alpha u\|_N \leqslant C_1 \sum_{|\alpha|=k} \sum_{j=1}^{m} \varepsilon^{-j}\|D^\alpha u\|_{N-j,\, r_1+\varepsilon}.$$

Using (4.18), for $j < k$ we obtain

$$\sum_{|\alpha|=k} \|D^\alpha u\|_{N,\, r_1} \leqslant C_1 \sum_{j=1}^{m} \varepsilon^{-j} \sum_{|\beta|=k-j} \|D^\beta u\|_{N,\, r_1+\varepsilon} \leqslant C_1 \sum_{j=1}^{m} \varepsilon^{-j}\varepsilon^{-(k-j)} A^{k-j+1}.$$

The last sum does not exceed $\varepsilon^{-k}A^{k+1}$ if $A > C_1 + 1$. Thus inequality (4.18) is proved.

Assuming $r_1 = r$ and $\varepsilon = r/j$ we obtain from (4.18)

$$\sum_{|\alpha|=j} \|D^\alpha u\|_{N,\, r} \leqslant j^j r^{-j} A^{j+1}$$

for all j. By Theorem 3.5 the estimate

$$\max_{|x-x_0|\leqslant r/2} |D^\alpha u(x)| \leqslant C\|D^\alpha u\|_{N,\, r},$$

is valid. From the estimate it follows that

$$\sum_{|\alpha|=j} \max_{|x-x_0|\leqslant r/2} |D^\alpha u(x)| \leqslant C j^j r^{-j} A^{j+1}.$$

Therefore the function u is analytic in the ball $K_{r/2}$ and the theorem is proved.

Remark. If $P(x, D)$ is an elliptic operator with analytic coefficients and the function $P(x, D)u$ is analytic in the subdomain $\omega \subset \Omega$ for for $u \in \mathscr{D}'(\Omega)$, then u is analytic in ω. This assertion is proved using the same method as in Theorem 4.6. This method was suggested by C. Morrey and L. Nirenberg.[1]

4.10. The results of Petrowsky, who has completely described a class of equations of the type $P(D)u = 0$, having solutions all of which are analytic, were further developed by Hörmander, who succeeded in describing a similar class of equations, every solution of which is an infinitely differentiable function.

Definition 4.4. An operator $P(D)$ is called *hypoelliptic in* Ω if for every subdomain $\omega \subset \Omega$ each generalized solution of the equation $P(D)u = 0$ in ω is an infinitely differentiable function.

Theorem 4.7. Each of the following conditions is necessary and sufficient for the operator $P(D)$ to be hypoelliptic:

a) if $d(\xi)$ is the distance from $\xi \in \mathbb{R}_n$ to the surface $\{\zeta, \zeta \in \mathbb{C}^n, P(\zeta) = 0\}$, then $d(\xi) \to \infty$ as $\xi \to \infty$.

b) There exist positive constants C and a such that $d(\xi) \geqslant |C\xi|^a$.

c) For all $\alpha \neq 0$ $\lim\limits_{\xi \to \infty} \dfrac{P^{(\alpha)}(\xi)}{P(\xi)} = 0$.

d) There exist positive constants C and a such that

$$\left| \frac{P^{(\alpha)}(\xi)}{P(\xi)} \right| \leqslant C |\xi|^{-|\alpha| a}$$

if $\xi \in \mathbb{R}_n$ and ξ is sufficiently large. (Note that $P^{(\alpha)}(\xi) = \partial_\xi^\alpha P(\xi)$.)

We shall not prove this statement. The proofs of this and the next theorem are available in Hörmander.[3]

Although the Definition 4.4 relates only to a homogeneous equation, from the conditions of Theorem 4.7 it follows that for every subdomain $\omega \subset \Omega$ the solution of an inhomogeneous equation $P(D)u = f$ in the class $\mathscr{D}'(\Omega)$ is a smooth function in ω provided f is smooth in ω.

Theorem 4.8. If the operator P(D) is hypoelliptic and $P(D)u \in C^\infty(\omega)$, where ω is a subdomain in Ω and $u \in \mathscr{D}'(\Omega)$, then $u \in C^\infty(\omega)$.

§5. Distributions on Manifolds

5.1. In studying boundary problems for partial differential equations, one often has to consider distributions defined on a manifold which bounds an open set in $\mathbb{R}^n$ and the differential operators acting on this manifold.

Since only linear equations are discussed in this book, we shall not try to find the exact conditions of smoothness for the coefficients of equations and manifolds, but shall consider only infinitely differentiable manifolds. In this section we define an infinitely differentiable manifold (see De Rham[1]).

Definition 5.1. An *n-dimensional manifold* is a topological space in which every point has a neighborhood homeomorphic to some neighborhood of the origin in $\mathbb{R}^n$. A manifold Ω is called an *infinitely differentiable* manifold if there is given on it an atlas consisting of a denumerable family of open sets Ω_ν covering Ω and a family of homeomorphisms f_ν of the set Ω_ν in a neighborhood ω_ν of the origin in $\mathbb{R}^n$, the mapping

$$f_\nu f_\mu^{-1}\colon f_\mu(\Omega_\nu \cap \Omega_\mu) \to f_\nu(\Omega_\nu \cap \Omega_\mu)$$

(of one neighborhood of the origin in $\mathbb{R}^n$ onto the other) being infinitely differentiable for all ν and μ.

The sets Ω_ν are called *coordinate neighborhoods* and f_ν are called the *coordinate mappings*.

Two atlases are called *equivalent* if they define the same C^∞ structure. A class of equivalent atlases is called a C^∞ *differentiable structure* on a manifold.

A function $a\colon \Omega \to \mathbb{C}$ is called *infinitely differentiable* if all of the functions

$$a|_{\Omega_\nu} \circ f_\nu^{-1}\colon \omega_\nu \to \mathbb{C}$$

are infinitely differentiable.

A manifold is called a *closed manifold* if it is compact and has no boundary.

5.2. Definition 5.2. A *distribution* u on the manifold Ω with an atlas $\{\Omega_\nu, \omega_\nu, f_\nu\}$ is a set of distributions $u_\nu \in \mathscr{D}'(\omega_\nu)$ which are compatible; that is, they satisfy

$$u_\mu = u_\nu (f_\nu f_\mu^{-1} x) \text{ in } \omega_\nu \cap f_\nu (\Omega_\mu)$$

for all μ and ν. Two distributions given on Ω by different but equivalent atlases are called *equivalent* if the distributions $u_\nu \in \mathscr{D}'(\omega_\nu)$ and $u'_\nu \in \mathscr{D}'(\omega'_\nu)$ form a compatible set corresponding to the atlas obtained by combining the given atlases. Let us denote by $\mathscr{D}'(\Omega)$ a family of equivalent distributions.

In further discussion we shall identify equivalent distributions and call the whole class of equivalent distributions the distribution. To show that Definition 5.2 is correct we must explain how a distribution given in some domain of the space $\mathbb{R}^n$ varies when independent variables are transformed. Let $y = F(x)$ be such a transformation where the functions F and F^{-1} are infinitely differentiable. Let $J(y) = |\partial F^{-1}(y)/\partial y|$ be the Jacobian. Let Ω' be the image of the domain Ω in this mapping.

If $u \in \mathscr{D}'(\Omega')$, then its pullback $v = u \cdot F^* u$ for a mapping F is a distribution $u \in \mathscr{D}'(\Omega)$ which is defined as follows:

$$v(\varphi) = u((\varphi \circ F^{-1}) | J |), \quad \varphi \in C_0^\infty (\Omega).$$

It is not hard to see that this definition agrees with the classical rule of change of variables under the integral sign when

$$u(\varphi) = \int_\Omega u(x) \varphi(x) \, dx \text{ and } u \in L_1^{\text{loc}} (\Omega).$$

From the definition it is clear that the mapping $u \to u \cdot F$ is continuous as a mapping of $\mathscr{D}'(\Omega')$ into $\mathscr{D}'(\Omega)$. From (1.3) it thus follows that the chain rule can be extended

$$D_j (u \circ F) = \sum_{k=1}^{n} (D_k u \circ F) \frac{\partial F_k}{\partial x^j}.$$

Similarly, we obtain

$$(au) \circ F = (a \circ F)(u \circ F) \quad \text{if} \quad u \in \mathscr{D}'(\Omega').$$

This enables the differential operator $P(x, D)$ on a manifold to be defined as a set of differential operators defined in the domains $\omega_\nu \subset \mathbb{R}^n$ and matched with each other such that

$$(P_\nu v) \circ (f_\nu f_\mu^{-1}) = P_\mu (v \circ f_\nu f_\mu^{-1}).$$

As above, we shall now introduce an equivalence relation of such sets which enables the differential operator to be determined independent of the selection of systems of coordinate neighborhoods and coordinate mappings.

This definition is convenient for applications since it makes it possible to define a differential operator on a manifold starting from the coordinate neighborhoods. It is equivalent to the following, formally briefer, definition in which the operator on a manifold is taken as a starting object.

Definition 5.3. A linear operator $P: C^\infty(\Omega) \to C^\infty(\Omega)$ defined on a manifold Ω is called a *differential operator* if for every coordinate neighborhood Ω_ν the operator

$$P_\nu f_\nu^{-1} \colon C^\infty (\omega_\nu) \to C^\infty (\omega_\nu)$$

is a differential operator.

5.3. A differential operator $P(x, D)$ is local in the sense that

$$\operatorname{supp} Pu \subset \operatorname{supp} u, \quad u \in C^\infty (\Omega).$$

This proves to be the defining property of differential operators. More precisely, we have the following

<u>Theorem 5.1.</u> Let Ω be a domain in $\mathbb{R}^n$ and let P be a continuous linear mapping $C_0^\infty(\Omega) \to C^\infty(\Omega)$ which is local. Then for every compact subset $K \subset \Omega$ we can find $m \in \mathbb{Z}_+$ and $a_\alpha \in C^\infty(K)$ such that $P(x, D)u = \sum_{|\alpha| \leqslant m} a_\alpha(x) \times D^\alpha u(x)$ for all $u \in C^\infty(K)$.

<u>Proof.</u> Let x_0 be an arbitrary point belonging to K. We now consider a functional F defined for $u \in C_0^\infty(K)$ such that

$$F(u) = (Pu)(x_0).$$

According to Definitions 1.3 and 1.5, F is a distribution and supp F consists of one point x_0. Hence by Theorem 1.3 there exist numbers $m(x_0)$, $a_\alpha(x_0)$ for $|\alpha| \leqslant m(x_0)$ such that

$$F = \sum_{|\alpha| \leqslant m(x_0)} a_\alpha(x_0) D^\alpha \delta(x - x_0),$$

that is,

$$Pu(x) = \sum_{|\alpha| \leqslant m(x)} a_\alpha(x) D^\alpha u(x).$$

Here the number m(x) may depend on compact K, but is bounded on every compact set. In order to prove that $a_\alpha \in C^\infty(K)$ it is sufficient to take a function $h \in C_0^\infty(K)$ equal to 1 in $K' \subset K$ and consider the function Pu for $u = hQ$, where Q are all possible polynomials of degree not higher than m. Since $Pu \in C^\infty(K')$, it follows that $a_\alpha \in C^\infty(K)$. As K' is an arbitrary compact set lying within K, we have $a_\alpha \in C^\infty(K)$.

5.4. Let P be a linear differential operator of order m in a domain $\Omega \subset \mathbb{R}^n$ and let F be a diffeomorphic mapping of Ω onto Ω'. Thus the operator P is transformed into a differential operator P' of the same order. Indeed, this order obviously cannot increase; it also cannot decrease, as this would mean that it increases in an inverse mapping.

Another invariant of the mapping is the characteristic form of the operator P(x, D), that is, the principal part of the polynomial $P(x, \xi)$ in the variables ξ. This is the polynomial $P_m(x, \xi)$ of degree m, homogeneous in ξ, appearing in $P(x, \xi)$, if we assume that this function is defined on the cotangent space of Ω. We recall that a topological space $T(\Omega)$ whose points are vectors tangent to Ω is called a tangent bundle.

Locally, within a coordinate neighborhood, the tangent bundle is isomorphic to the direct product $\omega_\nu \times \mathbb{R}_n$ and we can take (x, ξ), where $x \in \omega_\nu$ and $\xi \in \mathbb{R}_n$, as the coordinate of a point of the tangent bundle. Changing to other coordinates $y = F(x)$, the point (x, ξ) of the tangent bundle transforms to $(y, F'(x)\xi)$. This law of transformation of coordinates agrees with the rule used to transform the points where $a = (a_1, \ldots, a_n)$ and a_j are coefficients of the first order operator $\sum_{j=1}^n a_j(x) D_j$. That is why sections of the tangent bundle $T(\Omega)$ are usually identified with vector fields on Ω.

The adjoint space of the tangent bundle is called the cotangent bundle and is denoted by $T^*\Omega$. It is not hard to see that upon changing to new coordinates $y = F(x)$ in a coordinate neighborhood Ω_ν, the point (x, ξ) changes to $(y, {}^tF'(x)^{-1}\xi)$, where ${}^tF'(x)$ is the transpose of the Jacobian matrix $\| \partial F_i / \partial x^j \|$. This agrees with the law of transformation of the space of points $(x, a(x))$, where $a = (a_1, \ldots, a_n)$ and a_j are the coefficients of a differential form $\sum_{j=1}^n a_j(x) dx^j$.

The invariance of the characteristic form with respect to transformations of independent variables can be verified directly, since given the transformation $y = F(x)$, the characteristic polynomial $P_m(x, \xi)$ changes to $P_m(F^{-1}(y), {}^tF'(x)\eta)$. If P is a differential operator on a manifold Ω, then we consider a function $e^{-i\tau S}P(e^{i\tau S})$ where τ is a parameter, and expand it in powers of τ; $S \in C^\infty(\Omega)$. It is not hard to see that this function is a polynomial in τ. The highest degree of this polynomial over all S is called the order of the operator P. Here

$$e^{-i\tau S}P(e^{i\tau S}) = \tau^m P_m(x, \operatorname{grad} S) + \text{a polynomial of degree } m-1.$$

From the definition of a differential operator it follows that for an arbitrary function $S \in C^\infty(\Omega)$

$$P_m(x, \operatorname{grad} S) \circ f_\mu^{-1} = P_{\mu, m}(f_\mu^{-1}(x), \operatorname{grad}(S \circ f_\mu^{-1})), \quad x \in \omega_\mu.$$

Definition 5.4. A surface $S(x) = S(x_0)$ is called the *characteristic of a differential operator* P of order m if $S \in C^1(\Omega)$, $\operatorname{grad} S \neq 0$, and $P_m(x, \operatorname{grad} S) = 0$.

5.5. We now briefly recall the classical method of integrating the equation $P_m(x, \operatorname{grad} S) = 0$. Assume that the polynomial P_m has real coefficients in $C^\infty(\Omega)$ and that $d_\xi P_m(x_0, \xi^0) \neq 0$, where $\xi^0 = \operatorname{grad} S(x_0)$.

Differentiating the equation $P_m(x, \operatorname{grad} S) = 0$ with respect to x^j we obtain

$$P_{m(j)}(x, \operatorname{grad} S) + \sum_{k=1}^{n} P_m^{(k)}(x, \operatorname{grad} S) \frac{\partial^2 S}{\partial x^j \partial x^k} = 0, \tag{5.1}$$

where we have used the notation

$$P_{m(j)}(x, \xi) = \frac{\partial P_m(x, \xi)}{\partial x^j}, \quad P_m^{(k)}(x, \xi) = \frac{\partial P_m(x, \xi)}{\partial \xi_k}.$$

Let us consider the system of ordinary differential equations

$$\frac{dx^k}{dt} = P_m^{(k)}(x, \operatorname{grad} S), \qquad k = 1, \ldots, n. \tag{5.2}$$

Since $\sum_{k=1}^{n} |P_m^{(k)}(x, \operatorname{grad} S)| \neq 0$, one and only one integral curve without singularities passes through every point in Ω. From (5.1) it follows that on this curve

$$\frac{d}{dt}\left(\frac{\partial S}{\partial x^j}\right) = -P_{m(j)}(x, \operatorname{grad} S).$$

Writing $\xi = \operatorname{grad} S$ and considering ξ as a function of t along the curve $x = x(t)$, we obtain the Hamiltonian equations

$$\frac{dx^k}{dt} = P_m^{(k)}(x, \xi), \quad \frac{d\xi_k}{dt} = -P_{m(k)}(x, \xi), \qquad k = 1, \ldots, n.$$

Definition 5.5. An integral curve $x = x(t)$, $\xi = \xi(t)$ of the system

$$\frac{dx^k}{dt} = \frac{\partial f(x, \xi)}{\partial \xi_k}, \quad \frac{d\xi_k}{dt} = -\frac{\partial f(x, \xi)}{\partial x^k}, \qquad k = 1, \ldots, n \tag{5.3}$$

is called a bicharacteristic of the function f. A bicharacteristic is called a null bicharacteristic if

$$f(x(t), \xi(t)) \equiv 0.$$

Also note that the function $f(x(t), \xi(t))$ always has a constant value, since

$$\frac{d}{dt} f(x(t), \xi(t)) = \sum_{k=1}^{n} \left(\frac{\partial f}{\partial x^k} \frac{dx^k}{dt} - \frac{\partial f}{\partial \xi_k} \frac{d\xi_k}{dt} \right) = 0$$

by (5.3).

5.6. Partition of Unity. In this book we shall often make use of the following

Theorem 5.2. Let $\{U_j\}$ be an open locally finite covering of a manifold Ω. Then there exists a finite or denumerable system of functions $\{\varphi_j\}$ with the following properties:

1°. $\varphi_j \geqslant 0$, $\sum \varphi_j = 1$;

2°. $\varphi_j \in C^\infty(\Omega)$, and the support of φ_j is compact and belongs to one of the U_i;

3°. supp φ_j lies in the domain of definition of some system of local coordinates.

The proof involves the following auxiliary assertions:

Lemma 5.1. Let A be a compact set in $\mathbb{R}^n$ and B be an open set containing A. Then there exists a function $h \in C_0^\infty(B)$ whose values in A are equal to unity.

Proof. Let $f(x) = e^{-1/x^2}$ when $x > 0$ and $f(x) = 0$ when $x \leqslant 0$. Let $f_1(x) = \int_{-\infty}^{x} f(t) f(1-t)\,dt \left(\int_{-\infty}^{\infty} f(t) f(1-t)\,dt \right)^{-1}$. Then the function $g(x) = f_1(x+1) - f_1(x)$ is an infinitely differentiable non-negative function, vanishing when $|x| > 1$. In this case $\sum_{j=-\infty}^{\infty} g(x-j) = 1$.

Let $x \in \mathbb{R}^n$. Then for $\alpha = (\alpha_1, \ldots, \alpha_n) \in \mathbb{Z}^n$ we can assume that $h_\alpha(x) = \prod_{j=1}^{n} g(x_j\varepsilon^{-1} - \alpha_j)$. We now obtain $\sum h_\alpha(x) = 1$, $h_\alpha \in C^\infty$, $h_\alpha \geqslant 0$ with supp h_α in a cube of edge length 2ε centered at $\alpha\varepsilon$.

Now let ε be so small that the diagonal of the cube of edge length 2ε is less than the distance between A and the boundary of the manifold B. Assume that $h(x) = \sum' h_\alpha(x)$, where summation is taken over those α for which supp h_α has common points with the manifold A. It is not difficult to see that h is the desired function.

Proof of Theorem 5.2. There exists a locally finite covering $\{V_i\}$ of the manifold Ω for which V_i belongs to one of the U_j and is such that $\overline{V}_i$ lies in the domain of definition of some local system of coordinates. Indeed, for every point of Ω there exists a neighborhood V' with compact closure $\overline{V'}$ confined in one of the U and simultaneously in the domain of definition of some system of coordinates. Thus there exists a finite or a denumerable covering of Ω by such sets. If the covering is not locally finite, it may be decreased by means of the following arguments.

We define by induction the compact sets $K_1, K_2, \ldots$ and a sequence of integers $n_1, n_2, \ldots$ such that $K_1 = \overline{V}'_1$, $n_1 = 1$. Further, let n_m be the least integer greater than n_{m-1} such that $K_{m-1} \subset \bigcup_{i=1}^{n_m} V'_i$ and $K_m = \bigcup_{i=1}^{n_m} \overline{V}'_i$.

Then $K_{m-1} \subset K_m$ and any compact set is in K_m for sufficiently large numbers m. Now let $V_i = V'_i$ if $i \leqslant n_2$ and $V_i = V'_i \cap (\Omega \setminus K_{m-1})$ if $n_m < i \leqslant n_{m+1}$ when $m > 1$; here C is the complement. We shall now show that the set V_i forms a covering of the manifold Ω. To do this it is sufficient to prove that

$$\bigcup_{i=1}^{n_{m+1}} V_i = \bigcup_{i=1}^{n_{m+1}} V'_i.$$

When m = 1 it is valid because $V_i = V'_i$ for $i \leqslant n_2$. Setting $\bigcup_{i=1}^{n_m} V_i = \bigcup_{i=1}^{n_m} V'_i = A$, since $K_{m-1} \subset \bar{A}$ we obtain $V_i \cup A = V'_i \cup A$ when $n_m < i \leqslant n_{m+1}$. Thus we obtain the desired result for m + 1.

The covering $\{V_i\}$ is locally finite because any compact set is confined in all K_m for sufficiently large m and for $i > n_{m+1}$ the sets V_i and K_m do not intersect.

We shall now construct another covering $\{W_i\}$ for which $\bar{W}_i \subset V_i$. By Lemma 5.1 we can find a function ψ_i in $C_0^\infty(V_i)$ with values in [0. 1] equal to 1 on W_i. The sum $\psi = \sum \psi_i$ is taken everywhere to be $\geqslant 1$. Now let $\varphi_i = \psi_i/\psi$. It is clear that $\varphi_i \in C^\infty$, supp $\varphi_i \subset V_i$ and $\varphi_i \geqslant 0$, $\sum \varphi_i = 1$. Thus, Theorem 5.2 is proved.

CHAPTER II. PSEUDODIFFERENTIAL OPERATORS

§1. Singular Integral Operators

1.1. The theory of integral operators of the type

$$Au(x) = \int K(x, x-y)\, u(y)\, dy \tag{1.1}$$

is closely associated with the theory of linear partial differential equations. In most cases the solution of a boundary problem for a differential equation can be represented as (1.1) and the boundedness of the operator in the corresponding function spaces is indicative of the smoothness of the solution of the boundary problem and its continuous dependence on the right-hand terms.

As a rule, the kernel K of the operator (1.1) is a smooth function when $x \neq y$ and has a singularity only at the point $x = y$. This is a *weak* singularity if

$$|K(x, x-y)| \leqslant C_x |x-y|^{-m}, \qquad m < n. \tag{1.2}$$

The integral operator A is called *singular* provided $K(x, tz) = t^{-n}K(x, z)$ for $t > 0$ and the condition

$$\int_{|y-x|=r} K(x, y)\, dS_y = 0, \qquad x \in \Omega \tag{1.3}$$

is fulfilled for $r \leqslant r_0$, where $r_0 = r_0(x)$ is a positive constant. This condition enables (1.1) to be taken in the sense of the Cauchy principal value, that is,

$$\lim_{\varepsilon \to +0} \int_{|x-y|>\varepsilon} K(x, x-y)\, u(y)\, dy.$$

The theory of singular operators has been developed by G. Giraud, S. G. Mikhlin, A. P. Calderon, and A. Zygmund. In this section we shall give only basic facts pertaining to the L_2-theory of these operators.

1.2. We shall first prove the boundedness of integral operators with a weak singularity.

Theorem 1.1. An operator $A\colon u \mapsto \int K(x, y)\, u(y)\, dy$ with a weak singularity of the kernel K is bounded in $L_2(\Omega)$, and

$$\|A\| \leqslant \max\left(\sup_{x\in\Omega} \int_\Omega |K(x, y)|\, dy,\ \sup_{y\in\Omega} \int_\Omega |K(x, y)|\, dx\right). \tag{1.4}$$

Proof. Denote by N the constant appearing in the right-hand side of (1.4). It is sufficient to prove that

$$\left|\iint K(x, y) f(x)\, g(y)\, dx\, dy\right| \leqslant N \|f\|_{L_2(\Omega)} \|g\|_{L_2(\Omega)}.$$

Applying the Cauchy-Schwarz-Bunyakovsky inequality, we obtain

$$\left|\iint K(x, y) f(x)\, g(y)\, dx\, dy\right| \leqslant \iint \sqrt{|K(x, y)|}\, |f(x)|\, \sqrt{|K(x, y)|}\, |g(y)|\, dx\, dy \leqslant$$

$$\left(\iint |K(x, y)| |f(x)|^2 dx\, dy\right)^{1/2} \left(\iint |K(x, y)| |g(y)|^2 dx\, dy\right)^{1/2} \leqslant$$
$$(N\int |f(x)|^2 dx)^{1/2} \left(N\int |g(y)|^2 dy\right)^{1/2} = N\|f\|_{L_2(\Omega)} \|g\|_{L_2(\Omega)}.$$

This proves the theorem.

1.3. We shall first prove the boundedness of singular integral operators in $L_2(\Omega)$ for a particular case of the Hilbert operator:

$$Hu(x) = \text{v. p.} \frac{1}{\pi} \int_{-\infty}^{\infty} \frac{u(y)}{x-y} dy, \text{ where v. p.} \int \frac{u(y)}{x-y} dy = \lim_{\substack{\varepsilon \to +0 \\ R \to +\infty}} \int_{\varepsilon < |x-y| < R} \frac{u(y)}{x-y} dy.$$

To do so, we consider the distribution

$$f(\varphi) = \left(\text{v. p.} \frac{1}{x}\right)(\varphi) = \lim_{\substack{\varepsilon \to +0 \\ R \to +\infty}} \int_{\varepsilon < |y| < R} \frac{\varphi(y)}{y} dy \tag{1.5}$$

in $\mathscr{D}'(\mathbb{R})$ and compute its Fourier transform. Note that

$$\lim_{\substack{\varepsilon \to +0 \\ R \to +\infty}} \int_{\varepsilon < |y| < R} \frac{e^{iy\xi}}{y} dy = \lim_{\substack{\varepsilon \to +0 \\ R \to +\infty}} i \int_{\varepsilon < |y| < R} \frac{\sin y\xi}{y} dy = i\pi \operatorname{sgn} \xi = \begin{cases} i\pi \text{ when } \xi > 0, \\ -i\pi \text{ when } \xi < 0. \end{cases}$$

By definition, if $\varphi \in \mathscr{S}$, then

$$\tilde{f}(\varphi) = f(\tilde{\varphi}) = \lim_{\substack{\varepsilon \to +0 \\ R \to +\infty}} \int_{\varepsilon < |y| < R} \frac{dy}{y} \int e^{-iy\xi} \varphi(\xi)\, d\xi =$$
$$\int \varphi(\xi) \left(\lim_{\substack{\varepsilon \to +0 \\ R \to +\infty}} \int_{\varepsilon < |y| < R} \frac{e^{-iy\xi}}{y} dy \right) d\xi = -i\pi \int \varphi(\xi) \operatorname{sgn} \xi\, d\xi,$$

and hence

$$\tilde{f}(\xi) = -i\pi \operatorname{sgn} \xi.$$

Since $Hu = \frac{1}{\pi} f * u$, we see that

$$\widetilde{Hu}(\xi) = -i (\operatorname{sgn} \xi)\, \tilde{u}(\xi),$$

and, by Parseval's identity,

$$\int |Hu(x)|^2 dx = (2\pi)^{-1} \int |\widetilde{Hu}(\xi)|^2 d\xi = (2\pi)^{-1} \int |\tilde{u}(\xi)|^2 d\xi = \int |u(x)|^2 dx,$$

that is, $\|Hu\|_{L_2} = \|u\|_{L_2}$. Thus the operator H is isometric in $L_2(\mathbb{R})$.

1.4. The L_2 boundedness of a general singular integral operator with kernel K dependent only on x − y can be proved by the same method, i.e., by using the proof of the boundedness of the Fourier transform of its kernel.

Let us now consider the operator

$$Au(x) = \lim_{\varepsilon \to +0} \int_{|x-y| > \varepsilon} K(x-y)\, u(y)\, dy = (\text{v. p. } K) * u \tag{1.6}$$

for $u \in \mathscr{D}'(\mathbb{R}^n)$, assuming that

$$K(tz) = t^{-n} K(z) \text{ when } t > 0, \quad \int_{|z|=1} K(z)\, dz = 0. \tag{1.7}$$

Elsewhere we shall prove

Proposition 1.1. If a function $K \in C^1(\mathbb{R}^n \setminus 0)$ satisfies (1.7), then the Fourier transform

$$\widetilde{(\text{v. p. } K)}(\xi) = p(\xi)$$

is a continuous function in $\mathbb{R}^n \setminus 0$ for which

$$p(t\xi) = p(\xi) \text{ when } t > 0, \qquad \int_{|\xi|=1} p(\xi)\, d\xi = 0. \tag{1.8}$$

First we shall prove the following auxiliary statement.

Lemma 1.1. If a function f in $L_1^{loc} \cap \mathcal{S}'$ is such that $\tilde{f} \in L_1^{loc}$ and f(x) depends only on $|x|$, then $\tilde{f}$ is a function of $|\xi|$.

Proof. Let first $f \in L_1(\mathbb{R}^n)$ and U be an arbitrary rotation about the origin in $\mathbb{R}^n$. Then

$$\tilde{f}(U\xi) = \int f(x) e^{-ix\cdot U\xi}\, dx = \int f(x) e^{-iU^{-1}x\cdot\xi}\, dx = \int f(Uy) e^{-iy\xi}\, dy = \int f(y) e^{-iy\xi}\, dy = \tilde{f}(\xi),$$

and this proves the statement.

In the general case let $h \in C_0^\infty(\mathbb{R}^n)$, $h(x) = 1$ when $|x| < 1$ and $h(x) = 0$ when $|x| > 2$, where h depends only on $|x|$. Assuming $f_\varepsilon(x) = h(\varepsilon x)f(x)$ it follows that $f_\varepsilon \in L_1$ and $f_\varepsilon \to f$ in $\mathcal{S}'$. Since, as seen above, $\tilde{f}_\varepsilon$ depends only on $|\xi|$, then $\tilde{f}$ is also dependent only on $|\xi|$. This proves the lemma.

Lemma 1.2. If $f(tx) = t^m f(x)$ when $t > 0$, $f \in L_1^{loc} \cap \mathcal{S}'$ and $\tilde{f}(\xi) \in L_1^{loc}$, then $\tilde{f}(t\xi) = t^{-n-m}\tilde{f}(\xi)$.

Proof. First let $f \in \mathcal{S}$. For $t > 0$ $\tilde{f}(t\xi) = \int f(x) e^{-ix\cdot t\xi}\, dx = t^{-n} \int f\left(\frac{y}{t}\right) e^{-iy\xi}\, dy = t^{-n-m}\int f(y) e^{-iy\xi}\, dy = t^{-n-m}\tilde{f}(\xi)$. In the general case let $\varphi \in$ s. Then for $t > 0$ $\int \tilde{f}(t\xi)\, \tilde{\varphi}(\xi)\, d\xi = \iint f(x) e^{-itx\xi}\tilde{\varphi}(\xi)\, dx\, d\xi = t^{-n} \int\int f\left(\frac{y}{t}\right) e^{-iy\xi}\tilde{\varphi}(\xi)\, dy d\xi = t^{-n-m} \iint f(y) e^{-iy\xi}\tilde{\varphi}(\xi)\, dy\, d\xi = t^{-n-m}\int \tilde{f}(\xi)\, \tilde{\varphi}(\xi)\, d\xi$, that is, $\tilde{f}(t\xi) = \tilde{f}(\xi)t^{-n-m}$. This proves the lemma.

Proof of Proposition 1.1. Note that $k(x) = k(\omega)|x|^{-n}$, where $\omega = |x|^{-1}$ is a point on a unit sphere Ω. By (1.7) we have $\int_\Omega k(\omega)\, d\omega = 0$.

Let $\varphi \in \mathcal{D}$ and $\varphi(x) = 0$ when $|x| > R$. Then

$$\varphi(x) = \varphi(0) + \sum_{j=1}^n x^j \varphi_j(x)$$

and $\varphi_j \in C_0^\infty(\mathbb{R}^n)$. Therefore

$$(\text{v. p. } K)(\varphi) = \lim_{\varepsilon\to+0} \int_{|x|>\varepsilon} K(x)\, \varphi(x)\, dx =$$

$$\lim_{\varepsilon\to+0} \int_{|x|>\varepsilon} \frac{k(\omega)}{|x|^n} \left[\varphi(0) + \sum_{j=1}^n x^j\varphi_j(x)\right] dx = \sum_{j=1}^n \int \frac{x^j}{|x|^n} k\left(\frac{x}{|x|}\right) \varphi_j(x)\, dx. \tag{1.9}$$

Let $h \in C_0^\infty(\mathbb{R}^n)$ and $h(x) = 1$ in a neighborhood of the origin. Then

$$\text{v. p. } K(x) = h(x)\, \text{v. p. } K(x) + [1 - h(x)]\, K(x),$$

where the first term on the right-hand side belongs to $\mathcal{E}'$, and the second to the space L_2. Therefore, $p(\xi) = \widetilde{\text{v. p. } K}$ is a function.

Setting $K_\delta(x) = k(\omega)|x|^{n-\delta}$ when $\delta > 0$, we obtain by (1.9),

$$h(x)\, K_\delta(x) \to h(x)\, \text{v. p. } K(x)$$

when $\delta \to +0$ in $\mathcal{E}'$. Besides, it is easy to see that $[1 - h(x)]K_\delta(x) \to [1 - h(x)]K(x)$ in L_2. Therefore, $K_\delta \to$ v.p.K in $H^{-\infty} = \bigcup_s H_s$. Since the Fourier transform $\tilde{K}_\delta(\xi)$ of the function $K_\delta(x)$ is, by Lemma 1.1, a homogeneous function of degree $-\delta$, the limiting function $p(\xi) = \widetilde{\text{v. p. } K}$ is homogeneous of degree zero.

Suppose $\varphi \in \mathcal{S}$ and $\varphi(\xi)$ depends only on $|\xi|$. Then $\tilde{\varphi} = \psi(x)$ depends, by Lemma 1.1, only on $|x|$ and $\psi \in \mathcal{S}$. Therefore,

$$(2\pi)^{-n} \int p(\xi)\, \varphi(\xi)\, d\xi = \text{v. p.} \int K(x)\, \psi(x)\, dx.$$

By (1.7) the right-hand side is equal to zero. Therefore,

$$\int p(\xi)\varphi(\xi)\,d\xi = 0$$

for all such functions φ. Whence it follows that

$$\int_{|\xi|=1} p(\xi)\,d\xi = 0.$$

It remains to prove that the function p is continuous in $\mathbb{R}_n\setminus 0$. The Fourier transform of the function h(x)v.p.K(x) is an infinitely differentiable function.

On the other hand

$$|\Delta[(1-h(x))K(x)]| \leqslant C(1+|x|)^{-n-2},$$

so that the Fourier transform of this function, equal to $|\xi|^2\widetilde{(1-h)K}$, is continuous and bounded in $\mathbb{R}_n$. Therefore the function $\widetilde{(1-h)K}$ is continuous when $\xi \neq 0$ and Proposition 1.1 is proved.

Proposition 1.2. The operator A defined by formula (1.6) is bounded in L_2.

Proof. Applying Parseval's identity and Proposition 1.1, we obtain

$$\|Au\|^2 = (2\pi)^{-n}\|\widetilde{Au}(\xi)\|^2 = (2\pi)^{-n}\|p(\xi)\tilde{u}(\xi)\|^2 \leqslant (2\pi)^{-n}C\|\tilde{u}(\xi)\|^2 = C\|u(x)\|^2,$$

as p is a continuous function when $|\xi| = 1$ and $|p(\xi)| \leqslant \max_{|\xi|=1}|p(\xi)| = C$. The proposition is proved.

§2. Definitions and Basic Properties of Pseudo-Differential Operators

2.1. At present several different definitions of pseudo-differential operators are available. Initially these operators were called singular integro-differential operators and were defined by

$$Au(x) = \sum_{|\alpha|\leqslant m} A_\alpha(x)D^\alpha u(x),$$

where $A_\alpha(x)$ are singular operators. Such operators are useful in studies of boundary problems for elliptic differential equations.

In the expositions of M. I. Vishik and G. I. Eskin, pseudo-differential operators are called "operators in convolutions" and are defined as

$$Au(x) = \int K(x, x-y)u(y)\,dy,$$

where $K(x, z) \in \mathscr{D}'(\Omega\times\Omega)$. Such operators are also useful in studies of boundary problems for elliptic differential equations.

The term "pseudo-differential operator" was first used by K. Friedrichs and P. Lax[1] and J. J. Kohn and L. Nirenberg[1] for operators of the type

$$Au(x) = (2\pi)^{-n}\int a(x, \xi)\tilde{u}(\xi)e^{i(x,\xi)}\,d\xi, \tag{2.1}$$

where $\tilde{u}(\xi)$ is the Fourier transform of function u. The function a is called the symbol of the operator. It is easy to see that if $a(x, \xi) = \sum_{|\alpha|\leqslant m} a_\alpha(x)\xi^\alpha$, then the operator A is a differential operator.

In the general case certain restrictions are imposed on a, for example, in the form of inequalities

$$|D_\xi^\alpha D_x^\beta a(x, \xi)| \leqslant C_{\alpha,\beta,K}(1+|\xi|)^{m-|\alpha|}, \qquad x\in K, \tag{2.2}$$

which are satisfied for all α and β and an arbitrary compact set $K \subset \Omega$ such that $a \in C^\infty(T^*\Omega)$. The number m in this case is called the order of operator A. The class of functions a satisfying (2.2) in Ω is denoted by $S^m(\Omega)$. By $S_0^m(\Omega)$ we denote a subclass of symbols in $S^m(\Omega)$ which are equal to zero outside a certain compact subset of the domain Ω. Of course, in specific problems it is sufficient that inequalities of the type (2.2) are satisfied for $|\alpha + \beta| \leqslant N$ with some N. However, it is convenient to assume that these inequalities hold for all α and β.

We could define the class of pseudo-differential operators as a minimal class of linear operators containing all differential operators and the operators "inverse" to elliptic differential operators, so that for each elliptic differential operator P in this class there exists an operator Q such that $\varphi + T\varphi = PQ\varphi$, $\varphi \in C_0^\infty(\omega)$, where T is a compact operator.

Elsewhere we shall see that a class of pseudo-differential operators is naturally obtained from a class of differential operators by means of transformations that transform every function $\varphi \in C_0^\infty(\Omega)$ into a value of a solution of the hyperbolic equation

$$\frac{\partial u}{\partial t} = P(t, x, D_x)u \text{ for } t = 1,$$

if $u(0, x) = \varphi(x)$.

Each of these definitions is used to suit a specific problem.

2.2. We shall first consider operators of the type (2.1) with symbols satisfying (2.2), and study their basic properties.

To start with we shall prove the boundedness of these operators.

Theorem 2.1. The operator A defined by (2.1) with the symbol a belonging to the class $S_0^m(\mathbb{R}^n)$ carries $C_0^\infty(\mathbb{R}^n)$ into itself. This operator is bounded as an operator: $H_s \to H_{s-m}$ for all real s.

Proof. Let $u \in C_0^\infty(\mathbb{R}^n)$. Then, by Theorem 2.3 from Chapter I,

$$|\tilde{u}(\xi)| \leqslant C_N(1+|\xi|)^{-N}$$

for any N. By (2.2)

$$|D_x^\beta a(x, \xi)\tilde{u}(\xi)e^{i(x,\xi)}| \leqslant C_{\beta,N}(1+|\xi|)^{m+|\beta|-N}, \qquad x \in K,$$

where K is an arbitrary compact set in $\mathbb{R}^n$. These estimates reveal that the integral obtained after differentiating the integrand in (2.1) converges uniformly. Hence $Au \in C^\infty(\mathbb{R}^n)$ if $u \in C_0^\infty(\mathbb{R}^n)$. Since $a(x, \xi) = 0$ for $x \in \mathbb{R}^n \setminus K_0$, where K_0 is a compact set, $Au \in C_0^\infty(\mathbb{R}^n)$.

To prove the boundedness of the operator A we shall compute the Fourier transform of the function Au(x). We now have

$$|\widetilde{Au}(\eta)| = \int \tilde{a}(\eta-\xi, \xi)\tilde{u}(\xi)\,d\xi.$$

From (2.2) it follows that for any N

$$|\tilde{a}(\eta-\xi, \xi)| \leqslant C'_N(1+|\xi-\eta|^2)^{-N}(1+|\xi|^2)^{m/2}.$$

Therefore we must estimate the L_2 norm of the function

$$(1+|\eta|^2)^{\frac{s-m}{2}}\int \tilde{a}(\eta-\xi, \xi)\tilde{u}(\xi)\,d\xi =$$

$$(1+|\eta|^2)^{\frac{s-m}{2}}\int (1+|\xi|^2)^{-\frac{s}{2}}\tilde{a}(\eta-\xi, \xi)\left[(1+|\xi|^2)^{\frac{s}{2}}\tilde{u}(\xi)\right]d\xi$$

in terms of the L_2 norm of the function $(1 + |\xi|^2)^{s/2}\tilde{u}(\xi)$.

By Theorem 1.1 it is sufficient to prove that the integrals in ξ or η of the function

$$(1+|\eta|^2)^{(s-m)/2}(1+|\xi|^2)^{-s/2}|\tilde{a}(\eta-\xi,\ \xi)|$$

do not exceed the constant C. In view of the aforementioned estimates on the function $\tilde{a}$, this function may be estimated in terms of

$$C'_N\left(\frac{1+|\eta|^2}{1+|\xi|^2}\right)^{(s-m)/2}(1+|\xi-\eta|^2)^{-N}.$$

Note that the following estimates follow from the triangle inequality:

$$1+|\eta|^2\leqslant 2(1+|\xi-\eta|^2)(1+|\xi|^2),\qquad 1+|\xi|^2\leqslant 2(1+|\xi-\eta|^2)(1+|\eta|^2),$$

and therefore

$$\left(\frac{1+|\eta|^2}{1+|\xi|^2}\right)^{(s-m)/2}\leqslant 2^{|(s-m)/2|}(1+|\xi-\eta|^2)^{|(s-m)/2|}.$$

If N is selected so that

$$2N>n+|s-m|,$$

then

$$\int\left(\frac{1+|\eta|^2}{1+|\xi|^2}\right)^{(s-m)/2}(1+|\xi-\eta|^2)^{-N}\,d\xi\leqslant C$$

and the integral in η is also estimated. Thus, the theorem is proved.

2.3. The operator (2.1) can, of course, be written without using the Fourier transform. Substituting $\tilde{u}(\xi)=\int u(y)e^{-iy\xi}\,dy$, we obtain

$$Au(x)=(2\pi)^{-n}\int\int a(x,\ \xi)\,u(y)\,e^{i(x-y,\ \xi)}\,dy\,d\xi=\int K(x,\ x-y)\,u(y)\,dy,$$

where we have formally used the notation

$$K(x,\ x-y)=(2\pi)^{-n}\int a(x,\ \xi)\,e^{i(x-y,\ \xi)}\,d\xi.$$

Note that for $x\neq y$ and any N

$$e^{i(x-y,\ \xi)}=|x-y|^{-2N}(-\Delta_\xi)^N e^{i(x-y,\ \xi)}.$$

Let $h\in C_0^\infty(\mathbb{R}_n)$ and $h(\xi)=1$ when $|\xi|\leqslant 1$. Then

$$K(x,\ z)=(2\pi)^{-n}\lim_{t\to+\infty}\int h\left(\frac{\xi}{t}\right)a(x,\ \xi)\,e^{iz\xi}\,d\xi$$

and therefore

$$|z|^{2N}K(x,\ z)=(2\pi)^{-n}\lim_{t\to+\infty}\int h\left(\frac{\xi}{t}\right)a(x,\ \xi)(-\Delta_\xi)^N e^{iz\xi}\,d\xi.$$

Integrating by parts we obtain

$$|z|^{2N}K(x,\ z)=(2\pi)^{-n}\lim_{t\to+\infty}\int(-\Delta_\xi)^N\left[h\left(\frac{\xi}{t}\right)a(x,\ \xi)\right]e^{iz\xi}\,d\xi=$$

$$(2\pi)^{-n}\int(-\Delta_\xi)^N a(x,\ \xi)\,e^{iz\xi}\,d\xi,$$

provided N is sufficiently large. We now see that $|z|^{2N}K(x,\ z)\in C^l$ when N is so large that $l+m-2N<-n$. Therefore, the kernel $K(x,\ x-y)\in C^\infty$ when $x\neq y$, and has a singularity only when $x=y$.

Also note that in this case the symbol a satisfies (2.2) for any m, that is,

$$|D_\xi^\alpha D_x^\beta a(x,\ \xi)|\leqslant C_{\alpha,\beta,N}(1+|\xi|)^{-N}$$

for any α, β and N, the kernel

$$K(x,\ z)=(2\pi)^{-n}\int a(x,\ \xi)\,e^{iz\xi}\,d\xi$$

is an infinitely differentiable function when $(x, z) \in \Omega \times \Omega$ and $Au \in C^\infty$ for every distribution $u \in \mathscr{E}'(\Omega)$. In this case the order of the operator A equals $-\infty$ and we identify these operators as *smoothing operators*.

Let $h \in C_0^\infty(\mathbb{R})$ and $h(t) = 1$ when $|t| < 1$, $h(t) = 0$ when $|t| \geqslant 2$ and $h(t) \geqslant 0$. Then

$$Au(x) = \int K(x, x-y)\, u(y)\, dy = A_1 u(x) + A_2 u(x),$$

where

$$A_1 u(x) = \int h(|x-y|)\, K(x, x-y)\, u(y)\, dy.$$

The operator A_2 has a kernel equal to 0 in a neighborhood of the set $\{(x, y) \in \mathbb{R}^{2n},\ x = y\}$ and is therefore infinitely differentiable. As seen above, such an operator is a smoothing operator.

The operator A_1 is properly supported in the following sense. Recall that continuous mapping $f\colon X \to Y$, where X and Y are topological spaces, is called *proper* if for every compact set $K \subset Y$ its inverse image $f^{-1}(K)$ is a compact set in X. We call the set $M \subset \Omega \times \Omega$ a *proper* set if its canonical projections $\pi_1, \pi_2\colon M \to \Omega$ are proper mappings. The integral $Bu(x) = \int K(x, y)\, u(y)\, dy$ will be referred to as *properly supported* if the set supp K is proper.

Proposition 2.1. If the operator

$$Bu(x) = \int_\Omega K(x, y)\, u(y)\, dy$$

with a kernel from $\mathscr{D}'(\Omega\times\Omega)$, smooth off the diagonal, is properly supported, then it defines a mapping

$$B\colon C_0^\infty(\Omega) \to C_0^\infty(\Omega),$$

which may be extended to continuous mappings

$$B\colon C^\infty(\Omega) \to C^\infty(\Omega); \quad B\colon \mathscr{D}'(\Omega) \to \mathscr{D}'(\Omega); \quad B\colon \mathscr{E}'(\Omega) \to \mathscr{E}'(\Omega).$$

Proof. If $u \in C_0^\infty(\Omega)$, then according to the definition of a properly supported operator, the set

$$\{x \in \Omega,\ K(x, y)\, u(y) \not\equiv 0\}$$

is a compact set such that $Bu \in C_0^\infty(\Omega)$. For the same reason, if $u \in \mathscr{E}'(\Omega)$, then $Bu \in \mathscr{E}'(\Omega)$.

Formally, the adjoint operator B^* defined by $(Bu, v) = (u, B^*v)$, where $u, v \in C_0^\infty(\Omega)$, $(u, v) = \int u(x)\overline{v(x)}\, dx$, has a kernel $K(y, x)$ and therefore is also properly supported. It determines continuous mappings

$$B^*\colon C_0^\infty(\Omega) \to C_0^\infty(\Omega) \text{ and } B^*\colon \mathscr{E}'(\Omega) \to \mathscr{E}'(\Omega).$$

By duality this means that the mappings $B\colon \mathscr{D}'(\Omega) \to \mathscr{D}'(\Omega)$ and $B\colon C^\infty(\Omega) \to C^\infty(\Omega)$ are continuous. Thus the proposition is proved.

2.4. In further discussion we shall use

Lemma 2.1. Let $A = A(x, y, \xi, \eta)$ be an infinitely differentiable function in $\mathbb{R}^n\times\mathbb{R}^n\times\mathbb{R}_n\times\mathbb{R}_n$ such that

$$|D_\xi^\alpha D_\eta^{\alpha'} D_x^\beta D_y^{\beta'} A(x, y, \xi, \eta)| \leqslant C_{\alpha, \alpha', \beta, \beta'} (1+|\xi|)^{m-|\alpha|} (1+|\eta|)^{m'-|\alpha'|} \tag{2.3}$$

for all $\alpha, \alpha', \beta, \beta'$. Let $A(x, y, \xi, \eta) = 0$ be equal to 0 when $y \in \mathbb{R}^n \setminus K$, where K is a compact subset in $\mathbb{R}^n$. Then the function

$$a(x, \xi) = (2\pi)^{-n} \iint A(x, y, \xi, \eta)\, e^{i(x-y,\, \eta-\xi)}\, dy\, d\eta \tag{2.4}$$

is infinitely differentiable in $\mathbb{R}^n\times\mathbb{R}_n$ and the estimates

$$|D_\xi^\alpha D_x^\beta a(x, \xi)| \leqslant C_{\alpha, \beta} (1+|\xi|)^{m+m'-|\alpha|} \tag{2.5}$$

are valid for all α and β.

In addition, if we take

$$a_N(x, \xi) = \sum_{|\alpha|<N} \frac{i^{|\alpha|}}{\alpha!} D_y^\alpha D_\eta^\alpha A(x, y, \xi, \eta)\big|_{y=x,\ \eta=\xi}, \tag{2.6}$$

then the function $a - a_N$ is the symbol of an operator of order $m + m' - N$.

Proof. For fixed values of x, ξ, and η the function $A(x, y, \xi, \eta) \in C_0^\infty(\mathbb{R}^n)$ and therefore the integral in y converges absolutely, defining a function that can be estimated in terms of

$$C_l(1+|\xi-\eta|)^{-l}\ (1+|\xi|)^m(1+|\eta|)^{m'}$$

for any integer l. Therefore, the integral appearing in (2.4) converges and defines an infinitely differentiable function a.

According to Taylor's formula

$$A(x, y, \xi, \eta) = \sum_{|\alpha|<N} \frac{i^{|\alpha|}}{\alpha!} D_\eta^\alpha A(x, y, \xi, \xi)(\eta-\xi)^\alpha + r_N(x, y, \xi, \eta). \tag{2.7}$$

In integral a we replace the function A by the sum occurring in (2.7), and denote by a'_N the integral obtained:

$$a'_N(x, \xi) = (2\pi)^{-n} \int\int \sum_{|\alpha|<N} \frac{i^{|\alpha|}}{\alpha!} D_\eta^\alpha A(x, y, \xi, \xi)(\eta-\xi)^\alpha e^{i\langle x-y, \eta-\xi\rangle}\, dy\, d\eta.$$

As is known (see Example 2.1 in Chapter I),

$$(2\pi)^{-n}\int(\eta-\xi)^\alpha e^{i\langle x-y,\ \eta-\xi\rangle}\, d\eta = (-D_y)^\alpha \delta(x-y).$$

Therefore

$$a'_N(x, \xi) = \sum_{|\alpha|<N} \frac{i^{|\alpha|}}{\alpha!} D_\eta^\alpha D_y^\alpha A(x, x, \xi, \xi) = a_N(x, \xi).$$

From (2.3) and (2.6) it follows that

$$|D_\xi^\alpha D_x^\beta a_N(x, \xi)| \leqslant C'_{\alpha,\beta}(1+|\xi|)^{m+m'-|\alpha|},$$

and we have to prove similar inequalities for the function

$$a(x, \xi) - a_N(x, \xi) = (2\pi)^{-n}\int\int r_N(x, y, \xi, \eta)\, e^{i\langle x-y,\ \eta-\xi\rangle}\, dy\, d\eta. \tag{2.8}$$

We now write the remainder term in Taylor's formula (2.7) in the form

$$r_N(x, y, \xi, \eta) = \sum_{|\alpha|=N} \frac{N i^{|\alpha|}}{\alpha!} \int_0^1 (1-\theta)^{N-1} D_\eta^\alpha A(x, y, \xi, \xi+\theta(\eta-\xi))\, d\theta\, (\eta-\xi)^\alpha.$$

Writing $(\eta-\xi)^\alpha e^{i\langle x-y,\ \eta-\xi\rangle}$ in the form $(-D_y)^\alpha e^{i\langle x-y,\ \eta-\xi\rangle}$ and integrating by parts with respect to y we obtain

$$a(x, \xi) - a_N(x, \xi) =$$

$$(2\pi)^{-n} \sum_{|\alpha|=N} \frac{N i^N}{\alpha!} \int\int\int_0^1 (1-\theta)^{N-1} D_y^\alpha D_\eta^\alpha A(x, y, \xi, \xi+\theta(\eta-\xi))\, e^{i\langle x-y,\ \eta-\xi\rangle}\, d\theta\, dy\, d\eta. \tag{2.9}$$

By hypothesis $A = 0$ when $y \in \mathbb{R}^n \setminus K'$, where K' is a compact subset, so integration in y is taken over the compact set K'.

Representing the exponent in the form

$$e^{i\langle x-y,\ \eta-\xi\rangle} = (1-\Delta_y)\, e^{i\langle x-y,\ \eta-\xi\rangle}(1+|\xi-\eta|^2)^{-1} =$$

$$(1-\Delta_y)^l e^{i\langle x-y,\ \eta-\xi\rangle}(1+|\xi-\eta|^2)^{-l},$$

where $l > 0$ is an integer, we obtain

$$a(x, \xi) - a_N(x, \xi) = (2\pi)^{-n} \sum_{|\alpha| = N} \frac{Ni^N}{\alpha!} \times$$

$$\iint\int_0^1 (1-\theta)^{N-1} (1-\Delta_y)^l D_y^\alpha D_\eta^\alpha A(x, y, \xi, \xi+\theta(\eta-\xi)) e^{i(x-y,\, \eta-\xi)} (1+|\xi-\eta|^2)^{-l} \, d\theta\, dy\, d\eta.$$

By (2.3) the integrand does not exceed

$$C_1(1+|\xi|)^m (1+|\xi+\theta(\eta-\xi)|)^{m'-N} (1+|\xi-\eta|^2)^{-l} h_1(y),$$

where $h_1 \in C_0^\infty(\mathbb{R}^n)$, $h_1 = 1$ in a neighborhood of the compact set K'. Therefore

$$|a(x, \xi) - a_N(x, \xi)| \leqslant C_2 \int\int_0^1 (1+|\xi|)^m (1+|\xi+\theta(\eta-\xi)|)^{m'-N} (1+|\xi-\eta|^2)^{-l} \, d\theta\, d\eta.$$

Now divide the domain of integration into two subdomains:

$$\Omega_1 = \{(\theta, \eta);\ |\theta(\eta-\xi)| < |\xi|/2\}$$

and

$$\Omega_2 = \{(\theta, \eta);\ |\theta(\eta-\xi)| > |\xi|/2\}.$$

In Ω_1 the inequalities

$$|\xi| \leqslant 2|\xi+\theta(\eta-\xi)| \leqslant 3|\xi|$$

hold and therefore the integrand does not exceed

$$C_2'(1+|\xi|)^{m+m'-N}(1+|\xi-\eta|^2)^{-l}.$$

The integral in Ω_1 converges when $2l > n$ and does not exceed

$$C_3(1+|\xi|)^{m+m'-N}.$$

In Ω_2 the following inequalities

$$|\xi| \leqslant 2|\eta-\xi| \text{ and } 1+|\xi+\theta(\eta-\xi)| \leqslant 1+3|\eta-\xi|$$

are valid. Therefore, the integrand does not exceed

$$C_4(1+|\xi|)^{m+m'-N}(1+|\eta-\xi|^2)^{-l+|N-m'|/2}$$

because for $m' \geqslant N$

$$(1+|\xi+\theta(\eta-\xi)|)^{m'-N} \leqslant 3^{m'-N}(1+|\eta-\xi|)^{m'-N} \leqslant$$
$$3^{m'-N}(1+|\xi|)^{m'-N}(1+|\eta-\xi|)^{m'-N},$$

and for $m' < N$

$$(1+|\xi+\theta(\eta-\xi)|)^{m'-N} \leqslant 1 \leqslant (1+|\xi|)^{m'-N}(1+|\eta-\xi|)^{N-m'}2^{N-m'}.$$

Assuming $l = |N - m_2| + n + 1$ we obtain

$$|a(x, \xi) - a_N(x, \xi)| \leqslant C_5(1+|\xi|)^{m+m'-N}.$$

To estimate the derivative of $a - a_N$ it is sufficient to differentiate (2.9) with respect to x and ξ and to repeat the same estimates. As a result we obtain

$$|D_\xi^\alpha D_x^\beta [a(x, \xi) - a_N(x, \xi)]| \leqslant C'_{\alpha,\beta}(1+|\xi|)^{m+m'-N-|\alpha|}.$$

Thus, Lemma 2.1 is proved.

Remark 2.1. If under the assumption (2.3) we consider instead of (2.4) the integral

$$b(x, \xi) = (2\pi)^{-n} \iint A(x, y, \xi, \eta) e^{i(y-x,\, \eta-\xi)} dy, d\eta,$$

which differs from (2.3) in the sign of the exponent, then passing to complex conjugates it is easy to see that $b \in S^{m+m'}$ and $b - b_N \in S^{m+m'-N}$, where

$$b_N(x, \xi) = \sum_{|\alpha|<N} \frac{(-i)^{|\alpha|}}{\alpha!} D_y^\alpha D_\eta^\alpha A(x, y, \xi, \eta)\Big|_{\substack{\eta=\xi \\ y=x}}.$$

Remark 2.2. From the proof it is evident that (2.3) can be replaced by a weaker condition

$$|D_\xi^\alpha D_\eta^{\alpha'} D_x^\beta D_y^{\beta'} A(x, y, \xi, \eta)| \leqslant$$
$$C_{\alpha, \alpha', \beta, \beta'} (1+|\xi|)^{m-|\alpha|} (1+|\eta|)^{m'-|\alpha'|} (1+|\xi-\eta|)^{s_{\alpha, \alpha', \beta}}, \tag{2.10}$$

where $s_{\alpha,\alpha',\beta,\beta'}$ are numbers.

Indeed, nothing changes for sums of the form a_N because there $\xi = \eta$. When estimating the remainder term we have to increase the auxiliary parameter l by s, which is insignificant for the proof.

2.5. Let us now consider the composition of two operators A and B with symbols a and b belonging to classes $S_0^{m_1}(\Omega)$ and $S_0^{m_2}(\Omega)$.

Theorem 2.2. Let $a \in S_0^{m_1}(\Omega)$ and $b \in S_0^{m_2}(\Omega)$. The operator C which transforms a function $u \in C_0^\infty(\Omega)$ into a function $B \circ Au$, is a pseudo-differential operator of order $m_1 + m_2$ with the symbol c belonging to class $S_0^{m_1+m_2}(\Omega)$. Further for every N the function $c - c_N$, where

$$c_N(x, \xi) = \sum_{|\alpha|<N} \frac{i^{|\alpha|}}{\alpha!} D_\xi^\alpha b(x, \xi) D_x^\alpha a(x, \xi), \tag{2.11}$$

is a symbol of class $S_0^{m_1+m_2-N}$.

Proof. We compute formally

$$BAu = B(2\pi)^{-n} \int a(y, \xi) \tilde{u}(\xi) e^{i(y, \xi)} d\xi =$$

$$(2\pi)^{-2n} \iiint b(x, \eta) e^{ix\eta} e^{-iy\eta} a(y, \xi) \tilde{u}(\xi) e^{i(y, \xi)} d\xi\, dy\, d\eta = (2\pi)^{-n} \int c(x, \xi) \tilde{u}(\xi) e^{ix\xi} d\xi,$$

where

$$c(x, \xi) = (2\pi)^{-n} \iint b(x, \eta) a(y, \xi) e^{i(y-x, \xi-\eta)} dy\, d\eta. \tag{2.12}$$

Note that since $a = 0$ when $y \in \mathbb{R}^n \setminus K$, the integral (2.12) converges and determines an infinitely differentiable function in x and ξ.

It is not hard to see that Lemma 2.1 is applicable to the integral (2.12), and from this we can immediately obtain the desired statement.

Note that formula (2.11) can be regarded as a natural generalization of the Leibniz' formula from Chapter I.

Corollary 2.1. The operator $T = BA - C_0$, where C_0 is an operator with symbol equal to $b(x, \xi)a(x, \xi)$, is of order $m_1 + m_2 - 1$.

Corollary 2.2. The operator $T_1 = BA - AB - C_1$ is of order $m_1 + m_2 - 2$ if the symbol of the operator C_1 is equal to

$$c_1(x, \xi) = \frac{1}{i} \sum_{k=1}^n \left(\frac{\partial b}{\partial \xi_k} \frac{\partial a}{\partial x^k} - \frac{\partial a}{\partial \xi_k} \frac{\partial b}{\partial x^k} \right) = \frac{1}{i} \{b, a\}, \tag{2.13}$$

where $\{b, a\}$ is the Poisson bracket of functions b and a.

Corollary 2.3. If $b(x, \xi) = 0$ in the support of function a, then the pseudo-differential operator BA is of order $-\infty$.

Corollary 2.4. If P is a pseudo-differential operator of order m and T is a pseudo-differential operator of order $-\infty$, then PT and TP are pseudo-differential operators of order $-\infty$.

Proof. We can assume T to be an operator of order N for any integer N. Therefore, the operators PT and TP are of the order $m - N$ for any N.

But this implies that their order is equal to $-\infty$, and this proves the corollary.

2.6. Often it is convenient to consider operators of a somewhat more general form than (2.1), namely, operators such that

$$Au(x)=(2\pi)^{-n}\int\int b(x, y, \xi)\,u(y)\,e^{i(x-y,\,\xi)}\,dy\,d\xi, \tag{2.14}$$

where the function $b\in C^{\infty}(\Omega\times\Omega\times\mathbb{R}_n)$, $b=0$ when $x\in\mathbb{R}^n\setminus K$ or $y\in\mathbb{R}^n\setminus K'$ and

$$|D_\xi^\alpha D_x^\beta D_y^\gamma b(x, y, \xi)|\leqslant C_{\alpha,\beta,\gamma}(1+|\xi|)^{m-|\alpha|}. \tag{2.15}$$

Clearly this definition agrees with (2.1) when function b is independent of y.

The following theorem reveals that in reality the operators (2.14) are not more general than the operators (2.1).

Theorem 2.3. Suppose operator A is determined by formula (2.14) with symbol satisfying (2.15). Then the operator A is a pseudo-differential operator of order m with symbol belonging to $S_0^m(\Omega)$. If we assume

$$a_N(x, \xi)=\sum_{|\alpha|<N}\frac{i^{|\alpha|}}{\alpha!}D_y^\alpha D_\xi^\alpha b(x, y, \xi)\big|_{y=x},$$

then the function $a-a_N$ is the symbol of an operator of order $m-N$.

Proof. Formally we write the operator A as a pseudo-differential operator

$$Au(x)=(2\pi)^{-2n}\int\int\int b(x, y, \eta)\,\tilde{u}(\xi)\,e^{i(y,\,\xi)}e^{i(x-y,\,\eta)}\,d\xi\,dy\,d\eta=$$
$$(2\pi)^{-n}\int a(x, \xi)\,\tilde{u}(\xi)\,e^{i(x,\,\xi)}\,d\xi,$$

where it is assumed that

$$a(x, \xi)=(2\pi)^{-n}\int\int b(x, y, \eta)\,e^{i(x-y,\,\eta-\xi)}\,dy\,d\eta.$$

Applying Lemma 2.1 we obtain the desired statement.

2.7. When using integral estimates it is often necessary to perform "integration by parts," that is, to transfer a pseudo-differential operator from one co-factor in the integrand to another.

Definition 2.1. An operator P^* is said to be *formally adjoint* to an operator P if

$$\int Pu\bar{v}\,dx=\int u\overline{P^*v}\,dx,\quad u, v\in C_0^\infty(\Omega).$$

In Section 1.1 we have encountered this concept for a differential operator. The following theorem will enable us to compute the symbol of a formally adjoint operator for an arbitrary pseudo-differential operator P.

Theorem 2.4. Let A be a pseudo-differential operator of order m with symbol $a\in S_0^m(\Omega)$ and let

$$a_N^*(x, \xi)=\sum_{|\alpha|<N}\frac{i^{|\alpha|}}{\alpha!}D_\xi^\alpha D_x^\alpha\overline{a(x, \xi)}. \tag{2.16}$$

Operator A^* is a pseudo-differential operator of order m; its symbol a^* is a function such that for any N the operator with symbol a^*-a_N is of order $m-N$.

Proof. We compute formally

$$\int Au(x)\,\overline{v(x)}\,dx=(2\pi)^{-n}\int\int\int a(x, \xi)\,u(y)\,e^{i(x-y,\,\xi)}\,\overline{v(x)}\,dy\,d\xi\,dx=$$
$$\int u(y)\,\overline{A^*v(y)}\,dy,$$

where it is assumed that

$$A^*v(y) = (2\pi)^{-n} \int\int \overline{a(x, \xi)} e^{i\langle y-x, \xi\rangle} v(x)\, dx\, d\xi.$$

We now write this operator in the form (2.1)

$$A^*v(y) = (2\pi)^{-2n} \int\int\int \overline{a(x, \xi)} e^{i\langle y-x, \xi\rangle} \tilde{v}(\eta) e^{-i\langle x, \eta\rangle} d\eta\, dx\, d\xi =$$

$$(2\pi)^{-n} \int a^*(y, \eta) \tilde{v}(\eta) e^{i\langle y, \eta\rangle} d\eta,$$

where

$$a^*(y, \eta) = (2\pi)^{-n} \int\int \overline{a(x, \xi)} e^{i\langle y-x, \xi-\eta\rangle} dx\, d\xi.$$

The expression obtained satisfies the conditions of Lemma 2.1. Our statement is a simple corollary which follows from this lemma.

2.8. The fact that a kernel K(x, y) of a pseudo-differential operator turns out to be a smooth operator when $x \neq y$ results in the pseudo-local property of the considered operators.

Definition 2.2. A point x_0 is called a *singular point* of a distribution $u \in \mathscr{D}'(\Omega)$ if it has no neighborhood where u coincides with an infinitely differentiable function. The set of singular points of u is called the singular support and is denoted by sing supp u. An operator $P: \mathscr{E}'(\Omega) \to \mathscr{D}'(\Omega)$ is called *pseudo-differential* if sing supp Pu $\subset$ sing supp u for every $u \in \mathscr{E}'(\Omega)$. An operator $P: \mathscr{E}'(\Omega) \to \mathscr{D}'(\Omega)$ is called *hypoelliptic* if sing supp Pu $\supset$ sing supp u for every $u \in \mathscr{E}'(\Omega)$.

In practice the last definition agrees with the Definition 4.4 of Chapter I although formally it is more general.

Pseudo-differential operators are not local in the general case (see Theorem 5.1 from Chapter I) but have the weaker pseudo-local property.

Theorem 2.5. Pseudo-differential operators are pseudo-local.

Proof. Let $u \in \mathscr{E}'(\Omega)$, supp u $\subset$ K and u coincides with an infinitely differentiable function in a neighborhood ω of x_0. If $h \in C_0^\infty(\Omega)$, then $hu \in C_0^\infty(\Omega)$ and by Theorem 2.1 $P(hu) \in C^\infty(\Omega)$.

Let us assume that $h \in C_0^\infty(\omega)$ and h = 1 in a smaller neighborhood ω', and $g \in C_0^\infty(\omega')$. Then

$$gPu = gPhu + gP(1-h)u.$$

Function $gP(1-h)u$ is infinitely differentiable by Corollary 2.3 because $g(1-h) = 0$. Thus $gPu \in C^\infty$ and if $g \neq 0$ in ω'', then $Pu \in C^\infty(\omega'')$.

Since ω'' is an arbitrary subdomain compactly imbedded in ω, it follows that the function Pu is infinitely differentiable in ω. Hence sing supp Pu $\subset$ sing supp u. The theorem is proved.

2.9. In order to define a pseudo-differential operator on an arbitrary smooth manifold, we need first to study its dependence on the selection of a system of coordinates in the subdomain Ω of $\mathbb{R}^n$.

Theorem 2.6. The class of pseudo-differential operators with symbols in $S^m(\Omega)$ is invariant under diffeomorphisms $F: \Omega \to \Omega$ for which $|D^\alpha F(x)| \leqslant C_\alpha$ when $|\alpha| \geqslant 1$ and $|\partial F/\partial x| \geqslant c_0$, where $\partial F/\partial x$ is a Jacobi matrix, $c_0 = \text{const} > 0$, Ω is a domain in $\mathbb{R}^n$. Here, if A is a pseudo-differential operator with symbol $a \in S^m(\Omega)$, then the operator $A \circ F$ differs from the operator with symbol $a(F(x), {}^tF'(x)^{-1}\xi)$ by an operator of order $m - 1$.

The proof of this theorem is based on the following auxiliary proposition.

Lemma 2.2. An operator of the form

$$Au(x) = \int\int b(x, y, \xi) u(y) e^{i\langle x-y, \xi\rangle} dy\, d\xi,$$

where the function b is such that

$$|D_\xi^\alpha D_x^\beta D_y^\gamma b(x, y, \xi)| \leqslant C_{\alpha,\beta,\gamma,M}(1+|\xi|)^{m-|\alpha|}, \quad x \in M, \quad y \in M,$$

where M is an arbitrary compact set in $\mathbb{R}^n$ and $b(x, y, \xi) = 0$ when $|x - y| \leqslant c_0$, $c_0 = \text{const} > 0$, may be written as

$$Au(x) = \int K(x, y) u(y) dy,$$

where $K(x, y) \in C^\infty(\mathbb{R}^n \times \mathbb{R}^n)$. Every operator of the form $Bu(x) = \int K(x, y - x) u(y) dy$ with an infinitely differentiable kernel K such that

$$|D_x^\alpha D_z^\beta K(x, z)| \leqslant C_{\alpha,\beta,M,N}(1+|z|)^{-N}, \quad x \in M,$$

for all α, β, N, z and $M \Subset \mathbb{R}^n$ is a pseudo-differential operator of order $-\infty$.

Proof. Since the inequality $|x - y| \geqslant c_0$ is satisfied in the support of b, we can use the identity

$$e^{i(x-y,\xi)} = |x-y|^{-2l}(-\Delta_\xi)^l e^{i(x-y,\xi)}$$

and integrate by parts with respect to ξ. In so doing we obtain

$$Au(x) = \lim_{\varepsilon \to +0} \iint e^{-\varepsilon|\xi|} b(x, y, \xi)|x-y|^{-2l} u(y)(-\Delta_\xi)^l e^{i(x-y,\xi)} dy\, d\xi =$$

$$\lim_{\varepsilon \to +0} \iint (-\Delta_\xi)^l [b(x, y, \xi) e^{-\varepsilon|\xi|}] u(y) e^{i(x-y,\xi)} dy\, d\xi =$$

$$\iint (-\Delta_\xi)^l b(x, y, \xi)|x-y|^{-2l} u(y) e^{i(x-y,\xi)} dy\, d\xi = \int K(x, y) u(y) dy,$$

if $m - 2l < -n$. Then

$$K(x, y) = \int (-\Delta_\xi)^l b(x, y, \xi)|x-y|^{-2l} e^{i(x-y,\xi)} d\xi,$$

and the integral converges absolutely. If the number l is such that

$$2l > n + m + k,$$

then $K \in C^k$. Since l can be taken as arbitrarily large, K is an infinitely differentiable function.

An integral operator B can be represented as

$$Bu(x) = \int K(x, y-x) u(y) dy = (2\pi)^{-n} \iint K(x, y-x) \tilde{u}(\xi) e^{iy\xi} d\xi\, dy =$$

$$(2\pi)^{-n} \int b(x, \xi) \tilde{u}(\xi) e^{ix\xi} d\xi,$$

where we have formally assumed that

$$b(x, \xi) = \int K(x, y-x) e^{i(y-x,\xi)} dy = \int K(x, z) e^{i(z,\xi)} dz.$$

When $|\xi| \geqslant 1$, we have

$$e^{iz\xi} = |\xi|^{-2l}(-\Delta_z)^l e^{iz\xi}.$$

Integrating by parts we obtain

$$b(x, \xi) = |\xi|^{-2l} \int [(-\Delta_z)^l K(x, z)] e^{iz\xi} dz.$$

The assumptions on K imply that for $x \in M$ and all α and β

$$|D_\xi^\alpha D_x^\beta b(x, \xi)| \leqslant C_{\alpha,\beta,l,M}(1+|\xi|)^{-2l}.$$

The lemma is proved.

Proof of Theorem 2.6. Let $a \in S^m$ and

$$Au(x) = (2\pi)^{-n} \iint a(x, \xi) u(y) e^{i(x-y,\xi)} dy\, d\xi.$$

Let $h \in C_0^\infty(\mathbb{R})$ be a function such that $h(t) = 1$ when $|t| \leqslant 1$ and $h(t) = 0$ when $|t| \geqslant 2$. We may decompose A as $A = A_1 + A_2$ where

$$A_1 u(x) = (2\pi)^{-n} \iint a(x, \xi) h(2\varepsilon^{-1}|x-y|) u(y) e^{i(x-y,\xi)} dy\, d\xi.$$

Let now

$$x = F(x'),\ y = F(y'),\ u(F(y')) = v(y'),\ (A_1u)(F(x')) = w_1(x').$$

Then

$$w_1(x') = (2\pi)^{-n} \iint a(F(x'),\ \xi)\, h(2\varepsilon^{-1}|F(x') - F(y')|)\, v(y') \cdot$$
$$e^{i[F(x') - F(y')]\cdot\xi}\, |\partial F(y')/\partial y'|\, dy'\, d\xi, \tag{2.17}$$

where $|\partial F(y')/\partial y'|$ is the Jacobian. According to Hadamard's lemma

$$F(x') - F(y') = B(x',\ y')(x' - y'),$$

where

$$B(x',\ y') = \int_0^1 \partial F(y' + s(x' - y'))/\partial y'\, ds.$$

Note that $B(x',\ x') = \partial F(x')/\partial x'$, so that the matrix $B(x',\ y')$ is invertible if $|y' - x'|$ is sufficiently small, say $|y' - x'| < \varepsilon$.

Assuming $\eta = {}^tB(x',\ y')\xi$, where ${}^tB$ is the transpose of the matrix B, we have

$$w_1(x') = (2\pi)^{-n} \iint a(F(x'),\ {}^tB(x',\ y')^{-1}\eta)\, h(2\varepsilon^{-1}|F(x') - F(y')|) \cdot$$
$$v(y')\, e^{i(x' - y',\ \eta)} \frac{|\partial F(y')/\partial y'|}{|B(x',\ y')|}\, dy'\, d\eta. \tag{2.18}$$

Operator A_1 transforming v into w_1 is of the form (2.14) and is therefore pseudo-differential. Note that by Theorem 2.3 it differs from the operator with symbol

$$a\left(F(x'),\ \left(\frac{\partial F(x')}{\partial x'}\right)^{-1}\xi\right) \tag{2.19}$$

by an operator of order $m - 1$.

We now show that the operator A_2 is a pseudo-differential operator of order $-\infty$. For this, note that by Lemma 2.2 this is an integral operator

$$A_2u(x) = \int K(x,\ y - x)\, u(y)\, dy,$$

and

$$K(x,\ z) = \int a(x,\ \xi)[1 - h(2\varepsilon^{-1}|z|)]\, e^{-iz\xi}\, d\xi.$$

Since $z \neq 0$ in the support of the integrand, use can be made of the identity

$$e^{-iz\xi} = |z|^{-2l}(-\Delta_\xi)^{2l} e^{-iz\xi},$$

which after integration by parts gives

$$K(x,\ z) = |z|^{-2l} \int [(-\Delta_\xi)^l a(x,\ \xi)][1 - h(2\varepsilon^{-1}|z|)]\, e^{-iz\xi}\, d\xi.$$

From this formula it is evident that for all α, β and l

$$|D_x^\alpha D_z^\beta K(x,\ z)| \leq C_{\alpha,\beta,l,M}(1 + |z|)^{-2l} \text{ when } x \in M, \tag{2.20}$$

where M is an arbitrary compact set in $\mathbb{R}^n$.

The diffeomorphism F transforms the operator A_2 into the operator A_2' such that

$$A_2'v(x') = \int K(F(x'),\ F(y') - F(x'))\, v(y') \left|\frac{\partial F(y')}{\partial y'}\right| dy' = \int K'(x',\ y' - x')\, v(y')\, dy'.$$

Let us verify that the kernel

$$K'(x',\ z') = K(F(x'),\ F(x' + z') - F(x'))\, |\partial F(x' + z')/\partial x'|$$

satisfies the conditions of Lemma 2.2. Let $x' \in M$ (M is a compact set in $\mathbb{R}^n$). Note that

$$|F(x'+z')-F(x')| \geqslant c_0|z'|, \quad c_0=\text{const}>0,$$

$$|D_{x'}^{\alpha} D_{z'}^{\beta}\, \partial F(x'+z')/\partial x'| \leqslant C'_{\alpha,\beta}+C''_{\alpha,\beta}|z'|,$$

and therefore

$$|D_{x'}^{\alpha} D_{z'}^{\beta} K'(x', z')| \leqslant C'_{\alpha,\beta,N,M}(1+|z'|)^{-N},$$

for any arbitrary α, β, and N.

Thus, the conditions of Lemma 2.2 are satisfied for the operator A_2' and it is therefore a pseudo-differential operator of order $-\infty$.

We have already shown that operator A changes to a pseudo-differential operator $A_1 + A_2'$ as a result of a smooth nonsingular replacement of independent variables. Theorem 2.6 is thus proved.

2.10. Theorem 2.6 makes it possible to define effectively the principal symbol of the operator A as

$$a_0(x, \xi)=\lim_{t\to+\infty}\frac{a(x, t\xi)}{t^m}.$$

A function a_0 is called a *principal symbol* of operator A if the function $a(x, \xi) - h(|\xi|)a_0(x, \xi)$ is a symbol of the operator of order $m - 1$. Here, the function $h \in C_0^{\infty}(\mathbb{R})$ is such that $h(t) = 1$ when $|t| \geqslant 2$ and $h(t) = 0$ when $|t| \leqslant 1$.

From the definition it is clear that a_0 is a positive homogeneous function in ξ of degree m, that is, $a_0(x, t\xi) = t^m a_0(x, \xi)$ when $t > 0$.

If A is a pseudo-differential operator, then the principal symbol so defined agrees with the characteristic form of the operator A.

In this book we shall study operators whose properties are determined by the principal symbol a_0 and are independent of lower terms.

Corollary 2.5. From Theorem 2.6 it follows that the principal symbol of a pseudo-differential operator is defined invariantly under diffeomorphisms of the domain of Ω.

2.11. We shall now determine a pseudo-differential operator on the manifold Ω. Let $\{\Omega_\nu, \omega_\nu, f_\nu\}$ be an atlas on Ω. A linear operator P given on Ω is called a *pseudo-differential operator of order* m if every operator of the type $f_\nu^{-1}\circ P\circ f_\nu\colon C_0^{\infty}(\omega_\nu)\to C^{\infty}(\omega_\nu)$, where ω_ν is an open (not necessarily connected) set in $\mathbb{R}^n$ and f_ν is a diffeomorphism of ω_ν onto an open subset in Ω, is a pseudo-differential operator of order m.

Theorem 2.7. The principal symbol of a pseudo-differential operator on a manifold is defined invariantly as a function $a_0\colon T^*\Omega\setminus 0\to\mathbb{C}$.

Proof. This statement follows at once from the definition of a pseudo-differential operator on a manifold and Corollary 2.5.

§3. Elliptic Pseudo-Differential Operators

3.1. Elliptic pseudo-differential operators form a class of almost reversible elements in the algebra of pseudo-differential operators and are therefore of particular importance. As already noted, a class of pseudo-differential operators arises naturally from the closure of the algebra containing all differential operators and the operators inverse to elliptic differential operators. The methods of constructing such operators as (singular) integral operators are taken from classical analysis.

Note that for many elliptic operators defined on a closed manifold there does not exist an inverse operator, since an operator P may have a nontrivial kernel. However this kernel has always finite dimension and we can construct an "almost inverse" operator Q for which

$$PQ = I + T, \tag{3.1}$$

where I is the identity operator and T is an operator of order $-\infty$.

The operator Q satisfying (3.1) is called a *parametrix*.

We shall recall a classical method due to E. E. Levi for constructing a parametrix for an elliptic differential operator P. It consists of "freezing" the coefficients of the operator P. In so doing we first consider an equation with constant coefficients

$$P(x_0, D) E_0 = I, \tag{3.2}$$

and find the operator E_0. Then we look for the operator E_1 from the condition

$$P(x_0, D) E_1 = [P(x_0, D) - P(x, D)] E_0,$$

for the operator E_2 from the condition

$$P(x_0, D) E_2 = [P(x_0, D) - P(x, D)] E_1$$

and so on. If the series $E_0 + E_1 + \dots$ converges then summing these equalities gives

$$P(x, D)[E_0 + E_1 + \dots] = I.$$

If x varies in a sufficiently small neighborhood of the point x_0, then we can in fact prove that this series converges. In fact, to do so we cannot in general solve even equation (3.2); we can only find the solution to the equation

$$P(x_0, D) E_0 = I + T_0,$$

where T_0 is an operator with a finite-dimensional range. The operator E_1 is then found from the formula

$$E_1 f = E_0 [P(x_0, D) - P(x, D)] E_0 f,$$

and we obtain

$$P(x_0, D) E_1 = (I + T_0)[P(x_0, D) - P(x, D)] E_0 =$$
$$[P(x_0, D) - P(x, D)] E_0 + T_1$$

and so on. The main difficulty of this method is the proof of convergence. It is true that we can break the construction into a finite number of steps and obtain a solution for the equation

$$P(x, D) E' = I + T' + T_0,$$

where T' is not a smoothing operator but has a small norm. However, here also E' is represented as an awkward expression involving repeated integration.

The method of constructing a parametrix in the theory of pseudo-differential operators does not differ in principal from Levi's parametrix method. This theory, however, saves us from the need to prove the convergence of the successive approximations and enables us to construct a parametrix in a rather nice form.

3.2. Definition 3.1. A pseudo-differential operator P of order m with symbol $p(x, \xi)$ is called *elliptic* in Ω if there are constants $c_1 > 0$ and $C_2 > 0$ such that

$$|p(x, \xi)| \geqslant c_1 (1 + |\xi|)^m \text{ when } x \in \Omega, \ |\xi| \geqslant C_2. \tag{3.3}$$

It is not hard to see that this definition agrees with Definition 4.2 of Chapter I for differential operators.

Definition 3.2. A pseudo-differential operator $Q \in S^{-m}(\Omega)$ is called a *right* (correspondingly *left*) *parametrix* for the operator P in the domain $\Omega' \Subset \Omega$ if

$$Ph Q\varphi = \varphi + T\varphi \text{ (correspondingly } QhP\varphi = \varphi + T'\varphi), \tag{3.4}$$

where T and T' are operators of order $-\infty$, $h \in C^\infty(\Omega)$, $h = 1$ in the domain Ω'.

The construction of a parametrix is based on formula (2.11) obtained in Theorem 2.2. To prove this we shall need the following:

Lemma 3.1. (L. Hörmander). Let $a_j \in S^{m-j}(\Omega)$ for $j = 0, 1, 2, \ldots$. Then there exists a symbol $a \in S^m(\Omega)$ such that for any N

$$a - \sum_{j=0}^{N} a_j \in S^{m-N-1}(\Omega).$$

Proof. Let $\{M_k\}$ be a sequence of compact subsets in Ω such that $M_k \subset M_{k+1}$ and $\lim_{k\to\infty} M_k = \Omega$.

Let $h \in C^\infty(\mathbb{R}_n)$ and $h(\xi) = 0$ when $|\xi| < 1$ and $h(\xi) = 1$ when $|\xi| \geqslant 2$. We shall now construct an increasing sequence t_j such that $t_j \to +\infty$ and

$$|D_\xi^\alpha D_x^\beta [h(\xi t_j^{-1}) a_j(x, \xi)]| \leqslant 2^{-j}(1+|\xi|)^{m-j-|\alpha|+1} \tag{3.5}$$

when $x \in M_k$, $|\alpha+\beta|+k \leqslant j$, $\xi \in \mathbb{R}_n$.

Then $a(x, \xi) = \sum_{j=0}^{\infty} h(\xi t_j^{-1}) a_j(x, \xi)$ is the desired function. Indeed, at every point (x, ξ) only a finite number of addends is different from zero and therefore $a \in C^\infty(\Omega \times \mathbb{R}_n)$. If K is an arbitrary compact set in Ω, then we can find k for which $K \subset M_k$. Let $x \in K$ and ξ be an arbitrary point from $\mathbb{R}_n$. We can now find j_0 such that $t_{j_0} > |\xi|$. Therefore

$$a(x, \xi) = \sum_{j=0}^{j_0-1} h(\xi t_j^{-1}) a_j(x, \xi).$$

For arbitrary multi-indices α and β we have

$$|D_\xi^\alpha D_x^\beta a(x, \xi)| \leqslant \sum_{j=0}^{j_0-1} |D_\xi^\alpha D_x^\beta [h(\xi t_j^{-1}) a_j(x, \xi)]| \leqslant$$

$$\sum_{j=1}^{j_0-1} 2^{-j}(1+|\xi|)^{m-j-|\alpha|+1} + C_{\alpha\beta}^0 (1+|\xi|)^{m-|\alpha|} \leqslant C'_{\alpha\beta}(1+|\xi|)^{m-|\alpha|}.$$

Thus we have proved that $a \in S^m(\Omega)$.

In exactly the same way it is verified that

$$a(x, \xi) - \sum_{j=0}^{N} a_j(x, \xi) =$$

$$\sum_{j=0}^{N} [h(\xi t_j^{-1}) - 1] a_j(x, \xi) + \sum_{j=N+1}^{\infty} h(\xi t_j^{-1}) a_j(x, \xi) \in S^{m-N-1}(\Omega).$$

3.3. We shall first construct a right parametrix. By assumption we have

$$|p(x, \xi)| \geqslant c_1(1+|\xi|)^m \text{ when } |\xi| \geqslant c_2.$$

Let $\chi \in C^\infty(\mathbb{R}_n)$, $\chi(\xi) = 0$ when $|\xi| \leqslant c_2$ and $\chi(\xi) = 1$ when $|\xi| \geqslant c_2 + 1$. Further let $h \in C_0^\infty(\mathbb{R}^n)$ and $h = 1$ in a neighborhood of $\overline{\Omega'}$.

If Q_0 is an operator with symbol $q_0(x, \xi) = \chi(\xi)P^{-1}(x, \xi)$ (so that

$q_0(x, \xi) = 0$ when $|\xi| < C_2$), then its order is equal to $-m$; the order of the operator

$$R_1 = PhQ_0 - hI$$

is equal to -1. Let r_1 be its symbol and

$$q_1(x, \xi) = -r_1(x, \xi)\, p^{-1}(x, \xi)\, \chi(\xi).$$

Then the order of the operator Q_1 with this symbol is equal to $-m - 1$, and the order of the operator

$$R_2 = Ph\,(Q_0 + Q_1) - hI$$

is equal to -2, since

$$R_2 = hI + R_1 + PhQ_1 - hI = R_1 + PhQ_1.$$

Applying the same reasoning, we find a sequence of operators Q_0, Q_1, ... where the operator Q_j is of order $-m - j$, and the order of the operator

$$R_j = Ph\,(Q_0 + Q_1 + \ldots + Q_{j-1}) - hI$$

is equal to $-j$.

According to Lemma 3.1 there exists an operator $Q \in S_0^{-m}(\Omega)$ such that the order of the operator $Q - (Q_0 + Q_1 + \ldots + Q_{j-1})$ is equal to $-m - j$. This implies that the operator

$$R = PhQ - hI$$

is of order $-j$ for any j; that is, it is a smoothing operator.

3.4. The left parametrix Q' is constructed in a like manner. The operator Q_0' with symbol $\chi(\xi)p(x, \xi)^{-1}$ is of order $-m$, and the order of the operator

$$R_1' = Q_0'hP - hI$$

is equal to -1. Let us denote its symbol by r_1' and assume

$$q_1'(x, \xi) = -r_1'(x, \xi)\, \chi(\xi)\, p^{-1}(x, \xi).$$

Then the order of the operator Q_1' with this symbol is equal to $-m - 1$, and that of the operator

$$R_2' = (Q_0' + Q_1')\, hP - hI$$

is equal to -2, since

$$R_2' = hI + R_1' + Q_1'hP - hI = R_1' + Q_1'hP.$$

Applying the same reasoning we find a sequence of operators Q_0', Q_1', ... in which the operator Q_1' is of order $-m - j$, and the order of the operator

$$R_j' = (Q_0' + \ldots + Q_{j-1}')\, hP - hI$$

is equal to $-j$.

According to Lemma 3.1 there exists an operator $Q' \in S_0^{-m}(\Omega)$ such that the order of the operator $Q' - (Q_0' + \ldots + Q_j')$ is equal to $-m - j$, and therefore the order of the operator

$$R' = Q'hP - hI$$

is equal to $-j$ for any j so that this is a smoothing operator.

We now show that the operator $(Q - Q')h$ is also a smoothing operator.

In fact

$$Q'h\,(PhQ) = Q'h\,(hI + R) = Q'h^2 + T,$$

$$(Q'hP)\,hQ = (hI + R')\,hQ = h^2Q + T',$$

where T and T' are smoothing operators.

It is seen that $Q'h^2 - h^2Q$ is a smoothing operator; that is, $q' - q \in S^{-\infty}(\Omega')$.

Thus the operators Q and Q' coincide in Ω'. We now prove the following theorem.

Theorem 3.1. Let a pseudo-differential operator P be an elliptic operator of order m in Ω. Then for every function $h \in C_0^\infty(\Omega)$ there exists a pseudo-differential operator Q of order $-m$ such that

$$QhP = h \cdot I + T_1, \text{ and } PhQ = h \cdot I + T_2,$$

where T_1 and T_2 are smoothing operators.

3.5. Let us now consider an elliptic operator P on a closed manifold

Definition 3.3. A pseudo-differential operator P: $C^\infty(\Omega) \to C^\infty(\Omega)$ defined on a closed smooth manifold is called *elliptic* if there exists an atlas $\{\Omega_\nu, \omega_\nu, f_\nu\}$ on Ω such that for every operator $P_\nu = f_\nu^{-1}Pf_\nu$ is elliptic in ω_ν.

Theorem 2.7 implies that a principal symbol P^0, if it exists, is defined on $T^*\Omega\backslash 0$ and is a positively homogeneous function of ξ in local coordinates (x, ξ). Therefore, considering the bundle $S^*\Omega$ of unit spheres in the space $T^*\Omega$ we can find, by Definitions 3.3 and 3.1, a constant $c_0 > 0$ for which

$$|p^0| \geqslant c_0 \quad \text{on} \quad S^*\Omega.$$

This inequality can, of course, be taken as the definition of an elliptic operator assuming that there exists a principal symbol. Fortunately, inequality (3.3) is also invariant with respect to diffeomorphisms of the manifold Ω, by Theorem 2.6.

Let $\{\Omega_\nu, \omega_\nu, f_\nu\}$ be an atlas of the manifold Ω, $\nu = 1, \ldots, N$. To every map of the atlas there corresponds a pseudo-differential operator P_ν defined on functions with support lying in $\omega_\nu = f_\nu(\Omega_\nu) \subset \mathbb{R}^n$.

Let functions $\varphi_\nu: \Omega \to \mathbb{R}$ form a partition of unity corresponding to the given atlas such that $\varphi_\nu \in C_0^\infty(\Omega_\nu)$, $\varphi_\nu \geqslant 0$, $\sum \varphi_\nu \equiv 1$ and for every ν the support of φ_ν lies in some coordinate neighborhood. Let ψ_ν be a smooth function whose support also lies in some coordinate neighborhood, and for which $\psi_\nu = 1$ in a neighborhood of the support of φ_ν, so that $\varphi_\nu\psi_\nu = \varphi_\nu$.

If $u \in C^\infty(\Omega)$, then $u = \sum \varphi_\nu u$, and

$$Pu = \sum_\nu \psi_\nu P(\varphi_\nu u) = \sum_\nu \psi_\nu f_\nu^*(P_\nu u_\nu) + T_1 u,$$

where $u_\nu \in C_0^\infty(\omega_\nu)$, $f_\nu^2 u_\nu = \varphi_\nu u$ and T_1 is an operator of order $-\infty$.

By Theorem 3.1 there exists an operator of order $-m$ such that $Q_\nu\varphi_\nu P = \varphi_\nu \cdot I + T_\nu$ and T_ν is an operator of order $-\infty$.

Let now

$$Q = \sum \psi_\mu Q_\mu \varphi_\mu, \quad Q' = \sum_\mu \varphi_\mu Q_\mu \psi_\mu.$$

If $u \in C^\infty(\Omega)$, then

$$QPu = \sum_\mu \psi_\mu Q_\mu \varphi_\mu \left(\sum_\nu \psi_\nu P \varphi_\nu\right) u + T_2' u = \sum_{\mu,\nu} \psi_\mu Q_\mu \varphi_\mu P \psi_\nu \varphi_\nu u + T_2 u =$$

$$\sum_{\mu,\nu} \psi_\mu (\varphi_\mu I + T_\mu) \varphi_\nu u + T_2 u = \sum_{\mu,\nu} \varphi_\mu \varphi_\nu u + \sum_{\mu,\nu} \psi_\mu T_\mu \varphi_\nu u + T_2 u = u + T_3 u,$$

where T_2, T_3, T_4 are operators of order $-\infty$ on Ω.

On the other hand, by Theorem 3.1

$$P\varphi_\nu Q_\nu = \varphi_\nu I + T'_\nu,$$

where T'_ν is an operator of order $-\infty$. Therefore, if $u \in C^\infty(\Omega)$, then

$$PQ'u = \sum_\mu \varphi_\mu P \psi_\mu \sum_{\mu'} \varphi_{\mu'} Q_{\mu'} \psi_{\mu'} u + T'_1 u =$$

$$\sum_{\mu,\nu} \varphi_\mu \psi_\mu P \varphi_\nu Q_\nu \psi_\nu u + T'_2 u = \sum_{\mu,\nu} \varphi_\mu (\varphi_\nu I + T'_\nu) \psi_\nu u + T'_2 u =$$

$$\sum_{\mu,\nu} \varphi_\mu \varphi_\nu u + \sum_{\mu,\nu} \varphi_\mu T'_\nu \psi_\nu u + T'_2 u = u + T'u,$$

where T'_1, T'_2, T' are operators of order $-\infty$ on Ω. As above, it can be shown that $Q - Q'$ is a smoothing operator. Thus we have proved

Theorem 3.2. If P is an elliptic pseudo-differential operator of order m on a smooth closed manifold Ω, then there exists a pseudo-differential operator Q of order $-m$ which is a right and left parametrix of operator P.

3.6. We shall now define the Sobolev spaces $H_s(\Omega)$ for a smooth compact manifold Ω.

We shall do this first for $s = 0$. A space $H_0(\Omega) = L_2(\Omega)$ can be defined by means of an arbitrary atlas $\{\Omega_\nu, \omega_\nu, f_\nu\}$. Namely, we shall say that $u \in L_2(\Omega)$ if the norm

$$\| u \|_0 = \left(\sum_\nu \int_{\omega_\nu} | u_\nu |^2 \, dx \right)^{1/2}$$

is finite. Here, $u_\nu = (f_\nu)_*(\varphi_\nu u)$ and the functions φ_ν form a partition of unity corresponding to our atlas as in Section 3.5.

It is not hard to see that the space $L_2(\Omega)$ is independent of the selection of atlas although the norm may be different in another atlas. In this case, however, there always exists a constant C for which

$$C^{-1} \| u \|_0 \leqslant \| u \|'_0 \leqslant C \| u \|_0,$$

where $\|\cdot\|_0$ and $\|\cdot\|'_0$ are norms corresponding to the atlases

$$\{\Omega_\nu, \omega_\nu, f_\nu\} \text{ and } \{\Omega'_\nu, \omega'_\nu, f'_\nu\}.$$

Definition 3.4. The space $H_s(\Omega)$ on a smooth closed manifold Ω consists of functions u such that $Pu \in L_2(\Omega)$ for any pseudo-differential elliptic operator P of order s. A norm in this space may be introduced as

$$\| u \|_s = \left(\sum_\nu \| u_\nu \|_s^2 \right)^{1/2},$$

where $u_\nu = (f_\nu)_*(\varphi_\nu u)$ and $\| u_\nu \|_s^2 = \int (1 + |\xi|^2)^s | \tilde{u}_\nu(\xi) |^2 \, d\xi$.

The definition is based on the following:

Lemma 3.2. Let $u \in \mathscr{E}'(\Omega)$, where Ω is a domain in $\mathbb{R}^n$ and $\omega \Subset \Omega$. Let $Pu \in H_s(\omega)$, where P is a certain elliptic operator of order m. Then $u \in H^{loc}_{s+m}(\omega)$.

Proof. Let $h \in C_0^\infty(\omega)$ and Q be an operator of order $-m$ such that $QhPu = hu + Tu$. Then

$$hu = QhPu - Tu \in H_{s+m}(\Omega),$$

since Q is an operator of order $-m$ and $Tu \in C^\infty(\Omega)$. As h is an arbitrary function from $C_0^\infty(\omega)$ it follows that $u \in H^{loc}_{s+m}(\omega)$. The lemma is proved.

Since the class of elliptic operators of order s is invariant with respect to diffeomorphisms of the set Ω, the spaces $H_s(\Omega)$ are also invariant with respect to these diffeomorphisms.

As for the space $L_2(\Omega)$, passage to another atlas leads to the replacement of the norm by another norm equivalent to the former.

Using a partition of unity, we can obtain from Lemma 3.2 the following statement.

Theorem 3.3. If $u \in \mathscr{D}'(\Omega)$, where Ω is a closed smooth manifold, and $Pu \in H_s(\Omega)$, where P is an elliptic pseudo-differential operator of order m, then $u \in H_{s+m}(\Omega)$ and there exists a constant $C = C(s)$ for which

$$\|\varphi\|_{s+m} \leqslant C(\|P\varphi\|_s + \|\varphi\|_{s+m-1}), \qquad \varphi \in C^\infty(\Omega). \tag{3.6}$$

Proof. The first statement follows immediately from Lemma 3.2 since $H_{s+m}(\Omega) = H^{loc}_{s+m}(\Omega)$. The second statement follows from Theorem 3.2, since

$$\varphi = QP\varphi + T\varphi$$

and therefore

$$\|\varphi\|_{s+m} \leqslant \|QP\varphi\|_{s+m} + \|T\varphi\|_{s+m} \leqslant C(\|P\varphi\|_s + \|\varphi\|_{s+m-1}).$$

Another corollary of Lemma 3.2 is

Theorem 3.4. Elliptic operators are hypoelliptic.

Proof. Let $u \in \mathscr{E}'(\Omega)$ and $Pu \in C^\infty(\omega)$, where ω is a subdomain in Ω. By Lemma 3.2, $u \in H^{loc}_s(\omega)$ for any s and therefore $u \in C^{s-\left[\frac{n}{2}\right]-1}(\omega)$, when $s > \left[\frac{n}{2}\right] + 1$, by Sobolev's Theorem 3.5 of Chapter I. Hence $u \in C^\infty(\omega)$. The theorem is proved.

3.7. In this section we shall study the solvability of elliptic equations on a smooth closed manifold.

Theorem 3.5. Let Ω be a smooth closed manifold and $d\mu$ a positive infinitely differentiable measure on Ω. Then the space $N(\Omega) = \{v \in \mathscr{D}'(\Omega),$ $P^*v = 0\}$, where P is an elliptic pseudo-differential operator of order m and P^* is the operator adjoint to P with respect to the measure $d\mu$, is finite dimensional and all of its elements belong to $C^\infty(\Omega)$.

The equation $Pu = f$, where $f \in \mathscr{D}'(\Omega)$, has a solution if and only if $\int f\bar{v}\,d\mu = 0$ for all $v \in N(\Omega)$. If this condition is satisfied and $f \in H_s(\Omega)$, then there exists a solution u belonging to the class $H_{s+m}(\Omega)$ for which

$$\|u\|_{s+m} \leqslant C\|f\|_s$$

for some constant C independent of f.

Proof. According to Theorems 2.4 and 2.6, the operator P^* differs from the operator with symbol $p(x, \xi)$ by an operator of order $m - 1$. Therefore, the operator P^* is also elliptic and all solutions of the equation $P^*v = 0$ are infinitely differentiable by Theorem 3.4.

By Theorem 3.3 the inequality $\|v\|_1 \leqslant C\|v\|_0$ is satisfied for these solutions and therefore the set of solutions v satisfying the condition $\|v\|_0 \leqslant 1$ is compact in the $L_2(\Omega)$ norm. From Kolmogorov's theorem it follows that $\dim N(\Omega) < \infty$.

We further note that $\int u\,(\overline{P^*v})\,d\mu = \int f\bar{v}\,d\mu$. Hence the condition $\int f\bar{v}\,d\mu = 0$ for all $v \in N(\Omega)$ is necessary for the solvability of the equation $Pu = f$.

To prove the sufficiency of this condition it may be noted that for functions $\varphi \in C_0^\infty(\Omega)$ satisfying the orthogonality condition for the subspace $N(\Omega)$, the equality

$$\|\varphi\|_{-s} \leqslant C_1 \|P^*\varphi\|_{-s-m} \tag{3.7}$$

is valid.

Indeed, if it is false, then there exists a sequence of functions $\varphi_k \in C^\infty(\Omega)$ for which

$$\|\varphi_k\|_{-s} = 1, \ \|P^*\varphi_k\|_{-s-m} \leqslant k^{-1}, \ \int \varphi_k \bar{v}\, d\mu = 0 \text{ for } v \in N(\Omega).$$

But by Theorem 3.3

$$\|\varphi_k\|_{-s} \leqslant C_2 (\|P^*\varphi_k\|_{-s-m} + \|\varphi_k\|_{-s-1}).$$

The sequence $\{\varphi_k\}$ is compact in $H_{-s-1}(\Omega)$ and therefore, replacing it by a sequence, we may assume that $\{\varphi_k\}$ converges in H_{-s-1}. But

$$\|\varphi_k - \varphi_l\|_{-s} \leqslant C_2 (\|P^*\varphi_k\|_{-s-m} + \|P^*\varphi_l\|_{-s-m} + \|\varphi_k - \varphi_l\|_{-s-1}),$$

and, therefore, $\{\varphi_k\}$ converges to $\varphi \in H_{-s}(\Omega)$ in $H_{-s}(\Omega)$. Here, $P^*\varphi = 0$ and $\|\varphi\|_{-s} = 1$; that is, $\varphi \in N(\Omega)$ and $\varphi \neq 0$, which contradicts the condition that $\int \varphi \bar{v}\, d\mu = 0$ for $v \in N(\Omega)$.

The obtained inequality (3.7) indicates that the functional $\int f\bar{\varphi}\, d\mu$ of function $P^*\varphi$ is bounded in H_{-s-m}, since, if φ is orthogonal to $N(\Omega)$,

$$\left|\int f\bar{\varphi}\, d\mu\right| \leqslant C_1 \|f\|_s \|\varphi\|_{-s} \leqslant C_1 C_2 \|P^*\varphi\|_{-s-m} \|f\|_s.$$

By the Riesz representation theorem, there exists an element $u \in H_{s+m}$ for which

$$\int f\bar{\varphi}\, d\mu = \int u\, (\overline{P^*\varphi})\, d\mu,$$

and $\|u\|_{s+m} \leqslant C_1 C_2 \|f\|_s$. Function u is the required solution. The theorem is proved.

3.8. In this section we prove that an elliptic operator $P\colon H_s(\Omega) \to H_{s-m}(\Omega)$ on a closed set is a Fredholm operator for any real s. The latter implies that the range of the operator P is closed and the spaces $\ker P = \{u \in H_s(\Omega),\ Pu = 0\}$ and $\ker P^* = \{u \in H_{m-s}(\Omega),\ P^*u = 0\}$ have finite dimension.

<u>Theorem 3.6.</u> Let Ω be a smooth closed manifold and P be an elliptic pseudo-differential operator of order m on Ω.

We assume that $t^{-m}a(x, t, \xi) \to a_0(x, \xi)$ when $t \to +\infty$ uniformly for $x \in \omega_\nu$, $|\xi| = 1$. Then for any real s, $P\colon H_s(\Omega) \to H_{s-m}(\Omega)$ is a Fredholm operator; $\operatorname{ind} P \equiv \dim \ker P - \dim \ker P^* = 0$ if $n > 2$.

<u>Proof.</u> From Theorem 3.4 it follows that the spaces ker P and ker P* consist of smooth functions and are independent of s. We have noted in the proof of Theorem 3.5 that $N(\Omega) = \ker P^*$ is a finite-dimensional space. In exactly the same manner it is proved that the space ker P is finite dimensional. In proving Theorem 3.5 we found the set $P(H_s(\Omega))$ to consist of functions $f \in H_{s-m}(\Omega)$ orthogonal to $N(\Omega)$. Denoting by $H_t^{\perp}(\Omega)$ the orthogonal complement of $N(\Omega)$ in $H_t(\Omega)$, Theorem 3.5 reveals that $H_{s+m}^{\perp}(\Omega)$ is isomorphic to $H_s^{\perp}(\Omega)$ and this isomorphism is such that $\|u\|_{s+m} \leqslant C\|Pu\|_s$ for $u \in H_{s+m}^{\perp}(\Omega)$. Hence it follows that the image $P(H_{s+m}(\Omega))$ forms a closed subspace of finite codimensionality in $H_s(\Omega)$.

The number

$$\operatorname{ind} P = \dim \ker P - \dim \ker P^*$$

is called the *index* and represents an integral characteristic for an elliptic operator on a compact manifold.

As is seen, this number is independent of s. To prove the identity $\operatorname{ind} P = 0$ for $n \geqslant 3$, some auxiliary statements are needed. We shall first list them without proofs.

3.9. Let E be a Banach space and A: $E \to E$ be a continuous linear operator. Let ind A = dim N(A) − dim N(A^*), where N(A) is the kernel of the operator A and N(A^*) is the kernel of the adjoint operator A^*. It is assumed that the values dim N(A) and dim N(A^*) are finite and that the operator A has a closed range, in other words, that A is a Fredholm operator.

Lemma 3.3. If A and B are Fredholm operators, then BA is also a Fredholm operator and ind BA = ind A + ind B.

Lemma 3.4. If T is a compact operator, then I + T is a Fredholm operator and

$$\operatorname{ind}(I+T)=0.$$

3.10. Conclusion of the Proof of Theorem 3.6. Using the lemmas, we can find the index of the operator P.

Let Q be a pseudo-differential operator of order −m/2 with symbol $p(x, \xi)^{-1/2}$ when $|\xi| \geqslant C_2$, where C_2 is the constant which appeared in Definition 3.1 of an elliptic operator. Note that $|\xi| \geqslant C_2$ is a simply connected set if the dimensionality n of the manifold is greater than 2. Therefore the function $p(x, \xi)^{-1/2}$ is well defined on this manifold for fixed x if the function has a fixed value at one point. This enables one to determine the function $p^{-1/2}$ within the confines of every simply connected coordinate neighborhood when $|\xi| \geqslant C_2$. Now we can determine the operator on the whole manifold with the use of a partition of unity. The operator Q^*QP is of order 0, and

$$\operatorname{ind} Q^*QP = \operatorname{ind} P + \operatorname{ind} Q^*Q,$$

by Lemma 3.3. Since Q^*Q is a self-adjoint operator, its index is equal to 0. Hence, the indices of the operators P and Q^*QP agree. By Theorems 2.2 and 2.4, the operator Q^*QP differs from the identity operator by an operator T of order −1. Thus

$$\operatorname{ind} P = \operatorname{ind}(Q^*QP) = \operatorname{ind}(I+T).$$

But by Lemma 3.4, ind (I + T) = 0. The theorem is proved.

3.11. Proof of Lemma 3.3. Let $F' = \ker B \cap \operatorname{coker} A$. Recall that $\operatorname{coker} A = E \ominus (AE)$. Note that the subspace ker B is isomorphic to coker B^* and ker A^* is isomorphic to coker A, so that dim ker B = dim coker B^* and dim coker A = dim ker A^*.

Now it can be easily verified that

$$\dim\ker BA = \dim\ker A + \dim\ker B - \dim F'.$$

Similarly, $\dim\ker A^*B^* = \dim\ker A^* + \dim\ker B^* - \dim F'^*$ where $F'^* = \ker A^* \cap \operatorname{coker} B^*$. Since F' and F'^* are isomorphic and finite dimensional, $\dim F^* = \dim F'$ and hence

$$\operatorname{ind} BA = \dim\ker BA - \dim\ker (BA)^* = \dim\ker A -$$

$$\dim\ker A^* + \dim\ker B - \dim\ker B^* = \operatorname{ind} A + \operatorname{ind} B.$$

The proof is complete.

Proof of Lemma 3.4. 1°. First assume that T is a finite dimensional operator, that is, dim Im T < +∞.

Now write E as $E_1 + E_2$ so that E_1 is a closed subspace in E, $E_1 \subset \ker T$, $E_2 \supset \operatorname{Im} T$, $\dim E_2 < \infty$. It is clear that I + T = I on E_1 and $(I + T)E_2 \subset E_2$. The operator I + T is a Fredholm operator since $\ker(I + T) \subset E_2$ and $\operatorname{Im}(I + T) \supset E_1$. Since on E_1 this operator coincides with I, its index agrees with that of the restriction of the operator I + T to the finite dimensional space E_2. Therefore, by a known theorem of linear algebra, this index is equal to 0.

2°. Let us verify that $I + T$ is a Fredholm operator. Since on ker $(I + T)$ the operator I coincides with $-T$, it is compact and therefore $\dim \ker(I + T) < \infty$. Similarly, $\dim \ker(I + T^*) < \infty$. It now remains to prove that $\operatorname{Im}(I + T)$ is closed.

Let $\{x_n\} \in E$, and assume that the sequence $y_n = (I + T)x_n$ converges to y in E as $n \to \infty$. By F we denote the closed complement to ker $(I + T)$ in E. We may, of course, take $x_n \in F$ for all n.

The sequence $\{x_n\}$ is bounded. Otherwise we would have that $\|x_n\| \to \infty$, and assuming $x_n' = x_n/\|x_n\|$, $y_n' = (I + T)x_n'$, we would reach a contradiction: $\|x_n'\| = 1$ and $\|y_n'\| \to 0$, so $\{Tx_n'\}$ is a compact set, and from the convergence of $\{y_n'\}$ it would follow that the sequence $\{x_n'\}$ is also compact.

Passing to a subsequence we can let $x_n' \to x$, but $\|x\| = 1$ and $(I + T)x = 0$, which contradicts the condition $x_n \in F$.

Thus, $\{x_n\}$ is bounded and we know that $\lim_{n\to\infty} Tx_n$ exists, and therefore $\lim_{n\to\infty} x_n = y - \lim Tx_n = x$. Then $(I + T)x = y$, so $y \in \operatorname{Im}(I + T)$. Hence $\operatorname{Im}(I + T)$ is closed and $I + T$ is a Fredholm operator.

3°. We shall now show that $\operatorname{ind}(I + T) = 0$. To do so it suffices to show that $\operatorname{ind}(I + \varepsilon T)$ does not vary when ε increases from 0 to 1. The latter follows from the continuous dependence of ind A_ε on ε if A_ε is continuously dependent on ε in the operator norm. We shall prove this property.

Let A be a Fredholm operator. Then there exists $\varepsilon > 0$ such that ind $(A + B) = \operatorname{ind} A$ when $|B| < \varepsilon$.

Indeed, let R be an operator such that

$$RA = I - P_1, \quad AR = I - P_2,$$

where P_1 and P_2 are finite-dimensional projection operators onto ker A and onto $E \ominus \operatorname{Im} A$, respectively. If E_1 is the closed complement of ker A in E and E_2 is the complement of Im A, then the operator R is defined from the conditions $RA = I$ on E_1 and $R = 0$ on E_2.

Assume $\varepsilon = 1/\|R\|$. Then $R(A + B) = I - P_1 + RB$, $(A + B)R = I - P_2 + BR$. The operators $I + RB$ and $I + BR$ have inverses since $\|BR\| < 1$. Hence

$$(I + RB)^{-1} R (A + B) = I - P_1, \quad (A + B) R (I + BR)^{-1} = I - P_2.$$

From these equations it follows that $\ker(A + B) \subset \ker(I - P_1)$ and $\operatorname{Im}(A + B) \supset \operatorname{Im}(I - P_2)$ so that $A + B$ is a Fredholm operator. Now note that $\operatorname{ind}(I + BR) = \operatorname{ind}(I + RB) = \operatorname{ind}(I - P_1) = \operatorname{ind}(I - P_2) = 0$. Therefore, by Lemma 3.3, $\operatorname{ind}(A + B) = -\operatorname{ind} R = \operatorname{ind} A$, and Lemma 3.4 is proved.

§4. Canonical Transformations

4.1. In this section we show that the set of transformations preserving the class of pseudo-differential operators may be widened compared to the set of Theorem 2.6. This considerably facilitates the task of reducing (pseudo-)differential operators to a simpler form and thereby widens the scope of the study of general pseudo-differential operators.

Definition 4.1. A closed nonsingular differential 2-form ω^2 is called a *symplectic structure* on a manifold M of even dimensionality $2n$. A manifold equipped with a symplectic structure is called a symplectic manifold.

Recall that the form ω^2 is called closed if $d\omega^2 = 0$. The form ω^2 is called nonsingular if for every nonzero vector $\xi \in TM$ we can find a vector $\eta \in TM$ such that $\omega^2(\xi, \eta) \neq 0$.

Let Ω be a smooth (C^∞) n-dimensional manifold and let $M = T^*\Omega$. When $\Omega = \mathbb{R}^n$ the form $\omega^2 = dx \wedge d\xi = \sum dx^j \wedge d\xi_j$ is obviously a symplectic structure for $T^*\Omega$. It is not hard to see that for an arbitrary manifold Ω the symplectic structure on $T^*\Omega$ may be defined by the form ω^2 which appears as $dx \wedge d\xi$ in every system of local coordinates. In this case, $\omega^2 = d\omega^1$, where ω^1 is a form which in local coordinates is given by $\omega^1 = \xi dx$.

The form ω^2 enables one to establish an isomorphism between $T^*(T^*\Omega)$ and $T(T^*\Omega)$ in the following manner. If $\alpha \in T(T^*\Omega)$, then by ω'_α we denote the following 1-form:

$$\omega'_\alpha(\beta) = \omega^2(\beta, \alpha), \quad \beta \in T(T^*\Omega).$$

It is not hard to show that to every 1-form $\omega \in T^*(T^*\Omega)$ there corresponds a unique element $\alpha \in T(T^*\Omega)$ satisfying this identity, so that the mapping I: $T^*(T^*\Omega) \to T(T^*\Omega)$, which is an isomorphism, is defined. From the definition it follows that in local coordinates the mapping I is defined by the matrix $\begin{pmatrix} 0 & -E \\ E & 0 \end{pmatrix}$, where E is a unit matrix.

If a real-valued function $H \in C^\infty(T^*\Omega)$, then its differential dH defines an element belonging to $T^*(T^*\Omega)$. The vector field IdH is called the Hamiltonian vector field, and H, the Hamiltonian function. In local coordinates this vector field is the phase velocity of the Hamiltonian canonical equations:

$$\dot{x} = \partial H(x, \xi)/\partial\xi, \quad \dot{\xi} = -\partial H(x, \xi)/\partial x.$$

The integral curves of this system are called the bicharacteristics of H.

Definition 4.2. The mapping Φ: $T^*\Omega \to T^*\Omega$ is called *canonical* provided it retains its symplectic structure.

4.2. We shall now consider some examples of canonical transformations of the space $T^*\Omega$ when Ω is a domain in $\mathbb{R}^n$.

Example 4.1. Replacement of variables $y = F(x)$ gives rise to a canonical transformation of $T^*\Omega$:

$$\Phi\colon (x, \xi) \longmapsto (y, \eta), \text{ where } y = F(x), \ \xi = {}^tF'(x)\eta. \tag{4.1}$$

In this case

$$d\eta \wedge dy = d\eta \wedge F'(x)\,dx = {}^tF'(x)\,d\eta \wedge dx;$$

$$d\xi \wedge dx = ({}^tF'(x)\,d\eta + {}^tF''(x)\,\eta\,dx) \wedge dx = {}^tF'(x)\,d\eta \wedge dx,$$

so that Φ is canonical.

Example 4.2. Let $(y, \eta) = \Phi(x, \xi)$ where

$$y^1 = \xi_1, \quad \eta_1 = -x^1, \quad y' = x', \quad \eta' = \xi',$$
$$y' = (y^2, \ldots, y^n), \quad \eta' = (\eta_2, \ldots, \eta_n). \tag{4.2}$$

It is clear that $d\eta \wedge dy = -dx^1 \wedge d\xi_1 + d\xi' \wedge dx' = d\xi \wedge dx$.

Example 4.3. Let S be a real-valued function on $T^*\Omega$ such that $\det S_{x\xi} \neq 0$. Let $(y, \eta) = \Phi(x, \xi)$, where

$$\xi = \partial S(x, \eta)/\partial x, \quad y = \partial S(x, \eta)/\partial\eta. \tag{4.3}$$

Clearly this transformation is well-defined and is nonsingular. Here

$$d\xi \wedge dx = \frac{\partial^2 S}{\partial x^2}\,dx \wedge dx + \frac{\partial^2 S}{\partial x\,\partial\eta}\,d\eta \wedge dx = \frac{\partial^2 S}{\partial x\,\partial\eta}\,d\eta \wedge dx,$$

$$d\eta \wedge dy = d\eta \wedge \frac{\partial^2 S}{\partial\eta\,\partial x}\,dx + d\eta \wedge \frac{\partial^2 S}{\partial\eta^2}\,d\eta = \frac{\partial^2 S}{\partial x\,\partial\eta}\,d\eta \wedge dx.$$

Hence this transformation is also canonical.

Example 4.4. Let

$$(y, \eta) = \Phi_t(x, \xi) = (x(t), \xi(t)),$$

where $(x(t), \xi(t))$ is a solution of the Hamiltonian system

$$\dot{x} = H_\xi, \quad \dot{\xi} = -H_x,$$

for which $((x, 0), \xi(0)) = (x, \xi)$. Then

$$\frac{d}{dt}(dx \wedge d\xi)(t) = d\dot{x} \wedge d\xi + dx \wedge d\dot{\xi} =$$

$$dH_\xi \wedge d\xi - dx \wedge dH_x = H_{\xi x}\, dx \wedge d\xi + H_{\xi\xi}\, d\xi \wedge d\xi - dx \wedge H_{xx}\, dx - dx \wedge H_{x\xi}\, d\xi = 0.$$

Hence the mapping Φ_t is canonical for all t; in other words, the Hamiltonian phase flow retains its symplectic structure.

4.3. Let f and g be functions belonging to $C^\infty(T^*\Omega)$.

Definition 4.3. The derivative of a function f with respect to the direction of the phase flow for a Hamiltonian function g is called the *Poisson bracket* $\{f, g\}$. (This concept has been used earlier, see (2.13).)

In local coordinates

$$\{f, g\} = \frac{d}{dt} f(x(t), \xi(t)),$$

where

$$\dot{x} = g_\xi(x, \xi), \quad \dot{\xi} = -g_x(x, \xi).$$

Therefore

$$\{f, g\} = \sum_{j=1}^{n} \left(\frac{\partial f}{\partial x^j} \frac{\partial g}{\partial \xi_j} - \frac{\partial f}{\partial \xi_j} \frac{\partial g}{\partial x^j} \right).$$

From the definition it is seen that a Poisson bracket may be defined by the form ω^2 so that $\{f, g\} = \omega^2(Idf, Idg)$.

Proposition 4.1. The transformation Φ: $T^*\Omega \to T^*\Omega$, where $\Omega \subset \mathbb{R}^n$, is canonical if and only if the Poisson bracket $\{f, g\}$ of any two functions $f, g \in C^\infty(T^*\Omega)$ is invariant under the transformation.

Proof. Since the Poisson bracket is defined by the symplectic structure, it must be invariant under canonical transformations. The converse can be readily obtained from the identity

$$\{f, g\} = \omega^2(I\, df, I\, dg).$$

The differentials of coordinate functions x^i, ξ_j form a basis (set) in $T(T^*\Omega)$. Therefore, the invariance of the form ω^2 follows from the invariance of the Poisson bracket, and this implies that the given mapping is canonical.

4.4. The following assertion shows that in fact all the canonical transformations are included in the transformations considered in Examples 4.2 and 4.3. Further, we note that transformation (4.3) with function S linear in η is a transformation of the form (4.1).

Proposition 4.2. If Φ: $(x, \xi) \mapsto (y, \eta)$ is a canonical transformation defined in a sufficiently small neighborhood ω of a point (x_0, ξ^0) of the cotangent bundle $T^*\Omega$, then Φ is representable in this neighborhood as the composition of a finite number of transformations of the form (4.1) with $F(x) = Ax$, (4.2), or (4.3). Here $|\partial^2 S/\partial x^i \partial \xi_j - \delta^i_j| < \varepsilon$, where $\varepsilon \to 0$ together with the diameter of the given neighborhood.

Proof. Let $y = \Phi_1(x, \xi)$, $\eta = \Phi_2(x, \xi)$ in the transformation. The rank of the matrices $\left\| \frac{\partial \Phi_1}{\partial x}, \frac{\partial \Phi_1}{\partial \xi} \right\|$ and $\left\| \frac{\partial \Phi_2}{\partial x}, \frac{\partial \Phi_2}{\partial \xi} \right\|$ is equal to n. By transformations of the type (4.2) and renumbering of the variables, renumbering being accomplished by transformations of the type (4.1), one can assume that $\det\left(\frac{\partial \Phi_1}{\partial x}\right) \neq 0$ and $\det\left(\frac{\partial \Phi_2}{\partial \xi}\right) \neq 0$ in ω.

As immediately follows from the definition, the linear part of a canonical transformation which is computed at an arbitrary point again defines a canonical transformation. Let T_1 be a canonical transformation of the type

$$y = Ax, \quad \eta = A^{*-1}\xi,$$

where $A^* = \partial \Phi_2(x_0, \xi^0)/\partial \xi$. It is not hard to see that this is a transformation of type (4.3) for a function $S(x, \eta) = (Ax, \eta)$.

The linear part of the transformation $T_1\Phi$ at (x_0, η^0) will be described by the matrix

$$\begin{pmatrix} I + B_1 A_1 & B_1 \\ A_1 & I \end{pmatrix},$$

where A_1, B_1 are quadratic symmetric matrices of order n. Let T_2 be a transformation with a generating function $x\eta - \frac{1}{2}(B_1\eta, \eta)$, so that $\eta = \xi$, $y = x - B_1\xi$. The linear part of the transformation $T_2T_1\Phi$ at (x_0, ξ^0) is then described by the matrix

$$\begin{pmatrix} I & 0 \\ A_1 & I \end{pmatrix}.$$

Note that the transformations T_1 and T_2 are defined in the whole space $T^*\mathbb{R}^n$ so that the transformation $T_2T_1\Phi$ has meaning. Now we show that this transformation is defined by the generating function $S(x, \eta)$, and

$$S(x, \eta) = x\eta + O(|x - x_0|^2 + |\eta - \eta^0|^2)$$

when $x \to x_0$ and $\eta \to \eta^0$.

Indeed, this transformation takes the form

$$y^j = \varphi^j(x, \xi), \quad \eta_j = \psi_j(x, \xi), \quad j = 1, \ldots, n.$$

Solving these equations relative to the variables y, ξ (this is possible by the Implicit Function Theorem) in some neighborhood of the point (x_0, ξ^0), we obtain

$$y^j = \alpha^j(x, \eta), \quad \xi_j = \beta_j(x, \eta), \quad j = 1, \ldots, n.$$

It is clear that

$$\frac{\partial}{\partial x^j} = \sum_{k=1}^{n} \left(\frac{\partial \alpha^k}{\partial x^j} \frac{\partial}{\partial y^k} + \frac{\partial \beta_k}{\partial x^j} \frac{\partial}{\partial \xi_k} \right), \quad \frac{\partial}{\partial \eta_j} = \sum_{k=1}^{n} \left(\frac{\partial \alpha^k}{\partial \eta_j} \frac{\partial}{\partial y^k} + \frac{\partial \beta_k}{\partial \eta_j} \frac{\partial}{\partial \xi_k} \right), \quad j = 1, \ldots, n.$$

Therefore

$$\{f, g\}(x, \xi) = -\sum_{j,k=1}^{n} \left(\frac{\partial \alpha^k}{\partial x^j} \frac{\partial f}{\partial \xi_j} \frac{\partial g}{\partial y^k} + \frac{\partial \beta_k}{\partial x^j} \frac{\partial f}{\partial \xi_j} \frac{\partial g}{\partial \xi_k} - \frac{\partial \alpha^k}{\partial x^j} \frac{\partial f}{\partial y^k} \frac{\partial g}{\partial \xi_j} - \frac{\partial \beta_k}{\partial x^j} \frac{\partial f}{\partial \xi_k} \frac{\partial g}{\partial \xi_j} \right),$$

$$\{f, g\}(y, \eta) = -\sum_{j,k=1}^{n} \left(\frac{\partial \alpha^k}{\partial \eta_j} \frac{\partial f}{\partial y^k} \frac{\partial g}{\partial y^j} + \frac{\partial \beta_k}{\partial \eta_j} \frac{\partial f}{\partial \xi_k} \frac{\partial g}{\partial y^j} - \frac{\partial \alpha^k}{\partial \eta_j} \frac{\partial f}{\partial y^j} \frac{\partial g}{\partial y^k} - \frac{\partial \beta_k}{\partial \eta_j} \frac{\partial f}{\partial y^j} \frac{\partial g}{\partial \xi_k} \right).$$

Since the transformation is canonical by hypothesis, we have $\{f, g\}(x, \xi) = \{f, g\}(y, \eta)$ for any functions f and g. Assuming, for example,

$f = \xi_j$, $g = y^k = \varphi^k(x, \xi)$, we obtain $\frac{\partial \alpha^k}{\partial x^j} = \frac{\partial \beta_j}{\partial \eta_k}$. Similarly, choosing f and g, we can verify that

$$\frac{\partial \beta_k}{\partial x^j} = \frac{\partial \beta_j}{\partial x^k}, \quad \frac{\partial \alpha^k}{\partial \eta_j} = \frac{\partial \alpha^j}{\partial \eta_k}, \qquad j, k = 1, \ldots, n.$$

From these identities it follows that there exists a smooth function $S(x, \eta)$ such that

$$\alpha^k(x, \eta) = \frac{\partial S(x, \eta)}{\partial \eta_k}, \quad \beta_k(x, \eta) = \frac{\partial S(x, \eta)}{\partial x^k}, \qquad k = 1, \ldots, n.$$

In this case

$$\frac{\partial^2 S(x_0, \eta^0)}{\partial x^j \partial \eta_k} = \frac{\partial \alpha^k(x_0, \eta^0)}{\partial x^j} = \delta_j^k,$$

so that the inequality

$$\det \left\| \frac{\partial^2 S(x, \eta)}{\partial x^j \partial \eta_k} \right\| \neq 0$$

is satisfied in some neighborhood of the point (x_0, η^0).

Thus, we have proved that mapping $T_3 = T_2 T_1 \Phi$ is of the form (4.3) with some generating function $S(x, \eta)$. Since the mapping T_2^{-1} is a linear canonical transformation with generating function $x\eta + \frac{1}{2}(B_1\eta, \eta)$, and T_1^{-1} is a transformation with generating function $(A^{-1}x, \eta)$, the Proposition 4.2 is proved.

4.5. We now wish to establish a relationship between canonical transformations on the space $T^*\Omega$ and transformations of functions belonging to $C_0^\infty(\Omega)$. It can be easily seen that a transformation of type (4.2) implies a change from the function $u(x)$ to its partial Fourier transformation with respect to the variable x^1, that is, to the function $\tilde{u}(\eta_1, x')$. Such a transformation, from the viewpoint of the theory of pseudo-differential operators, is inadmissible since the variables x and ξ have different roles in this theory. We shall show that a transformation of type (4.3) is admissible provided the function S satisfies some additional conditions.

Definition 4.4. The real-valued function $S \in C^\infty(T^*\Omega)$ is called a *phase function* if

1°. $|D_\xi^\alpha D_x^\beta S(x, \xi)| \leq C_{\alpha,\beta}(1 + |\xi|)^{1-|\alpha|}$;

2°. $|\partial S(x, \xi)/\partial x| \geq C_1|\xi|$ when $|\xi| \geq C_2$, $C_1 = \text{const} > 0$.

Theorem 4.1. Let P be a pseudo-differential operator of order m and

$$\Phi u(x) = (2\pi)^{-n} \int a(x, \xi)\, \tilde{u}(\xi)\, e^{iS(x, \xi)}\, d\xi, \tag{4.4}$$

where S is a phase function and $a \in S^0(\mathbb{R}^n)$. Let the symbols p and *a* of these operators vanish when $x \in \mathbb{R}^n \setminus K$, where K is a bounded domain in $\mathbb{R}^n$. Then there exists a pseudo-differential operator Q of order m such that the order of the operator

$$T = P\Phi - \Phi Q$$

is equal to $-\infty$, and Q differs from the operator with symbol $q^0(y, \eta) = p(x, \xi)$, where the points (x, ξ) and (y, η) are related by formulas (4.3), by an operator of order $m - 1$.

Proof 1°. The operator $P\Phi$ is an operator of type (4.4) since

$$P\Phi u(x) = (2\pi)^{-2n} \iiint a(y, \xi)\, \tilde{u}(\xi)\, e^{iS(y, \xi)}\, d\xi\, e^{-iy\eta}\, dy\, p(x, \eta)\, e^{ix\eta}\, d\eta =$$

$$(2\pi)^{-n}\int k(x,\ \xi)\,\tilde{u}(\xi)\,e^{iS(x,\ \xi)}\,d\xi,$$

where we assumed that

$$k(x,\ \xi)=(2\pi)^{-n}\iint a(y,\ \xi)\,p(x,\ \eta)\,e^{i[S(y,\ \xi)-S(x,\ \xi)]+i(x-y)\eta}\,dy\,d\eta.$$

From conditions 1° and 2° of Definition 4.4 it follows that

$$\left|\int_0^1 \frac{\partial S(x+t(y-x),\ \xi)}{\partial x}\,dt\right|\geqslant\frac{C_1}{2}|\xi|\quad\text{for } |\xi|\geqslant 2C_2,$$

if $|x-y|<\varepsilon$ and ε is sufficiently small. Let $h(t)=1$ when $|t|\leqslant\frac{1}{2}$, $h(t)=0$ when $|t|\geqslant 1$. Now divide (4.5) into two integrals I_1+I_2, the integrands of which contain additional factors $h(|x-y|\varepsilon^{-1})$ and $1-h(|x-y|\varepsilon^{-1})$ compared to (4.5). Note that

$$e^{i[S(y,\ \xi)-S(x,\ \xi)]+i(x-y)\eta}=(A_0+|\eta-S_y(y,\ \xi)|^2+\Delta_y S(y,\ \xi)+|\eta|^2|x-y|^2)^{-1}\times$$
$$(A_0-\Delta_y-|\eta|^2\Delta_\eta)\,e^{i[S(y,\ \xi)-S(x,\ \xi)]+i(x-y)\eta}.$$

Now we choose the constant A_0 so that the inequality

$$F\equiv A_0+|\eta-S_y(y,\ \xi)|^2+\Delta_y S(y,\ \xi)+|\eta|^2|x-y|^2\geqslant c_0(1+|\xi|^2),\quad c_0>0$$

is fulfilled when $|x-y|\geqslant\varepsilon/2$. This is possible by the estimate

$$1+|\eta-S_y(y,\ \xi)|^2+|\eta|^2|x-y|^2\geqslant\frac{1}{5}\varepsilon^2|S_y(y,\ \xi)|^2+1\geqslant C_1(\varepsilon)(1+|\xi|^2),$$

since

$$|\Delta_y S(y,\ \xi)|\leqslant C(1+|\xi|)\leqslant\frac{1}{2}C_1(\varepsilon)(1+|\xi|^2)+C_2(\varepsilon).$$

It now remains to select A_0 so that $A_0>C_2(\varepsilon)+1$, and assume $C_0=\frac{1}{2}C_1(\varepsilon)$. Similarly

$$e^{i(x-y)\eta}=|x-y|^{-2}(-\Delta_\eta)\,e^{i(x-y)\eta}$$

Therefore, for any N and K

$$I_2=(2\pi)^{-n}\iint[(A_0-\Delta_y-|\eta|^2\Delta_\eta)(A_0+|\eta-S_y(y,\ \xi)|^2+\Delta_y S(y,\ \xi)+|\eta|^2|x-y|^2)^{-1}]^N\times$$
$$\{|x-y|^{-2k}[1-h(|x-y|\varepsilon^{-1})]\,a(y,\ \xi)(-\Delta_\eta)^k p(x,\ \eta)\}\,e^{i[S(y,\ \xi)-S(x,\ \xi)]+i(x-y)\eta}\,dy\,d\eta.$$

If $2k>m+n$, then the integral converges absolutely and uniformly. In this case $|I_2|\leqslant C(\varepsilon,\ N)(1+|\xi|^2)^{-N}$ for arbitrary ε and N. This shows that $I_2\in S^{-\infty}$.

We now show that $I_1\in S^m$. Note that

$$S(y,\ \xi)-S(x,\ \xi)=\frac{\partial S(x,\ \xi)}{\partial x}(y-x)+(A(x,\ y,\ \xi)(y-x),\ y-x),$$

where $A_{ij}(x,\ y,\ \xi)=\int_0^1(1-t)\frac{\partial^2 S(x+t(y-x),\ \xi)}{\partial x^i\,dx^j}\,dt.$ Recall that because of our choice of ε, the inequality

$$|1+A(x,y,\xi)|\geqslant C_3(1+|\xi|)$$

is satisfied for $|x-y|<\varepsilon$. Here $C_3=\text{const}>0$.

Further we shall note that the following estimates are satisfied for the vector-valued function A:

$$|D_\xi^\alpha D_x^\beta D_y^\gamma A(x,y,\xi)|\leqslant C_{\alpha,\beta,\gamma}(1+|\xi|)^{1-|\alpha|},$$

$$C_3(1+|\xi|)\leqslant 1+|A(x,y,\xi)|\leqslant(1+|\xi-\eta+A|)(1+|\xi-\eta|)$$

so that

$$(1 + |\xi - \eta + A|)^{-1} \leqslant C_4(1 + |\xi - \eta|)(1 + |\xi|)^{-1},$$

$$(1 + |\xi - \eta + A|)^{-1} \leqslant C_4(1 + |\xi - \eta|)(1 + |\eta|)^{-1}.$$

Therefore

$$|D_\xi^\alpha D_\eta^{\alpha'} D_x^\beta D_y^\beta\, a(y, \xi)\, p(x, \xi - \eta) + A(x, y, \xi)|$$
$$\leqslant C_{\alpha,\alpha',\beta,\beta'}(1 + |\xi|^{m-|\alpha|}(1 + |\eta|)^{-|\alpha'|}(1 + |\xi - \eta|)^{m+|\alpha+\beta+\beta'|+2|\alpha'|}.$$

Remark 2.2 guarantees that $k \in S^m$ and $k'(x, \xi) \equiv k(x, \xi) - a(x, \xi)\, p(x, \partial S(x, \xi)/\partial x) \in S^{m-1}$.

2°. Similarly we can consider the operator ΦQ which is also of type (4.4), since

$$\Phi Qu(x) = (2\pi)^{-2n} \int\int\int q(y, \xi)\, \tilde{u}(\xi)\, e^{iy\xi}\, d\xi\, e^{-iy\eta}\, dy\, a(x, \eta)\, e^{iS(x, \eta)}\, d\eta =$$

$$(2\pi)^{-n} \int k_1(x, \xi)\, \tilde{u}(\xi)\, e^{iS(x, \xi)}\, d\xi,$$

where we have assumed

$$k_1(x, \xi) = (2\pi)^{-n} \int\int q(y, \xi)\, a(x, \eta)\, e^{iy(\xi-\eta)+i[S(x, \eta)-S(x, \xi)]}\, dy\, d\eta. \tag{4.6}$$

Now we show that $k_1 \in S^m$. Let

$$S(x, \eta) - S(x, \xi) = B(x, \xi, \eta)(\eta - \xi),$$

where $B(x, \xi, \eta) = \int_0^1 \partial S(x, \xi + t(\eta - \xi))/\partial\xi \cdot dt$, so that $B(x, \xi, \xi) = \partial S(x, \xi)/\partial\xi$

and

$$|D_\xi^\alpha D_\eta^\beta D_x^\gamma B(x, \xi, \eta)| \leqslant C_{\alpha,\beta,\gamma}(1 + |\xi|)^{-|\alpha|} \cdot (1 + |\eta|)^{-|\beta|}(1 + |\xi - \eta|)^{|\alpha|+2|\beta|}.$$

Therefore replacing y by $y_1 = y - B + x$ we obtain

$$k_1(x, \xi) = (2\pi)^{-n} \int\int q(y + B(x, \xi, \eta) - x, \xi)\, a(x, \eta)\, e^{i(\eta-\xi, x-y)}\, d\eta\, dy,$$

and

$$|D_\xi^\alpha D_\eta^{\alpha'} D_x^\beta D_y^{\beta'} q(y + B(x, \xi, \eta) - x, \xi)\, a(x, \eta)| \leqslant$$

$$C_{\alpha,\alpha',\beta,\beta'}(1 + |\xi|)^{m-|\alpha|}(1 + |\eta|)^{-|\beta|}(1 + |\xi - \eta|)^{|\alpha+2\alpha'|}.$$

Using Remark 2.2, we find that $k_1 \in S^m$ and

$$k_1'(x, \xi) \equiv k_1(x, \xi) - a(x, \xi)\, q(\partial S(x, \xi)/\partial\xi, \xi) \in S^{m-1}.$$

3°. Thus we have shown that if

$$q_0(\partial S(x, \xi)/\partial\xi, \xi) = p(x, \partial S(x, \xi)/\partial x), \tag{4.7}$$

then the order of the operator $P\Phi - \Phi Q_0$ is equal to $m - 1$ and

$$(P\Phi - \Phi Q)\, u(x) = (2\pi)^{-n} \int p_1(x, \xi)\, \tilde{u}(\xi)\, e^{iS(x, \xi)}\, d\xi,$$

where $p_1 \in S^{m-1}$. We now find a symbol $q_1 \in S^{m-1}$ such that

$$(P\Phi - \Phi Q_0 - \Phi Q_1)\, u(x) = (2\pi)^{-n} \int p_2(x, \xi)\, \tilde{u}(\xi)\, e^{iS(x, \xi)}\, d\xi,$$

where $p_2 \in S^{m-2}$. To do so, it is sufficient to assume of course that

$$q_1(\partial S(x, \xi)/\partial\xi, \xi) = p_1(x, \partial S(x, \xi)/\partial x),$$

Applying the same reasoning we find a symbol $q_j \in S^{m-j}$ such that

$$[P\Phi - \Phi(Q_0 + Q_1 + \ldots + Q_j)]\, u(x) = (2\pi)^{-n} \int p_{j+1}(x, \xi)\, \tilde{u}(\xi)\, e^{iS(x, \xi)}\, d\xi$$

and $p_{j+1} \in S^{m-j-1}$. According to Lemma 3.1, there exists an operator with symbol $q(x, \xi) \sim q_0 + q_1 + \ldots + q_j + \ldots$ of class S^m which satisfies all the conditions of the theorem.

4.6. The following often proves useful.

Theorem 4.2. Let P be a pseudo-differential operator of order m and Φ be an operator defined by formula (4.4). Assume that the phase function also satisfies the following two conditions

$$\left|\det \frac{\partial^2 S(x, \xi)}{\partial x\, \partial \xi}\right| \geqslant c_0 > 0; \qquad \frac{\partial S(x, \xi)}{\partial x} = \frac{\partial S(x, \eta)}{\partial x} \Longleftrightarrow \xi = \eta \tag{4.8}$$

when $|\xi| \geqslant C_1$, $|\eta| \geqslant C_1$, and $x \in K$, where K is a bounded domain in $\mathbb{R}^n$. Let the symbols p and a of these operators vanish when $x \in \mathbb{R}^n \setminus K$. Then Q = $\Phi^* P \Phi$ is a pseudo-differential operator with symbol q of class S^m, and

$$q(y, \eta) - p(x, \xi)\,|a(x, \xi)|^2\,|\det S_{x\xi}|^{-1} \in S^{m-1},$$

where the points (x, ξ) and (y, η) are related by (4.3).

Proof. As in the proof of Theorem 4.1 we find that $P\Phi$ is an operator of type (4.4) with symbol $k(x, \xi)$ given by formula (4.5). Operator Q may therefore be represented as

$$Qu(x) = (2\pi)^{-2n} \int\int\int k(y, \xi)\, \tilde{u}(\xi)\, e^{iS(y, \xi)}\, d\xi\, \overline{a(y, \eta)}\, e^{-iS(y, \eta)}\, dy e^{i\eta x}\, d\eta =$$

$$(2\pi)^{-n} \int q(x, \xi)\, \tilde{u}(\xi)\, e^{ix\xi}\, d\xi,$$

where

$$q(x, \xi) = (2\pi)^{-n} \int\int k(y, \xi)\, \overline{a(y, \eta)}\, e^{i[S(y, \xi) - S(y, \eta)] + ix(\eta - \xi)}\, dy\, d\eta. \tag{4.9}$$

Let $S(y, \xi) - S(y, \eta) = B(y, \xi, \eta)(\xi - \eta)$, where

$$B(y, \xi, \eta) = \int_0^1 \frac{\partial S(y, \eta + t(\xi - \eta))}{\partial \xi}\, dt,$$

so that $B(y, \xi, \xi) = \partial S(y, \xi)/\partial \xi$. By the hypotheses imposed on S, there exists an $\varepsilon > 0$ such that

$$\left|\det\left(\frac{\partial B(y, \xi, \eta)}{\partial y}\right)\right| \geqslant \frac{c_0}{2} \quad \text{when } |\xi - \eta| \leqslant \varepsilon |\eta|.$$

As above, represent integral (4.9) as the sum of two integrals $I_1 + I_2$ by introducing under the integral sign the functions $h(|\xi - \eta|\varepsilon^{-1}|\eta|^{-1})$ and $1 - h(|\xi - \eta|\varepsilon^{-1}|\eta|^{-1})$, where $h \in C_0^\infty(\mathbb{R})$, $h(t) = 0$ when $|t| \geqslant 1$ and $h(t) = 1$ when $|t| \leqslant 1/2$. To do so we may assume, without loss of generality, that $a(y, \eta) = 0$ when $|\eta| \leqslant 1$.

In integral I_1 we shall pass to a new variable, assuming

$$z = B(y, \xi, \eta),$$

which is possible in our construction. Thus we obtain

$$I_1(x, \xi) = (2\pi)^{-n} \int\int h\left(\frac{|\xi - \eta|}{\varepsilon |\eta|}\right) \frac{k(y, \xi)\, \overline{a(y, \eta)}}{|\det(\partial B/\partial y)|}\, e^{i(x - z, \eta - \xi)}\, dz\, d\eta.$$

Applying Lemma 2.1 we obtain $I_1 \in S^m$ and $I_1 - q_0 \in S^{m-1}$ where

$$q_0\left(\frac{\partial S(x, \xi)}{\partial \xi}, \xi\right) = |a(x, \xi)|^2\, p\left(x, \frac{\partial S(x, \xi)}{\partial x}\right) \left|\det\left(\frac{\partial^2 S(x, \xi)}{\partial x\, \partial \xi}\right)\right|^{-1}.$$

This theorem will be proved if we show that $I_2 \in S^{-\infty}$. By assumption, a smooth function φ for which $\dfrac{\partial S(x, \varphi(x, \xi))}{\partial x} = \xi$ is defined, and therefore

$$|\xi - \eta| \leqslant C(1 + |\partial S(x, \xi)/\partial x - \partial S(x, \eta)/\partial x|).$$

Thus there exist constants A_0 and $c_0 > 0$ such that

$$|A_0+|S_y(y,\xi)-S_y(y,\eta)|^2+\Delta_y S(y,\xi)-\Delta_y S(y,\eta)|\geqslant c_0(1+|\xi-\eta|^2).$$

Hence

$$I_2(x,\xi)=(2\pi)^{-n}\int\int\left[1-h\left(\frac{|\xi-\eta|}{\varepsilon|\eta|}\right)\right]\{A_0-\Delta_y\,[A_0+|S_y(y,\xi)-S_y(y,\eta)|^2+$$

$$\Delta_y S(y,\xi)-\Delta_y S(y,\eta)]^{-1}\}^N\,[k(y,\xi)\overline{a(y,\eta)}]\cdot e^{i[S(y,\xi)-S(y,\eta)]+ix(\eta-\xi)}\,dy\,d\eta.$$

The integral in y is taken over a bounded region. Further, the expression under the integral sign is estimated via

$$C(\varepsilon,N)(1+|\eta-\xi|^2)^{-N}(1+|\xi|^2)^{m/2}.$$

Since $|\eta|\leqslant 2\varepsilon^{-1}|\xi-\eta|$, we have $|\xi|\leqslant(1+2\varepsilon^{-1})|\xi-\eta|$ so that the integrand may be estimated in terms of

$$C'(\varepsilon,N)(1+|\eta|^2)^{-(n+1)/2}(1+|\xi|^2)^{m/2+(n+1)/2-N}.$$

Similar estimates can be obtained for the derivatives $D_\xi^\alpha D_x^\beta I_2$. It is evident that $I_2\in S^{-\infty}$, and the theorem is proved.

Now note the following important corollary of Theorem 4.2.

Corollary 4.1. If the phase function S satisfies the conditions of Theorem 4.2, and if $a(x,\xi)=|\det S_{x\xi}|^{1/2}$, then

$$\Phi^*\Phi=I+B,$$

where B is a pseudo-differential operator with symbol of class S^{-1}.

4.7. To illustrate the application of Theorem 4.2 we shall show that a first-order operator of principal type with a real-valued symbol can be reduced to the form $Q=D_1$.

Definition 4.5. The operator P(x, D) is called an *operator of principal type* in Ω if there exists a principal symbol $p^0(x,\xi)$ for which the form $dp^0(x,\xi)$ is not proportional to the form ξdx at any point of $T^*\Omega\backslash 0$.

Recall that $p^0(x,\xi)$ is a positively homogeneous function in ξ of degree m so that by the Euler identity,

$$\sum_{j=1}^n \xi_j\frac{\partial p^0(x,\xi)}{\partial\xi_j}=mp^0(x,\xi),$$

$\operatorname{grad}_\xi p^0(x,\xi)\neq 0$ if $p^0(x,\xi)\neq 0$. Definition 4.2 therefore concerns only the behavior of the principal symbol p^0 at characteristic points and implies that if at some point $p^0(x,\xi)=0$, $\operatorname{grad}_\xi p^0(x,\xi)=0$, then the vector $\operatorname{grad}_x p^0(x,\xi)$ is not collinear to the vector ξ.

For the equation of principal type Pu = f we can always assume without loss of generality that m = 1, since elliptic operators of order 1 − m can be applied to both sides of the equation.

Assume that p^0 is a real-valued function and P is a first-order operator of principal type. Let $p^0(x_0,\xi^0)=0$ and $\operatorname{grad}_\xi p^0(x_0,\xi^0)\neq 0$. For definiteness we assume that $\partial_{\xi_1}p^0(x_0,\xi^0)\neq 0$. Now consider the equation

$$p^0\left(x,\frac{\partial S(x,\eta)}{\partial x}\right)=\eta_1 \qquad (4.10)$$

with the initial condition

$$S=\sum_{j=2}^n \eta_j\left(x^j-x_0^j\right)\text{ when } x^1=x_0^1. \qquad (4.11)$$

Since the plane $x^1=x_0^1$ does not have characteristic directions in some neighborhood of (x_0,ξ^0), the Cauchy problem (4.10), (4.11) has a solution in some neighborhood of this point.

Further, replacing the vector η by the vector $t\eta$, when $t > 0$, gives that the function tS satisfies this new equation and the initial conditions. Therefore

$$S(x, t\eta) = tS(x, \eta), \quad t > 0,$$

so that S is a first-order positively homogeneous function of η.

Finally, we note that differentiating equation (4.10) with respect to η_1 gives $\sum_{j=1}^{n} \frac{\partial p^0}{\partial \xi_j}\left(x, \frac{\partial S}{\partial x}\right) \frac{\partial^2 S}{\partial x^j \partial \eta_1} = 1$. From condition (4.11) it follows that $\frac{\partial^2 S}{\partial x^j \partial \eta_1} = 0$ for $j = 2, \ldots, n$ if $x^1 = x_0^1$. Hence $\frac{\partial^2 S(x_0, \eta^0)}{\partial x^1 \partial \eta_1} = \left|\frac{\partial p^0(x_0, \xi^0)}{\partial \xi_1}\right|^{-1} \neq 0$ and $\det\left\|\frac{\partial^2 S(x_0, \eta^0)}{\partial x^i \partial \eta_j}\right\| = \frac{\partial^2 S(x_0, \eta^0)}{\partial x^1 \partial \eta_1} \neq 0$, where $\xi^0 = \frac{\partial S(x_0, \eta^0)}{\partial x}$. The equation $\frac{\partial S(x, \eta)}{\partial x} = \xi$ can be solved for η in some neighborhood of (x_0, ξ^0) by the Implicit Function Theorem. In this case, $|\eta| \leqslant C$ if, for example, $|\xi| \leqslant 1$. By homogeneity it follows that

$$|\eta| \leqslant C\left|\frac{\partial S(x, \eta)}{\partial x}\right|$$

for all η and x in a neighborhood of x_0.

Thus we have verified that all the conditions of Definition 4.2 are fulfilled with the exception of the smoothness condition. The constructed function S is smooth everywhere when $\xi \neq 0$. In order to remove this deficiency it is sufficient to multiply S by the function $h \in C^\infty$ equal to 0 when $|\xi| \leqslant 1/2$ and 1 when $|\xi| \geqslant 1$.

4.8. We shall now consider the case when $p^0(x_0, \xi^0) = 0$ and $\text{grad}_\xi\, p^0(x_0, \xi^0) = 0$, but the vector $\text{grad}_x\, p^0(x_0, \xi^0)$ does not vanish and is not collinear to the vector ξ^0.

Let for definiteness $\text{grad}_x\, p^0(x_0, \xi^0) = (a, 0, \ldots, 0)$ and let the vector ξ^0 have a coordinate ξ_2^0 different from 0. In this case it is convenient to use canonical transformations inverse to (4.3). We shall take the generating function of this transformation to have the form $S = S(y, \xi)$.

Let us now consider the Cauchy problem

$$p^0\left(\frac{\partial S}{\partial \xi}, \xi\right) = \frac{\partial S}{\partial y^1}; \quad S = \sum_{j=2}^{n} y^j \xi_j + \xi_1^2 (2\xi_2)^{-1} \tag{4.12}$$

for $y^1 = 0$.

It is not hard to see that this problem has a solution and

$$S(y, t\xi) = tS(y, \xi) \text{ when } t > 0.$$

At (x_0, ξ^0) by (4.12), the following identities are satisfied:

$$\frac{\partial^2 S}{\partial \xi_1 \partial y^1} = a(\xi_2^0)^{-1}; \quad \frac{\partial^2 S}{\partial y^j \partial \xi_i} = \delta_j^i, \quad i, j = 2, \ldots, n,$$

Thus

$$\det\left\|\frac{\partial^2 S}{\partial y^j \partial \xi_i}\right\| = a(\xi_2^0)^{-1} \neq 0.$$

We find that

$$|\xi| \leqslant C\left|\frac{\partial S(y, \xi)}{\partial y}\right|$$

in a conic neighborhood of (y_0, ξ^0).

Thus in this case also the operator may be reduced to the form D_1 +

(operator of zero order) in a small conic neighborhood of (x_0, ξ^0).

4.9. The results obtained earlier may be improved as follows: an operator of the type $D_1 + Q$, where Q is a pseudo-differential operator of zero order, is equivalent to an operator of the type $D_1 + T$, where T is a pseudo-differential operator of order $-\infty$.

Proposition 4.3. If Q is a pseudo-differential operator of zero order, then there exists an elliptic pseudo-differential operator A, also of zero order, such that

$$B(D_1+Q)A = D_1 + T, \tag{4.13}$$

where B is a parametrix of the operator A, and T is a pseudo-differential operator of order $-\infty$.

Proof. The symbol of the operator AD_1 is equal to $a\xi_1$.

The symbol of the operator $(D_1 + Q)A$ is equal to $(\xi_1 + q)a + \frac{\partial a}{\partial x^1} + \sum_{|\alpha|=1}^{N-1} \frac{i^{|\alpha|}}{\alpha!} D_\xi^\alpha q \cdot D_x^\alpha a + t_N$, where $t_N \in S^{-N}$.

The method for constructing the operator A is now evident. Let a_0 be a nontrivial solution of the equation $\frac{\partial a_0}{\partial x^1} + qa_0 = 0$, for example, $a_0 = \exp\left(-\int_0^{x^1} q\,dx^1\right)$, so that $a_0 \in S^0$. We shall find a function a_1 of class S^{-1} such that

$$\frac{\partial a_1}{\partial x^1} + qa_1 + \sum_{|\alpha|=1} iD_\xi^\alpha q \cdot D_x^\alpha a_0 = 0.$$

Applying the same reasoning we find a function a_j of class S^{-j} such that

$$\frac{\partial a_j}{\partial x^1} + qa_j + \sum_{|\alpha|+l=1} \frac{i^{|\alpha|}}{\alpha!} D_\xi^\alpha q D_x^\alpha a_l = 0.$$

From the given sequence $\{a_j\}$ the symbol $a \in S^0$ is constructed using Lemma 3.1.

Identity (4.13) is immediately verified.

4.10. Another way of transforming the symbol of a pseudo-differential operator consists of using the canonical transformations considered in Example 4.4. Let $H(x, \xi)$ be a smooth real-valued function of class S^1. By $U(t)$ we denote the operator which transforms the function φ belonging to $C_0^\infty(\mathbb{R}^n)$ to a solution $u(t, x)$ of the Cauchy problem

$$\frac{\partial u}{\partial t} = H(x, D_x)u, \quad u(0, x) = \varphi(x). \tag{4.14}$$

This problem is studied using well-known methods of geometrical optics. If $\varphi(x) = e^{ix\xi}$, then a good approximation to the solution can be obtained provided the solution is of the form $e^{iS(t,x,\xi)}a(t, x, \xi)$. The phase function S is obtained as the solution to the nonlinear Cauchy problem:

$$\frac{\partial S}{\partial t} = H\left(x, \frac{\partial S}{\partial x}\right), \quad S(0, x, \xi) = x \cdot \xi,$$

and the amplitude a is found by integrating along the bicharacteristic

$$\dot{x} = H_\xi(x, \xi), \quad \dot{\xi} = -H_x(x, \xi). \tag{4.15}$$

The operator $U(t)$ is bounded in every H_s space, as is its inverse. In this case

$$U(t)\varphi = (2\pi)^{-n}\int e^{iS(t,x,\xi)}a(t, x, \xi)\tilde{\varphi}(\xi)\,d\xi,$$

where T is an operator of order $-\infty$.

From Theorem 4.1 it follows that for every pseudo-differential operator P with symbol p in S^m there is a pseudo-differential operator Q(t) with symbol in S^m for which

$$PU(t) = U(t)Q(t) + T(t). \tag{4.16}$$

In this case, the operator Q differs by an operator of order $m - 1$ from an operator with symbol

$$q(t, y, \eta) = p(x(t, y, \eta), \xi(t, y, \eta)),$$

where $(x(t, y, \eta), \xi(t, y, \eta))$ is the solution of system (4.15) which satisfies the initial condition

$$x(0, y, \eta) = y, \qquad \xi(0, y, \eta) = \eta.$$

4.11. In studying necessary conditions for local solvability in Chapter VI and subelliptic operators in Chapter VIII we encounter "localized" estimates that are obtained from the ordinary ones by passing to differential operators with polynomial coefficients of the form $T_k^{(x,\xi)}p(x + y, \xi + D)$; here $T_k^{(x,\xi)}p$ is the k-th order expansion of function p at (x, ξ) by the Taylor formula. We shall show that such localized estimates are invariant under arbitrary canonical transformations on the space $T^*(\Omega)$.

We shall illustrate this by reproducing the results of L. Hörmander (see Ref. 6).

Theorem. Let P be a first-order pseudo-differential operator and

$$\|u\|_{-\sigma} \leqslant C(K)(\|Pu\|_0 + \|u\|_{-\sigma-1}), \qquad u \in C_0^\infty(K),$$

where K is a compact subset of Ω in $\mathbb{R}^n$. Then for every subset M belonging to $T^*\Omega\backslash 0$ there exists a constant C such that the inequality

$$\|\psi\| \leqslant C\lambda^\sigma\left\{\left\|\sum_{2j+|\alpha+\beta|<N} p_{(\beta)}^{j(\alpha)}(x, \xi)\frac{1}{\alpha!\,\beta!}y^\beta D^\alpha\psi(y)\lambda^{-j-\frac{|\alpha+\beta|}{2}}\right\|_0 + \lambda^{-N/2}\sum_{|\alpha+\beta|\leqslant N}\|y^\beta D^\alpha\psi\|\right\}, \qquad \psi \in \mathcal{S}, \tag{4.17}$$

is satisfied when $(x, \xi) \in M$ and $\lambda \geqslant 1$. Here $\|\cdot\|$ represents the norm in the space $L_2(\mathbb{R}^n)$.

We shall show that (4.17) is invariant under the canonical transformations on $T^*\Omega$.

Theorem 4.3. Let Φ be an arbitrary canonical transformation on $T^*\Omega$ for which the set M changes to some compact set M'. For all $(x, \xi) \in M$, condition (4.17) holds if and only if for all $(x', \xi') \in M'$ an analogous condition holds, where the functions $p_{(\beta)}^{j(\alpha)}$ have been replaced by the functions $q_{(\beta)}^{j(\alpha)}$. Here

$$q_{\alpha,\beta}^0(x', \xi') = p_{(\beta)}^{0(\alpha)}(x, \xi), \quad (x', \xi') = \Phi(x, \xi).$$

Proof. By Proposition 4.2 it suffices to prove the invariance of condition (4.17) under transformations (4.2) and (4.3).

1°. Substituting

$$\psi(y) = 2\pi\int\psi_1(\eta, y')e^{iy^1\eta_1}d\eta_1, \quad \text{where } \psi_1 \in \mathcal{S},$$

into (4.17), we obtain a similar inequality for function ψ_1, replacing $p^0(x, \xi)$ by $q^0(x, \xi) = p^0(\xi_1, -x^1, \xi')$. Indeed,

$$\|\psi\| = \|\psi_1\|, \quad \|y^\beta D^\alpha \psi\| = \|D_1^{\beta_1}(y')^{\beta'}(y^1)^{\alpha_1}(D')^{\alpha'}\psi_1\| \leqslant C \sum_{|\gamma+\delta|\leqslant|\alpha+\beta|} \|y^\delta D^\gamma \psi_1\|.$$

Similarly,

$$\|y^\beta D^\alpha \psi_1\| \leqslant C \sum_{|\gamma+\delta|\leqslant|\alpha+\beta|} \|y^\delta D^\gamma \psi\|.$$

Further we have

$$\left\| \sum_{2j+|\alpha+\beta|<N} p_{(\beta)}^{j(\alpha)}(x, \xi)\frac{1}{\alpha!\,\beta!} y^\beta D^\alpha \psi \lambda^{1-j-|\alpha+\beta|/2} \right\| =$$

$$\left\| \sum_{2j+|\alpha+\beta|<N} p_{(\beta)}^{j(\alpha)}(x, \xi)\frac{1}{\alpha!\,\beta!}(-D_1)^{\beta_1}(y')^{\beta'}(y^1)^{\alpha_1}(D')^{\alpha'}\psi_1\lambda^{1-j-|\alpha+\beta|/2} \right\| =$$

$$\left\| \sum_{2j+|\alpha+\beta|<N} p_{(\beta)}^{j(\alpha)}(x, \xi)\frac{1}{\alpha!\,\beta!}(y')^{\beta'} \times \right.$$

$$\left. (D')^{\alpha'} \sum_{\gamma=0}^{\beta_1} \frac{\beta_1!}{\gamma!\,(\beta_1-\gamma)!} D_1^\gamma (y^1)^{\alpha_1} D_1^{\beta_1-\gamma}\psi_1\lambda^{1-j-|\alpha+\beta|/2} \right\| = \sum_{2j+|\alpha+\beta|<N} q_{\alpha\beta}^j(\tilde{x}, \tilde{\xi})\frac{1}{\alpha!\,\beta!} y^\beta,$$

where

$$q_{\alpha\beta}^0(\tilde{x}, \tilde{\xi}) = p_{(\beta)}^{0(\alpha)}(x, \xi), \quad \tilde{x}^1 = \xi_1, \quad \tilde{\xi}_1 = -x^1, \quad \tilde{x}' = x', \quad \tilde{\xi}' = \xi'.$$

2°. Now let A be a nonsingular quadratic matrix and $\psi(y) = \psi_1(Ay)$. Then

$$\|\psi\| = |\det A|^{-1/2}\|\psi_1\|,$$

$$c^{-1}\sum_{|\alpha+\beta|\leqslant N}\|y^\beta D^\alpha\psi\| \leqslant \sum_{|\alpha+\beta|\leqslant N}\|y^\beta D^\alpha \psi_1\| \leqslant c\sum_{|\alpha+\beta|\leqslant N}\|y^\beta D^\alpha\psi\|,$$

where $c > 0$ is independent of ψ. Finally

$$\left\| \sum_{2j+|\alpha+\beta|<N} p_{(\beta)}^{j(\alpha)}(x, \xi)\frac{1}{\alpha!\,\beta!} y^\beta D^\alpha \psi\lambda^{1-j-|\alpha+\beta|/2} \right\| =$$

$$|\det A|^{-1/2}\left\| \sum_{2j+|\alpha+\beta|<N} p_{(\beta)}^{j(\alpha)}(x, \xi)\frac{1}{\alpha!\,\beta!}(A^{-1}z)^\beta\,(A^*D_z)^\alpha\psi_1\lambda^{1-j-|\alpha+\beta|/2} \right\| =$$

$$|\det A|^{-1/2}\left\| \sum_{|\alpha+\beta|<N} q_{(\beta)}^{0(\alpha)}(\tilde{x}, \tilde{\xi})\frac{1}{\alpha!\,\beta!} z^\beta D_z^\alpha \psi_1 \lambda^{1-j-|\alpha+\beta|/2} + \right.$$

$$\left. \sum_{j=1}^{N}\sum_{|\alpha+\beta|<N-j} q_{\alpha,\beta}^j(\tilde{x}, \tilde{\xi}) z^\beta D_z^\alpha\psi_1\lambda^{1-j-|\alpha+\beta|/2} \right\|,$$

where

$$q^0(\tilde{x}, \tilde{\xi}) = p^0(A^{-1}\tilde{x}, \ A^*\tilde{\xi}) = p^0(x, \xi),$$

so that the transformation corresponds to the linear canonical transformation

$$\tilde{x} = Ax, \quad \tilde{\xi} = A^{*-1}\xi.$$

3°. We now consider a canonical transformation of type (4.3) with generating function S for which $\partial^2 S(x_0, \xi^0)/\partial x^i \partial\xi_j = \delta_i^j$. In this case S can be conveniently replaced by $x\xi + S(x, \xi)$.

<u>Proposition 4.4.</u> A canonical transformation Φ with generating function $x\xi + S(x, \xi)$ may be represented as the product of the canonical transformation Φ_1 with the generating function

$$x\xi + S(x, \xi) - (\operatorname{grad}_x S(x_0, \xi^0), x) - (\operatorname{grad}_\xi S(x_0, \xi^0), \xi)$$

and two parallel translations T_1, T_2 in the $\mathbb{R}^n_x$ and $\mathbb{R}^n_\xi$ spaces, that is, canonical transformations with generating functions $x\xi + a\xi$ and $x\xi + bx$, where a and b are constant vectors.

Proof. To prove the proposition we represent Φ_1 as the product of two parallel translations and the transformation Φ. This is sufficient since the transformation inverse to a parallel translation is again a parallel translation.

The transformation Φ_1 takes the form

$$y = x + \partial S(x, \eta)/\partial\eta - \partial S(x_0, \xi^0)/\partial\eta, \qquad \xi = \eta + \partial S(x, \eta)/\partial x - \partial S(x_0, \xi^0)/\partial x.$$

Let $x_1 = x$, $\xi^1 = \xi + \partial S(x_0, \xi^0)/\partial x$. Let $x_2 = x_1 + \partial S(x_1, \xi^2)/\partial\xi$, $\xi^1 = \xi^2 + \partial S(x_1, \xi^2)/\partial x$. Finally, we assume that $y = x_2 - \partial S(x_0, \xi^0)/\partial\xi$, $\eta = \xi^2$. It is easy to see that $(x, \xi) \mapsto (y, \eta)$ coincides with transformation Φ_1. Proposition 4.4 is proved.

Proposition 4.5. The canonical transformation Φ with generating function $x\xi + S(x, \xi)$, where $\partial^2 S(x_0, \xi^0)/\partial x^i \partial\xi_j = 0$ may be represented as the composition of a canonical transformation Φ_1 with generating function

$$x\xi + S(x, \xi) - \frac{1}{2} \sum_{i,j=1}^{n} \frac{\partial^2 S(x_0, \xi^0)}{\partial x^i \partial x^j} \cdot x^i x^j - \frac{1}{2} \sum_{i,j=1}^{n} \frac{\partial^2 S(x_0, \xi^0)}{\partial \xi_i \partial \xi_j} \xi_i \xi_j$$

and canonical transformations with generating functions $x\xi + \frac{1}{2}(Bx, x)$ and $x\xi + \frac{1}{2}(A\xi, \xi)$, where A and B are quadratic matrices, and elementary transformations of the type (4.2).

Proof. Note that the product of the canonical transformation $(x, \xi) \mapsto (x, \xi + Bx)$ with generating function $x\xi^1 - \frac{1}{2}(Bx, x)$ and the transformation Φ such that

$$x_2 = x_1 + \partial S(x_1, \xi^2)/\partial\xi, \quad \xi^1 = \xi^2 + \partial S(x_1, \xi^2)/\partial x$$

gives a canonical transformation with generating function $S(x, \xi^2) - \frac{1}{2}(Bx, x)$ having the same second-order derivatives with respect to ξ as $S(x, \xi^2)$. Using elementary canonical transformations, one can switch the roles of the variables x and ξ and perform a similar operation. Proposition 4.5 is proved.

Proposition 4.6. Let Φ_1 be a canonical transformation with generating function $x\xi + S(x, \xi)$, where $S \in C_0^\infty(\mathbb{R}^n \times \mathbb{R}_n)$ and $|\partial^2 S(x, \xi)/\partial x^i \partial\xi_j| \leqslant \delta$. Here $i, j = 1, \ldots, n$ and δ is a sufficiently small number. Let $S_1 \in C_0^\infty(\mathbb{R}^n \times \mathbb{R}_n)$ and Φ_2 be a canonical transformation with generating function $x\xi + S(x, \xi) + \varepsilon S_1(x, \xi)$. Then we can find an ε_0 such that for every ε from $(0, \varepsilon_0)$ the function $T_\varepsilon \in C_0^\infty(\mathbb{R}^n \times \mathbb{R}_n)$ is defined such that $|D_\xi^\alpha D_x^\beta T_\varepsilon(x, \xi)| \leqslant C_{\alpha,\beta}$, where $C_{\alpha,\beta}$ are independent of ε, and the transformation $\Phi_2\Phi_1^{-1}$ is canonical with generating function $x\xi + \varepsilon T_\varepsilon(x, \xi)$.

Proof. The transformation $\Phi_2\Phi_1^{-1}$ is canonical and is locally defined by a smooth generating function. It is also clear that outside the supports of functions S and S_1 this function coincides with $x\xi$.

Let now Φ_1: $(x_0, \xi^0) \mapsto (x_1, \xi^1)$, Φ_2: $(x_0, \xi^0) \mapsto (x_2, \xi^2)$. We have

$$x_1 = x_0 + \frac{\partial S(x_0, \xi^1)}{\partial\xi}, \quad \xi^0 = \xi^1 + \frac{\partial S(x_0, \xi^1)}{\partial x};$$

$$x_2 = x_0 + \frac{\partial S(x_0, \xi^2)}{\partial\xi} + \varepsilon \frac{\partial S_1(x_0, \xi^2)}{\partial\xi}, \qquad \xi^0 = \xi^2 + \frac{\partial S(x_0, \xi^2)}{\partial x} + \varepsilon \frac{\partial S_1(x_0, \xi^2)}{\partial x}. \tag{4.18}$$

It follows that

$$x_2 = x_1 + \int_0^1 \frac{\partial^2 S(x_0, \xi^1 + t(\xi^2 - \xi^1))}{\partial\xi\, \partial\xi} dt\, (\xi^2 - \xi^1) + \varepsilon \frac{\partial S_1(x_0, \xi^2)}{\partial\xi},$$

$$\xi^1 = \xi^2 + \int_0^1 \frac{\partial^2 S(x_0, \xi^1 + t(\xi^2 - \xi^1))}{\partial x\, \partial \xi}\, dt\, (\xi^2 - \xi^1) + \varepsilon \frac{\partial S_1(x_0, \xi^2)}{\partial x}.$$

Therefore

$$\xi^2 - \xi^1 = -\varepsilon \left(I + \int_0^1 \frac{\partial^2 S(x_0, \xi^1 + t(\xi^2 - \xi^1))}{\partial x\, \partial \xi}\, dt \right)^{-1} \frac{\partial S_1(x_0, \xi^2)}{\partial x}.$$

If

$$x^2 = x^1 + \varepsilon \frac{\partial T_\varepsilon(x^1, \xi^2)}{\partial \xi}, \quad \xi^1 = \xi^2 + \varepsilon \frac{\partial T_\varepsilon(x^1, \xi^2)}{\partial x},$$

we obtain

$$\frac{\partial T_\varepsilon(x_1, \xi^2)}{\partial x} = \left(I + \int_0^1 \frac{\partial^2 S(x_0, \xi^1 + t(\xi^2 - \xi^1))}{\partial x\, \partial \xi}\, dt \right)^{-1} \frac{\partial S_1(x_0, \xi^2)}{\partial x},$$

$$\frac{\partial T_\varepsilon(x_1, \xi^2)}{\partial \xi} = -\int_0^1 \frac{\partial^2 S(x_0, \xi^1 + t(\xi^2 - \xi^1))}{\partial \xi\, \partial \xi}\, dt\, \frac{\partial T_\varepsilon(x_1, \xi^2)}{\partial x} + \frac{\partial S_1(x_0, \xi^2)}{\partial \xi}.$$

It is seen that the derivatives, of any order, of T with respect to x and ξ are bounded uniformly in ε. Proposition 4.6 is proved.

4°. Referring to Proposition 4.4 we shall check whether inequality (4.17) is invariant under parallel displacement. Because of the symmetry between the variables x and ξ it is sufficient to prove that this inequality is invariant under parallel displacement along x. To do so, it suffices to substitute $\psi(y) = \psi_1(y)^{iay}$.

Analogously, referring to Proposition 4.5, we consider a canonical transformation with generating function $x\xi + \frac{1}{2}(Ax, x)$. Thus we substitute $\psi(y) = \psi_1(y)\exp \frac{1}{2}(Ay, y)$ into (4.17). We have

$$\| y^\beta D^\alpha \psi \| = \| y^\beta (D + Ay)^\alpha \psi_1 \| + O\left(\sum_{|\gamma+\delta| \leqslant |\alpha+\beta|-2} \| y^\delta D^\gamma \psi \| \right),$$

so that

$$\left\| \sum_{2j+|\alpha+\beta|<N} p_{(\beta)}^{j(\alpha)}(x, \xi) \frac{1}{\alpha!\,\beta!} y^\beta D^\alpha \psi \lambda^{1-|\alpha+\beta|/2-j} \right\| =$$

$$\left\| \sum_{|\alpha+\beta|<N} p_{(\beta)}^{0(\alpha)}(x, \xi) \frac{1}{\alpha!\,\beta!} y^\beta (D + Ay)^\alpha \psi_1 \lambda^{1-|\alpha+\beta|/2} + \sum_{j \geqslant 1} a_{\alpha\beta}^j y^\beta D^\alpha \psi_1 \lambda^{-j-|\alpha+\beta|/2} \right\| =$$

$$\left\| \sum_{|\alpha+\beta|<N} q_{(\beta)}^{0(\alpha)}(x, \xi) \frac{1}{\alpha!\,\beta!} y^\beta D^\alpha \psi_1 \lambda^{1-|\alpha+\beta|/2} + \ldots \right\|,$$

where $q^0(x, \xi) = p^0(x, \xi + Ax)$. Obviously

$$\| \psi \| = \| \psi_1 \|, \quad \| y^\beta D^\alpha \psi \| \leqslant C \sum_{|\gamma+\delta| \leqslant |\alpha+\beta|} \| y^\delta D^\gamma \psi_1 \|,$$

$$\| y^\beta D^\alpha \psi_1 \| \leqslant C \sum_{|\gamma+\delta| \leqslant |\alpha+\beta|} \| y^\delta D^\gamma \psi \|.$$

5°. Finally we consider the most difficult case, that is, a canonical transformation which is locally defined by the generating function $x\xi + S(x, \xi)$, where grad $S(x_0, \xi^0) = 0$ and $D_\xi^\alpha D_x^\beta S(x_0, \xi^0) = 0$ for $|\alpha + \beta| = 2$. By Propositions 4.2, 4.4, and 4.5, Theorem 4.3 will then be proved. Note that in a ρ-neighborhood of (x_0, ξ^0) the following inequalities

$$|S(x, \xi)| \leqslant \varepsilon\rho^2, \quad |\operatorname{grad} S(x, \xi)| \leqslant \varepsilon\rho,$$

$$|D_\xi^\alpha D_x^\beta S(x, \xi)| \leqslant \varepsilon \quad \text{when} \quad |\alpha + \beta| = 2,$$

are satisfied for small ρ, where $\varepsilon \to 0$ when $\rho \to 0$. Let $h \in C_0^\infty(\mathbb{R}^n \times \mathbb{R}_n)$ and $h = 0$ outside a ρ-neighborhood of (x_0, ξ^0) and $h = 1$ in a $\rho/2$-neighborhood of this point. We assume that $|D_\xi^\alpha D_x^\beta h(x, \xi)| \leqslant C\rho^{-|\alpha+\beta|}$ for $|\alpha + \beta| \leqslant 2$. Then $hS \in C_0^\infty(\mathbb{R}^n \times \mathbb{R}_n)$, $hS = S$ in a $\rho/2$-neighborhood of (x_0, ξ^0), and $|D_\xi^\alpha D_x^\beta hS(x, \xi)| \leqslant C\varepsilon$ for $1 \leqslant |\alpha + \beta| \leqslant 2$. Therefore, if ρ is sufficiently small, then

$$\det \| \delta_i^j + \partial^2 hS/\partial x^i \partial \xi_j \| = 1 + \varphi(x, \xi),$$

where $\varphi \in C_0^\infty(\mathbb{R}^n \times \mathbb{R}_n)$, $|\varphi(x, \xi)| \leqslant C_1\varepsilon$ and $\varphi = 0$ outside a ρ-neighborhood of the point (x_0, ξ^0).

This enables us to restrict ourselves to considering the case when $S \in C_0^\infty(\mathbb{R}^n \times \mathbb{R}_n)$, $D_\xi^\alpha D_x^\beta S(x_0, \xi^0) = 0$ for $|\alpha + \beta| \leqslant 2$ and $D_\xi^\alpha D_x^\beta S(x, \xi)$, when $|\alpha + \beta| = 2$, are small for all $(x, \xi) \in \mathbb{R}^n \times \mathbb{R}_n$.

6°. Proposition 4.7. Let

$$v(x) = Au(x) \equiv \int \tilde{u}(\xi) e^{i(x, \xi) + i\varepsilon\lambda S(x\lambda^{-1/2}, \xi\lambda^{-1/2})} d\xi,$$

where $\lambda \geqslant 1$, $|D_\xi^\alpha D_x^\beta S(x, \xi)| \leqslant C$ for $|\alpha+\beta| \leqslant 2$, $S \in C_0^\infty(\mathbb{R}^n \times \mathbb{R}_n)$. Then there exists a constant ε_0 such that the inequalities

$$\|v\|/2 \leqslant \|u\| \leqslant 2\|v\|, \quad u \in C_0^\infty(\mathbb{R}^n),$$

are valid for $\varepsilon \leqslant \varepsilon_0$.

Proof. We have

$$\|v\|^2 = \iiint \tilde{u}(\xi) \overline{\tilde{u}(\eta)}\, e^{ix(\xi-\eta) + i\varepsilon\lambda S(x\lambda^{-1/2}, \xi\lambda^{-1/2}) - i\varepsilon\lambda S(x\lambda^{-1/2}, \eta\lambda^{-1/2})} d\xi\, d\eta\, dx. \tag{4.19}$$

According to Hadamard's lemma, there exist functions $a^i \in C_0^\infty(\mathbb{R}^n \times \mathbb{R}_n \times \mathbb{R}_n)$ such that

$$S(x, \xi) - S(x, \eta) = \sum_{i=1}^n a^i(x, \xi, \eta)(\xi_i - \eta_i)$$

for all x, ξ, η. Therefore

$$\lambda[S(x\lambda^{-1/2}, \xi\lambda^{-1/2}) - S(x\lambda^{-1/2}, \eta\lambda^{-1/2})] = \sqrt{\lambda} \sum_{i=1}^n a^i(x\lambda^{-1/2}, \xi\lambda^{-1/2}, \eta\lambda^{-1/2})(\xi_i - \eta_i).$$

Replacing the variable x under the integral in (4.19) by $x - \varepsilon\sqrt{\lambda}a(x\lambda^{-1/2}, \xi\lambda^{-1/3}, \eta\lambda^{-1/2})$ gives

$$\|v\|^2 = \iiint \tilde{u}(\xi) \overline{\tilde{u}(\eta)}\, e^{i(x, \xi-\eta)} [1 + \varepsilon\varphi(x, \xi, \eta, \varepsilon)]\, d\xi\, d\eta\, dx =$$

$$\|u\|^2 + \varepsilon \iint \tilde{u}(\xi) \overline{\tilde{u}(\eta)}\, \tilde{\varphi}(\eta - \xi, \xi, \eta, \varepsilon)\, d\xi\, d\eta,$$

where $1 + \varepsilon\varphi = \det \| \delta_i^j + \varepsilon\, \partial a^i(x\lambda^{-1/2}, \xi\lambda^{-1/2}, \eta\lambda^{-1/2})/\partial x^j \|^{-1}$ and $\tilde{\varphi}(\zeta, \xi, \eta, \varepsilon)$ is the Fourier transform of $\varphi(x, \xi, \eta, \varepsilon)$ in the first argument. Select ε_0 so small that the inequalities

$$\varepsilon \max_\xi \int |\tilde{\varphi}(\eta - \xi, \xi, \eta, \varepsilon)|\, d\eta < 1/2,$$

$$\varepsilon \max_\eta \int |\tilde{\varphi}(\eta - \xi, \xi, \eta, \varepsilon)|\, d\xi < 1/2$$

are satisfied. Note that the last integrals can be estimated, for example, by the following rather crude method. Let $h \in C_0^\infty(\mathbb{R}^n)$, $0 \leqslant h(x) \leqslant 1$, $h(x) = 1$ when $|x| \leqslant 1$. Then

$$\int |\tilde{\varphi}(\eta - \xi, \xi, \eta)|\, d\xi = \int \left| \int \varphi(x, \xi, \eta) e^{i(x, \xi-\eta)} dx \right| d\xi =$$

$$\int \left| \int h((\xi - \eta)\lambda^{1/2}) \varphi(x, \xi, \eta) e^{i(x, \xi-\eta)} dx \right| d\xi +$$

$$\int \left| \int [1 - h((\xi - \eta)\lambda^{1/2})] \varphi(x, \xi, \eta) e^{i(x, \xi-\eta)} dx \right| d\xi =$$

$$\int \left| \int h((\xi - \eta)\lambda^{1/2}) \varphi(x, \xi, \eta) e^{i(x, \xi-\eta)} dx \right| d\xi +$$

$$\int |\xi-\eta|^{-2n}[1-h((\xi-\eta)\lambda^{1/2})]\left|\int \Delta_x^n \varphi(x,\ \xi,\ \eta) e^{i(x,\ \xi-\eta)}\,dx\right| d\xi.$$

The first integral on the right-hand side can be easily estimated by replacing the integrand by $\max\limits_{x,\xi,\eta} |\varphi(x,\ \xi,\ \eta)|$, x by $y\lambda^{1/2}$, $\xi\lambda^{1/2}$ by ξ^1, and $\eta\lambda^{1/2}$ by η^1, since it is taken over a bounded region. In the second integral the integration with respect to y of a bounded function is taken over a bounded region, and the integrals taken over ξ^1 converge. Application of Shur's lemma leads to the estimate

$$|\|v\|^2-\|u\|^2|\leqslant \|u\|^2/2.$$

It follows that

$$\|u\|^2/2\leqslant \|v\|^2\leqslant 3\|u\|^2/2,$$

which completes the proof.

7°. Proposition 4.8. Suppose the conditions of Proposition 4.7 are satisfied and let

$$w(x)=Bu(x)\equiv \int b(x,\ \xi)\,\tilde u(\xi)\,e^{ix\xi+ie\lambda S(x\lambda^{-1/2},\ \xi\lambda^{-1/2})}\,d\xi,$$

where $|b(x,\ \xi)|\leqslant C$ and $\lambda\geqslant 1$. Then $\|w\|\leqslant C_1\|u\|$, where the constant C_1 is independent of u and ε.

Proof. This proposition is proved in exactly the same manner as the first statement in Proposition 4.7.

8°. Proposition 4.9. Suppose the conditions of Proposition 4.8 are satisfied and $\varepsilon\leqslant\varepsilon_0$, where ε_0 is sufficiently small. Then

$$\sum_{|\alpha+\beta|=l}\|x^\beta D^\alpha v\|\leqslant C_1\sum_{|\delta+\gamma|\leqslant l}\|x^\delta D^\gamma u\|,$$

$$\sum_{|\alpha+\beta|=l}\|x^\beta D^\alpha u\|\leqslant C_2\sum_{|\delta+\gamma|\leqslant l}\|x^\delta D^\gamma v\|,$$

where the constants C_1 and C_2 are independent of u.

Proof. Notice that

$$D_j v(x)=\int[\xi_j+\varepsilon\sqrt{\lambda}\,\partial S(x\lambda^{-\frac12},\ \xi\lambda^{-\frac12})/\partial x^j]\,\tilde u(\xi)\,e^{i(x,\ \xi)+ie\lambda S(x\lambda^{-\frac12},\ \xi\lambda^{-\frac12})}d\xi=A(D_j u)+$$

$$\varepsilon\sqrt{\lambda}\int\partial S(x\lambda^{-\frac12},\ \xi\lambda^{-\frac12})/\partial x^j\,\tilde u(\xi)\,e^{i(x,\ \xi)+ie\lambda S(x\lambda^{-\frac12},\ \xi\lambda^{-\frac12})}d\xi.$$

By Proposition 4.2 we assume that

$$\frac{\partial S}{\partial x^j}(x\lambda^{-1/2},\ \xi\lambda^{-1/2})=\sum_{k=1}^n a_{jk}(x,\ \xi,\ \lambda)\,x^k\lambda^{-1/2}+\sum_{k=1}^n b_j^k(x,\ \xi,\ \lambda)\,\xi_k\lambda^{-1/2},$$

where a_{jk} and $b_j^k\in C_0^\infty(\mathbb{R}^n\times\mathbb{R}_n)$ for $\lambda\geqslant 1$. It is clear that

$$\left\|\varepsilon\sqrt{\lambda}\int b_j^k(x,\ \xi,\ \lambda)\,\xi_k\lambda^{-1/2}\tilde u(\xi)\,e^{i(x,\ \xi)+ie\lambda S(x\lambda^{-1/2},\ \xi\lambda^{-1/2})}d\xi\right\|\leqslant C\varepsilon\|D_k u\|.$$

On the other hand,

$$\int a_{jk}(x,\ \xi,\ \lambda)\,x^k\tilde u(\xi)\,e^{i(x,\ \xi)+ie\lambda S(x\lambda^{-1/2},\ \xi\lambda^{-1/2})}d\xi=$$

$$-\int[D_{\xi_k}a_{jk}\tilde u(\xi)+ia_{jk}\tilde u(\xi)\,\varepsilon\sqrt{\lambda}\,D_{\xi_k}S(x\lambda^{-1/2},\ \xi\lambda^{-1/2})]\,e^{i(x,\ \xi)+ie\lambda S(x\lambda^{-1/2},\ \xi\lambda^{-1/2})}d\xi.$$

Let

$$D_{\xi_k}S(x\lambda^{-1/2},\ \xi\lambda^{-1/2})=\sum_{l=1}^n c_l^k(x,\ \xi,\ \lambda)\,\lambda^{-1/2}x^l+\sum_{l=1}^n d^{kl}(x,\ \xi,\ \lambda)\,\xi_l\lambda^{-1/2},$$

where c_l^k, d^{kl} are uniformly bounded in $C_0^\infty(\mathbb{R}^n\times\mathbb{R}_n)$ for all $\lambda\geqslant 1$. Then

$$\sqrt{\lambda}\int a_{jk}\tilde{u}(\xi)\,D_{\xi_k}S(x\lambda^{-1/2},\ \xi\lambda^{-1/2})\,e^{i(x,\,\xi)+ie\lambda S(x\lambda^{-1/2},\ \xi\lambda^{-1/2})}\,d\xi=$$

$$\int a_{jk}\left[\sum_{l=1}^{n}c_l^k x^l+\sum_{l=1}^{n}d^{kl}\xi_l\right]\tilde{u}(\xi)\,e^{i(x,\,\xi)+ie\lambda S(x\lambda^{-1/2},\ \xi\lambda^{-1/2})}\,d\xi,$$

and we can repeat the argument. As a result we obtain

$$D_j v=A(D_j u)+\sum_{k=1}^{\infty}\int\sum_{l=1}^{n}\left(A_k^l\xi_l+B_{lk}D_{\xi_l}+C_{lk}\right)\tilde{u}(\xi)\,e^{i(x,\,\xi)+ie\lambda S(x\lambda-1/2,\ \xi\lambda-1/2)}\,d\xi,$$

where $|A_k^l|+|B_{lk}|+|C_{lk}|\leqslant C(\varepsilon nM)^k$. Here $M=\max(|a_{jk}|,\ |b_{jk}|,\ |c_{jk}|,\ |d_{jk}|)$. If $\varepsilon nM<1/2$, then the series converges uniformly and we obtain, by Proposition 4.8,

$$\|D_j v\|\leqslant C\left(\|D_j u\|+\varepsilon\|u\|+\varepsilon\sum_{k=1}^{n}\|x^k u\|+\varepsilon\sum_{k=1}^{n}\|D_k u\|\right),$$

Application to Proposition 4.7 gives

$$\|D_j u\|\leqslant C\left(\|D_j v\|+\varepsilon\|u\|+\varepsilon\sum_{k=1}^{n}\|x^k u\|+\varepsilon\sum_{k=1}^{n}\|D_k u\|\right).$$

Notice that

$$x^j v=\int\left[D_{\xi_j}e^{ix\xi}\right]\tilde{u}(\xi)\,e^{ie\lambda S(x\lambda^{-1/2},\ \xi\lambda^{-1/2})}\,d\xi=$$

$$-\int e^{ix\xi}\left[D_{\xi_j}+\varepsilon\sqrt{\lambda}\,\partial S(x\lambda^{-1/2},\ \xi\lambda^{-1/2})/\partial\xi_j\right]\tilde{u}(\xi)\,e^{ie\lambda S(x\lambda^{-1/2},\ \xi\lambda^{-1/2})}\,d\xi.$$

Arguing as above we obtain

$$\|x^j v\|\leqslant C\left(\|x^j u\|+\varepsilon\|u\|+\varepsilon\sum_{k=1}^{n}\|x^k u\|+\varepsilon\sum_{k=1}^{n}\|D_k u\|\right),$$

$$\|x^j u\|\leqslant C\left(\|x^j v\|+\varepsilon\|u\|+\varepsilon\sum_{k=1}^{n}\|x^k u\|+\varepsilon\sum_{k=1}^{n}\|D_k u\|\right).$$

Thus, we have proved that

$$\sum_{j=1}^{n}(\|x^j v\|+\|D_j v\|)\leqslant C\sum_{|\alpha+\beta|\leqslant 1}\|x^\beta D^\alpha u\|,$$

$$\sum_{j=1}^{n}(\|x^j u\|+\|D_j u\|)\leqslant C\sum_{|\alpha+\beta|\leqslant 1}\|x^\beta D^\alpha v\|,$$

which proves our assertion for $l=1$. It can also be proven for larger values of l. For example

$$D_j D_k v=\int\Big[\big(\xi_j+\varepsilon\sqrt{\lambda}\,\partial S(x\lambda^{-1/2},\ \xi\lambda^{-1/2})/\partial x^j\big)\cdot\big(\xi_k+\varepsilon\sqrt{\lambda}\,\partial S(x\lambda^{-1/2},\ \xi\lambda^{-1/2})/\partial x^k\big)+$$
$$\varepsilon\,\partial^2 S(x\lambda^{-1/2},\ \xi\lambda^{-1/2})/\partial x^j\,\partial x^k\Big]\tilde{u}(\xi)\,e^{i(x,\,\xi)+ie\lambda S(x\lambda^{-1/2},\ \xi\lambda^{-1/2})}\,d\xi=$$
$$\int\Bigg[\left(\xi_j+\varepsilon\sum_{s=1}^{n}a_{js}x^s+\varepsilon\sum_{s=1}^{n}b_j^s\xi_s\right)\left(\xi_k+\varepsilon\sum_{s=1}^{n}a_{ks}x^s+\varepsilon\sum_{s=1}^{n}b_k^s\xi_s\right)+$$
$$\varepsilon\,\partial^2 S(x\lambda^{-1/2},\ \xi\lambda^{-1/2})/\partial x^j\,\partial x^k\Bigg]\tilde{u}(\xi)\,e^{i(x,\,\xi)+ie\lambda S(x\lambda^{-1/2},\ \xi\lambda^{-1/2})}\,d\xi.$$

Replacing every x^s by the derivative of $-ie^{i(x,\xi)}$ with respect to ξ_s and integrating by parts, we obtain the desired result. This completes the proof of Proposition 4.9.

9°. Completion of the Proof of Theorem 4.3. We substitute

$$\psi(y)=\int\tilde{\varphi}_1(\eta)\,e^{i(y,\,\eta)+ie\lambda S(x+y\lambda^{-1/2},\ \xi+\eta\lambda^{-1/2})}\,d\eta$$

into (4.17), assuming that $D_\xi^\alpha D_x^\beta S(x,\ \xi)=0$ for $|\alpha+\beta|\leqslant 2$, $S\in C_0^\infty(\mathbb{R}^n\times\mathbb{R}_n)$.

and ε is a sufficiently small number. From Proposition 4.7 it follows that

$$\|\psi\| \leqslant 2\|\psi_1\| \leqslant 4\|\psi\|.$$

Further, from Proposition 4.9 it follows that

$$\sum_{|\alpha+\beta|\leqslant N} \|y^\beta D^\alpha \psi\| \leqslant C_1 \sum_{|\alpha+\beta|\leqslant N} \|y^\beta D^\alpha \psi_1\| \leqslant C_2 \sum_{|\alpha+\beta|\leqslant N} \|y^\beta D^\alpha \psi\|.$$

In addition

$$y^\beta D^\alpha \psi = \int [D_\eta - \varepsilon\lambda^{1/2}\,\partial S(x+y\lambda^{-1/2},\ \xi+\eta\lambda^{-1/2})/\partial\xi]^\beta \times$$
$$[\eta + \varepsilon\lambda^{1/2}\,\partial S(x+y\lambda^{-1/2},\ \xi+\eta\lambda^{-1/2})/\partial x]^\alpha\, \tilde{\psi}_1(\eta) \times$$
$$e^{i(y,\eta)+i\varepsilon\lambda S(x+y\lambda^{-1/2},\ \xi+\eta\lambda^{-1/2})}\,d\eta + O\Big(\sum_{|\gamma+\delta|\leqslant|\alpha+\beta|-2} \|y^\delta D^\gamma \psi\|\Big),$$

since the inequality

$$|\lambda^{1/2} D_\eta^\alpha D_y^\beta S(x+y\lambda^{-1/2},\ \xi+\eta\lambda^{-1/2})| \leqslant C_{\alpha\beta}\lambda^{1/2-|\alpha+\beta|/2}$$

is valid for all α and β.

Note that

$$\sum_{2j+|\alpha+\beta|<N} p^{j\,(\alpha)}_{(\beta)}(x,\ \xi)\,\frac{1}{\alpha!\beta!}\,y^\beta D^\alpha \psi \lambda^{1-|\alpha+\beta|/2} =$$
$$\lambda \sum_{2j<N} T^{N-2j}_{(x,\ \xi)} p^j(y\lambda^{-1/2},\ D\lambda^{-1/2})\,\psi = \lambda \int T^N_{(x,\ \xi)} p^0 (D_\eta\lambda^{-1/2} -$$
$$\varepsilon\,\partial S(x+y\lambda^{-1/2},\ \xi+\eta\lambda^{-1/2})/\partial\xi,\ \eta\lambda^{-1/2} + \varepsilon\,\partial S(x+y\lambda^{-1/2},\ \xi+\eta\lambda^{-1/2})/\partial x)\cdot$$
$$\tilde{\psi}_1(\eta)\, e^{i(y,\eta)+i\varepsilon\lambda S(x+y\lambda^{-1/2},\ \xi+\eta\lambda^{-1/2})}\,d\eta +$$
$$\sum_{j=1}^{N} \sum_{|\alpha+\beta|<N-2j} a_{j\alpha\beta}\, y^\beta D^\alpha \psi \lambda^{-j-\frac{|\alpha+\beta|}{2}} + O\Big(\lambda^{1-\frac{N}{2}} \sum_{|\alpha+\beta|\leqslant N} \|y^\beta D^\alpha \psi\|\Big). \tag{4.20}$$

Now, without varying the order of the last term, we can replace grad $S(x+y\lambda^{-1/2},\ \xi+\eta\lambda^{-1/2})$ by the N-th order Taylor expansion with respect to y and η. Integrating by parts, we replace every $y^\beta e^{i(y,\eta)}$ by $D_\eta^\beta E^{i(y,\eta)}$. This requires the replacement of every y by $D_\eta - \varepsilon\lambda^{1/2}\,\partial S(x+y\lambda^{-1/2},\ \xi+\eta\lambda^{-1/2})/\partial\xi$. Thus, terms with second-order derivatives with respect to η can be estimated by $O(\|y^\beta D^\alpha \psi\|\lambda^{-|\alpha+\beta|/2})$. Since $D_\xi^\alpha D_x^\beta S(x,\ \xi) = 0$ for $|\alpha + \beta| \leqslant 2$ and the terms containing $y^\beta D^\alpha \psi$ with $|\alpha + \beta| \geqslant N + 1$ relate to the remainder term, we only have to perform integration by parts a finite number of times in order to eliminate y from the integrand in (4.20).

The remaining integral then takes the form

$$\lambda \int T^N_{(x,\ \xi)} q^0 \Big(D_\eta \lambda^{-\frac{1}{2}},\ \eta\lambda^{-\frac{1}{2}}\Big)\, \tilde{\psi}_1(\eta)\, e^{i(y,\eta)+i\varepsilon\lambda S\left(x+y\lambda^{-\frac{1}{2}},\ \xi+\eta\lambda^{-\frac{1}{2}}\right)}\,d\eta, \tag{4.21}$$

where $q^0(x,\ \xi) = p^0(x - \varepsilon\partial_\xi S(x - \varepsilon\partial_\xi S\ (x - \ldots\xi),\ \xi),\ \ \xi + \varepsilon\partial_x S(x - \varepsilon\partial_\xi S(x - \ldots,\ \xi),\ \xi))$ or, if $\bar{x} = x - \varepsilon\partial_\xi S(\bar{x},\ \xi)$ and $\bar{\xi} = \xi + \varepsilon\partial_x S(\bar{x},\ \xi)$, then $q^0(x,\ \xi) = p^0(x,\ \bar{\xi})$. Integral (4.21) can now be represented as

$$\lambda \int F_{z\to\eta}\Big\{T^N_{(x,\ \xi)} q^0 \Big(z\lambda^{-\frac{1}{2}},\ D_z\lambda^{-\frac{1}{2}}\Big)\psi(z)\Big\} e^{i(y,\eta)+i\varepsilon\lambda S\left(x+y\lambda^{-\frac{1}{2}},\ \xi+\eta\lambda^{-\frac{1}{2}}\right)}\,d\eta,$$

where $F_{z\to\eta} f(z) = \int f(z) e^{-i(z,\eta)}\,dz$. By Proposition 4.7, the norm in L_2 of this integral is estimated from above and below by

$$\lambda \| T^N_{(x,\ \xi)} q^0 (z\lambda^{-1/2},\ D_z\lambda^{-1/2}) \psi(z) \|.$$

The terms of (4.20) with the coefficients $a_{j\alpha\beta}$ also permit such transformations; they vary in a more complicated way and an explicit expression cannot be obtained for them. We can only assert that

$$\lambda \sum_{2j+|\alpha+\beta|<N} p^{j\,(\alpha)}_{(\beta)}(x,\ \xi) \frac{1}{\alpha!\beta!} y^\beta D^\alpha \psi \lambda^{1-j-|\alpha+\beta|/2} =$$

$$\lambda \int \{T^N_{(x,\ \xi)} q^0 (D_\eta \lambda^{-1/2},\ \eta\lambda^{-1/2}) + \sum b_{j\alpha\beta} \eta^\alpha D^\beta_\eta \lambda^{-j-|\alpha+\beta|/2}\} \tilde{\psi}_1(\eta) \times$$

$$e^{i(y,\ \eta)+i\varepsilon\lambda S(x+y\lambda^{-1/2},\ \xi+\eta\lambda^{-1/2})} d\eta + O\Big(\lambda^{1-N/2} \sum_{|\alpha+\beta|\leqslant N} \| y^\beta D^\alpha \psi \|\Big).$$

Our assertion now follows from this by Proposition 4.7 and also from the estimates of Proposition 4.8. However, note that the theorem has been proven only for the case when the generating function is of the form $x\xi + \varepsilon S(x,\ \xi)$.

We shall now show that this suffices to prove Theorem 4.3. To do so we consider a family Φ_t of canonical transformations generated by the functions $x\xi + tS(x,\ \xi)$ for $0 \leqslant t \leqslant 1$. Let M be the set of points t in [0, 1] for which the transformation Φ_t is permissible, that is, such that condition (4.17) is invariant under the canonical transformation Φ_t. It has thus been proven that $M \supset [0,\ \varepsilon_0]$ when ε_0 is small, that is, for $\varepsilon_0 > 0$. Theorem 4.3 states that the point $t = 1$ also belongs to M.

<u>10°. Proposition 4.10.</u> The set M coincides with the interval [0, 1].

<u>Proof.</u> We first prove that M is an open set in [0, 1]. Indeed, if $t_1 \in M$ and $t_1 < 1$, then, by Proposition 4.6 the transformation $\Phi_{t_1+\varepsilon}\Phi^{-1}_{t_1}$ for $0 \leqslant \varepsilon < \varepsilon_0$ is defined by the generating function $x\xi + \varepsilon T_\varepsilon(x,\ \xi)$, where

$$|D^\alpha_\xi D^\beta_x T_\varepsilon(x,\ \xi)| \leqslant C_{\alpha,\ \beta},\quad 0 \leqslant \varepsilon \leqslant \varepsilon_0,$$

and is permissible by what has been proven above. Multiplication from the right of this transformation by Φ_{t_1} is also permissible. This is $\Phi_{t_1+\varepsilon}$.

Now we shall prove that M is closed in [0, 1]. Let $t_n \in M$ and $t_n \to t_0$. By Proposition 4.6 Φ_{t_0} is the product of the transformation Φ_{t_n} and the transformation defined by a generating function of the type $x\xi + (t_0 - t_n)T_{t_0-t_n}(x,\ \xi)$, the derivatives of the last function being bounded uniformly with respect to n when $n \geqslant n_0$. Hence, the last transformation is permissible. Therefore Φ_{t_0} is also permissible.

Since the set M is both open and closed in [0, 1], it necessarily coincides with [0, 1]. Proposition 4.10 is proven.

With this the proof of Theorem 4.3 is completed.

<u>§5. Gårding's Inequality</u>

Gårding's classical inequality was obtained for an elliptic equation of the type $P(x,\ D)u = f(x)$, where P is a differential operator. In so doing the characteristic form $P^0(x,\ \xi)$ was assumed to satisfy the inequality

$$\operatorname{Re} p^0(x,\ \xi) \geqslant c_0 |\xi|^m,\quad x \in \Omega,\quad \xi \in \mathbb{R}_n,\quad c_0 = \text{const} > 0. \tag{5.1}$$

Let $\varepsilon > 0$ be a small number and let K be a compact set in Ω such that the distance between K and $\partial\Omega$ is more than ε. From the covering of the compact set K by balls of radius ε we choose a finite covering by balls $\omega_1,\ \ldots,\ \omega_N$ and construct a partition of unity $\sum \varphi_j^2(x) = 1$ such that $\varphi_j \in C_0^\infty(\omega_j)$.

Note that

$$\varphi_j Pu - P^0\varphi_j u = Qu, \tag{5.2}$$

where Q is an operator of order $m-1$. If ε is sufficiently small, then the inequality

$$([P^0(x, D) - P^0(x_j, D)]\varphi_j u, \varphi_j u) \leqslant \frac{c_0}{3}\|\varphi_j u\|^2_{m/2}, \quad j = 1, \ldots, N, \tag{5.3}$$

where $x_j \in \operatorname{supp} \varphi_j$, is satisfied.

In Section 1 (para. 4) we proved Gårding's inequality for differential operators $P^0(x_j, D)$ with constant coefficients. Recalling that

$$\operatorname{Re}(P^0(x_j, D)\varphi_j u, \varphi_j u) = \operatorname{Re}\int p^0(x_j, \xi)|\widetilde{\varphi_j u}(\xi)|^2 d\xi \geqslant$$
$$c_0\int|\xi|^m|\widetilde{\varphi_j u}(\xi)|^2 d\xi \geqslant c_0\|\varphi_j u\|^2_{m/2} - C_1\|\varphi_j u\|^2_{m/2-1}. \tag{5.4}$$

and combining the inequalities (5.2), (5.3), and (5.4) we can write the estimates

$$\operatorname{Re}\int Pu\cdot\bar{u}\,dx = \operatorname{Re}\int\sum_j \varphi_j Pu\overline{\varphi_j u}\,dx = \operatorname{Re}\sum_j\int P^0\varphi_j u\cdot\overline{\varphi_j u}\,dx + \operatorname{Re}\sum_j\int Q_j u\cdot\overline{\varphi_j u}\,dx \geqslant$$
$$\operatorname{Re}\sum_j\int P^0(x_j, D)\varphi_j u\cdot\overline{\varphi_j u}\,dx - \frac{c_0}{3}\sum_j\|\varphi_j u\|^2_{m/2} -$$
$$\sum_j\|\varphi_j u\|_{m/2}\|u\|_{m/2-1} \geqslant c_0\sum_j\|\varphi_j u\|^2_{m/2} - C_1\sum\|\varphi_j u\|^2_{m/2-1} -$$
$$\frac{c_0}{3}\sum\|\varphi_j u\|^2_{m/2} - \frac{c_0}{3}\sum\|\varphi_j u\|^2_{m/2} - C_2\|u\|^2_{m/2-1} \geqslant$$
$$\frac{c_0}{3}\sum\|\varphi_j u\|^2_{m/2} - C_3\|u\|^2_{m/2-1} \geqslant \frac{c_0}{3}\|u\|^2_{m/2} - C_4\|u\|^2_{m/2-1}$$

valid for $u \in C_0^\infty(K)$.

Now we can write Gårding's inequality as

$$\operatorname{Re}\int Pu\cdot\bar{u}\,dx \geqslant \frac{c_0}{3}\|u\|^2_{m/2} - C\|u\|^2_{m/2-1}, \quad u \in C_0^\infty(K). \tag{5.5}$$

5.2. A simple analysis of the proof indicates that by varying ε, instead of the constant $c_0/3$, one can obtain $c_0(1-\delta)$ for any $\delta > 0$ in the final form of the inequality. Obviously, in this context the constant C can also vary.

However, this constant can also be made equal to c_0. This was proved by L. Hörmander for pseudo-differential elliptic operators.

<u>Theorem 5.1.</u> Let $P(x, D)$ be a pseudo-differential operator of order m with principal symbol $p^0 \in S^m(\Omega)$ and

$$\operatorname{Re} p^0(x, \xi) \geqslant c_0|\xi|^m, \quad c_0 = \text{const}, \quad x \in \Omega. \tag{5.6}$$

Then the inequality

$$\operatorname{Re}\int Pu\cdot\bar{u}\,dx \geqslant c_0\|u\|^2_{m/2} - C_K\|u\|^2_{m/2-1}, \quad u \in C_0^\infty(K), \tag{5.7}$$

where K is an arbitrary compact set in Ω, is valid.

The proof is based on the following simple inequality.

<u>Lemma 5.1.</u> If $f \in C^2(\mathbb{R}^n)$, $f \geqslant 0$, and

$$\sum_{j=1}^n \sup\left|\frac{\partial^2 f}{(\partial x^j)^2}\right| = M, \text{ then } \sum_{j=1}^n\left|\frac{\partial f}{\partial x^j}\right|^2 \leqslant 2Mf(x).$$

<u>Proof.</u> To prove this lemma it is sufficient to show that the inequality $f^2_{x^j}(x) \leqslant 2f(x)\max|f''_{x^j x^j}|$ holds when $j = 1, \ldots, n$, that is, to prove the assertion for $n = 1$.

Let $f \in C^2(\mathbb{R})$ and x be an arbitrary point. Then

$$0 \leqslant f(x+t) \leqslant f(x) + tf'(x) + \frac{1}{2}t^2 \max|f''|.$$

Substituting $t = -f'(x)/\max|f''|$ in this inequality, we obtain

$$0 \leqslant f(x) - \frac{1}{2}f'^2(x)/\max|f''|$$

and the lemma is proved.

Proof of Theorem 5.1. Let $P^0(x, D)$ be an operator with symbol $\operatorname{Re} p^0(x, \xi)h(\xi) = p_1^0(x, \xi)$, where $h \in C^\infty$, $h(\xi) = 0$ when $|\xi| \leqslant 1$, and $h(\xi) = 1$ when $|\xi| \geqslant 2$. Note that

$$\operatorname{Re}\int Pu\bar{u}\,dx = \operatorname{Re}\int P^0 u\bar{u}\,dx + O(\|u\|^2_{(m-1)/2}).$$

Also if A is an operator with symbol $\operatorname{Im} p^0(x, \xi)h(\xi)$, then the operator $P - P^0 - iA$ is of order $m - 1$ and

$$\operatorname{Re} i\int Au\bar{u}\,dx = i\int (A - A^*)u\cdot\bar{u}\,dx = O(\|u\|^2_{(m-1)/2})$$

by Theorem 3.4.

Let first $c_0 = 0$.

We now consider a covering of the space $\mathbb{R}_n$ with balls K_ξ for $\xi \in \mathbb{R}_n$ so that the ball K_ξ centered at the point ξ has radius $(1 + |\xi|^2)^{1/4}/2$. Choose a locally finite subcovering such that every point of the space $\mathbb{R}_n$ is covered by not more than $c(n)$ balls. Number these balls $K_1, \ldots, K_j, \ldots$ and construct functions $h_j(\xi)$ such that $h_j \in C_0^\infty(\mathbb{R}_n)$, $h_j \geqslant 0$, $h_j(\xi) = 1$ when $\xi \in K_j$ and $h_j = 0$ outside the ball which is concentric to K_j' and has double the radius. Assume that

$$|D^\alpha h_j(\xi)| \leqslant C_\alpha(1 + |\xi|)^{-|\alpha|/2}$$

and C is independent of j. Clearly such functions can be obtained from one of the standard functions h using translations and dilations.

Now assume that $\varphi_j(\xi) = h_j(\xi)\left[\sum_{l=1}^{\infty} h_l^2(\xi)\right]^{-1/2}$. Then

$$\sum \varphi_j^2(\xi) = 1, \quad \varphi_j \in C_0^\infty(\mathbb{R}_n) \text{ and } |D^\alpha \varphi_j| \leqslant C'_\alpha \lambda_j^{-|\alpha|/2},$$

where the constants C'_α are independent of j, $\lambda_j = 1 + |\xi^j|$, and ξ^j is the center of the ball K_j.

For $u \in C_0(K)$ we have

$$\operatorname{Re}\int P^0 u\cdot\bar{u}\,dx = \operatorname{Re}\sum_j \int \varphi_j(D)P^0 u\cdot\overline{\varphi_j(D)u}\,dx =$$

$$\operatorname{Re}\sum_j \int P^0(x, D)\varphi_j u\cdot\overline{\varphi_j u}\,dx + \operatorname{Re}\sum_j \int [\varphi_j(D), P^0]u\cdot\overline{\varphi_j u}\,dx,$$

where $[\varphi_j, P^0] = \varphi_j P^0 - P^0\varphi_j$ is the commutator of the operators φ_j and P^0. Passing to Fourier transforms we can write

$$\int [\varphi_j, P^0]u\overline{\varphi_j u}\,dx = (2\pi)^{-n}\int [\varphi_j(\xi)\widetilde{P^0 u}(\xi) - \widetilde{P^0(\varphi_j u)}(\xi)]\overline{\widetilde{\varphi_j u}(\xi)}\,d\xi.$$

Since

$$\widetilde{P^0 v}(\xi) = (2\pi)^{-n}\int \tilde{b}(\eta - \xi, \eta)\tilde{v}(\eta)\,d\eta,$$

where $b(y, \eta) = \operatorname{Re} p^0(y, \eta)h(\eta)$ and $\tilde{b}(\zeta, \eta)$ is the Fourier transform of $b(y, \eta)$ in y, we have

$$\int [\varphi_j, P^0]u\overline{\varphi_j u}\,dx = (2\pi)^{-2n}\iint [\varphi_j(\xi) - \varphi_j(\eta)]\tilde{b}(\eta - \xi, \eta)\tilde{u}(\eta)\overline{\varphi_j(\xi)\tilde{u}(\xi)}\,d\xi\,d\eta. \quad (5.8)$$

Note that for any N

$$|\tilde{b}(\eta-\xi,\ \eta)| \leqslant C_N(1+|\eta-\xi|^2)^{-N/2}(1+|\eta|^2)^{m/2} \leqslant$$
$$2^{\frac{m}{4}} C_N (1+|\eta-\xi|^2)^{-\frac{N}{2}+\frac{|m-1|}{4}}(1+|\eta|^2)^{\frac{m+1}{4}}(1+|\xi|^2)^{\frac{m-1}{4}}, \tag{5.9}$$

since

$$1+|\eta|^2 \leqslant 2(1+|\xi|^2)(1+|\eta-\xi|^2), \quad (1+|\eta|^2)^{-1} \leqslant 2(1+|\xi|^2)^{-1}(1+|\eta-\xi|^2).$$

On the other hand, when $|\xi - \eta| > |\eta|/2$, the inequalities

$$\sum_j |\varphi_j(\xi)-\varphi_j(\eta)|^2 \leqslant 4 \leqslant 8|\xi-\eta||\eta|^{-1}$$

and, for $|\xi - \eta| < |\eta|/2$,

$$\sum_j |\varphi_j(\xi)-\varphi_j(\eta)|^2 = \sum_j \left| \sum_{l=1}^n \frac{\partial \varphi_j}{\partial \eta_l}(\eta+\theta_j(\xi-\eta))(\xi_l-\eta_l)\right|^2 \leqslant$$
$$C_1(1+|\eta|)^{-1}|\xi-\eta|^2$$

hold.

Therefore, for $|\eta| \geqslant 1$ the estimate

$$\sum_j |\varphi_j(\xi)-\varphi_j(\eta)|^2 \leqslant C_2(1+|\eta|^2)^{-1/2}(1+|\xi-\eta|^2) \tag{5.10}$$

is valid.

Combining (5.8), (5.9), and (5.10) gives

$$\sum_j \left| \int [\varphi_j,\ P^0] u \overline{\varphi_j u}\, dx \right| \leqslant$$
$$C'_N \iint (1+|\eta-\xi|^2)^{-\frac{N}{2}+\frac{|m-1|}{4}}(1+|\eta|^2)^{\frac{m+1}{4}}(1+|\xi|^2)^{\frac{m-1}{4}}|\tilde{u}(\eta)| \times$$
$$\sum_j \overline{|\varphi_j(\xi)\tilde{u}(\xi)|}\,|\varphi_j(\xi)-\varphi_j(\eta)|\, d\xi\, d\eta \leqslant A\|u\|^2_{(m-1)/2}.$$

Here we have used Theorem 1.1 and denoted by A the value

$$\max_\eta \int |K(\xi,\ \eta)|\, d\xi = \max_\xi \int |K(\xi,\ \eta)|\, d\eta,$$

where $K(\xi,\ \eta) = (1+|\eta-\xi|^2)^{-N/2+|m-1|/4+1}$.

It is clear that these integrals converge for $N > |m - 1|/2 + n + 2$.

Thus, we have shown that

$$\mathrm{Re}\int Pu\bar{u}\, dx \geqslant \mathrm{Re}\sum_j \int P^0\varphi_j u \cdot \overline{\varphi_j u}\, dx - C_3\|u\|^2_{(m-1)/2}. \tag{5.11}$$

Now denote by $R_j(x, D)$ the operator

$$P^0(x,\ D) - p^0(x,\ \xi^j) - \sum_{l=1}^n \frac{\partial p^0(x,\ \xi^j)}{\partial \xi_l}(D_l - \xi_l^j),$$

whose symbol $R_j(x, \xi)$ and derivatives $D_x^\beta R_j(x, \xi)$ can be estimated by $C_\beta(1 + |\xi|)^{m-1}$ if $\xi, \xi^j \in \operatorname{supp} \varphi_j$.

Let us estimate the integral

$$\mathrm{Re}\int R_j(x,\ D)\varphi_j u \overline{\varphi_j u}\, dx = \mathrm{Re}\iint \tilde{R}_j(\eta-\xi,\ \xi)\varphi_j(\xi)\tilde{u}(\xi)\overline{\varphi_j(\eta)\tilde{u}(\eta)}\, d\xi\, d\eta.$$

As above, using the estimates

$$|\tilde{R}_j(\eta-\xi,\ \xi)| \leqslant C_N''(1+|\eta-\xi|)^{-N}(1+|\xi|)^{m-1} \leqslant$$
$$C_N'''(1+|\eta-\xi|)^{-N+|m-1|/2}(1+|\xi|)^{(m-1)/2}(1+|\eta|)^{(m-1)/2},$$

we can check that for $N > n + |m - 1|/2$

$$\left|\operatorname{Re}\int R_j\varphi_j u\overline{\varphi_j u}\,dx\right| \leqslant C_4\|\varphi_j u\|^2_{(m-1)/2},$$

so that

$$\sum_j\left|\operatorname{Re}\int R_j\varphi_j u\overline{\varphi_j u}\,dx\right| \leqslant C_5\|u\|^2_{(m-1)/2}.$$

Using Lemma 5.1 we can affirm that

$$\left|\frac{\partial p_1^0(x,\ \xi)}{\partial\xi_l}\right|^2 \leqslant C_6 p_1^0(x,\ \xi)(1+|\xi|)^{m-2}, \qquad l=1,\ \ldots,\ n.$$

Therefore

$$\left|\int\frac{\partial p_1^0(x,\ \xi^j)}{\partial\xi_l}(D_l-\xi_l^j)\,\varphi_j u\cdot\overline{\varphi_j u}\,dx\right| \leqslant$$
$$\left(\int\left|\frac{\partial p_1^0(x,\ \xi^j)}{\partial\xi_l}\varphi_j u\right|^2 dx\right)^{1/2}\left(\int\left|(D_l-\xi_l^j)\,\varphi_j u\right|^2 dx\right)^{1/2} \leqslant$$
$$C_6(1+|\xi^j|)^{(m-2)/2}\left(\int p_1^0(x,\ \xi^j)\,|\varphi_j u|^2\,dx\right)^{1/2}\left(\int(\xi_l-\xi_l^j)^2\,|\varphi_j(\xi)\,\tilde{u}(\xi)|^2\,d\xi\right)^{1/2} \leqslant$$
$$\frac{1}{2n}\int p_1^0(x,\ \xi^j)\,|\varphi_j u|^2\,dx + C_7(1+|\xi^j|)^{m-1}\int|\varphi_j(\xi)\,\tilde{u}(\xi)|^2\,d\xi.$$

Returning to inequality (5.11) we see that

$$\operatorname{Re}\int Pu\cdot\bar{u}\,dx \geqslant \operatorname{Re}\sum_j\int p_1^0(x,\ \xi^j)\,|\varphi_j u|^2\,dx - \frac{1}{2}\sum_j\int p_1^0(x,\ \xi^j)\,|\varphi_j u|^2\,dx -$$
$$C_7 n\sum_j(1+|\xi^j|)^{m-1}\int|\varphi_j u|^2\,dx - C_3\|u\|^2_{(m-1)/2} \geqslant$$
$$\frac{1}{2}\operatorname{Re}\sum\int p_1^0(x,\ \xi^j)\,|\varphi_j u|^2\,dx - C_8\|u\|^2_{(m-1)/2}, \tag{5.12}$$

since

$$\frac{1}{2}(1+|\xi|) \leqslant 1+|\xi^j| \leqslant 2(1+|\xi|) \quad \text{for} \quad \xi\in\operatorname{supp}\varphi_j.$$

Since by hypothesis $p_1^0 \geqslant 0$, it follows from (5.12) that

$$\operatorname{Re}\int Pu\cdot\bar{u}\,dx \geqslant -C_8\|u\|^2_{(m-1)/2}.$$

In the general case, if c_0 is any number, we consider an operator $Q = P - c_0\Lambda^m$, where Λ is an operator with symbol $(1 + |\xi|^2)^{1/2}$. Since $\operatorname{Re} q^0(x, \xi) \geqslant 0$, the inequality

$$\operatorname{Re}\int Qu\bar{u}\,dx \geqslant -C\|u\|^2_{(m-1)/2}$$

holds by what has been proven above. From this inequality it follows that

$$\operatorname{Re}\int Pu\bar{u}\,dx \geqslant c_0\|u\|^2_{m/2} - C\|u\|^2_{(m-1)/2}.$$

This proves the theorem.

5.3. An interesting refinement of the Gårding-Hörmander inequality (5.7) is given in C. Fefferman and D. H. Phong.[1]

Theorem 5.2. Let $P(x, D)$ be a pseudo-differential operator of order m with symbol $p(x, \xi)$ of class $S^m(\Omega)$, and

$$\operatorname{Re} p(x,\ \xi) \geqslant c_0(1+|\xi|)^m, \qquad c_0=\text{const}, \quad x\in\Omega, \quad |\xi| \geqslant c_1.$$

Then the inequality

$$\operatorname{Re}\int Pu\bar{u}\,dx \geq c_0\|u\|^2_{m/2} - C_K\|u\|^2_{(m-2)/2}, \qquad u \in C_0^\infty(K),$$

holds, where K is a subdomain in Ω, $K \subset\subset \Omega$.

Due to limitations of space we shall not prove this theorem here. Various generalizations of the theorem are found in C. Fefferman and D. H. Phong.[2]

§6. Generalizations

6.1. L. Hörmander proposed considering the $S^m_{\rho,\delta}$ classes as more general classes of pseudo-differential operators of type (2.1), but with symbols satisfying the conditions

$$|D^\alpha_\xi D^\beta_x a(x, \xi)| \leq C_{\alpha,\beta,K}(1+|\xi|)^{m-|\alpha|\rho+|\beta|\delta}, \qquad x \in K, \tag{6.1}$$

instead of conditions (2.2), which are obtained for $\delta = 0$ and $\rho = 1$.

Generally it is assumed that $0 \leq \delta \leq \rho \leq 1$. Smooth functions a satisfying condition 6.1 form the class $S^m_{\rho,\delta}(\Omega)$.

The majority of the results of Section 2 hold in this case.

Example 6.1. Let P(D) be a hypoelliptic differential operator and let a smooth function $h(\xi)$ be such that $h = 0$ in a neighborhood of the zeros of P and $h = 1$ in a neighborhood of infinity. Then $p = hp^{-1}$ is an infinitely differentiable function such that

$$|D^\alpha_\xi p(\xi)| \leq C_\alpha(1+|\xi|)^{m-|\alpha|\rho},$$

where $\rho = \text{const}$, $0 < \rho \leq 1$.

After a change of variables $y = f(x)$ the operator changes to an operator with symbol $q(y, \eta) = p({}^t f'^{-1}(x)\eta)$ and

$$|D^\alpha_\eta D^\beta_y q(y, \eta)| \leq C_{\alpha,\beta}(1+|\eta|)^{m-|\alpha|\rho+|\beta|(1-\rho)}.$$

Example 6.2. Let $\theta(\xi) \in C_0^\infty$ and $\theta(\xi) = 1$ when $|\xi| \leq 1$ and $\theta(\xi) = 0$ when $|\xi| \geq 2$. Let $z_1, \ldots, z_l, \ldots$ be all points in $\mathbb{R}^n$ with integer coordinates. Assume that

$$\varphi_l(\xi) = \theta(\xi - z_l)\left[\sum_j \theta^2(\xi - z_j)\right]^{-1/2}.$$

Then $\varphi_l \in C_0^\infty(\mathbb{R}^n)$ and $\sum \varphi_l^2(\xi) = 1$. The function

$$f(x, \xi) = \sum_{l,m=1}^{\infty} A_{lm}\varphi_l^2(\xi|\xi|^{-\rho})\varphi_m^2(x|\xi|^\delta),$$

where A_{lm} are arbitrary constants, is infinitely differentiable and belongs to $S^0_{\rho,\delta}$ if $|A_{lm}| \leq C$.

6.2. The first proposition of Theorem 2.1 is, obviously, preserved. The second one holds when $0 \leq \delta \leq \rho \leq 1$, except for the case where $\delta = \rho = 1$. This will be proven in the following paragraph.

In Theorems 2.2, 2.4, and 2.5 it is natural to assume that $0 \leq \delta < \rho \leq 1$. Then the terms of (2.11) are of decreasing order in ξ. The case $\delta = \rho$ will be considered in the following paragraph.

Lemma 2.1 is retained if condition (2.3) is replaced by the condition

$$|D^\alpha_\xi D^{\alpha'}_\eta D^\beta_x D^{\beta'}_y A(x, y, \xi, \eta)| \leq$$
$$C_{\alpha,\alpha',\beta,\beta'}(1+|\xi|)^{m-|\alpha|\rho+|\beta|\delta}(1+|\eta|)^{m'-|\alpha'|\rho+|\beta'|\delta}$$

when $0 \leq \delta < \rho \leq 1$. Then $a \in S^{m+m'}_{\rho,\delta}$.

Theorem 2.3 holds if (2.15) is replaced by

$$|D_\xi^\alpha D_x^\beta D_y^\gamma b(x, y, \xi)| \leqslant C_{\alpha,\beta,\gamma}(1+|\xi|)^{m-|\alpha|\rho+|\beta+\gamma|\delta}$$

for $0 \leqslant \delta < \rho \leqslant 1$.

Theorem 2.6 holds when $0 \leqslant \delta < \rho \leqslant 1$, $1 - \rho \leqslant \delta$.

In these cases the proofs of the aforementioned theorems are practically unchanged.

6.3. Almost all of the theorems of this chapter hold for systems of pseudo-differential operators and for the operators acting on cross sections of vector bundles on a smooth manifold Ω. Only Corollary 2.3 fails to hold when $ba \neq ab$.

6.4. A new class of pseudo-differential operators $S^{M,m}(\Phi, \varphi)$, much broader than $S^m_{\rho,\delta}$ and retaining most of the properties of operators in $S^m_{\rho,\delta}$, was defined by R. Beals.[2]

The functions Φ and φ are called *weight functions*. They are smooth functions in $\mathbb{R}^n \times \mathbb{R}_n$ with non-negative real values. Their ratio is assumed to be independent of x. We also assume that

1° $\Phi\varphi \geqslant 1 \geqslant \varphi$;

2° $|\Phi^{(\alpha)}_{(\beta)}| \leqslant C_{\alpha\beta}\Phi^{1-|\alpha|}\varphi^{|\beta|}$, $|\varphi^{(\alpha)}_{(\beta)}| \leqslant C_{\alpha\beta}\Phi^{-|\alpha|}\varphi^{1-|\beta|}$;

3° $|R^{(\alpha)}| \leqslant C_\alpha R^{1-\delta}\Phi^{-|\alpha|}$, $\alpha \neq 0$, $\delta > 0$.

The class $S^{M,m}$, where M and m are real numbers, is the set of symbols a for which

$$|a^{(\alpha)}_{(\beta)}(x, \xi)| \leqslant C_{\alpha,\beta}\Phi^{M-|\alpha|}\varphi^{m-|\beta|}.$$

Let $L^{M,m}$ be a set of operators of type (2.1) with symbols in $S^{M,m}$. Let $H^{M,m}$ be the Sobolev space consisting of distributions $u \in \mathcal{S}'$ for which $Au \in L_2$ for any operator A in $L^{M,m}$.

It can be proven (see R. Beals[2]) that

1° if $A \in L^{M,m}$, $B \in L^{K,k}$, then $AB \in L^{M+K,m+k}$;

2° if $A \in L^{M,m}$, then $A^* \in L^{M,m}$;

3° if $A \in L^{M,m}$, then the mapping A: $H^{M+K,m+k} \to H^{K,k}$ is continuous for all K and k.

It is remarkable that this class can be characterized internally.

Let $L_jB = [B, ix^j] = iBx^j - ix^jB$, $M_jB = [D_j, B] = D_jB - BD_j$ and let

$$B^{(\alpha)}_{(\beta)} = L_1^{\alpha_1} \ldots L_n^{\alpha_n} M_1^{\beta_1} \ldots M_n^{\beta_n} B.$$

Theorem (R. Beals). A linear operator B: $\mathcal{S} \to \mathcal{S}'$ belongs to $L^{0,0}$ if and only if for all α, β in $(\mathbb{Z}_+)^n$ the operator $B^{(\alpha)}_{(\beta)}$ can be extended to a continuous mapping of the space $H^{-|\alpha|,-|\beta|}$ into L_2.

6.5. Operators of type (2.1) with symbols $a(x, \xi)$, satisfying conditions of the form

$$|D_\xi^\alpha D_x^\beta a(x, \xi)| \leqslant C_{\alpha,\beta}(1+|x|+|\xi|)^{m-|\alpha|},$$

are considered by A. Grossman, G. Loupias and E. M. Stein[1] and in the books of F. A. Berezin and M. A. Shubin[1] and V. P. Maslov.[2]

These are natural operators for applications to quantum mechanics. Here the variables x and ξ play the role of coordinates and momenta. All results of this chapter are carried over without any difficulty to such operators. The spaces are also convenient since arbitrary canonical transformations are permitted, as the variables x and ξ are equivalent.

In these circumstances it is often convenient to consider operators of the type

$$Wu(x) = (2\pi)^{-n} \iint e^{i(x-y)\xi} a\left(\frac{x+y}{2}, \xi\right) u(y)\, dy\, d\xi$$

due to H. Weyl, instead of operators of the type (2.1). It is not difficult to verify that W differs from the operator A, with symbol $a(x, \xi)$, by an operator of order $m - 1$.

The Weyl calculus of pseudo-differential operators, suited for the space $L_2(\Omega)$, is developed in L. Hörmander.[16]

6.6. All results of this chapter are carried over without any difficulty to operators of the type

$$A_\tau u(x) = (2\pi)^{-n} \int a(\tau, x, \xi)\, \tilde{u}(\xi)\, e^{i(x, \xi)}\, d\xi,$$

where $a \in C^\infty(\mathbb{R}_+ \times \mathbb{R}^n \times \mathbb{R}_n)$ and

$$|D_\xi^\alpha D_x^\beta a(\tau, x, \xi)| \leqslant C_{\alpha, \beta, K} (1 + |\xi| + \tau)^{m - |\alpha|}, \quad x \in K, \tag{6.2}$$

for all α and $\beta \in \mathbb{Z}_+^n$, $\tau \geqslant \tau_0$. In doing so, instead of the ordinary Sobolev spaces H_s, it is natural to consider the $H_{s,\tau}$ spaces in which the norm is determined by the formula

$$\|u\|_{s,\tau}^2 = \int (1 + |\xi|^2 + \tau^2)^s |\tilde{u}(\xi)|^2\, d\xi.$$

We shall use Weyl operators in Chapter IX for determining Carleman estimates.

§7. A Class of Pseudo-Differential Operators

7.1. In this section we consider a class of pseudo-differential operators which will be used further.

Definition 7.1. A symbol $a \in C^\infty(\Omega \times \mathbb{R}_n)$ belongs to $S_\delta^m(\Omega)$ if for all α, β and every compact set $K \subset \Omega$ the inequalities

$$|D_\xi^\alpha D_x^\beta a(x, \xi)| \leqslant C_{\alpha, \beta, K} (1 + |\xi|)^{m + \delta(|\beta| - |\alpha|)}, \quad x \in K, \tag{7.1}$$

where $\delta = \text{const}$, $-\infty < \delta < 1$, are valid.

As in Section 2, we start with an auxiliary statement.

Lemma 7.1. Let $A \in C^\infty(\Omega \times \Omega \times \mathbb{R}_n \times \mathbb{R}_n)$ and

$$|D_\xi^\alpha D_x^\beta D_\eta^{\alpha'} D_y^{\beta'} A(x, y, \xi, \eta)| \leqslant$$
$$C_{\alpha, \alpha', \beta, \beta'} (1 + |\xi|)^{m + \delta(|\beta| - |\alpha|)} (1 + |\eta|)^{m' + \delta(|\beta'| - |\alpha'|)}. \tag{7.2}$$

Then the function

$$a(x, \xi) = \iint A(x, y, \xi, \eta)\, e^{i(x-y, \eta-\xi)}\, dy\, d\eta \tag{7.3}$$

is infinitely differentiable and belongs to $S_\delta^{m+m'}$.

Proof. We use the identity

$$e^{i(x-y, \eta-\xi)} = [1 + (1 + |\eta|)^{2\delta} |x - y|^2]^{-N} [1 - (1 + |\eta|)^{2\delta} \Delta_\eta]^N \times$$
$$[1 + (1 + |\eta|)^{-2\delta} |\eta - \xi|^2]^{-M} [1 - (1 + |\eta|)^{-2\delta} \Delta_y]^M e^{i(x-y, \eta-\xi)}$$

and integrate by parts with respect to η and y so that

$$a(x, \xi) = \iint B(x, y, \xi, \eta)\, e^{i(x-y, \eta-\xi)}\, dy\, d\eta,$$

where

$$B(x, y, \xi, \eta) = [1 - (1 + |\eta|)^{-2\delta} \Delta_y]^M [1 + (1 + |\eta|)^{-2\delta} |\eta - \xi|^2]^{-M} \times$$

$$[1-(1+|\eta|)^{2\delta}\Delta_\eta]^N[1+(1+|\eta|)^{2\delta}|x-y|^2]^{-N}e^{i(x-y,\,\eta-\xi)}.$$

Letting $y = x + z(1 + |\eta|)^{-\delta}$, we see that

$$|a(x,\xi)| \leqslant \iint |(1-\Delta_z)^M[1+(1+|\eta|)^{-2\delta}|\eta-\xi|^2]^{-M}\times$$
$$[1-(1+|\eta|)^{2\delta}\Delta_\eta]^N[1+|z|^2]^{-N}A(x,y,\xi,\eta)|(1+|\eta|)^{-n\delta}dz\,d\eta \leqslant$$
$$C\int[1+(1+|\eta|)^{-2\delta}|\eta-\xi|^2]^{-M}(1+|\xi|)^m(1+|\eta|)^{m'-n\delta}d\eta, \qquad (7.4)$$

provided N is selected so that $2N > n$.

To estimate the integral (7.4) we divide the range of integration into three zones.

In the region where $1 + |\eta| > 2(1 + |\xi|)$ the integrand is estimated via

$$(1+|\xi|)^{-l}(1+|\eta|)^{m'-n\delta+|m|+l-2M(1-\delta)}$$

for any $l > 0$. When $2M(1-\delta) > l + |m| + m' - n\delta + n$, the integral is estimated via $C_1(1 + |\xi|)^{-l}$.

If $1 + |\xi| > 2(1 + |\eta|)$, then $|\eta - \xi|^2(1 + |\eta|)^{-2\delta} \geqslant C_2(1 + |\xi|)^{2(1-\delta)}$ and the integral over this region does not exceed the value

$$C_3(1+|\xi|)^{m+|m'|+n-2(1-\delta)M}.$$

Finally, the integral over a domain for which $1 + |\xi| < 2(1 + |\eta|) < 4(1 + |\xi|)$ is estimated via

$$C_4(1+|\xi|)^{m+m'-n\delta}\int\frac{d\eta}{[1+(1+|\xi|)^{-2\delta}|\eta-\xi|^2]^M},$$

which is finite when $2M > n$ and does not exceed $C_5(1 + |\xi|)^{m+m'}$.

Thus for $\delta < 1$ we have an estimate for the integral (7.4): $|a(x,\xi)| \leqslant C_6(1 + |\xi|)^{m+m'}$.

The derivatives $D_\xi^\alpha D_x^\beta a(x, \xi)$ are estimated in the same way. Now we have

$$D_\xi^\alpha D_x^\beta a(x,\xi) = \sum_{\substack{\alpha'+\alpha''=\alpha\\ \beta'+\beta''=\beta}} C_{\alpha',\alpha'',\beta',\beta''}\iint D_\xi^{\alpha'}D_x^{\beta'}A(x,y,\xi,\eta)(-D_y)^{\beta''}(-D_\eta)^{\alpha''}\times$$
$$e^{i(x-y,\,\eta-\xi)}dy\,d\eta = \sum_{\substack{\alpha'+\alpha''=\alpha\\ \beta'+\beta''=\beta}} C_{\alpha',\alpha'',\beta',\beta''}\iint D_\xi^{\alpha'}D_\eta^{\alpha''}D_x^{\beta'}D_y^{\beta''}A(x,y,\xi,\eta)e^{i(x-y,\,\eta-\xi)}dy\,d\eta,$$

and we obtain integrals of the same type as (7.3). Clearly, they can be estimated in the same manner as integral (7.3). The lemma is proven.

7.2. We now show that the theorems of Section 2 hold for operators in S_δ^m.

<u>Theorem 7.1.</u> If pseudo-differential operators A and B have symbols a and b belonging to S_δ^m and $S_\delta^{m'}$ and $\delta < 1$, $h \in C_0^\infty(\Omega)$, then the operator $C = BhA$ is pseudo-differential and its symbol C belongs to $S_\delta^{m+m'}$.

<u>Proof.</u> As in the proof of Theorem 2.2, we consider the integral

$$c(x,\xi) = (2\pi)^{-n}\iint b(x,\eta)h(y)a(y,\xi)e^{i(y-x,\,\xi-\eta)}dy\,d\eta.$$

Using Lemma 7.1, we immediately obtain the desired result.

<u>Theorem 7.2.</u> An operator of the type

$$Au(x) = (2\pi)^{-n}\iint b(x,y,\xi)u(y)e^{i(x-y,\,\xi)}dy\,d\xi,$$

where the function $b \in C^\infty(\Omega\times\Omega\times\mathbb{R}_n)$ is chosen so that

$$|D_\xi^\alpha D_x^\beta D_y^\gamma b(x, y, \xi)| \leqslant C_{\alpha,\beta,\gamma}(1+|\xi|)^{m+\delta(|\beta+\gamma|-|\alpha|)}, \quad \delta<1$$

is pseudo-differential with symbol belonging to S_δ^m.

Proof. As in the proof of Theorem 2.3, we consider the integral

$$a(x, \xi) = (2\pi)^{-n}\iint b(x, y, \eta)\, e^{i(x-y,\, \eta-\xi)}\, dy\, d\eta,$$

to which Lemma 7.1 applies.

Theorem 7.3. If A is a pseudo-differential operator with symbol belonging to $S_\delta^m(\Omega)$ and $\delta < 1$, then the formal adjoint operator A^* is pseudo-differential with symbol from $S_\delta^m(\Omega)$.

Proof. As in the proof of Theorem 2.4, it is sufficient to consider the integral

$$a^*(x, \xi) = (2\pi)^{-n}\iint a(y, \eta)\, e^{i(x-y,\, \eta-\xi)}\, dy\, d\eta.$$

Using Lemma 7.1, we obtain the result.

7.3. Now we show that the operators of the given class are bounded.

Theorem 7.4. If A is a pseudo-differential operator with symbol a in S_δ^m, $\delta < 1$ and $h \in C_0^\infty(\Omega)$, then there exists a constant $C = C(s)$ such that

$$\|hAu\|_s \leqslant C\|u\|_{s+m}, \quad u \in C_0^\infty(\mathbb{R}^n). \tag{7.5}$$

The proof is based on the following lemma, due to M. Kotlar.

Lemma 7.2. Let H be a Hilbert space, X be a space upon which a measure is defined, and let A(x) be a measurable function of $x \in X$ with values in the space L(H) of operators bounded in H for which

$$\int \|A(x)A^*(y)\|^{1/2}\, dy \leqslant C, \quad \int \|A^*(x)A(y)\|^{1/2}\, dy \leqslant C.$$

Then the integral $A = \int A(x)\, dx$ weakly converges and $\|A\| \leqslant C$.

Proof. We first consider the case when $\|A(x)\| \leqslant M$ and the measure m of X is finite. From the inequality

$$\|A_1A_2\ldots A_{2n}\| \leqslant \|A_1\|^{1/2}\|A_1A_2\|^{1/2}\ldots\|A_{2n-1}A_{2n}\|^{1/2}\|A_{2n}\|^{1/2}$$

it follows that

$$\|A\|^{2n} = \|A^*A\|^n \leqslant \int \|A^*(x_1)\|^{1/2}\|A^*(x_1)A(x_2)\|^{1/2}\|A(x_2)A^*(x_3)\|^{1/2}\ldots$$

$$\|A^*(x_{2n-1})A(x_{2n})\|^{1/2}\|A(x_{2n})\|^{1/2}\, dx_1\ldots dx_{2n} \leqslant mMC^{2n-1}.$$

Hence $\|A\| \leqslant C(mM/C)^{1/2n}$ and as $n \to \infty$ we find $\|A\| \leqslant C$. In the general case, if Y is an arbitrary subset in X of finite measure on which the norm $\|A(x)\|$ is bounded, then for any $u, v \in H$ we have $\left|\int_Y (A(x)u, v)\, dx\right| \leqslant C\|u\|\|v\|$. This proves the lemma.

Proof of Theorem 7.4. 1°. First we let $s = 0$ and $m = 0$. We have

$$hAu(x) = (2\pi)^{-n}\iint h(x)\, a(x, \xi)\, u(y)\, e^{i(x-y)\xi}\, dy\, d\xi =$$

$$(2\pi)^{-n}\iint b(x, y, \xi)\, u(y)\, e^{i(x-y)\xi}\, dy\, d\xi,$$

where we assume

$$b(x, y, \xi) = [1+(-\Delta_\xi)^N(1+|\xi|^2)^{\delta N}]\,[1+|x-y|^{2N}(1+|\xi|^2)^{\delta N}]^{-1}h(x)\, a(x, \xi).$$

The operator A is thus represented in the form $A = \int A(\xi)\, d\xi$, where $A(\xi)$ is an operator with kernel $(2\pi)^{-n}e^{i(x-y)\xi}b(x, y, \xi)$ and

$$|b(x, y, \xi)| \leqslant C[1+|x-y|^{2N}(1+|\xi|^2)^{\delta N}]^{-1}.$$

We see that $A(\xi)$ is a bounded operator if $2N > n$ by Theorem 1.1.

2°. To apply Lemma 7.2, the norms $\|A(\xi)A^*(\eta)\|$ and $\|A^*(\xi)A(\eta)\|$ have to be estimated. The operator $A(\xi)A^*(\eta)$ has kernel

$$c(x, y, \xi, \eta) = (2\pi)^{-2n}\int e^{i(x-z, \xi)} b(x, z, \xi) e^{-i(y-z, \eta)} \overline{b(y, z, \eta)}\, dz.$$

We shall represent $e^{i(z, \eta-\xi)}$ as $L^k e^{i(z, \eta-\xi)}$, where $L = [1+(1+|\xi|^2+|\eta|^2)^{-\delta}|\eta-\xi|^2]^{-1}[1-(1+|\xi|^2+|\eta|^2)^{-\delta}\Delta_z]$ and integrate by parts. Note that $|{}^tL^k[b(x, z, \xi)\times \overline{b(y, z, \eta)}]| \leqslant C[1+(1+|\xi|^2+|\eta|^2)^{-\delta}|\eta-\xi|^2]^{-k}[1+|x-z|^{2N}(1+|\xi|^2)^{\delta N}]^{-1}[1+|y-z|^{2N}\times (1+|\eta|^2)^{\delta N}]^{-1}$. By Theorem 1.1

$$\|A(\xi)A^*(\eta)\| \leqslant C(1+|\xi|^2)^{-n\delta/2}(1+|\eta|^2)^{-n\delta/2}[1+(1+|\xi|^2+|\eta|^2)^{-\delta}|\eta-\xi|^2]^{-k}. \quad (7.6)$$

Also the norm of the operator $A(\xi)^*A(\eta)$ having kernel

$$(2\pi)^{-2n}\int e^{i(x-z)\xi}\overline{b(z, x, \xi)}\, e^{i(z-y, \eta)} b(z, y, \eta)\, dz$$

is estimated so that

$$\|A(\xi)^* A(\eta)\| \leqslant C(1+|\xi|^2)^{-n\delta/2}(1+|\eta|^2)^{-n\delta/2}[1+(1+|\xi|^2+|\eta|^2)^{-\delta}|\eta-\xi|^2]^{-k}. \quad (7.7)$$

3°. Let us first estimate $\int \|A(\xi)A^*(\eta)\|^{1/2} d\eta$. To do so, we divide the region of integration into three parts: $\Omega_1 = \{\eta;\ |\eta| \leqslant (1+|\xi|)/2\}$, $\Omega_2 = \{\eta;\ |\eta| \geqslant 2(1+|\xi|)\}$, $\Omega_3 = \{\eta;\ (1+|\xi|)/2 < |\eta| < 2(1+|\xi|)\}$.

In Ω_1 we have

$$1+(1+|\xi|^2+|\eta|^2)^{-\delta}|\eta-\xi|^2 \geqslant c_0(1+|\xi|^2)^{1-\delta}, \qquad c_0 = \text{const} > 0.$$

Therefore

$$\int_{\Omega_1}\|A(\xi)A^*(\eta)\|^{1/2} d\eta \leqslant C_1\int_{\Omega_1}(1+|\xi|^2)^{-k(1-\delta)/2} d\eta \leqslant C_2(1+|\eta|^2)^{[n-k(1-\delta)]/2}. \quad (7.8)$$

In Ω_2 we have

$$1+(1+|\xi|^2+|\eta|^2)^{-\delta}|\eta-\xi|^2 \geqslant c_0(1+|\eta|^2)^{1-\delta}$$

and therefore

$$\int_{\Omega_2}\|A(\xi)A^*(\eta)\|^{1/2} d\eta \leqslant C_3\int_{\Omega_2}(1+|\eta|^2)^{-k(1-\delta)/2} d\eta. \quad (7.9)$$

Finally,

$$\int_{\Omega_3}\|A(\xi)A^*(\eta)\|^{1/2} d\eta \leqslant C_4\int_{\Omega_3}[1+(1+|\xi|^2)^{-\delta}|\eta-\xi|^2]^{-k/2}(1+|\xi|^2)^{-n\delta/2} d\eta. \quad (7.10)$$

The integrals on the right-hand sides of (7.8), (7.9), and (7.10) are uniformly bounded in ξ if $n < k(1-\delta)$.

By (7.7) the integral $\int \|A(\xi)^* A(\eta)\|^{1/2} d\eta$ is estimated in the same way. Hence the theorem is proven for $s = m = 0$.

4°. In the general case, let Λ be an operator with symbol $(1 + |\xi|^2)^{1/2}$ so that $\|u\|_s = \|\Lambda^s u\|_0$. Then inequality (7.5) can be written as

$$\|\Lambda^s h A u\|_0 \leqslant C\|\Lambda^{s+m}u\|_0, \qquad u \in C_0^\infty(\mathbb{R}^n).$$

Clearly the estimate holds only when it is valid for all $u \in \mathcal{S}$. Further, if $u \in \mathcal{S}$, then $v = \Lambda^{s+m}u \in \mathcal{S}$. Hence, (7.5) is equivalent to the inequality

$$|\Lambda^s h A \Lambda^{-s-m} v|_0 \leqslant C\|v\|_0. \quad (7.11)$$

But by Theorem 7.1, the operator $\Lambda^s h A \Lambda^{-s-m}$ has symbol in S^0_δ; inequality (7.11) therefore holds by the above theorem.

Remark 7.1. Note that we actually make use only of finite smoothness of the symbol of the operator A. For example, in the case when $m = 0$ and s = 0, it suffices to assume that there exist derivatives $\partial_\xi^\alpha \partial_x^\beta a(x, \xi)$ for $|\alpha| \leqslant 2N$ and $|\beta| \leqslant 2k$. Here N and k are integers such that $2N > n$, $k(1 - \delta) > n$, and the estimates

$$|\partial_\xi^\alpha \partial_x^\beta a(x, \xi)| \leqslant C(1+|\xi|)^{\delta(|\beta|-|\alpha|)}$$

hold for the derivatives.

CHAPTER III. BOUNDARY PROBLEMS FOR ELLIPTIC EQUATIONS

In this chapter we shall show how boundary problems for linear elliptic differential equations of arbitrary order are solved. The general method is based on the fact that these equations can be approximated by appropriate equations with constant coefficients, for which the task of describing correct boundary problems reduces to a simple algebraic problem. Another approach, based on the reduction of a problem to a solution of a pseudo-differential equation on the boundary of the domain, is discussed in Section 3. A boundary problem is correct, or properly posed, only when the pseudo-differential equation is elliptic on the boundary.

§1. Equations with Constant Coefficients in a Half Space

1.1. We first consider a boundary problem of the type

$$P(D)u = f(x) \text{ when } x^n \geq 0, \quad n \geq 3 \tag{1.1}$$

with conditions

$$B_j(D)u = g_j(x) \text{ when } x^n = 0, \quad j = 1, \ldots, \mu. \tag{1.2}$$

Here, $P(\xi)$ is a homogeneous elliptic polynomial of order m such that the inequality

$$|P(\xi)| \geq c_0 |\xi|^m, \quad c_0 = \text{const} > 0, \quad \xi \in \mathbb{R}_n$$

is satisfied, and μ stands for the number of roots λ of the equation $P(\xi', \lambda) = 0$ with positive imaginary parts. This number is independent of vector $\xi' = (\xi_1, \ldots, \xi_{n-1})$ varying in the space $\mathbb{R}_{n-1}$, since, by the ellipticity condition, the equation $P(\xi', \lambda) = 0$ cannot have real roots when $\xi' \neq 0$. The order m_j of the differential operator B_j is assumed not to exceed m $- 1$.

The boundary problem (1.1)-(1.2) is not correct for arbitrary operators $B_j(D)$. Suppose the polynomials $B_j(\xi)$ are homogeneous in ξ, of degree m_j. Further we see that these polynomials satisfy certain conditions known as *Lopatinski conditions*.

Let us describe these conditions. To do so, we consider the following equation with constant coefficients

$$P(\xi', D_n)v = 0;$$

its general solution is written as

$$v = \sum_{j=1}^{r_0} \sum_{l=0}^{\nu_j - 1} c_{jl}(\xi')(x^n)^l e^{ix^n \lambda_l(\xi')},$$

where $\lambda_1, \ldots, \lambda_r$ are the roots of the characteristic equation, their multiplicities being equal to $\nu_1, \ldots, \nu_r$ and $\nu_1 + \ldots + \nu_r = m$. We are interested only in those exponents which are Fourier transforms in the variables x' of distributions in the half space $x^n > 0$. By this, we select roots $\lambda_1, \ldots, \lambda_{r_0}$ at which $\text{Im}\, \lambda_j > 0$. Substituting

$$v(\xi', x^n) = \sum_{j=1}^{r_0} \sum_{l=0}^{\nu_j - 1} c_{jl}(\xi')(x^n)^l e^{ix^n \lambda_l(\xi')} \tag{1.3}$$

into

$$B_j(\xi', D_n)v = g_j(\xi'), \qquad j = 1, \ldots, \mu,$$

we obtain a linear system of μ equations for μ functions c_{jl}, where $\mu = \nu_1 + \ldots + \nu_{r_0}$ is the number of roots of the characteristic equation with positive real parts, taking into account multiplicity. According to the Lopatinski condition, this system of linear equations should have a unique solution for every real vector $\xi' \neq 0$ and every function $g_1, \ldots, g_\mu$. This condition is fulfilled if and only if the determinant of the matrix of order μ, composed of the coefficients of c_{jl}, is different from zero.

Example 1.1. For an elliptic equation of second order with real coefficients, μ is always equal to unity and therefore one boundary condition $B(x, D)u = g(x)$ should be given; here the order of the operator B is equal to m. If $m = 0$ it takes the form $bu = g(x)$ and the Lopatinski condition implies that $b \neq 0$. This problem is known as the *Dirichlet problem*.

If $m = 1$, then the condition is written as $\sum_1^n b_j D_j u = g$, and the Lopatinski condition reduces to the inequality $b_n \neq 0$.

We shall say that the problem (1.1)-(1.2) is well posed provided the following conditions are fulfilled:

1°. The space of solutions u of the boundary problem

$$P(D)u = 0 \text{ when } x^n \geqslant 0; \quad B_j(D)u = 0 \text{ when } x^n = 0; \quad j = 1, \ldots, \mu \tag{1.4}$$

from the $H_m(\mathbb{R}^n_+)$ class has a finite dimension and there exists a constant C such that

$$\|v\|_m \leqslant C\left(\|P(D)v\|_0 + \sum_{j=1}^{\mu} \|B_j(D)v\|_{m-m_j-1/2} + \|v\|_0\right), \tag{1.5}$$

where $v \in H_m(\mathbb{R}^n_+)$.

2°. In the space of vector-valued functions $C_0^\infty(\overline{\mathbb{R}^n_+}) \times \prod_{j=1}^{\mu} C_0^\infty(\mathbb{R}^{n-1})$ there exists a subspace L, the complement of which has finite dimension and is such that if $(f, g_1, \ldots, g_\mu) \in L$, then the problem (1.1)-(1.2) has a solution $u \in H_m(\mathbb{R}^n_+)$.

3°. The set $\{(P(D)u, \gamma B_1(D)u, \ldots, \gamma B_\mu(D)u);\ u \in H_m(\mathbb{R}^n_+)\}$ is closed in $\{H_0(\mathbb{R}^n_+) \times H_{m-m_1-1/2}(\mathbb{R}^{n-1}) \times \ldots \times H_{m-m_\mu-1/2}(\mathbb{R}^{n-1})\}$. Here γ is the restriction to the plane $x^n = 0$.

1.2. Theorem 1.1. In order for the problem (1.1)-(1.2) to be well posed it is necessary and sufficient that the Lopatinski condition be fulfilled.

Proof. We shall prove that the *a priori* estimate (1.5) cannot hold when the Lopatinski condition is not satisfied.

Suppose this condition is violated when $\xi' = \xi_0' \neq 0$ so that there exists a function $v(x^n)$ which decreases as $x^n \to +\infty$ and satisfies the equation

$$P(\xi_0', D_n)v = 0 \text{ when } x^n > 0$$

and the conditions

$$B_j(\xi_0', D_n)v = 0 \text{ when } x^n = 0, \qquad j = 1, \ldots, \mu.$$

Assume that

$$u_\lambda(x) = \varphi_1(x')\varphi_2(x^n)e^{i\lambda(x', \xi_0')}v(\lambda x^n),$$

where $\lambda>0$, $\varphi_1\in C_0^\infty(\mathbb{R}^{n-1})$, $\varphi_2\in C_0^\infty(\mathbb{R})$ and $|\varphi_1\varphi_2|\leqslant 1$; the function $\varphi_1\varphi_2$ is equal to unity in some neighborhood ω of the origin in $\mathbb{R}^n$.

Now insert u_λ in (1.5)

$$\|u_\lambda\|_m\leqslant C\left(\|Pu_\lambda\|_0+\sum_{j=1}^{\mu}\|B_ju_\lambda\|_{m-m_j-1/2}+\|u\|_0\right).$$

Let $\alpha=(\alpha_1,\dots\ \alpha_{n-1},\ 0)$ and $|\alpha|=m$. Then

$$\|D^\alpha u_\lambda\|_0^2\geqslant\lambda^{2m-1}\int|\varphi_1(x')|^2dx'\int|v(t)|^2dt-C_1\lambda^{2m-3}\geqslant c_0^2\lambda^{2m-1},$$

if λ is sufficiently small. Therefore

$$\|u_\lambda\|_m\geqslant c_0\lambda^{m-1/2},\qquad c_0>0.$$

On the other hand, it is clear that $\|u_\lambda\|\leqslant$ const.

Since $P(D)[e^{i\lambda(x',\xi_0')}v(\lambda x^n)]=0$ we obtain

$$\|P(D)u_\lambda\|_0\leqslant C_2\lambda^{m-1}.$$

Similarly, from the identity $B_j(D)[e^{i\lambda(x',\xi_0')}v(\lambda x^n)]=0$ when $x^n=0$, it follows that

$$\|B_j(D)u_\lambda\|_{m-m_j-1/2}^{x^n=0}\leqslant C_3\lambda^{m-1}.$$

Thus, if (1.5) is satisfied for all functions u_λ, then

$$c_0\lambda^{m-1/2}\leqslant(C_2+C_3)\lambda^{m-1}+C_4.$$

This is impossible. The contradiction obtained reveals that (1.5) cannot be fulfilled if the Lopatinski condition is violated.

1.3. Let $u\in H_m(\mathbb{R}_+^n)$ be a solution of problem (1.4). Let $v(\xi',x^n)$ be the Fourier transform of u in the variable x'. Then by (1.4) we have

$$P(\xi',D_n)v=0 \text{ when } x^n>0 \text{ and } B_j(\xi',D_n)v=0$$

when $x^n=0$, $j=1,\dots,\mu$. The general solution of the equation $P(\xi',D_n)v=0$ in $L_2(\mathbb{R}_+^n)$ looks like (1.3). Therefore, from the Lopatinski con-ition it follows that $v=0$ and hence $u=0$.

We next prove that the problem (1.1)-(1.2) has a solution $u\in H_m(\mathbb{R}_+^n)$ if the Lopatinski condition is satisfied and $f\in L_2(\mathbb{R}_+^n)$, $g_j\in H_{m-m_j-1/2}(\mathbb{R}^{n-1})$, and

$$\|u\|_m\leqslant c\left(\|f\|_0+\sum_{j=1}^{\mu}\|g_j\|_{m-m_j-1/2}+\|u\|_0\right).$$

Extending the function f up to a function in $L_2(\mathbb{R}^n)$ we obtain the elliptic equation $P(D)u=f$ with constant coefficients in $\mathbb{R}^n$ considered in Section 4.3 of Chapter I. Such an equation always has a solution u_1 in $H_m(\mathbb{R}^n)$. The substitution $v=u-u_1$ reduces the problem (1.1)-(1.2) to the problem

$$P(D)v=0\quad\text{when}\quad x^n>0,\qquad B_j(D)v=h_j\quad\text{when}\quad x^n=0,$$

where $h_j=g_j-\gamma B_j(D)u_1\in H_{m-m_j-1/2}(\mathbb{R}^{n-1})$ and

$$\|h_j\|_{m-m_j-1/2}\leqslant\|g_j\|_{m-m_j-1/2}+C_1\|f\|_0.$$

Applying a Fourier transformation in x' and assuming

$$\tilde v(\xi',x^n)=\int v(x)e^{-ix'\xi'}dx',\quad \tilde h_j(\xi')=\int h_j(x')e^{-ix'\xi'}dx',$$

we obtain an ordinary differential equation

$$P(\xi',D_n)\tilde v=0\quad\text{when}\quad x^n>0 \tag{1.6}$$

with the condition

$$B_j(\xi', D_n)\tilde{v} = \tilde{h}_j \text{ when } x^n = 0, \; j = 1, \ldots, \mu. \tag{1.7}$$

By the Lopatinski condition, the linear space of solutions $\tilde{v}$ of equation (1.5) which decrease exponentially as $x^n \to +\infty$ has μ dimensions and the mapping

$$\tilde{v} \to \{\gamma B_1(\xi', D_n)\tilde{v}, \ldots, \gamma B_\mu(\xi', D_n)\tilde{v}\}$$

is one-to-one. Hence, the problem (1.6)-(1.7) has a unique solution in the given class. Since this solution and its derivatives decrease exponentially, the inequality

$$\sum_{j=0}^{m}\int_0^\infty |D_n^j\tilde{v}|^2\,dx^n + \sum_{j=0}^{m-1}|D_n^j\tilde{v}(\xi', 0)|^2 \leqslant C_2(\xi')\sum_{j=1}^{\mu}|\tilde{h}_j(\xi')|^2$$

holds so that

$$\sum_{j=0}^{m}\int_0^\infty |D_n^j\tilde{v}|^2\,dx^n + \sum_{j=0}^{m-1}|D_n^j\tilde{v}(\xi', 0)|^2 \leqslant C(\xi')\left[\int_0^\infty |\tilde{f}(\xi', x^n)|^2\,dx^n + \sum_{j=1}^{\mu}|\tilde{g}_j(\xi')|^2\right],$$

where the function $C(\xi')$ is continuously dependent on ξ'. If $C_0 = \max C(\xi')$ when $|\xi'| = 1$, then it can be concluded, using homogeneity, that

$$\sum_{j=0}^{m}|\xi'|^{2(m-j)}\int_0^\infty |D_n^j\tilde{v}(\xi', x^n)|^2\,dx^n + \sum_{j=0}^{m-1}|\xi'|^{2(m-j)-1}|D_n^j\tilde{v}(\xi', 0)|^2 \leqslant$$

$$C_0\left\{\int_0^\infty |\tilde{f}(\xi', x^n)|^2\,dx^n + \sum_{j=1}^{\mu}|\xi'|^{2(m-m_j)-1}|\tilde{g}_j(\xi')|^2\right\}.$$

Adding the integral $\int|\tilde{v}|^2\,dx^n$ to the left and right-hand sides and integrating the inequality with respect to ξ', we obtain the estimate

$$\|v\|_m^2 + \|v(x', 0)\|_{m-1/2}^2 \leqslant C_0\left(\|f\|_0^2 + \sum_{j=1}^{\mu}\|g_j\|_{m-m_j-1/2}^2 + \|v\|_0^2\right) \tag{1.8}$$

for $m > m_j$.

This shows that the boundary problem (1.1)-(1.2) is well posed.

1.4. The boundary problem (1.1)-(1.2) has a special form; it is of interest only as a first step to constructing a parametrix of a general problem for an equation with variable coefficients.

By H we denote the direct product of spaces $L_2(\mathbb{R}_+^n)$ and $H_{m-m_j-1/2}(\mathbb{R}^{n-1})$ when $j = 1, \ldots, \mu$. Let $A\colon H_m(\mathbb{R}_+^n) \to H$ so that $Au = (Pu, \gamma B_1 u, \ldots, \gamma B_\mu u)$. Then a *left* (respectively, *right*) parametrix of the boundary problem

$$Pu = f \text{ in } \mathbb{R}_+^n, \quad B_1 u = g_1, \ldots, B_\mu u = g_\mu \text{ when } x^n = 0$$

is an operator $E\colon H \to H_m(\mathbb{R}_+^n)$ such that

$$EAu = u + Tu, \quad u \in H_m(\mathbb{R}_+^n) \text{(respectively, } AEf = f + T_1 f, \quad f \in H)$$

where T and T_1 are compact operators in the spaces $H_m(\mathbb{R}_+^n)$ and H, respectively. If the operator E is such that T and $T_1 = 0$, then it is called a fundamental solution of the boundary problem.

The above-mentioned consideration makes it possible to construct a fundamental solution, although a parametrix suffices for our purpose. Using a Fourier transformation and solving a boundary problem for an ordinary differential equation, we construct an operator $E\colon H \to H_m(\mathbb{R}_+^n)$ for which $AEf = f$. On the other hand, assuming $f = Au$, where $u \in H_m(\mathbb{R}_+^n)$ we obtain $AEAu = AU$, and, since A is an invertible operator, $EAu = u$. This implies that E is a fundamental solution.

§2. Boundary Problem for an Equation with Variable Coefficients

2.1. Let us now consider a boundary problem for the elliptic operator $P(x, D)$ of order m in an arbitrarily bounded domain Ω of the space $\mathbb{R}^n$ with smooth boundary $\partial\Omega$. Assume the boundary $\partial\Omega$ to be composed of a finite number of connected components.

The number of boundary conditions μ is determined for every connected component of the boundary as the number of roots τ of the equation $P_m(x, \xi_x + \tau\nu_x) = 0$ with positive imaginary parts, where ν_x stands for the unit vector normal to the boundary and directed into the domain Ω, and the vector ξ_x is tangent to the boundary at a point x. Since the equation cannot have real roots, μ is neither dependent on x, which moves along a connected set, nor on the vector ξ_x.

We say that the boundary conditions $B_j(x, D)u = f_j(x)$ for $x \in \partial\Omega$, $j = 1, \ldots, \mu$, satisfy the Lopatinski condition at a point x_0 provided the conditions are satisfied for a system of operators $(P(x_0, D), B_1(x_0, D), \ldots, B\mu(x_0, D))$ in a local system of coordinates in which x_0 coincides with the origin, and the equation of a boundary portion in a neighborhood of x_0 is of the form $x^n = 0$.

Theorem 2.1. Let the Lopatinski condition be satisfied for a boundary problem

$$P(x, D)u = f \text{ in } \Omega, \tag{2.1}$$

$$B_j(x, D)u = g_j \text{ when } x \in \partial\Omega,\ j = 1, \ldots, \mu, \tag{2.2}$$

and let $f \in L_2(\Omega)$, $g_j \in H_{m-m_j-1/2}(\partial\Omega)$. There exists a right parametrix $E\colon\ H \to H_m(\Omega)$, where $H = H_0(\Omega) \times H_{m-m_1-1/2}(\partial\Omega) \times \ldots \times H_{m-m_\mu-1/2}(\partial\Omega)$, and a constant C such that if $u = E(f, g_1, \ldots, g_\mu)$, then

$$\|u\|_m \leqslant C\left(\|f\|_0 + \sum_{j=1}^{\mu} \|g_j\|_{m-m_j-1/2} + \|u\|_0\right). \tag{2.3}$$

Proof. Let $\varepsilon > 0$ be a positive number and $\delta = \delta(\varepsilon)$ be such that the values of the coefficients of the operators P_m and B_j^0 differ by no more than $\varepsilon/2$ at every two points belonging to Ω, the distance between which is less than δ.

Suppose $\varphi_j \in C_0^\infty(\mathbb{R}^n)$, $\varphi_j \geqslant 0$, $\sum \varphi_j(x) = 1$, the diameters of the supports of functions φ_j do not exceed $\delta/2$, and at every point no more than N φ_j-functions are different from zero. Also consider another set of functions $\{\psi_j\}$, where $\psi_j \geqslant 0$, $\psi_j \in C_0^\infty(\mathbb{R}^n)$, diam supp $\psi_1 < d/2$, and $\psi_j = 1$ in supp φ_j. Denote supp ψ_j by ω_j.

Without loss of generality we may assume that in a neighborhood ω_j having common points with the boundary $\partial\Omega$ it is possible to introduce local coordinates $y^1, \ldots, y^n$ in which the part of the boundary $\partial\Omega$ falling in ω_j has an equation $y^n = 0$ and ω_j lies in the range $y^n > 0$.

If neighborhood ω_j has no common points with the boundary $\partial\Omega$, then we choose an arbitrary point x_j belonging to ω_j and construct a parametrix E_j for operator $P(x_j, D)$ with constant coefficients in the entire space.

If neighborhood ω_j intersects $\partial\Omega$, then we choose a point x_j from $\omega_j \cap \partial\Omega$ and a system of local coordinates $y^1, \ldots, y^n$ so that x_j is their origin, the boundary $\omega_j \cap \partial\Omega$ has an equation $y^n = 0$ and the domain ω_j lies in the half-space $y^n > 0$.

Let us now construct a parametrix for the boundary problem

$$P(x_j, D)u = f \text{ when } y^n > 0, \quad B_l(x_j, D)u = g_l \text{ when } y^n = 0,\ l = 1, \ldots, \mu,$$

as in Section 1. Going back to the previous coordinates x, we obtain an operator E_j with the property that if

$$u_j = E_j(\psi_j f,\ \psi_j g_1,\ \ldots,\ \psi_j g_\mu),$$

then

$$P_m(x_j,\ D)\,u_j = \psi_j f + T_j(f,\ g_1,\ \ldots,\ g_\mu),$$

$$B_l^0(x_j,\ D)\,u_j = \psi_j g_l + T_{lj}(f,\ g_1,\ \ldots,\ g_\mu),\quad l = 1,\ \ldots,\ \mu,$$

where

$$T_j \in L(H,\ H_1(\Omega)),\ T_{lj} \in L(H,\ H_{m-m_l+1/2}(\partial\Omega)).$$

We now assume $E_0 = \sum \varphi_j E_j \psi_j$. This is a linear operator and $E_0 \in L(H, H_m(\Omega))$. Further, if $F = (f, g_1, \ldots, g_\mu) \in H$ we assume $u = E_0 F$. Then

$$P(x,\ D)\,u = \sum P(x,\ D)\,\varphi_j E_j \psi_j F = \sum \varphi_j P(x,\ D)\,E_j \psi_j F + T_1 F =$$

$$\sum \varphi_j P_m(x_j,\ D)\,E_j \psi_j F + \sum \varphi_j\,[P(x,\ D) - P_m(x_j,\ D)]\,E_j \psi_j F + T_1 F =$$

$$\sum \varphi_j \psi_j f + T_2 F + T_3 F = f + T_2 F + T_3 F,$$

where T_1 and T_2 are operators belonging to $L(H, H_1(\Omega))$, $T_3 \in L(H, L_2(\Omega))$, and $\|T_3\| \leqslant C_1 \varepsilon$, C_1 being independent of ε.

Further, at points of the boundary $\partial\Omega$

$$B_l(x,\ D)\,u = \sum B_l(x,\ D)\,\varphi_j E_j \psi_j F = \sum \varphi_j B_l(x,\ D)\,E_j \psi_j F + T_{l1} F =$$

$$\sum \varphi_j B_l^0(x_j,\ D)\,E_j \psi_j F + \sum \varphi_j\,[B_l(x,\ D) - B_l^0(x_j,\ D)]\,E_j \psi_j F + T_{l1} F =$$

$$\sum \varphi_j \psi_j g_l + T_{l2} F + T_{l3} F = g_l + T_{l2} F + T_{l3} F,$$

where T_{l1} and T_{l2} are operators belonging to $L(H, H_{m-m_l+1/2}(\partial\Omega))$, $T_{l3} \in L(H, H_{m-m_l-1/2}(\partial\Omega))$, and $\|T_{l3}\| \leqslant C_1\varepsilon$, C_1 being independent of ε.

Thus, if $A \in L(H_m(\Omega), H)$ and $Au = (Pu, \gamma B_1 u, \ldots, \gamma B_\mu)$, then

$$AE_0 F = F + TF + T'F,$$

where $T \in L(H, H')$, $T' \in L(H, H)$, $H' = H_1(\Omega) \times H_{m-m_l+1/2}(\partial\Omega) \times \ldots \times H_{m-m_\mu+1/2}(\partial\Omega)$ and $\|T'\| \leqslant C_0 \varepsilon$. It follows that for sufficiently small ε the operator $E = E_0(I + T')^{-1}$ is a right parametrix of the boundary problem.

Inequality (2.3) follows immediately from the construction of the parametrix, since

$$E \in L(H,\ H_m(\Omega)).$$

Theorem 2.1 is proven.

Remark 2.1. From the proof it is clear that the operators $P - P_m$ and $B_l - B_l^0$ need not be differential operators. If they belong to $L(H_m(\Omega), H_1(\Omega))$ and $L(H_m(\Omega), H_{m-m_l+1/2}(\partial\Omega))$, respectively, then the above construction of the parametrix holds.

Remark 2.2. From the proof it is clear that the construction is also valid when the operator P is defined on a compact n-dimensional manifold with boundary, the boundary being a smooth manifold of n − 1 dimensions, and the operators B_l are defined on this boundary manifold.

2.2. We now show that the boundary problem (2.1)-(2.2) can be also studied when $f \in H_{s-m}(\Omega)$, $g_l \in H_{s-m_l-1/2}(\partial\Omega)$ for any real s, $s > \max m_j + 1/2$.

Theorem 2.2. The parametrix constructed earlier is an operator from

$$H^{(s)}(\Omega) = H_s(\Omega) \times H_{s+m-m_1-1/2}(\partial\Omega) \times \ldots \times H_{s+m-m_\mu-1/2}(\partial\Omega)$$

to $H_{s+m}(\Omega)$ for any real s, $s > \max m_j + 1/2 - m$. In this case, if $u = E(f, g_1, \ldots, g_\mu)$, then

$$\|u\|_{s+m} \leqslant C(s)\,[\|f\|_s + \Sigma \|g_j\|_{s+m-m_j-1/2} + \|u\|_s].$$

Proof. 1°. We first consider the problem (1.1)-(1.2) in a half-space

Let $F = (f, g_1, \ldots, g_\mu)$ be a vector in $H^{(s)}(\Omega)$. Let Λ_s be a pseudo-differential operator acting on a function of $n-1$ variables $x^1, \ldots, x^{n-1}$, with symbol $(1 + \xi_1^2 + \ldots + \xi_{n-1}^2)^{s/2}$. $\Lambda_s F$ is contained in H and therefore $E\Lambda_s F \in H_m(\Omega)$. However, $\Lambda_s A = A\Lambda_s$ and therefore $E\Lambda_s = \Lambda_s E$ so that $E = \Lambda_s^{-1} E \Lambda_s$. Hence the function $u = EF$ is such that $\Lambda_s u \in H_m(\mathbb{R}_+^n)$.

Let $0 < s \leqslant 1$. Then it follows that $D_n^j D'^\alpha u \in H_s(\mathbb{R}_+^n)$ for $|\alpha| \leqslant m - j$, $m > j \geqslant 1$. By the identity

$$Pu = f \in H_s(\mathbb{R}_+^n)$$

it follows that $D_n^m u \in H_s(\mathbb{R}_+^n)$, and therefore $u \in H_{m+s}(\mathbb{R}_+^n)$.

If $1 < s \leqslant 2$, then, by what has been proven, $\Lambda_{s-1} u \in H_{m+1}(\mathbb{R}_+^n)$. Repeating the same arguments we find that $u \in H_{m+s}(\mathbb{R}_+^n)$, and so on.

Thus the operator $E \in L(H^{(s)}(\mathbb{R}_+^n), H_{s+m}(\mathbb{R}_+^n))$, which proves the theorem.

2°. In an arbitrary domain $E_0 = \Sigma \varphi_j E_j \psi_j$, where $E_j \in L(H^{(s)}(\Omega), H_{s+m}(\Omega))$, as in the proof of Theorem 2.1. Hence $E \in L(H^{(s)}(\Omega), H_{s+m}(\Omega))$. Since $E = E_0(1 + T')^{-1}$, where $T' \in L(H^{(s)}(\Omega), H^{(s)}(\Omega))$, $\|T'\| \leqslant C_0\varepsilon$ (see the proof of Theorem 2.1), we find that $E \in L(H^{(s)}(\Omega), H_{s+m}(\Omega))$. The theorem is proven.

Corollary 2.1. If the Lopatinski condition is satisfied, then the solutions of the homogeneous boundary problem $Pu = 0$ in Ω; $B_j u = 0$ on Γ, $j = 1, \ldots, \mu$ form a finite-dimensional subspace of $C^\infty(\overline{\Omega})$.

Proof. If E is a parametrix, then $EA = I + T$, and if $Au = 0$, then $u = -Tu$, where T is an operator of order -1. It follows that the unit sphere in the space of solutions of the homogeneous boundary problem is compact and therefore this is a finite-dimensional space. From the identity $u = -Tu$ it follows that if $u \in H_s(\Omega) \cap \ker A$, then $u \in H_{s+1}(\Omega)$ and therefore $u \in \cap H_s(\Omega)$, that is $u \in C^\infty(\overline{\Omega})$. This proves the corollary.

Corollary 2.2. If the Lopatinski condition is satisfied, then the range of the operator A: $H_{s+m}(\Omega) \to H^{(s)}(\Omega)$ is closed.

Proof. If $u \in H_{s+m}(\Omega)$, then by Theorem 2.2 $EAu = u + Tu$ so that

$$\|u\|_{s+m} \leqslant C(\|Au\|_{H^{(s)}} + \|u\|_s) \leqslant C_1\left(\|Pu\|_s + \sum_{j=1}^{\mu} \|B_j u\|_{s+m-m_j-1/2} + \|u\|_s\right).$$

From this we see that if $u_k \in H_{s+m}(\Omega)$, $Pu_k = f_k \in H_s(\Omega)$, $B_j u_k = g_{jk} \in H_{s+m-m_j-1/2}(\partial\Omega)$ and $f_k \to f$ in $H_s(\Omega)$, $g_{jk} \to g_j$ in $H_{s+m-m_j-1/2}(\partial\Omega)$ as $k \to \infty$, then $\{u_k\}$ converges in $H_{s+m}(\Omega)$ to a function u and $u \in R(A)$. This proves the corollary.

Corollary 2.3. If the Lopatinski condition is satisfied, then there exist functions $v_1, \ldots, v_N$ in $C^\infty(\partial\Omega)$ and $w_{j1}, \ldots, w_{jN}$ of $C^\infty(\partial\Omega)$ for $j = 1, \ldots, \mu$, such that if $f \in H_s(\Omega)$,

$$g_j \in H_{s+m-m_j-1/2}(\partial\Omega),\ s > \max m_j + 1/2 - m \text{ and}$$

$$\int_\Omega f v_l\, dx + \sum_{j=1}^{\mu} \int_{\partial\Omega} g_j w_{jl}\, dx = 0, \quad l = 1, \ldots, N,$$

then there exists a solution of the boundary problem (2.1)-(2.2) $u \in H_{s+m}(\Omega)$ and

$$\|u\|_{s+m} \leqslant C(\|f\|_s + \Sigma \|g_j\|_{s+m-m_j-1/2}).$$

Proof. From $AE = I + T'$, where $T' \in L(H^{(s)}(\Omega), H^{(s+1)}(\Omega)$, it follows that $E^*A^* = I + T'^*$. Therefore, as above, we find that dim ker $A^* < \infty$ and ker $A^* \subset \bigcap_s H^{(s)}(\Omega)$. The basis $G_1, \ldots, G_N$ in ker A^* determines the desired vector-valued function $G_l = (v_l, w_{1l}, \ldots, w_{\mu l})$, $l = 1, \ldots, N$. The solvability conditions of the equation $Au = F$ are equivalent to the orthogonality conditions $(F, G_l) = 0$, $l = 1, \ldots, N$, since the range of the operator A is closed (see, for example, Browder[1]).

Corollary 2.4. Let the Lopatinski condition be satisfied and $u \in H_{s_0}(\Omega)$, where $s_0 > \max m_j + 1/2$, be a solution to (2.1)-(2.2). If $s > s_0$ and $f \in H_{s-m}(\Omega)$, $g_j \in H_{s-m_j-1/2}(\partial\Omega)$, then $u \in C^\infty(\bar{\Omega})$.

Proof. Since $Au = (f, g_1, \ldots, g_\mu)$ we obtain immediately that $u + Tu = E(f, g_1, \ldots, g_\mu) \in H_s(\Omega)$. As the operator T belongs to $L(H_t(\Omega), H_{t+1}(\Omega))$ for all $t \geqslant s_0$, it follows that $u \in H_s(\Omega)$, and the corollary is proven.

§3. Reduction of a Boundary Problem to a Pseudo-Differential Equation

3.1. In this section we describe another approach to studying a boundary problem for elliptic equations. As an example we shall consider second-order equations. This approach resembles in many ways the method of potentials and enables the solution of a boundary problem to be reduced to the solution of an integro-differential equation on a boundary manifold. In doing so, the Lopatinski conditions prove to be the ellipticity conditions for the equation.

We now consider a boundary problem

$$Lu \equiv \sum_{j,\,l=1}^{n} a_{jl}(x)\, D_j D_l u + \sum_{j=1}^{n} b_j(x)\, D_j u + c(x)\, u = f(x), \quad x \in \Omega, \tag{3.1}$$

$$Bu(x) \equiv \sum_{j=1}^{n} \alpha_j(x)\, D_j u + \beta(x)\, u = g(x), \quad x \in \partial\Omega. \tag{3.2}$$

All the coefficients of operators L and B are assumed to be smooth complexed-valued functions defined on $\mathbb{R}^n$,

$$\sum_{j,\,l=1}^{n} a_{jl}(x)\,\xi_j\xi_l \geqslant c_0 |\xi|^2, \text{ where } c_0 = \text{const} > 0, \quad \sum_{j=1}^{n} |\alpha_j|^2 + |\beta|^2 \neq 0.$$

3.2. If a smooth function v is defined on $\bar{\Omega}$, we denote by v^0 a function equal to v in Ω and equal to zero outside Ω. By v_0 we denote the trace of function v on $\partial\Omega$, and by v_1, the trace of the derivative of v along the direction of the outward normal on $\partial\Omega$.

The derivatives of v^0 are smooth outside $\partial\Omega$ but may contain distributions of the type $\delta(\partial\Omega)$ and its derivatives. In order to compute these derivatives, it is convenient to straighten a portion of the boundary $\partial\Omega$ in a neighborhood of a given point.

Let x_0 be a point on the boundary $\partial\Omega$, $x^n = 0$ be the equations of the boundary manifold in the neighborhood ω of x_0. The domain $\Omega \cap \omega$ lies in the half-space $x^n > 0$. If v is a smooth function in $\bar{\Omega}$ equal to 0 outside ω, then

$$D_n v^0 = (D_n v)^0 - i\delta(x^n) \otimes v_0(x'), \quad D_j v^0 = (D_j v)^0 \text{ for } j = 1, \ldots, n-1.$$

The second derivatives are computed similarly:

$$D_n^2 v^0 = (D_n^2 v)^0 - \delta(x^n) \otimes v_1(x') - \delta'(x^n) \otimes v_0(x'),$$

$$D_n D_j v^0 = (D_n D_j v)^0 - i\delta(x^n) \otimes D_j v_0(x'), \quad j = 1, \ldots, n-1,$$

$$D_jD_lv^0=(D_jD_lv)^0, \quad j,\ l=1,\ \dots,\ n-1.$$

Let the operator L have the following form in the chosen system of coordinates:

$$L(x,\ D)=D_n^2+2\sum_{j=1}^{n-1}a_{jn}(x)D_jD_n+\sum_{j,\ l=1}^{n-1}a_{jl}(x)D_jD_l+\sum_{j=1}^{n}b_j(x)D_j+c(x).$$

Then from the formulas given earlier it is seen that

$$Lv^0(x)=(Lv)^0(x)-\delta(x^n)\otimes v_1(x')-\delta'(x^n)\otimes v_0(x')-$$

$$2i\sum_{j=1}^{n-1}a_{jn}(x)\,\delta(x^n)\otimes D_jv_0(x')-i\delta(x^n)\otimes v_0(x')\cdot b_n(x). \tag{3.3}$$

We apply the parametrix E for the operator L defined in the entire space $\mathbb{R}^n$ to both sides of (3.3). Let

$$Ev(x)=(2\pi)^{-n}\int e(x,\ \xi)\,\tilde{v}(\xi)\,e^{ix\xi}\,d\xi.$$

Then

$$E(\delta(x^n)\otimes w(x'))=(2\pi)^{-n}\int e(x,\ \xi)\,\tilde{w}(\xi')\,e^{ix\xi}\,d\xi=(2\pi)^{-n+1}\int k(x,\ \xi')\,\tilde{w}(\xi')\,e^{ix'\xi'}\,d\xi',$$

$$E(\delta'(x^n)\otimes w(x'))=(2\pi)^{-n}\int e(x,\ \xi)\,(i\xi_n)\,\tilde{w}(\xi')\,e^{ix\xi}\,d\xi=(2\pi)^{-n+1}\int k_1(x,\ \xi')\,\tilde{w}(\xi')\,e^{ix'\xi'}\,d\xi',$$

where

$$k(x,\ \xi)=\frac{1}{2\pi}\int e(x,\ \xi)\,e^{ix^n\xi_n}\,d\xi_n, \qquad k_1(x,\ \xi')=-\frac{1}{2\pi i}\int \xi_n e(x,\ \xi)\,e^{ix^n\xi_n}\,d\xi_n.$$

As in Lemma 3.1 of Chapter II, the symbol $e(x,\ \xi)$ is constructed as an asymptotic decreasing power series in $|\xi|$. Since $L(x,\ \xi)$ is a second-degree polynomial in ξ, the principal parts with symbols k and k_1 are computed easily. In fact, when $|\xi|>1$ the principal symbol e^0 of the parametrix E takes the form

$$e^0(x,\ \xi)=\left[\xi_n^2+2\xi_n\sum_{j=1}^{n-1}a_{jn}(x)\,\xi_j+\sum_{j,\ l=1}^{n-1}a_{jl}(x)\,\xi_l\xi_j\right]^{-1}=[(\xi_n-\alpha)(\xi_n-\beta)]^{-1},$$

where $\alpha(x,\ \xi')$ and $\beta(x,\ \xi')$ are the roots of polynomial $(e^0)^{-1}$. Thus

$$k^0(x,\ \xi')=\frac{1}{2\pi}\int\frac{e^{ix^n\xi_n}}{(\xi_n-\alpha)(\xi_n-\beta)}\,d\xi_n, \qquad k_1^0(x,\ \xi')=-\frac{1}{2\pi i}\int\frac{\xi_n e^{ix^n\xi_n}}{(\xi_n-\alpha)(\xi_n-\beta)}\,d\xi_n.$$

According to Jordan's lemma, for $x^n>0$ the contour of integration can be deformed so that it will lie in the upper half-plane $\operatorname{Im}\xi_n\geqslant 0$. The contour Γ is made up of a segment of the real axis and a semicircle and contains those points $\xi_n=\alpha$ and $\xi_n=\beta$ for which the real parts are positive. These integrals, when $x^n\to+0$, are computed using residues.

The boundary problem (3.1)-(3.2) can be well posed only when $\mu=1$, and we should assume, for example, that $\operatorname{Im}\alpha<0$ and $\operatorname{Im}\beta>0$. This happens, for instance, if the coefficients a_{jl} are real. If $\operatorname{Im}\alpha<0$ and $\operatorname{Im}\beta>0$ for all $x=(x',\ 0)$ and $|\xi'|>1$, then

$$k^0(x,\ \xi')=\frac{i}{\beta-\alpha}, \quad k_1^0(x,\ \xi')=\frac{-\beta}{\beta-\alpha}. \tag{3.4}$$

Thus, applying the operator E to both sides of (3.3) and assuming $x^n=0$, we obtain

$$u_0+(Tu)_0=(Ef^0)_0-Ku_1-K_1u_0-2iK\left(\sum_{j=1}^{n-1} a_{jn}D_ju_0\right)-iK(b_nu_0). \tag{3.5}$$

Since, by (3.4), $|k^0(x, \xi')| \geqslant C(1+|\xi'|)^{-1}$ when $|\xi'| \geqslant C_1$, the operator K is an elliptic operator of order -1 and equation (3.5) may be solved for u_1 so that $u_1 = Au_0 + A_1f$ and the principal symbol of A takes the form $i(-\alpha-2\sum a_{jn}\xi_j)$. Substituting the result into the boundary condition (3.2):

$$\frac{1}{i}\alpha_n(x)\,u_1+\sum_{i=1}^{n-1}\alpha_j(x)\,D_ju_0+\beta(x)\,u_0=g,$$

gives an equation for u_0 of the form $Pu_0 = F$, where F is a known function that can be expressed in terms of f and g. The last equation will be elliptic if

$$\alpha_n(x)\left(-\alpha-2\sum a_{jn}\xi_j\right)+\sum\alpha_j(x)\,\xi_j\neq 0, \qquad \xi\in\mathbb{R}_{n-1}.$$

If a_{jn} and α_j have real values, then this condition is satisfied when $\alpha_n(x) \neq 0$ since $\mathrm{Im}(2\beta-\alpha) \neq 0$.

Using a partition of unity, the problem can similarly be reduced in the entire range to an equation on a boundary manifold. For equation (3.1) with real coefficients and a real boundary condition (3.2), the problem is elliptic, provided the coefficient appearing in (3.2) for the derivative with respect to the direction of the normal to $\partial\Omega$ is different from zero.

3.3. A similar construction can be applied for the boundary problem (2.1)-(2.2) for an elliptic equation of order m. In this case, μ conditions are given at the boundary, which determine the relations between the functions $u_0, u_1, \ldots, u_{m-1}$ that are traces of the function u and its $(m-1)$th normal derivatives. The missing $m-\mu$ equations are obtained from an equality of the type (3.3) after applying the operator E to both sides and differentiating it $m-\mu-1$ times. The Lopatinski conditions in this case are equivalent to the ellipticity condition of the obtained system of pseudo-differential equations at the boundary.

This method can be applied to studying solvability conditions and to prove theorems on the smoothness of solutions of a boundary problem. For example, if the boundary problem (3.1)-(3.2) satisfies the Lopatinski condition, then from Theorem 3.6 of Chapter II it follows that the index of the problem is equal to zero when $n \geqslant 3$, that is, the dimensions of the kernel and the cokernel agree.

CHAPTER IV. THE OBLIQUE DERIVATIVE PROBLEM

Here, we shall consider a special type of problem for an elliptic second-order equation. As a matter of fact, this problem is essentially similar to the problem (3.1)-(3.2) of Chapter III, but with the difference that the Lopatinski condition may be violated at certain points of the boundary manifold. On reducing it to a pseudo-differential equation on a boundary manifold we obtain an equation of principal type which is not elliptic. In the general case, the kernel and cokernel of such an equation have infinite dimension. In the following chapters we shall show that not only is the oblique derivative problem an interesting example of an operator of principal type, but the methods described in this chapter may also be used for studying general operators of principal type.

§1. Statement of the Problem. Examples

1.1. Let Ω be a bounded domain in $\mathbb{R}^n$, where $n \geqslant 3$, with a smooth boundary Γ. In this domain we consider the equation

$$Lu \equiv \sum_{i,j=1}^{n} a^{ij}(x) \frac{\partial^2 u}{\partial x^i \partial x^j} + \sum_{i=1}^{n} b^i(x) \frac{\partial u}{\partial x^i} + c(x) u = f(x) \tag{1.1}$$

with smooth real coefficients. Equation (1.1) is assumed to be elliptic, that is

$$\sum_{i,j=1}^{n} a^{ij}(x) \xi_i \xi_j \geqslant c_0 |\xi|^2, \quad c_0 = \text{const} > 0.$$

We are given a smooth field $l = \sum_{j=1}^{n} \alpha^j(x) \frac{\partial}{\partial x^j}$ on the boundary Γ; the field may be tangent to the boundary at points forming disjoint smooth submanifolds $\Gamma_{01}, \ldots, \Gamma_{0N}$ of dimension $n - 2$, but is not tangent to the submanifolds themselves. Also the field l can be tangent to the boundary at arbitrary submanifolds of lower dimensions which are at a positive distance from $\Gamma_{01}, \ldots, \Gamma_{0N}$.

The problem consists in solving (1.1) with the condition that

$$Bu \equiv lu + k(x) u = g(x) \quad \text{on} \quad \Gamma, \tag{1.2}$$

where l is the smooth field $l = \sum_{j=1}^{n} \alpha^j(x) \partial/\partial x^j$ and k is a smooth function on Γ.

Note that the substitution $u = av$ (a is chosen in a particular manner) reduces the problem to the case $k = 0$. For this it is sufficient that the equality $la + k = 0$ be fulfilled. Here a is such that $c \geqslant a(x) \geqslant c^{-1} > 0$; this function does not affect the validity of all other arguments. Therefore, in future discussion we shall assume that $k = 0$ and $Bu = lu$.

1.2. Depending on the structure of l in a neighborhood of every submanifold Γ_{0j}, this submanifold may be placed into one of the three classes.

Let ν be a unit vector pointing in the outer normal direction to Γ. Every manifold Γ_{0j} is orientable in Γ by means of l. Let $b = (l, \nu)$ be the scalar product of two fields and let A be an arbitrary point on Γ_{0j}. If in some neighborhood of a point on Γ, b assumes negative values on the negative side of Γ_{0j} and positive values on the positive side, then we say that point A is in class I. If, on the contrary, b assumes positive values on the negative side of Γ_{0j} and negative values on the positive side, then we place A in class II. Finally, if b retains its sign in some neighborhood of point A, then A is placed into class III.

It is not hard to see that any two points belonging to one submanifold Γ_{0j} always belong to the same class, since l cannot be tangent to the manifold Γ_{0j} and (l, ν) is different from zero outside Γ_{0j} in some neighborhood of Γ_{0j} on Γ.

Representing the field given on $\Gamma \setminus \bigcup \Gamma_{0j}$ by arrows directed into the domain Ω, we find the arrows to be directed towards submanifolds of class I and away from submanifolds of class II. In the theory of Brownian motion, in accord with this interpretation, manifolds of class I are called *attracting* manifolds, and of class II, *repellent* manifolds. Class III manifolds are termed *neutral*.

It is important to note that the addition of $\varepsilon\nu$, ε being arbitrarily small, to l yields, for a suitable choice of the sign of ε, a problem satisfying the Lopatinski condition provided there are no class I and class II submanifolds on Γ. If there is any such submanifold, smooth regularization of the problem is impossible.

If we extend l to a neighborhood of the boundary Γ in Ω, then class I manifolds Γ_0 are characterized by the property that the integral curves of the extended field in a neighborhood of Γ_0 have both ends on Γ, while for class II manifolds Γ_0 the integral curves of the extended field go into Ω in both directions. Finally, for class III manifolds, one end of every integral curve lies on Γ and the integral curve in the other direction goes into Ω.

1.3. The following examples illustrate specific properties of the manifolds of various classes.

Example 1.1. (A. V. Bitsadze). Consider in $\mathbb{R}^3$ a ball $\Omega = \{x;\ (x^1)^2 + (x^2)^2 + (x^3)^2 < 1\}$ and the Laplace equation $\Delta u = 0$ in Ω with boundary condition $D_1 u = f$ on $\partial\Omega = \Gamma$.

In this case there is one special submanifold Γ_0 on Γ which belongs to class I. We shall find a solution to the boundary problem satisfying the additional condition

$$u = g \quad \text{on} \quad \Gamma_0. \tag{1.3}$$

The function $v = D_1 u$ satisfies the Laplace equation in Ω and the condition $v = f$ on Γ. This enables the function v to be defined uniquely in Ω.

Let us consider the boundary problem

$$\frac{\partial^2 u}{\partial (x^2)^2} + \frac{\partial^2 u}{\partial (x^3)^2} = -\frac{\partial v}{\partial x^1}, \quad u = g \quad \text{on} \quad \Gamma_0$$

in the equatorial plane $x^1 = 0$. Since v is already defined, these equalities enable u to be determined uniquely in that part of the equatorial plane which lies in Ω.

From the known values of u in the equatorial plane and of $v = D_1 u$ in Ω, u can be uniquely defined everywhere in Ω. Thus, the boundary problem has a unique solution if condition (1.3) is added. Without this condition the problem has an infinite-dimensional kernel.

Example 1.2. We now consider the boundary problem

$$\Delta u = 0 \text{ in } \Omega, \quad D_1 u = f \text{ on } \Gamma \tag{1.4}$$

when Ω, a domain in $\mathbb{R}^3$, is the exterior of a unit ball so that

$$\Omega = \{x;\ x \in \mathbb{R}^3,\ (x^1)^2 + (x^2)^2 + (x^3)^2 > 1\}.$$

Add to (1.4) the conditions at infinity

$$\lim_{x \to \infty} u(x) = 0, \quad \lim_{x \to \infty} D_1 u(x) = 0.$$

As above, solving the Dirichlet problem we find $v = D_1 u$ in Ω. Thereafter, we find u in Ω by the formula

$$u(x) = \begin{cases} \int_{-\infty}^{x^1} iv(t, x^2, x^3)\,dt, & x^1 < 0, \\ \int_{+\infty}^{x^1} iv(t, x^2, x^3)\,dt, & x^1 > 0. \end{cases}$$

In order that u be continuous in Ω it is necessary and sufficient that $\int_{-\infty}^{\infty} v\,dx^1 = 0$ when $(x^2)^2 + (x^3)^2 = 1$, since $\int_{-\infty}^{\infty} v\,dx^1$ is a harmonic function when $(x^2)^2 + (x^3)^2 > 1$. Using the integral representation of the solution of the Dirichlet problem

$$v(x) = \iint_\Gamma K(x, y) f(y)\,dy$$

and assuming $K_1(x^2, x^3, y) = \int_{-\infty}^{\infty} K(x, y)\,dx^1$ when $(x^2)^2 + (x^3)^2 \geqslant 1$ the solvability condition is written as

$$\iint K_1(x^2, x^3, y) f(y)\,dS_y = 0, \quad \text{if} \quad (x^2)^2 + (x^3)^2 = 1. \tag{1.5}$$

Thus, the cokernel has infinite dimension; that is, problem (1.4) can be solved only when the function f is orthogonal to an infinite-dimensional subspace of functions.

§2. Auxiliary Constructions

2.1. In this section we obtain some simple results which will be useful in the proofs that follow.

Proposition 2.1. To every point belonging to Γ_0 there exists a system of coordinates $y^1, \ldots, y^n$ with the origin at the point and such that l coincides with the field $\partial/\partial y^1$ in some neighborhood of the point; Γ_0 is defined by the equations $y^1 = 0$, $y^n = 0$, and Γ is defined by the equation $y^n = f(y^1, \ldots, y^{n-1})$ ($f \geqslant 0$ if Γ_0 is of class I and $f \leqslant 0$ if Γ_0 is of class II).

Proposition 2.2. There exists a partition of unity $\sum_{j=1}^{N+1} \psi_j(x) = 1$ such that $\psi_j \in C^\infty(\bar{\Omega})$, $\psi_{N+1} = 0$ in a neighborhood of all manifolds Γ_0, and, in the support of each of the remaining functions, a system of coordinates $y^1, \ldots, y^n$ satisfying the conditions of Proposition 2.1 and $l\psi_j = 0$ for $j = 1, \ldots, N$ is defined in some neighborhood of the manifold Γ_0.

Proposition 2.3. Let $\Omega' \subset \Omega$ and let the intersection Γ' of the boundary of Ω' with Γ be a smooth manifold. Let $\Gamma' \subset \Gamma'' \subset \Gamma$. Denote by B the trace operator on Γ, or the operator of the form $\sum_{j=1}^{n} A^j \frac{\partial}{\partial x^j} + A^0$ in which the normal derivative coefficient is different from 0. Let m be on the order of B (m = 0 or 1). Then the following assertions are valid.

I. If $u \in H_t(\Omega)$, $Lu = f \in H_s(\Omega)$ and $Bu = g \in H_{s-m+3/2}(\Gamma'')$, where $s > m - 3/2$, then $u \in H_{s+2}(\Omega')$ and there exists a constant C independent of u for which

$$\|u\|_{s+2}^{\Omega'} \leqslant C(\|Lu\|_s + \|g\|_{s-m+3/2}^{\Gamma''} + \|u\|_0).$$

II. If $f \in H_s(\Omega)$, $g \in H_{s-m+3/2}(\Gamma'')$, then there exists a function $v \in H_{s+2}(\Omega')$ such that $Lv = f - f_1$ in Ω', $Bv = g - g_1$ on Γ' and there exists a constant C independent of f and g for which

$$\|f_1\|_{s+1}^{\Omega'} + \|g_1\|_{s-m+5/2}^{\Gamma'} \leqslant C(\|f\|_s + \|g\|_{s-m+3/2}^{\Gamma''}).$$

2.2. Proof of Proposition 2.1. Let P be an arbitrary point belonging to Γ_0. Choose a system of coordinates $s^1, \ldots, s^n$ with the origin at P for which l in some neighborhood of P coincides with $\partial/\partial s^1$.

The points lying on the normals to the manifold Γ, drawn at all points of Γ_0, form a manifold N transverse to Γ of the form $s^1 = \varphi(s^2, \ldots, s^n)$. Assume that

$$t^1 = s^1 - \varphi(s^2, \ldots, s^n), \quad t^j = s^j, \quad j = 2, \ldots, n.$$

In the $t^1, \ldots, t^n$ coordinates the origin coincides with P, l takes the form $\partial/\partial t^1$, and Γ_0 is defined by the equations: $t^1 = 0$, $t^n = \psi(t^2, \ldots, t^n)^1$.

We now choose coordinates $y^1, \ldots, y^n$ so that

$$y^j = t^j, \quad j = 1, \ldots, n-1; \quad y^n = t^n - \psi(t^2, \ldots, t^{n-1}).$$

These coordinates satisfy all the conditions.

2.3. Proof of Proposition 2.2. At every point $P \in \Gamma_0$ there exists a neighborhood ω_p in which the system of coordinates $y^1, \ldots, y^n$, described in Proposition 2.1, is defined. Every such neighborhood contains cubes $\{y;\ |y^j| \leqslant \alpha,\ j = 1, \ldots, n\}$, where α stands for a positive number depending on P. The cubes cover Γ_0. Choose a finite subcovering $\omega_1, \ldots, \omega_N$. In each of the chosen cubes ω_j we construct a function $\theta_j \in C_0^\infty(\omega_j)$ positive in ω_j and such that $\partial\theta_j/\partial y^1 = 0$ when $(y^1)^2 + (y^n)^2 \leqslant \alpha^2/4$. In the x coordinates, to function θ_j there corresponds a function $\varphi_j(x)$ positive inside ω_j and such that $l\varphi_j = 0$ in some neighborhood of the manifold Γ_0.

We now assume that $\psi_j(x) = \varphi_j(x)\left(\sum_{k=1}^{N} \varphi_k(x)\right)^{-1}$ at those points where $\sum_{k=1}^{N} \varphi_k(x) \neq 0$ and $\psi_j(x) = 0$ at other points. Assuming $\psi_{N+1}(x) = 1 - \sum_{j=1}^{N} \psi_j(x)$ we obtain the desired partition of unity.

2.4. Proof of Proposition 2.3.

I. Let $h \in C^\infty(\bar{\Omega})$ be a function such that $h = 1$ in a neighborhood of Ω' and $h = 0$ in a neighborhood of the manifold $\Gamma \setminus \Gamma''$. Then

$$L(hu) = hf + L_1(u) \text{ in } \Omega, \quad B(hu) = hg + B_1(u) \text{ on } \Gamma,$$

where L_1 and B_1 are operators of order 1 and $m - 1$, respectively. Thus, $L(hu) = f_1 \in H_{t-1}(\Omega) \cup H_s(\Omega)$,

$$B(hu) = g_1 \in H_{s-m+3/2}(\Gamma) \cup H_{t-m+1/2}(\Gamma).$$

According to Theorem 2.2 of Chapter III, $hu \in H_{t+1}(\Omega) \cup H_{s+2}(\Omega)$ and the estimate

$$\|h_1 u\|_{t''} \leqslant C(\|h_1 f\|_s + \|h_1 g\|_{s-m+3/2}^{\Gamma} + \|u\|_{t''-1})$$

is valid, where $t^1 = \min(t+1, s+2)$. If $t \geqslant s+1$, then the theorem is proven.

If $t < s + 1$, we consider a function $h_1 \in C_0^\infty(\bar{\Omega})$ equal to 1 in some neighborhood of Ω' and equal to 0 outside the set on which $h = 1$. Then $L(h_1u) = h_1f + L_2(hu)$ in Ω and $B(h_1u) = h_1g + B_2(hu)$ on Γ, where L_2 and B_2 are operators of order 1 and $m - 1$, respectively. Therefore

$$L(h_1u) = f_2 \in H_t(\Omega) \cup H_s(\Omega), \quad B(h_1u) = g_2 \in H_{t-m+3/2}(\Gamma) \cup H_{s-m+3/2}(\Gamma).$$

By Theorem 2.2 of Chapter II we find that $h_1u \in H_{t+2}(\Omega) \cup H_{s+2}(\Omega)$ and the inequality

$$\|hu\|_{t''} \leqslant C(\|hf\|_s + \|hg\|_{s-m+3/2}^{\Gamma} + \|u\|_{t'-1})$$

holds, where $t'' = \min(t + 2, s + 2)$.

If $t \geqslant s$, then the theorem is proven. When $t < s$ we consider a new function $h_2 \in C^\infty(\bar{\Omega})$ equal to 0 outside the set on which $h = 1$ and equal to 1 in some neighborhood of Ω', and so on.

II. Let us extend g to a function belonging to $H_{s-m+3/2}(\Gamma)$. The boundary problem $Lv = f$ in Ω, $Bv = g$ on Γ may not have a solution since the functions (f, g) must be orthogonal to the cokernel, that is, to the complement to the range of the operator (L, B). By Corollary 2.3 of Chapter III, the cokernel consists of infinitely differentiable functions and the projection (f_1, g_1) of the vector (f, g) onto this cokernel is such that

$$\|f_1\|_t + \|g_1\|_{t-m+3/2}^{\Gamma} \leqslant C_t(\|f\|_s + \|g\|_{s-m+3/2}^{\Gamma})$$

for any real t. The boundary problem

$$Lv = f - f_1 \text{ in } \Omega, \quad Bv = g - g_1 \text{ on } \Gamma$$

can be solved and the solution v belongs to the space $H_{s+2}(\Omega)$.

§3. Manifolds of Class I

3.1. It will be assumed throughout this section that Γ contains one smooth submanifold Γ_0 on which the field l is tangent to the boundary Γ and that the manifold belongs to class I. We shall show that in this case the problem (1.1)-(1.2) becomes a Fredholm problem on adding the condition

$$u = u_0 \quad \text{on } \Gamma_0. \tag{3.1}$$

Thus, (1.1)-(1.2) is found to have an infinite-dimensional kernel isomorphic to the space of functions u_0 on Γ_0 accurate up to a finite-dimensional subspace. Also for any s the problem (1.1)-(1.2) can have a solution not belonging to $H_s(\Omega)$ even when f and g are (infinitely) smooth.

3.2. Let P be an arbitrary point of the submanifold Γ_0 and let $y^1, \ldots, y^n$ be a system of coordinates in a neighborhood U_P of this point, as in Proposition 2.1.

By d we denote the diameter of a neighborhood U_P and assume d to be a small number. Every point of U_P can be connected with the plane $y^1 = 0$ by a segment of a line parallel to Oy^1, completely lying in Ω and having length $\leqslant Cd$.

Lemma 3.1. To every $\varepsilon > 0$ one can find d such that if the diameter of the neighborhood U_P is equal to d, $u \in H_s(\Omega)$ for $s \geqslant 0$ and $u(x) = 0$ outside U_P, then

$$\left\| \int_0^{x^1} u(t, x')\, dt \right\|_s \leqslant \varepsilon \|u\|_s. \tag{3.2}$$

Proof. Let $\int_0^{x^1} u(t, x')\, dt = v(x)$. We define the norm $\|w\|_{0,s}$ to be

$\left(\int \|w\|_s'^2\,dx^1\right)^{1/2}$, where $\|\cdot\|_s'$ is the norm in $H_s(\mathbb{R}^{n-1})$. Clearly

$$\|v\|_s \leqslant C\left(\sum_{j=2}^{n} \|D_j v\|_{0,\,s-1} + \|D_1 v\|_{s-1} + \|v\|_0\right).$$

The estimate (3.2), when s = 0, follows from the Cauchy-Schwarz-Bunjakovski inequality if $\varepsilon = d$. Similarly

$$\sum_{j=2}^{n} \|D_j v\|_{0,\,s-1} \leqslant C_1 d \|u\|_s.$$

According to Theorem 3.12 of Chapter I (which can be applied to u after extending it to the entire space, the norm being retained in H_s), the estimate

$$\|u\|_{s-1} \leqslant \delta \|u\|_s + C_\delta \|v\|_0$$

holds for any $\delta > 0$.

Thus, combining all of these inequalities we obtain the estimate

$$\|v\|_s \leqslant C\,(C_1 d \|u\|_s + \delta \|u\|_s + C_\delta \|v\|_0 + \|v\|_0) \leqslant$$

$$(CC_1 d + \delta) \|u\|_s + C\,(C_\delta + 1)\, d \|u\|_0 \leqslant [CC_1 d + \delta + C\,(C_\delta + 1)\, d] \|u\|_s.$$

If $\delta < \varepsilon/2$ and d is so small that $dC(C_1 + (C_\delta + 1)) < \varepsilon/2$, then it follows that $\|v\|_s \leqslant \varepsilon \|u\|_s$. This proves the lemma.

3.3. Lemma 3.2. Let U_P be a neighborhood of point $P \in \Gamma_0$, with diameter d, and let $y = (y^1, \ldots, y^n)$ be the coordinates in U_P as in Proposition 2.1. Let $Au = (Ll - lL)u$ and $R(f, g)$ be a parametrix in the Dirichlet problem for the operator L in the domain U_P. Let

$$Sw = \int_0^{y^1} w\,(t,\ y')\,dt, \quad B_0 w = R\,(Aw,\ 0)$$

(that is $LBw = Aw + T_1 Aw$ in U_P; $B_0 w = T_2 Aw$ on ∂U_P, where T_1 and T_2 are operators of order -1). If d is sufficiently small, then

$$B_0 S w = B_1 w + B_1' w, \tag{3.3}$$

$$S B_0 w = B_2 w + B_2' w, \tag{3.4}$$

where B_1 and B_2 act from $H_s(U_P)$ into $H_s(U_P)$ and have norm <1/2 (under proper selection of the norm in $H_s(U_P)$), and the operators B_1' and B_2' take elements of $H_s(U_P)$ into $H_{s+1}(U_P)$.

Proof. Without loss of generality U_P may be assumed to be a convex domain with a smooth boundary. The operator A is a homogeneous differential operator of second order, since in (3.3)-(3.4) the operators B_j' of order -1 correspond to lower-order terms.

A change of independent variables $y = dz$ takes the domain U_P into a domain of unit diameter for which the operators B_0 and S are bounded:

$$\|B_0\| \leqslant C_1, \quad \|S\| \leqslant C_2, \tag{3.5}$$

where the constants C_1 and C_2 depend only on the maximum values of the coefficients of operators L and l and on the maximum values of the first derivatives of these coefficients.

For brevity we assume s to be a positive integer so that

$$\|u\|_s^2 = \sum_{j=0}^{s} \|D^j u\|_0^2,$$

where

$$\|D^j u\|_0^2 = \int_{U_P} \sum_{|\alpha| = j} |D^\alpha u|^2\,dy.$$

Let L' be the image of L under the transformation y = zd. Estimates (3.5) are retained if we consider the operator d^2L' for which the parametrix is equal to $R_d = d^{-2}R'u$, where R' corresponds to the parametrix R for operator L in z-coordinates.

It follows that the estimates

$$\|D^sR'(Aw,\ 0)\|_0+d^{-s}\|R'(Aw,\ 0)\|_0=d^{-s}\|D_z^sR_d(Aw,\ 0)\|_0+d^{-s}\|R_d(Aw,\ 0)\|_0\leqslant$$
$$Cd^{-s}(\|D_z^sw\|_0+\|w\|_0)\leqslant C_1(\|D_y^sw\|_0+\|d^{-s}w\|_0)$$

are valid.

Similarly it is verified that

$$\|D^sSw\|_0+d^{-s}\|Sw\|_0\leqslant C_2d(\|D^sw\|_0+d^{-s}\|w\|_0).$$

Now let d be so small that $C_1C_2d < 1/2$ and the sum $\|D^sw\|_0 + d^{-s}\|w\|_0$ is the norm of the space H_s. Then we have

$$\|B_1\|<1/2,\ \|B_2\|<1/2.$$

This proves the lemma.

3.4. Lemma 3.3. If d is sufficiently small and $u \in H_{s+1}(\Omega)$ is equal to 0 outside U_d, where U_d is a neighborhood of a point $P \in \Gamma_0$, with diameter d, then the inequality

$$\|u\|_s+\|lu\|_s+\|u\|_s^{\Pi}\leqslant$$
$$C(\|Lu\|_{s-2}+\|lLu\|_{s-2}+\|Lu\|_{s-2}^{\Pi}+\|lu\|_{s-1/2}^{\Gamma}+\|u\|_{s-1/2}^{\Gamma_0}+\|u\|_0) \tag{3.6}$$

holds when s > 1/2, where Π stands for the intersection of the plane $y^1 = 0$ with the neighborhood U_d, and C is a constant independent of u.

Proof. Let u(x) = v(y), y being the coordinates of Proposition 2.1. Assume Lu = f and u = w when $y^1 = 0$. For $y^1 = 0$, L is represented as $L_1+L_2\left(\frac{\partial}{\partial y^1}\right)$, where L_1 is an elliptic second-order operator acting on the variables $y^2, \ldots, y^n$, and L_2 is a first-order operator.

The function w, that is, the trace of the function u on plane $y^1 = 0$, can be regarded as a solution to the Dirichlet problem:

$$L_1w=f-L_2\left(\frac{\partial v}{\partial y^1}\right)\ \text{in}\ \Pi,\qquad w=u_0\ \text{when}\ y^n=0.$$

Therefore, when s > 1/2, we have the estimate

$$\|w\|_s^{\Pi}\leqslant C_1\left(\|f_1\|_{s-2}^{\Pi}+\left\|\frac{\partial v}{\partial y^1}\right\|_{s-1}^{\Pi}+\|u_0\|_{s-1/2}^{\Gamma_0}+\|u\|_0\right). \tag{3.7}$$

Differentiating Lv = f with respect to y^1 we obtain

$$L\left(\frac{\partial v}{\partial y^1}\right)=\frac{\partial f}{\partial y^1}+Av,$$

where A is a differential operator of second order. By g we denote the trace of function $\partial v/\partial y^1$ on Γ.

Using again the estimate of the solution to the Dirichlet problem, we obtain

$$\left\|\frac{\partial v}{\partial y^1}\right\|_s\leqslant C_2\left(\left\|\frac{\partial f}{\partial y^1}\right\|_{s-2}+\|Av\|_{s-2}+\|g\|_{s-1/2}^{\Gamma}+\|v\|_0\right)\leqslant$$
$$C_3\left(\left\|\frac{\partial f}{\partial y^1}\right\|_{s-2}+\|v\|_s+\|g\|_{s-1/2}^{\Gamma}+\|v\|_0\right). \tag{3.8}$$

Applying Lemma 3.1 we find that for sufficiently small d

$$\|v\|_s \leqslant \|v-w\|_s + \|w\|_s \leqslant \varepsilon \|\partial v/\partial y^1\|_s + \|w\|_s.$$

Combining this inequality with (3.7) and (3.8) we obtain, for $C_3 d < 1/2$, the inequality

$$\left\|\frac{\partial v}{\partial y^1}\right\|_s \leqslant C_4 \left(\left\|\frac{\partial f}{\partial y^1}\right\|_{s-2} + \|g\|^{\Gamma}_{s-1/2} + \|u_0\|^{\Gamma}_{s-1/2} + \|f\|^{\Pi}_{s-2} + \|v\|_0\right).$$

Thereafter, from (3.7) and (3.8) we obtain the estimates for $\|v\|_s$ and $\|v\|^{\Pi}_s = \|w\|_s$. Lemma 3.3 is proven.

3.5. Lemma 3.4. Let $u \in H_{s+1}(\Omega)$ and $u = 0$ outside a d-neighborhood of a class I manifold Γ_0. If d is sufficiently small and $s > 1/2$, then there exists a constant C such that

$$\|u\|_s + \|lu\|_s + \|u\|^N_s \leqslant C(\|Lu\|_{s-2} + \|lLu\|_{s-2} + \|Lu\|^N_{s-2} + \|lu\|^{\Gamma}_{s-1/2} + \|u\|^{\Gamma_0}_{s-1/2} + \|u\|_0),$$

where N is a manifold defined by the equation $y^1 = 0$ in the system of y coordinates of Proposition 2.1.

Proof. We make use of the special partition of unity of Proposition 2.2. Inequality (3.6) holds for every function $u\psi_k$.

Clearly

$$\sum_k \|L(u\psi_k)\|_{s-2} \leqslant \sum_k \|\psi_k Lu\|_{s-2} + C_1\|u\|_{s-1} \leqslant C_2(\|Lu\|_{s-2} + \|u\|_{s-1}),$$

$$\sum_k \|lL(u\psi_k)\|_{s-2} \leqslant \sum_k \|l\psi_k Lu\|_{s-2} + \sum_k \|lA^{(k)}u\|_{s-2} \leqslant$$

$$C_3 \sum_k \|\psi_k lLu\|_{s-2} + \sum_k \|A^{(k)} lu\|_{s-2} + C_3\|u\|_{s-1} \leqslant C_4(\|lLu\|_{s-2} + \|lu\|_{s-1} + \|u\|_{s-1});$$

$$\sum_k \|L(u\psi_k)\|^N_{s-2} = \sum_k \|\psi_k Lu\|^N_{s-2} + \sum_k \|A^{(k)}u\|^N_{s-2} \leqslant$$

$$C_5(\|Lu\|^N_{s-2} + \|u\|_{s-1/2}) \leqslant C_5\|Lu\|^N_{s-2} + \varepsilon\|u\|_s + C_\varepsilon\|u\|_{s-1}.$$

Therefore

$$\|u\|_s + \|lu\|_s + \|u\|^N_s \leqslant \sum_k C(\|u\psi_k\|_s + \|l(u\psi_k)\|_s + \|\psi_k u\|^N_s) \leqslant$$

$$C(\|Lu\|_{s-2} + \|lLu\|_{s-2} + \|Lu\|^N_{s-2} + \|g\|^{\Gamma}_{s-1/2} + \|u_0\|^{\Gamma_0}_{s-1/2} + \|u\|_{s-1}).$$

This proves the lemma.

3.6. Theorem 3.1. If $u \in H_{s+1}(\Omega)$, $d > 0$ is a sufficiently small number and $s > 1/2$, then there exists a constant $C > 0$, independent of u, such that

$$C^{-1}(\|u\|_s + \|l(hu)\|_s + \|hu\|^N_s) \leqslant$$

$$\|f\|_{s-2} + \|l(hf)\|_{s-2} + \|hf\|^N_{s-2} + \|g\|^{\Gamma}_{s-3/2} + \|hg\|^{\Gamma}_{s-1/2} +$$

$$\|u_0\|^{\Gamma_0}_{s-1/2} + \|u\|_0 \leqslant C(\|u\|_s + \|l(hu)\|_s + \|hu\|^N_s), \tag{3.9}$$

where $f = Lu$, $g = lu$ on Γ, $u_0 = u$ on Γ_0, $h \in C^\infty(\Omega)$ and $h = 1$ in a d/2-neighborhood of manifold Γ_0 and $h = 0$ outside a d-neighborhood of this manifold.

Proof. Let us denote by Ω_δ, where $\delta > 0$, the set of points in the region Ω which lie at a distance $>\delta$ from Γ_0, and let Γ'_δ denote the intersection $\Gamma \cap \Omega_\delta$.

Since the field l is not tangent to the boundary on $\Gamma'_{d/4}$, the estimate

$$\|(1-h)u\|_s+\|u\|_s^{G'_{d/2}}+\|u\|_{s-1/2}^{\Gamma'_{d/2}}\leqslant C_1\left(\|Lu\|_{s-2}^{G'_{d/4}}+\|lu\|_{s-3/2}^{\Gamma'_{d/4}}+\|u\|_0^{G'_{d/4}}\right)$$

holds by Proposition 2.3.

On the other hand, by Lemma 3.4 we have

$$\|hu\|_s+\|l(hu)\|_s+\|hu\|_s^N\leqslant C_2\left(\|hLu\|_{s-2}+\|lhLu\|_{s-2}+\|hLu\|_{s-2}^N+\right.$$

$$\left.\|hlu\|_{s-1/2}^{\Gamma}+\|u\|_{s-1/2}^{\Gamma_0}+\|u\|_0+\|u\|_s^{G'_{d/2}}+\|u\|_{s-1/2}^{\Gamma'_{d/2}}\right).$$

Therefore

$$\|u\|_s+\|l(hu)\|_s+\|hu\|_s^N\leqslant\|hu\|_s+\|(1-h)u\|_s+\|l(hu)\|_s+\|hu\|_s^N\leqslant$$

$$C_3\left(\|f\|_{s-2}+\|hf\|_{s-2}^N+\|l(hf)\|_{s-2}+\|g\|_{s-3/2}^{\Gamma}+\|hg\|_{s-1/2}^{\Gamma}+\|u_0\|_{s-1/2}^{\Gamma_0}+\|u\|_0\right).$$

This proves the theorem.

Corollary 3.1. The solution space of the homogeneous problem

$$Lu=0 \text{ in } \Omega,\ lu=0 \text{ on } \Gamma,\ u=0 \text{ on } \Gamma_0$$

has finite dimension.

Corollary 3.2. Denote by $\Pi_s(\Omega)$ the space of functions with finite norm

$$\|u\|_{\Pi_s(\Omega)}=\|u\|_s+\|l(hu)\|_s+\|hu\|_s^N$$

and by Γ_s the space of functions defined on Γ with finite norm

$$\|u\|_{\Gamma_s}=\|u\|_s^{\Gamma}+\|hu\|_{s+1}^{\Gamma}.$$

Then the range of the operator

$$u\longmapsto(Lu,\ lu|_{\Gamma},\ u|_{\Gamma_0})$$

from $\Pi_s(\Omega)$ into $\Pi_{s-2}(\Omega)\times\Gamma_{s-3/2}\times H_{s-1/2}(\Gamma_0)$ is closed.

3.7. We shall now show that the estimate (3.9) cannot be improved in H_s, that is, an estimate of the form

$$\|u\|_{s+\delta}\leqslant C\left(\|f\|_{s-1}+\|g\|_{s-1/2}^{\Gamma}+\|u_0\|_{s-1/2}^{\Gamma_0}+\|u\|_0\right)$$

is impossible for any $\delta>0$.

Example 3.1. The functions $u_m=(x^1+ix^2)^m x^3$ are harmonic functions in $\mathbb{R}^3$. They may be considered solutions of the boundary problem:

$$\Delta u=0 \text{ in } \Omega,\quad \frac{\partial u}{\partial x^3}=g \text{ on } \Gamma,\quad u=0 \text{ on } \Gamma_0,$$

where $\Omega=\{x;\ x\in\mathbb{R}^3,\ (x^1)^2+(x^2)^2+(x^3)^{2p}\leqslant 1\}$, F is the boundary of the domain Ω; Γ_0 is a set of points in Γ at which $x^3=0$.

Let us now prove the impossibility of the estimate

$$\|u\|_{s+\delta}\leqslant C\left(\|g\|_{s-1/2}^{\Gamma}+\|u\|_0\right) \tag{3.10}$$

with a constant C independent of u.

First we shall compute $\|u\|_k$ for an integer $k\geqslant 0$. We have

$$\sum_{|\alpha|=k}|D^\alpha u|^2=\left(\frac{m!}{(m-k)!}\right)^2[(x^1)^2+(x^2)^2]^{m-k}(x^3)^2+\left[\frac{m!}{(m-k+1)!}\right]^2[(x^1)^2+(x^2)^2]^{m-k+1}.$$

Furthermore

$$\int_\Omega[(x^1)^2+(x^2)^2]^{m-k}(x^3)^2dx=2\pi\int_0^1 r^{2(m-k)+1}dr\int_{-(1-r^2)^{1/2p}}^{(1-r^2)^{1/2p}}(x^3)^2dx^3=$$

$$\frac{4\pi}{3}\int_0^1 r^{2(m-k)+1}(1-r^2)^{3/2p}\,dr = \frac{2\pi}{3}\int_0^1 t^{m-k}(1-t)^{3/2p}\,dt =$$

$$\frac{2\pi}{3}\mathrm{B}\left(m-k+1,\ \frac{3}{2p}+1\right) = \frac{2\pi}{3}\ \frac{\Gamma(m-k+1)\,\Gamma\left(\frac{3}{2p}+1\right)}{\Gamma\left(m-k+\frac{3}{2p}+2\right)} \approx c_0 m^{-1-\frac{3}{2p}},$$

where c_0 is a positive constant independent of m and $m \to +\infty$.

On the other hand, we compute similarly

$$\int_\Omega ((x^1)^2+(x^2)^2)^{m-k+1}\,dx = 2\pi\int_0^1 r^{2(m-k+1)}\,dr \int_{-(1-r^2)^{1/2p}}^{(1-r^2)^{1/2p}} dx^3 =$$

$$(4\pi)\int_0^1 r^{2(m-k+1)+1}(1-r^2)^{1/2p}\,dr = 2\pi\int_0^1 t^{m-k+1}(1-t)^{1/2p}\,dt =$$

$$2\pi\mathrm{B}\left(m-k+2,\ \frac{1}{2p}+1\right) = 2\pi\ \frac{\Gamma(m-k+2)\,\Gamma\left(\frac{1}{2p}+1\right)}{\Gamma\left(m-k+3+\frac{1}{2p}\right)} \approx C_1 m^{-1-1/2p},$$

where C_1 is a positive constant independent of m.

It is seen that for integers k

$$\|u\|_k^\Omega \geqslant C_2 m^{k-1/2-3/4p}, \quad C_2>0.$$

Now we estimate the norm $\|g\|_l^\Gamma$ for $l \geqslant 0$. Since

$$\|g\|_{l-1/2}^\Gamma \leqslant \|D_3 u\|_l,$$

we shall first estimate $\|D_3 u\|_l$ for an integer l. We have

$$\sum_{|\alpha|=l} \|D^\alpha D_3 u\|^2 = \left[\frac{m!}{(m-l)!}\right]^2 [(x^1)^2+(x^2)^2]^{m-l}.$$

Therefore

$$\|D_3 u\|_l^2 \approx C_3 m^{2l-1-1/2p}.$$

Interpolating, one can obtain an analogous equality for all real $l \geqslant 0$. Therefore

$$\|g\|_{l-1/2} \leqslant 2\sqrt{C_3}\, m^{l-1/2-1/4p}.$$

Thus inequality (3.10) reduces to the inequality

$$m^{s+\delta-1/2-3/4p} \leqslant C_4\,(m^{s-1/2-1/4p} + m^{-1/2-3/4p}),$$

which may not be fulfilled, however, for all m if $\delta > 1/2p$. Since p can be arbitrarily large, estimate (3.10) does not hold for $\delta > 0$.

3.8. We now prove a theorem on the smoothness of the solution.

<u>Theorem 3.2.</u> If Γ_0 belongs to class I and

$$u \in H_s(\Omega), \quad Lu \in H_s(\Omega), \quad lu \in H_{s+1/2}(\Gamma), \quad u \in H_{s+1/2}(\Gamma_0),$$

where $s > 0$, then $u \in H_{s+1}(\Omega)$ and there exists a constant C, independent of u, such that

$$\|u\|_{s+1} \leqslant C\left(\|Lu\|_s + \|lu\|_{s+1/2}^\Gamma + \|u\|_{s+1/2}^{\Gamma_0} + \|u\|_s\right).$$

<u>Proof.</u> Let $h \in C^\infty(\overline{\Omega})$ be a function equal to 1 in a d/2-neighborhood $\Omega_{d/2}$ of Γ_0 and $h(x) = 0$ outside a d-neighborhood Ω_d of the manifold. Since the Lopatinski conditions are fulfilled outside Γ_0, $u \in H_{s+2}(\Omega\setminus\Omega_{d/4})$. The support of $Lhu - hLu$ lies in $\Omega\setminus\Omega_{d/2}$ so $L(hu) \in H_s(\Omega)$. Similarly, $l(hu) \in H_{s+1/2}(\Gamma)$. Thus, in proving the theorem the support of u may be assumed to lie in Ω_d.

Let d be so small that $l\psi_k = 0$ in Ω_d, where ψ_k are the functions constructed in Proposition 2.2. Let us denote by ψ one of the functions ψ_k and show that $\psi u \in H_{s+1}(\Omega)$ and

$$\|\psi u\|_{s+1} \leqslant C(\|Lu\|_s + \|lu\|^{\Gamma}_{s+1/2} + \|u_0\|^{\Gamma_0}_{s+1/2} + \|u\|_s).$$

Obviously, this suffices to prove the theorem. Thus, the support of u may be assumed to lie in a small neighborhood of P in which the coordinates $y^1, \ldots, y^n$ of Proposition 2.1 are defined.

We denote Lu by f, lu on Γ by g, and u on Γ_0 by u_0. According to the conditions of the theorem

$$Ll(u\psi) = \psi lf + Au \in H_{s-2}(\Omega), \qquad l(u\psi) = \psi lu \in H_{s+1/2}(\Gamma).$$

By Corollary 2.4 of Chapter III (on the smoothness of solutions of the Dirichlet problem), it follows that $l(u\psi) \in H_s(\Omega)$ so that $lu \in H_s$ in some neighborhood of Γ_0.

Note that

$$Ll(u\psi) = \psi lf + A(\psi u) + A_1 u + A_2(lu) = F + A(\psi u), \tag{3.11}$$

where A is a second-order differential operator; A_1 and A_2 are first-order operators; $F = \psi lf + A_1 u + A_2 lu$ belongs to $H_{s-1}(\Omega)$.

Let $u(x) = v(y)$, where y represents coordinates as in Proposition 2.1. The function $w(y) = v(0, y^2, \ldots, y^n) \in H_{s-1/2}(U)$ and when $y^1 = 0$

$$L_1 w = f + L_2(\partial v/\partial y^1),$$

where L_1 is an elliptic second-order operator, and the order of L_2 is equal to 1. Therefore, when $y^1 = 0$

$$L_1(\psi w) = \psi f + L_2\left(\frac{\partial \psi v}{\partial y^1}\right) + A_1' w + A_0'\left(\frac{\partial v}{\partial y^1}\right) = F_1 + A_1' w + L_2\left(\frac{\partial(\psi v)}{\partial y^1}\right), \tag{3.12}$$

where

$$F_1 = [\psi f + A_0'(\partial v/\partial y^1)]_{y^1=0} \in H_{s-1/2}(\Pi).$$

The order of the operator A_j is equal to j. Here Π stands for the part of the plane $y^1 = 0$ lying in U.

If R_1 is a parametrix of the Dirichlet problem for L in the domain U, then

$$R_1(Lz, z|_{\partial U}) = z + T_1 z, \tag{3.13}$$

where the order of the operator T_1 is equal to -1 and

$$R_1 \in L(H_{l-2}(U) \times H_{l-1/2}(\partial U), H_l(U)).$$

Let R_2 be a parametrix of the Dirichlet problem for L_1 in the domain Π:

$$R_2(L_1 z, z|_{\partial\Pi}) = z + T_2 z, \tag{3.14}$$

the order of the operator T_2 is equal to -1, and

$$R_2 \in L(H_{l-2}(\Pi) \times H_{l-1/2}(\partial\Pi), H_l(\Pi)).$$

Since $l(\psi v) = \psi g$ on $\Gamma \cap \partial U$ and $I(\psi v) = 0$ on the remainder of the boundary ∂U it follows from (3.11) and (3.13) that

$$\frac{\partial \psi_1 v}{\partial y^1} + T_1\left(\frac{\partial \psi v}{\partial y^1}\right) = R_1(F + A(\psi u), \tilde{g}) = F_2 + A_0(\psi u), \tag{3.15}$$

where $\tilde{g}$ stands for the values $l(\psi v)$ on ∂U;

$$F_2 = R_1(F, \tilde{g}) \in H_{s+1}(\Omega) \text{ and } A_0 \in L(H_s(U), H_s(U)).$$

Similarly, from (3.12) and (3.14) it follows that

$$\psi w + T_2\psi w = R_2\left(F_1 + A_1' w + L_2\left(\frac{\partial\psi v}{\partial y^1}\right),\ \tilde{u}_0\right) = F_3 + A_{-1/2}\left(\frac{\partial\psi v}{\partial y^1}\right), \tag{3.16}$$

where $\tilde{u}_0$ are the values of function ψw on $\partial\Pi$ so that $\tilde{u}_0 = \psi u_0$ on $\partial\Pi \cap \Gamma$ and $\tilde{u} = 0$ on the remainder of $\partial\Pi$, the order of the operator $A_{-1/2}$ being equal to $-1/2$, $F_3 = R_2(F_1 + A_1 w,\ \tilde{u}_0) \in H_{s+1/2}(\Pi)$, since $F_1 \in H_{s-1/2}(\Pi)$, $\tilde{u}_0 \in H_{s+1/2}(\partial\Pi)$ and $w \in H_{s-1/2}(\Pi)$.

From the equality

$$\psi v = \psi w + S\left(\frac{\partial\psi v}{\partial y^1}\right), \tag{3.17}$$

where S is a bounded operator, and equalities (3.15) and (3.16), it follows that

$$\frac{\partial\psi v}{\partial y^1} + T_1\left(\frac{\partial\psi v}{\partial y^1}\right) = F_2 + A_0(\psi w) + A_0 S\left(\frac{\partial\psi v}{\partial y^1}\right) =$$

$$F_2 + A_0F_3 + A_0A_{-1/2}\left(\frac{\partial\psi v}{\partial y^1}\right) - A_0T_2(\psi w) + A_0S\left(\frac{\partial\psi v}{\partial y^1}\right).$$

Therefore

$$\frac{\partial\psi v}{\partial y^1} + A'_{-1/2}\left(\frac{\partial\psi v}{\partial y^1}\right) = F_4 + A_0S\left(\frac{\partial\psi v}{\partial y^1}\right),$$

where $F_4 = F_2 + A_0F_3 - A_0T_2(\psi w) \in H_{s+1/2}(U)$, the order of $A'_{-1/2}$ being equal to $-1/2$, so that $A'_{-1/2}\left(\frac{\partial\psi v}{\partial y^1}\right) \in H_{s+1/2}(U)$.

As per Lemma 3.2, A_0S is represented as a sum $B_1 + B_1'$, where $B_1'\left(\frac{\partial\psi v}{\partial y^1}\right) \in H_{s+1}(U)$ and the norm of B_1 does not exceed 1/2. Thus, $I - B_1$ is invertible and

$$\frac{\partial\psi v}{\partial y^1} = (I - B_1)^{-1}\left[F_4 + B_1'\left(\frac{\partial\psi v}{\partial y^1}\right) - A'_{-1/2}\left(\frac{\partial\psi v}{\partial y^1}\right)\right]. \tag{3.18}$$

Hence, $\partial\psi v/\partial y^1 \in H_{s+1/2}(U)$.

From (3.16) we see that $\psi w \in H_{s+1/2}(\Pi)$, and from (3.17), that $\psi u \in H_{s+1/2}(U)$. This implies that $\psi u \in H_{s+1/2}(\Omega)$. In order to prove that $\psi u \in H_{s+1}(\Omega)$ we need to repeat the same argument starting from (3.16). Indeed, since $\psi w \in H_{s+1/2}(\Omega)$ and $F_3 \in H_{s+1/2}(\Pi)$ we see that $F_4 \in H_{s+1}(U)$. Since $A'_{-1/2}(\partial\psi v/\partial y^1) \in H_{s+1}(U)$, it follows from (3.18) that $\partial\psi v/\partial y^1 \in H_{s+1}(U)$ and from (3.16) that $\partial w \in H_{s+1}(\Pi)$; therefore, (3.17), $\psi v \in H_{s+1}(U)$. Thus, Theorem 3.2 is proven.

Remark. It is clear from the proof that the conditions of the theorem may be relaxed; namely, it may be assumed that $f \in H_{s-1}(\Omega)$, $l(hf) \in H_{s-1}(\Omega)$, $g \in H_{s-1/2}(\Gamma)$, $hg \in H_{s+1/2}(\Gamma)$, $u_0 \in H_{s+1/2}(\Gamma_0)$. Here, as above, h stands for a function in $C^\infty(U)$ which is equal to 0 outside some neighborhood of the manifold Γ_0 and is equal to 1 in a smaller neighborhood of this manifold. In this case, we find that not only $u \in H_{s+1}(\Omega)$ but also $l(hu) \in H_{s+1}(N)$ and $hu \in H_{s+1}(N)$ so that the estimates for the lower and upper bounds are valid for u.

3.9. Now we shall prove the theorem on the solvability of the boundary problem (1.1), (1.2), (3.1).

Theorem 3.3. Let $f \in H_{s-1}(\Omega)$, $g \in H_{s-1/2}(\Gamma)$, $u_0 \in H_{s-1/2}(\Gamma_0)$, where $s > 1/2$. Then there exists a finite-dimensional subspace N_0 of space $H_{s-1}(\Omega) \times H_{s-1/2}(\Gamma) \times H_{s-1/2}(\Gamma_0)$ such that if $(f,\ g,\ u_0)$ is orthogonal to N, then the problem (1.1), (1.2), (3.1) has a solution u in $H_s(\Omega)$.

This theorem is a corollary of the following more precise statement.

Theorem 3.4. Let $h \in C^\infty(\bar{\Omega})$ be a function equal to 1 in a d/2-neighborhood of manifold Γ_0 and equal to 0 outside a d-neighborhood. Let

$$f \in H_{s-2}(\Omega), \quad l(hf) \in H_{s-2}(\Omega), \quad hf \in H_{s-2}(N),$$

$$g \in H_{s-3/2}(\Omega), \quad hg \in H_{s-1/2}(\Gamma), \quad u_0 \in H_{s-1/2}(\Gamma_0),$$

where $s > 1/2$, and let (f, g, u_0) be orthogonal to a subspace N_0 of the finite-dimensional space $H_{s-1}(\Omega) \times H_{s-1/2}(\Gamma) \times H_{s-1/2}(\Gamma_0)$. Then there exists a solution u to the problem (1.1), (1.2), (3.1) belonging to $H_s(\Omega)$, for which $l(hu) \in H_s(\Omega)$ and $hu|_N \in H_s(N)$. There also exists a constant C, independent of u, for which

$$\|u\|_s + \|l(hu)\|_s + \|hu\|_s^N \leqslant C(\|f\|_{s-2} + \|l(hf)\|_{s-2} +$$

$$\|hf\|_{s-2}^N + \|g\|_{s-3/2}^\Gamma + \|hg\|_{s-1/2}^\Gamma + \|u_0\|_{s-1/2}^{\Gamma_0}).$$

Proof of Theorem 3.4. Let us introduce the spaces $\Pi_s(\Omega)$ with norms

$$\|u\|_{\Pi_s} = \|u\|_s + \|l(hu)\|_s + \|u\|_s^N$$

and $\Gamma_s(\Gamma)$ with norms

$$\|u\|_{\Gamma_s} = \|u\|_s^\Gamma + \|hu\|_{s+1}^\Gamma.$$

Let first s be sufficiently large, say $s \geqslant 5$. We now consider the partition of unity $\sum \varphi_j(x) = 1$, introduced in Proposition 2.2, and the analogous functions φ_j in $C^\infty(\Omega)$ such that $\varphi_i\psi_i \equiv \varphi_i$.

By Proposition 2.3, para. II, there exists an operator R_{N+1}: $H_{s-2}(\Omega) \times H_{s-3/2}(\Gamma) \to H_s(\Omega)$, for which

$$\varphi_{N+1} L R_{N+1}(\psi_{N+1} f, \psi_{N+1} g) = \varphi_{N+1} f + T_{N+1}(f, g),$$

$$\varphi_{N+1} l R_{N+1}(\psi_{N+1} f, \psi_{N+1} g) = \varphi_{N+1} g + T'_{N+1}(f, g) \quad \text{on } \Gamma,$$

where

$$T_{N+1} \in L(H_{s-2}(\Omega) \times H_{s-3/2}(\Gamma), \Pi_{s-1}(\Omega)),$$

$$T'_{N+1} \in L(H_{s-2}(\Omega) \times H_{s-3/2}(\Gamma), \Gamma_{s-1/2}(\Gamma)).$$

Let now $1 \leqslant i \leqslant N$. We construct the operator

$$R_i \in L(\Pi_{s-2}(U_i) \times \Gamma_{s-3/2}(\Gamma_i) \times H_{s-1/2}(\Gamma_0 \cap \partial U_i), \Pi_s(\Omega)),$$

where $U_i = \operatorname{supp} \psi_i$, $\Gamma_i = \partial U_i \cap \Gamma$, for which

$$\varphi_i L R_i(\psi_i f, \psi_i g, \psi_i u_0) = \varphi_i f + T_i(f, g, u_0),$$

$$\varphi_i l R_i(\psi_i f, \psi_i g, \psi_i u_0)|_\Gamma = \varphi_i g + T'_i(f, g, u_0),$$

$$\varphi_i R_i(\psi_i f, \psi_i g, \psi_i u_0)|_{\Gamma_0} = \varphi_i u_0 + T''_i(f, g, u_0),$$

where

$$T_i: \Pi_{s-2}(\Omega) \times \Gamma_{s-3/2}(\Gamma) \times H_{s-1/2}(\Gamma_0) \to \Pi_{s-1}(\Omega),$$

$$T'_i: \Pi_{s-2}(\Omega) \times \Gamma_{s-3/2}(\Gamma) \times H_{s-1/2}(\Gamma_0) \to \Gamma_{s-1/2}(\Gamma),$$

$$T''_i: \Pi_{s-2}(\Omega) \times \Gamma_{s-3/2}(\Gamma) \times H_{s-1/2}(\Gamma_0) \to H_{s+1/2}(\Gamma_0)$$

are bounded operators.

We next verify that the operator $R = \sum_{i=1}^{N+1} \varphi_i R_i \psi_i$ is a parametrix of the problem. Indeed

$$LR(f, g, u_0) = \sum L\varphi_i R_i(\psi_i f, \psi_i g, \psi_i u_0) = \sum \varphi_i L R_i(\psi_i f, \psi_i g, \psi_i u_0) + T(f, g, u_0) =$$

$$\sum \varphi_i f + T_1(f, g, u_0) = f + T_1(f, g, u_0),$$

where T and T_1 are bounded operators in the space

$$\Pi_{s-2}(\Omega)\times\Gamma_{s-3/2}(\Gamma)\times H_{s-1/2}(\Gamma_0) \text{ in } \Pi_{s-1}(\Omega).$$

Furthermore, at boundary points

$$lR(f,\ g,\ u_0)=\sum l\varphi_i R_i(\psi_i f,\ \psi_i g,\ \psi_i u_0)=\sum \varphi_i l R_i(\psi_i f,\ \psi_i g,\ \psi_i u_0)+T'(f,\ g,\ u_0)=$$

$$\sum \varphi_i\psi_i g+T_1'(f,\ g,\ u_0)=g+T_1'(f,\ g,\ u_0),$$

where T' and T_1' are bounded operators in the space

$$\Pi_{s-2}(\Omega)\times\Gamma_{s-3/2}(\Gamma)\times H_{s-1/2}(\Gamma_0) \text{ in } \Gamma_{s-1/2}(\Gamma).$$

To check the boundedness of operator T' we use the fact that $l\psi_i = 0$ in Ω_d so that the support of T'(f, g, u_0) lies outside Ω_d.

Finally, note that on Γ_0

$$R(f,\ g,\ u_0)=\sum \varphi_i R_i(\psi_i f,\ \psi_i g,\ \psi_i u_0)=\sum \varphi_i\psi_i u_0+T''(f,\ g,\ u_0)=u_0+T''(f,\ g,\ u_0),$$

where T'' is a bounded operator from $\Pi_{s-2}(\Omega)\times\Gamma_{s-3/2}(\Gamma)\times H_{s-1/2}(\Gamma_0)$ to $H_{s+1/2}(\Gamma_0)$.

The parametrix so constructed enables the boundary problem to be reduced to a second-order Fredholm problem. Indeed, assuming u = R($\tilde{f}$, $\tilde{g}$, $\tilde{u}_0$) we see that u is a solution of the boundary problem if and only if

$$\tilde{f}+T_1(\tilde{f},\ \tilde{g},\ \tilde{u}_0)=f,\quad \tilde{g}+T_1'(\tilde{f},\ \tilde{g},\ \tilde{u}_0)=g,\quad \tilde{u}_0+T''(\tilde{f},\ \tilde{g},\ \tilde{u}_0)=u_0.$$

Therefore, the theorem follows immediately from the Fredholm theorem. Since T_1, T_1', T'' are smoothing operators, the kernel of the adjoint operator consists of infinitely differentiable functions and has finite dimension. Thus, the solvability conditions are that the vector (f, g, u) be orthogonal to a finite number of functions of the type (F, G, U_0) belonging to $C^\infty(\Omega)\times C^\infty(\Gamma)\times C^\infty(\Gamma_0)$.

Thus, to prove the theorem it remains to construct an operator R_i for i = 1,..., N. For brevity we shall drop the index i and assume the functions f, g, u_0 to be different from 0 only in a small neighborhood U of point P lying on Γ_0.

By R_D we denote the parametrix for the Dirichlet problem for L in the region U and by R_D', the parametrix for the Dirichlet problem for L_1 which is obtained from L by neglecting the derivatives with respect to y^1 in the range Π. The operators R_D and R_D' are such that

$$LR_D(f,\ g)=f+T_1(f,\ g),\quad R_D(f,\ g)|_{\partial U}=g+T_2(f,\ g);$$

$$L_1R_D'(f,\ g)=f+T_1'(f,\ g),\quad R_D'(f,\ g)|_{\partial\Pi}=g+T_2'(f,\ g),$$

and for every $s\geqslant 2$

$$T_1\times T_2\in L(H_{s-2}(U)\times H_{s-1/2}(\partial U),\ H_{s-1}(U)\times H_{s+1/2}(\partial U));$$

$$T_1'\times T_2'\in L(H_{s-2}(\Pi)\times H_{s-1/2}(\partial\Pi),\ H_{s-1}(\Pi)\times H_{s+1/2}(\partial\Pi)).$$

If u is the solution of the problem, then, as shown above,

$$Llu=lf+A_2u \quad \text{in} \quad U,$$

$$L_1w=f+A_1(lu) \ \text{ on } \ \Pi,\quad u=w+Slu \quad \text{in} \quad U,$$

where A_1 and A_2 are first- and second-order differential operators, respectively, and $S\in L(H_s,\ H_s)$.

For functions $z\in H_s(U)$ we define the operators $B_1z = sR_D(A_2z,\ 0)$ and $B_2z = R_D'(A_1R_D(A_2z,\ 0),\ 0)$.

According to Lemma 3.2, B_1 is a sum of $B_1' + B_1''$, where B_1' has a norm $\leqslant 1/2$ in the spaces

$$H_s(U),\ H_{s-1}(U),\ H'_{s-2}(U),\ H_{s-3}(U)$$

if the diameter of the region U is sufficiently small and B_1'' is of order -1. The order of B_2 does not exceed $-1/2$.

Let us show that the image of space $H_s(U)$ under the mapping $I - B_1 - B_2$ has finite dimension and the functions in $H_s(U)$ which are orthogonal to this image belong to $H_{s+3}(U)$. Indeed, if

$$((I - B_1 - B_2)u,\ v)_s = 0, \quad \forall u \in H_s(U),$$

where $(,)_s$ is the scalar product in $H_s(U)$, then

$$(I - B_1^* - B_2^*)v = 0, \tag{3.19}$$

where B_1^* and B_2^* are operators adjoint to the operators B_1 and B_2 relative to the scalar product in $H_s(U)$. The operator B_1^* is bounded in $H_{s+k}(U)$ for $k = 1, 2, 3$ and has norm $\leqslant 1/2$; operator $B_2^* \in L(H_{s+k}(U), H_{s+k+1/2}(U)$. It follows that the solutions of equation (3.19) form a finite-dimensional subspace and belong to $H_{s+3}(U)$.

On the other hand, every element of the form $w = (I - B_1 - B_2)v$, where $v \in H_s(U)$, may have a finite number of inverse images under the mapping $I - B_1 - B_2$. We adopt the convention that Φw stands for an inverse image that has a minimum norm. If w is orthogonal to all elements of the form $(I - B_1 - B_2)v$, then we assume $\Phi w = 0$.

Now suppose

$$R(\psi f,\ \psi g,\ \psi u_0) = \Phi(R'_D(\psi f + A_1 R_D(l\psi f,\ \psi\tilde{g}),\ \psi\tilde{u}_0) + SR_D(l\psi f,\ \psi\tilde{g})),$$

where $\psi\tilde{g}$ and $\psi\tilde{u}_0$ are functions obtained from ψg and ψu_0 after assuming them to be zero over the boundary ∂U. We shall show that R is the desired parametrix.

Let $v = R(\psi f, \psi g, \psi u_0)$. Clearly $v \in H_{s-1}(U)$. We shall prove that $v \in \Pi_s(U)$. We have

$$(I - B_1 - B_2)v = R'_D(\psi f + A_1 R_D(l\psi f,\ \psi\tilde{g}),\ \psi\tilde{u}_0) +$$

$$SR_D(l\psi f,\ \psi\tilde{g}) + T_0(\psi f,\ \psi g,\ \psi u_0),$$

where $T_0(\psi f, \psi g, \psi u_0) \in \Pi_{s+2}(U)$ (even $H_{s+3}(U)$). From this and the definition of B_1 and B_2 it follows that

$$v = R'_D(\psi f + A_1 R_D(B\psi f + A_2 v,\ \psi\tilde{g}),\ \psi\tilde{u}_0) +$$

$$SR_D(l\psi f + A_2 v,\ \psi\tilde{g}) + T_0(\psi f,\ \psi g,\ \psi u_0).$$

Hence

$$lv = R_D(l\psi f + A_2 v,\ \psi\tilde{g}) + lT_0(\psi f,\ \psi g,\ \psi u_0).$$

We see that $lv \in H_{s-1}(U)$ and that

$$l(Lv - \psi f) = T_1(f,\ g,\ u_0) \in H_s(U), \tag{3.20}$$

$$(lv - \psi g)|_\Gamma = T'(f,\ g,\ u_0) \in H_{s+3/2}(U). \tag{3.21}$$

Furthermore

$$(Lv - \psi f)_{y^1=0} = T_2(f,\ g,\ u_0) \in H_s(\Pi), \tag{3.22}$$

$$v|_{\Gamma_0} - \psi u_0 = T''(f,\ g,\ u_0) \in H_{s+1/2}(\Gamma_0). \tag{3.23}$$

We find from (3.20) and (3.22) that

$$Lv - \psi f = T(f,\ g,\ u_0) \in H_s(U). \tag{3.24}$$

So far we have made use of the fact that $f \in H_{s-2}(\Omega)$ and $g \in H_{s-3/2}(\Gamma)$. But it is also known that

$$l(hf) \in H_{s-2}(\Omega), \quad hf|_{y^1=0} \in H_{s-2}(\Pi), \quad hg \in H_{s-1/2}(\Gamma).$$

It follows from (3.24) and (3.21) that $v \in H_s$ outside a d/2-neighborhood of the manifold Γ_0. Hence

$$l(hv)|_\Gamma - hg \in H_{s-1/2}(\Gamma \cap \partial U).$$

Since $L(hv) - hf \in H_{s-1}(U)$, we have

$$L(hv) \in \Pi_{s-2}(U), \quad l(hv)|_\Gamma \in \Gamma_{s-3/2}(\Gamma), \quad hv|_{\Gamma_0} \in H_{s-1/2}(\Gamma_0).$$

Arguing as in the proof of the smoothness theorem, we conclude that $v \in \Pi_s(U)$.

Equations (3.21), (3.23), and (3.24) reveal that

$$\varphi Lv - \varphi f = \varphi T(f, g, u_0) \in \Pi_{s-1}(\Omega), \qquad (\varphi lv - \varphi g)|_\Gamma = \varphi T'(f, g, u_0) \in \Gamma_{s-1/2}(\Gamma),$$

$$(\varphi v - \varphi u_0)|_{\Gamma_0} = \varphi T''(f, g, u_0) \in H_{s+1/2}(\Gamma_0).$$

Hence, R satisfies all of the required conditions and the theorem is proven for $s \geqslant 5$.

To prove the theorem for $1 \leqslant s < 5$ the data of the problem have to be approximated by smooth functions. Note that if a solution u of the problem (1.1), (1.2), (3.1) is orthogonal to the solution of a homogeneous problem, then inequality (3.9) holds without the term $\|u\|_0$ in the middle part. If (f, g, u_0) satisfies the orthogonality conditions necessary for solvability, then the approximating functions are selected so that these orthogonality conditions are fulfilled for them also. Then from (3.9) follows the convergence of the solutions of the approximating problems to functions u from $H_s(\Omega)$, which will be the solution of the problem (1.1), (1.2), (3.1). Theorem 3.4 is proven.

Corollary 3.3. The range of the operator $u \longmapsto (Lu, Bu|_\Gamma, u|_{\Gamma_0})$ taking $\Pi_s(\Omega)$ into $\Pi_{s-2}(\Omega) \times \Gamma_{s-1/2}(\Gamma) \times H_{s-1/2}(\Gamma_0)$ has finite codimension equal to the dimension of the kernel of this operator.

The proof of this statement follows from the equivalence of the boundary problem to an integral equation of the form $(I + T)u = f$, where T is an operator of order (-1). We have proven this equivalence in the proof of Theorem 3.4.

§4. Manifolds of Class II

4.1. In this section we consider the case when there is a unique submanifold Γ_0 on Γ belonging to class II. For this case, as will be seen below, we do not have any existence theorem, but the kernel is finite-dimensional and the theorem on the smoothness of solutions is valid.

Theorem 4.1. Let $u \in H_s(\Omega)$, where $s > 1/2$, be a solution of the problem (1.1)-(1.2), where $f \in H_{s-1}(\Omega)$ and $g \in H_{s-1/2}(\Gamma)$. Then there exists a constant C independent of u such that

$$\|u\|_s \leqslant C\left(\|f\|_{s-1} + \|g\|^\Gamma_{s-1/2} + \|u\|_{s-1}\right).$$

The proof of this theorem is based on the following lemma.

Lemma 4.1. There exist positive constants ε_0 and K such that if $u \in H_s(\Omega)$, $\varepsilon < \varepsilon_0$, and $u = 0$ outside an ε-neighborhood of manifold Γ_0, then

$$\|u\|_s \leqslant K\varepsilon \|lu\|_s.$$

Proof of Lemma 4.1. Let P be a point on Γ_0, $y^1, \ldots, y^n$ be coordinates in a neighborhood of point P as in Proposition 2.1, and ψ_j be functions forming a partition of unity in accordance with Proposition 2.2. Since Γ_0 belongs to class II, the equation of the boundary Γ in a neighborhood of P takes the form $y^n = \omega(y^1, \ldots, y^{n-1})$ and $\omega = 0$ when $y^1 = 0$ and $\omega \leqslant 0$, $y^1\partial\omega/\partial y^1 \leqslant 0$ in this neighborhood. Without loss of generality it may be assumed that $u = 0$ outside the support of some function ψ_j; that is, $u = 0$ when $(y^1)^2 + (y^n)^2 \geqslant \varepsilon^2$, and also in that case when even for one value of j, $2 \leqslant j \leqslant n - 1$, the inequality $|y^s| \geqslant \alpha$ is satisfied. By Q we denote the support of u. Let Q_1 be the part of Q in which $y^1 < 0$ and let Q_2 be the part of Q in which $y^1 > 0$. In the domain Q_1 we have

$$u = \int_{-\varepsilon}^{y^1} \frac{\partial u(t, y')}{\partial t}\, dt.$$

It is clear that $\| u \|_s^{Q_1} \leqslant C_1\varepsilon \| \partial u/\partial y^1 \|_s^{Q_1}$. Similarly $\| u \|_s^{Q_2} \leqslant C_1\varepsilon \| \partial u/\partial y^1 \|_s^{Q_2}$. Since $u \in H_s$ it follows that

$$\| u \|_s^Q \leqslant C_1\varepsilon \left\| \frac{\partial u}{\partial y^1} \right\|_s^Q.$$

Returning to x-variables we obtain

$$\| u \|_s \leqslant C_2\varepsilon \| lu \|_s.$$

In the general case, when the support of u is not a small neighborhood of P, we have

$$\| u \|_s = \| \sum \psi_j u \|_s \leqslant \sum \| \psi_j u \|_s \leqslant C_2\varepsilon \sum \| l\psi_j u \|_s \leqslant K\varepsilon \| lu \|_s.$$

This proves the lemma.

4.2. <u>Proof of Theorem 4.1.</u> First let $u = 0$ outside Ω_ε, and let $\varepsilon > 0$ be sufficiently small (below we shall indicate how small it should be). The function lu satisfies the equation

$$L(lu) = lf + A(u),$$

and on Γ

$$lu = g \in H_{s-1/2}(\Gamma).$$

Using an *a priori* estimate to solve the Dirichlet problem, we obtain

$$\begin{aligned} \| lu \|_s &\leqslant C_1 (\| lf \|_{s-2} + \| Au \|_{s-2} + \| g \|_{s-1/2}^{\Gamma} + \| lu \|_{s-2}) \leqslant \\ &C_2 (\| f \|_{s-1} + \| u \|_s + \| g \|_{s-1/2}^{\Gamma} + \| u \|_{s-1}). \end{aligned}$$

According to Lemma 4.1, we have

$$\| u \|_s \leqslant K\varepsilon \| lu \|_s.$$

Combining these estimates, we claim that

$$\| lu \|_s \leqslant C_2 (\| f \|_{s-1} + \| g \|_{s-1/2}^{\Gamma} + \| u \|_{s-1}) + C_2K\varepsilon \| lu \|_s.$$

If ε is so small that $C_2K\varepsilon < 1/2$, then

$$\| lu \|_s \leqslant 2C_2 (\| f \|_{s-1} + \| g \|_{s-1/2}^{\Gamma} + \| u \|_{s-1})$$

and

$$\| u \|_s \leqslant 2C_2K\varepsilon (\| f \|_{s-1} + \| g \|_{s-1/2}^{\Gamma} + \| u \|_{s-1}). \tag{4.1}$$

The theorem has thus been proven for the case when $u = 0$ outside Ω_ε.

In the general case, we shall represent u as

$$u = h_\varepsilon u + (1 - h_\varepsilon) u$$

where $h_\varepsilon \leqslant C^\infty(\bar{\Omega})$, $h_\varepsilon = 0$ outside Ω_ε and $h_\varepsilon = 1$ in $\Omega_{\varepsilon/2}$. Clearly

$$L(h_\varepsilon u) = h_\varepsilon f + A_1 u.$$

Here A_1 is a first-order operator whose coefficients are equal to 0 in $\Omega_{\varepsilon/2}$. Similarly, on Γ

$$l(h_\varepsilon u) = h_\varepsilon g + ul(h_\varepsilon)$$

and $l(h_\varepsilon) = 0$ in $\Omega_{\varepsilon/2}$. By property I of Proposition 2.3, we have

$$\|A_1 u\|_s + \|ul(h_\varepsilon)\|^\Gamma_{s-1/2} \leqslant C_\varepsilon(\|f\|_{s-1} + \|g\|^\Gamma_{s-1/2} + \|u\|_{s-1}).$$

By the same proposition

$$\|(1-h_\varepsilon)u\|_s \leqslant C(\|f\|_{s-1} + \|g\|^\Gamma_{s-1/2} + \|u\|_{s-1}).$$

Replacing u by $h_\varepsilon u$ in (4.1) we obtain

$$\|h_\varepsilon u\| \leqslant C_3(\|h_\varepsilon f\|_{s-1} + \|A_1 u\|_{s-1} + \|h_\varepsilon g + ul(h_\varepsilon)\|^\Gamma_{s-1/2} + \|u\|_{s-1}) \leqslant$$
$$C_4(\|f\|_{s-1} + \|g\|^\Gamma_{s-1/2} + \|u\|_{s-1}).$$

Therefore

$$\|u\|_s \leqslant \|h_\varepsilon u\|_s + \|(1-h_\varepsilon)u\|_s \leqslant C_5(\|f\|_{s-1} + \|g\|^\Gamma_{s-1/2} + \|u\|_{s-1}).$$

This proves the theorem.

Corollary 4.1. The space of solutions of the homogeneous problem (1.1), (1.2) which belong to $H_s(\Omega)$ when $s > 1/2$ has finite dimension.

Remark 4.1. It is evident from the proof that there are estimates for lower and upper bounds:

$$C^{-1}(\|u\|_s + \|l(hu)\|_s) \leqslant \|f\|_{s-2} + \|l(hf)\|_{s-2} + \|g\|^\Gamma_{s-3/2} +$$

$$\|hg\|^\Gamma_{s-1/2} + \|u\|_{s-1} \leqslant C(\|u\|_s + \|l(hu)\|_s),$$

where $s > 1/2$ and the constant C is independent of u.

Corollary 4.2. If Π_s is a space of functions in Ω with finite norm $\|u\|_{\Pi_s} = \|u\|_s + \|l(hu)\|_s$ and Γ_s is a space of functions defined on Γ with norm $\|g\|_{\Gamma_s} = \|g\|^\Gamma_s + \|hg\|^\Gamma_{s+1}$, then the range of the operator $u \longmapsto (Lu, lu|_\Gamma)$ taking $\Pi_s(\Omega)$ into $\Pi_{s-2}(\Omega) \times \Gamma_{s-3/2}$ is closed.

4.3. We now prove a theorem on the smoothness of solution of the problem (1.1), (1.2).

Theorem 4.2. If $u \in H_s(\Omega)$ is a solution of the problem (1.1), (1.2), where $f \in H_s(\Omega)$, $g \in H_{s+1/2}(\Gamma)$, and $s > -1/2$, then $u \in H_{s+1}(\Omega)$.

Proof. We use the norm $\|\cdot\|_{s,\delta}$ defined in Section 3.11 of Chapter I for functions in $H_{s-2}(\mathbb{R}^n)$. If u is defined only in the domain Ω, then its norm $\|u\|_{s,\delta}$ is defined as a lower bound of such norms for functions U which are extensions of u to the entire space.

Repeating the proof of Lemma 4.1 it is not hard to verify that the inequality

$$\|u\|_{s,\delta} \leqslant K\varepsilon\|lu\|_{s,\delta}$$

is valid; here K is independent of ε and δ.

Similarly, repeating the arguments of Section 3.2, it may be shown that for a solution u of the boundary problem

$$Lu = f \text{ in } \Omega, \quad u = \varphi \text{ on } \Gamma,$$

the estimate

$$\|u\|_{s,\delta} \leqslant C(\|f\|_{s-2,\delta} + \|\varphi\|_{s-1/2,\delta} + \|u\|_{s-2,\delta}), \quad s > 1/2,$$

with the constant C independent of δ, holds.

Thus, for a function v equal to lu in Ω which is a solution of the Dirichlet problem

$$Lv = lf + Au \text{ in } \Omega, \quad v = g \text{ on } \Gamma,$$

the inequality

$$\| v \|_{s+1,\delta} \leqslant C_1 (\| f \|_{s,\delta} + \| u \|_{s+1,\delta} + \| g \|^{\Gamma}_{s+1/2,\delta} + \| v \|_{s-1,\delta})$$

is satisfied provided $s > -1/2$. Estimating $\|u\|_{s+1,\delta}$ by $K\varepsilon \|v\|_{s+1,\delta}$ we obtain the estimate $\| u \|_{s+1,\delta} \leqslant C_2 \varepsilon (\| f \|_{s,\delta} + \| g \|^{\Gamma}_{s+1/2,\delta} + \| u \|_{s,\delta})$ for small ε.

Since the right-hand side of this inequality is bounded uniformly in δ as $\delta \to 0$, it follows that the norm $\|u\|_{s+1,\delta}$ is uniformly bounded in δ and therefore $u \in H_{s+1}$. This proves the theorem.

Remark 4.2. From the proof it also follows that $lu \in H_{s+1}(\Omega)$. A more precise assertion about smoothness than Theorem 4.2 may be obtained using the spaces $\Pi_s(\Omega)$ and $\Gamma_s(\Gamma)$ introduced in Corollary 4.2. Namely, if $u \in \Pi_{s-1}(\Omega)$, $f \in \Pi_{s-2}(\Omega)$, $g \in \Gamma_{s-3/2}(\Gamma)$, then $u \in \Pi_s(\Omega)$.

4.4. The following example reveals that Theorem 4.1 cannot be strengthened in the sense that the norm $\|u\|_s$ in the left-hand side cannot be replaced by the norm $\|u\|_{s+\rho}$ for any $\rho > 0$.

Example 4.1. Let Ω be the region formed by the intersection of a neighborhood of the origin in $\mathbb{R}^3$ with the half-space $x^3 > 0$. Consider the sequence of infinitely differentiable functions

$$u_m(x) = e^{(ix^2 - x^3)m} \varphi(rm^{1/(p+1)}),$$

where p is an odd natural number; $r = \sqrt{(x^1)^2 + (x^2)^2 + (x^3)^2}$ and $\varphi \in C^\infty$ when $r \geqslant 0$, $\varphi(r) = 0$ when $r \geqslant 1$, and $\varphi(r) = 1$ when $r \leqslant 1/2$.

We shall now verify that for integers $s \geqslant 0$

$$\| u_m \|_s \geqslant C m^{s-1/2-1/(p+1)}, \quad C = \text{const} > 0.$$

Indeed,

$$\sum_{|\alpha| = s} | D^\alpha u_m |^2 = m^{2s} e^{-2mx^3} \varphi^2 (rm^{1/(p+1)}) + O(e^{-2mx^3} m^{2s-2+2/(p+1)}).$$

Therefore,

$$\| u_m \|_s^2 = m^{2s} \int e^{-2mx^3} \varphi^2 (rm^{1/(p+1)})\, dx + O(m^{2s-2p/(p+1)}) \int\limits_{r \leqslant m^{-1/(p+1)}} e^{-2mx^3}\, dx \geqslant$$

$$m^{2s} \int\limits_{2r \leqslant m^{-1/(p+1)}} e^{-2mx^3}\, dx + O(m^{2s-2p/(p+1)}) \int\limits_{r \leqslant m^{-1/(p+1)}} e^{-2mx^3}\, dx =$$

$$c_0 m^{2s-2/(p+1)-1} + O(m^{2s-3}) \geqslant \frac{1}{2} c_0 m^{2s-2/(p+1)-1}.$$

We now consider the norm $\| \Delta u_m \|_k$. Since

$$\Delta u_m = e^{(ix^2 - x^3)m} \left[\Delta \psi + 2im \frac{\partial \psi}{\partial x^2} - 2m \frac{\partial \psi}{\partial x^3} \right],$$

where $\psi(x) = \varphi(rm^{1/(p+1)})$, we have for integers k

$$\sum_{|\alpha| \leqslant k} | D^\alpha u_m |^2 \leqslant m^{2k+2+\frac{2}{p+1}} \sum_{j=2}^{3} | D_j \psi |^2 e^{-2mx^3} + O\left(e^{-2mx^3} m^{2k+\frac{4}{p+1}}\right).$$

Hence

$$\| \Delta u_m \|_k^2 \leqslant C_1 m^{2k+2+\frac{2}{p+1}} \int\limits_{r \leqslant m^{-1/(p+1)}} e^{-2mx^3}\, dx \leqslant C_2 m^{2k+1}.$$

A similar inequality for real $k > 0$ is obtained by applying interpolational inequalities (see J. Lions and E. Magenes[1]).

Let us now consider the norm $\|lu\|^{\Gamma}_{n}$, where $lu = (x^1)^p \frac{\partial u}{\partial x^3} + \frac{\partial u}{\partial x^1}$. For $x^3 = 0$ we have

$$lu = -m(x^1)^p e^{imx^2}\varphi\left(rm^{\frac{1}{p+1}}\right) + m^{\frac{1}{p+1}} e^{imx^2} \frac{x^1}{\sqrt{(x^1)^2+(x^2)^2}} \varphi'\left(rm^{\frac{1}{p+1}}\right).$$

Then for $x^3 = 0$

$$\sum_{|\alpha|\leqslant n} |D^\alpha lu|^2 \leqslant m^{\frac{2}{p+1}+2n} \varphi'^2\left(rm^{\frac{1}{p+1}}\right) + C\sum_{j=1}^{n} m^{2+2(n-j)+\frac{2j}{p+1}} (x^1)^{2(p-j)} + C_1 m^{2n-2+\frac{4}{p+1}},$$

and therefore

$$\|lu\|^{\Gamma}_{n} \leqslant C_2 \sum_{j=0}^{n} m^{2n+2-2j+\frac{2j}{p+1}} \int_{r\leqslant m^{-\frac{1}{p+1}}} (x^1)^{2(p-j)}\,dx + m^{2n+\frac{2}{p+1}} \int_{r\leqslant m^{-\frac{1}{p+1}}} dx \leqslant C_3 m^{2n}.$$

A similar estimate

$$\|lu\|^{\Gamma}_{s} \leqslant C_3 m^{2s}$$

for $s > 0$, is obtained using interpolational inequalities.

The inequalities obtained reveal that

$$\|u_m\|_{s+\rho} \geqslant C m^{s+\rho-\frac{1}{2}-\frac{1}{p+1}},$$

while

$$\|\Delta u_m\|_{s-1} + \|lu_m\|^{\Gamma}_{s-1/2} + \|u_m\|_{s-1} \leqslant C m^{s-1/2}.$$

Clearly if p is so large that $1 < (p + 1)\rho$, then the estimate

$$\|u_m\|_{s+\rho} \leqslant C\left(\|\Delta u_m\|_{s-1} + \|lu_m\|^{\Gamma}_{s-1/2} + \|u_m\|_{s-1}\right)$$

may not be satisfied for all m.

§5. Manifolds of Class III

I. As already noted, this is the most simple manifold to be studied, since $l + \delta\nu$ is not tangent to the boundary for an appropriate sign of δ. The boundary problem (1.1)-(1.2) when there are no class I and class II manifolds on the boundary of Γ has a finite-dimensional kernel and cokernel, and the range of the corresponding operator is closed.

The method discussed in the previous two sections is applicable to the boundary problem (1.1)-(1.2).

As above, the proofs are based on the following lemma:

Lemma 5.1. Let the manifold Γ_0 belong to class III, and let the support of u be concentrated in a small neighborhood ω of this manifold. Then for every $\varepsilon > 0$ and $s \in \mathbb{R}$ we may find d such that if $\omega \subset \Omega_d$, where Ω_d is a set of points in Ω whose distance from Γ_0 does not exceed d, then

$$\|u\|_s \leqslant \varepsilon \|lu\|_s. \tag{5.1}$$

Proof. Using the partition of unity in Proposition 2.2 we reduce the proof to the case when the support of function u lies in such a small neighborhood ω of point $P \in \Gamma_0$ that the coordinates $y^1, \ldots, y^n$ of Proposition 2.1 are defined in it. Every point of this neighborhood may be connected with some point on the boundary Γ by a segment of a line of length $\leqslant d$, parallel to the Oy^1 axis.

The function $v = lu$ may be smoothly extended to the entire space $\mathbb{R}^n$; therefore, function u is continued beyond the limits of Ω as a solution of the equation $\partial u/\partial y^1 = v$. Inequality $\|u\|_s^\omega \leqslant \varepsilon\|v\|_s^\omega$ is easily verified, using the Cauchy-Schwarz-Bunyakovski inequality, when $d \leqslant \varepsilon$.

5.2. We first show that an *a priori* estimate is satisfied for the solution.

Theorem 5.1. If $u \in H_s(\Omega)$ and is a solution of the problem (1.1)-(1.2), then for $s > 3/2$ the inequality

$$\|u\|_s \leqslant C(\|f\|_{s-1} + \|g\|_{s-1/2}^{\Gamma} + \|u\|_0)$$

is valid, where the constant C is independent of u.

Proof. Let $lu = v$. Then v is the solution of the Dirichlet problem

$$Lv = lf + Au \quad \text{in} \quad \Omega, \qquad v = g \quad \text{on} \quad \Gamma.$$

Therefore, the inequality

$$\|v\|_s \leqslant C_1(\|lf\|_{s-2} + \|Au\|_{s-2} + \|g\|_{s-1/2} + \|u\|_0) \leqslant C_2(\|f\|_{s-1} + \|u\|_s + \|g\|_{s-1/2}) \tag{5.2}$$

is satisfied for $s > 1/2$.

By Lemma 5.1 $\|\psi_j u\|_s \leqslant K\varepsilon\|l\psi_j u\|_s$ when $j = 1, \ldots, N$. Here ψ_j are functions defined in Proposition 2.2. On the other hand,

$$\sum_1^N \|[l, \psi_j]u\|_s + \|\psi_{N+1}u\|_s \leqslant C_3(\|f\|_{s-2} + \|g\|_{s-3/2} + \|u\|_0)$$

by property I of Proposition 2.3, as the support of ψ_{N+1} and $l(\psi_j)$ is distinct from Γ_0. Thus,

$$\|u\|_s \leqslant \sum_1^{N+1} \|\psi_j u\|_s \leqslant K\varepsilon \sum_1^N \|l\psi_j u\|_s + C_\varepsilon(\|f\|_{s-2} + \|g\|_{s-3/2} + \|u\|_0) \leqslant$$

$$K_1\varepsilon\|v\|_s + C'_\varepsilon(\|f\|_{s-2} + \|g\|_{s-3/2} + \|u\|_0). \tag{5.3}$$

Now choose $\varepsilon > 0$ so that the inequality $C_2K\varepsilon < 1/2$ is satisfied, and fix this value. Combining (5.2) and (5.3) we obtain

$$\|u\|_s \leqslant C_4(\|f\|_{s-1} + \|g\|_{s-1/2} + \|u\|_0),$$

which proves the lemma.

Remark 5.1. As above, introducing spaces $\Pi_s(\Omega)$ with norm $\|u\|_s + \|hlu\|_s$ and $\Gamma_s(\Gamma)$ with norm $\|u\|_s^\Gamma + \|hu\|_{s+1}^\Gamma$, we have estimates *above and below*:

$$C^{-1}(\|f\|_{\Pi_{s-2}(\Omega)} + \|g\|_{\Gamma_{s-3/2}(\Gamma)} + \|u\|_0) \leqslant \|u\|_{\Pi_s(\Omega)} \leqslant$$

$$C(\|f\|_{\Pi_{s-2}(\Omega)} + \|g\|_{\Gamma_{s-3/2}(\Gamma)} + \|u\|_0).$$

Corollary 5.1. The space of solutions of the homogeneous problem (1.1)-(1.2) in $H_s(\Omega)$ for $s > 3/2$ is a finite-dimensional space.

Corollary 5.2. The intersection of the range of the operator $L \times l$: $H_s(\Omega) \to H_{s-2}(\Omega) \times H_{s-3/2}(\Gamma)$ with $H_{s-1}(\Omega) \times H_{s-1/2}(\Gamma)$ is closed.

5.3. We now prove a smoothness theorem.

Theorem 5.2. If $u \in H_s(\Omega)$, $f \in \Pi_{s-1}(\Omega)$, $g \in \Gamma_{s-1/2}(\Gamma)$, where $s > 1/2$ and u is a solution of the problem (1.1)-(1.2), then $u \in \Pi_{s+1}(\Omega)$ and the inequality

$$\|u\|_{\Pi_{s+1}(\Omega)} \leqslant C(\|f\|_{\Pi_{s-1}(\Omega)} + \|g\|_{\Gamma_{s-1/2}(\Gamma)} + \|u\|_0)$$

is valid.

Proof. By hypothesis $f \in H_{s-1}(\Omega)$ and $g \in H_{s-3/2}(\Gamma)$. By property I of Proposition 2.3, $u \in H_{s+1}$ outside an arbitrarily small neighborhood of Γ_0.

Recall that $h \in C^\infty(\bar{\Omega})$ is a function such that $h(x) = 1$ in a d/2-neighborhood of manifold Γ_0 and $h(x) = 0$ outside a d-neighborhood of this manifold. Note that

$$L(hu) - hLu \in H_s(\Omega),$$

since $Lh - hL$ is a first-order operator whose coefficients differ from 0 only in the region where $u \in H_{s+1}$. Therefore, $L(hu) \in \Pi_{s-1}(\Omega)$. Similarly, $l(hu) \in H_{s+1/2}(\Gamma)$.

Let $\{\psi_k,\ k = 1,\dots, N + 1\}$ be functions in the partition of unity constructed in Proposition 2.1. Obviously, to prove the theorem it suffices to show that $\psi_k u \in \Pi_{s+1}(\Omega)$ and that

$$\|\psi_k u\|_{\Pi_{s+1}(\Omega)} \leqslant C\left(\|Lu\|_{\Pi_{s-1}(\Omega)} + \|g\|_{\Gamma_{s-1/2}(\Gamma)} + \|u\|_0\right).$$

Let ψ be one of the functions ψ_k and let its support U be so small that the coordinates $y^1,\dots, y^n$ of Proposition 2.2 are defined in it. By hypothesis

$$Ll(\psi u) = \psi lf + Au \in H_{s-2}(\Omega), \tag{5.4}$$

$$l(\psi u) = \psi lu \in H_{s+1/2}(\Gamma), \tag{5.5}$$

since the functions $\psi_1,\dots, \psi_N$ are assumed to differ from 0 only in the domain where $h = 1$. According to a well-known theorem (from the theory of elliptic boundary problems) we conclude that $l(\psi u) \in H_s(\Omega)$, that is, $lu \in H_s$ in some neighborhood of Γ_0. Computing $Ll(\psi u)$ more accurately, we see that

$$Ll(\psi u) = \psi lf + A_2(\psi u) + A_1(u) + A_1'(lu) = F_1 + A_2(\psi u),$$

where A_1 and A_1' are first-order operators, A_2 is a second-order operator, and $F_1 \in H_{s-1}(\bar{\Omega})$.

Let R be a left parametrix for the Dirichlet problem in the region U so that

$$R(Lz,\ z)|_{\partial U} = z + T_1 z,$$

where T_1 is an operator of order -1 and

$$R \in L(H_{l-2}(U) \times H_{l-1/2}(\partial U)),\ H_l(U)).$$

From (5.4) and (5.5) it is clear that

$$l(\psi u) + T_1 l(\psi u) = R(F_1 + A_2(\psi u),\ \psi g) = F_2 + A_0(\psi u), \tag{5.6}$$

where $A_0 \in L(H_s(U),\ H_s(U))$ and $F_2 = R(F_1,\ \psi g) \in H_{s+1}(U)$.

Thus,

$$\psi u = SA_0(\psi u) + F_3,$$

where $F_3 \in H_{s+1}(U)$. According to Lemma 5.1, $\|S\| < \varepsilon$ if d is sufficiently small. Therefore, for small d the operator $I - SA_0$ is invertible so that $\psi u \in H_{s+1}(\Omega)$, and, by (5.6), $l(\psi u) \in H_{s+1}(\Omega)$, and

$$\|\psi u\|_{s+1} + \|l(\psi u)\|_{s+1} \leqslant C(\|f\|_{s-1} + \|lf\|_{s-1} + \|g\|_{s-1/2} + \|hg\|_{s+1/2} + \|u\|_0).$$

5.4. Theorem 5.3. If $f \in \Pi_{s-2}(\Omega)$, $g \in \Gamma_{s-3/2}(\Gamma)$, where $s > 1/2$ and the functions (f, g) satisfy solvability conditions of the form $F_i(f, g) = 0$, $i = 1,\dots, N$, where F_i are linear functionals in $\Pi_{s-2}(\Omega) \times \Gamma_{s-3/2}(\Gamma)$, then there is a solution u of the problem (1.1)-(1.2) in $\Pi_s(\Omega)$.

Proof. Let $\{\psi_j,\ j = 1,\dots, N + 1\}$ be the functions forming a partition of unity which appear in Proposition 2.2. Let $\{\varphi_j,\ j = 1,\dots, N + 1\}$ be an analogous system of functions and let supp φ_j lie in the set on which $\psi_j = 1$.

By property II of Proposition 2.3, there exists an operator $R \in L(H_{s-2}(\Omega)\times H_{s-3/2}(\Gamma), H_s(\Omega))$ for which

$$\varphi_{N+1}LR(\psi_{N+1}f, \psi_{N+1}g)=\varphi_{N+1}f+T_{N+1}(f, g) \text{ in } \Omega,$$

$$\varphi_{N+1}lR(\psi_{N+1}f, \psi_{N+1}g)=\varphi_{N+1}g+T'_{N+1}(f, g) \text{ on } \Gamma,$$

where

$$T_{N+1}\in L(H_{s-2}(\Omega)\times H_{s-3/2}(\Gamma), \Pi_{s-1}(\Omega)),$$

$$T'_{N+1}\in L(H_{s-2}(\Omega)\times H_{s-3/2}(\Gamma), \Gamma_{s-1/2}(\Gamma)).$$

Let now $1 \leqslant i \leqslant N$. We shall construct an operator $R_i \in L(\Pi_{s-2}(U_i)\times \Gamma_{s-3/2}(\Gamma_i), \Pi_s(\Omega))$, where $U_i = \operatorname{supp}\psi_i$ and $\Gamma_i = \partial U_i \cap \Gamma$, for which

$$\varphi_i LR_i(\psi_i f, \psi_i g)=\varphi_i f+T_i(f, g) \text{ in } \Omega,$$

$$\varphi_i lR_i(\psi_i f, \psi_i g)=\varphi_i g+T'_i(f, g) \text{ on } \Gamma,$$

and

$$T_i\in L(\Pi_{s-2}(\Omega)\times\Gamma_{s-3/2}(\Gamma), \Pi_{s-1}(\Omega)),$$

$$T'_i\in L(\Pi_{s-2}(\Omega)\times\Gamma_{s-3/2}(\Gamma), \Gamma_{s-1/2}(\Gamma)).$$

The operator

$$R=\sum_{i=1}^{N+1}\varphi_i R_i \psi_i$$

will then be a parametrix of our problem. Indeed,

$$LR(f, g)=\sum L\varphi_i R_i(\psi_i f, \psi_i g)=\sum \varphi_i LR_i(\psi_i f, \psi_i g)+T(f, g)=$$

$$\sum\varphi_i f+T_1(f, g)=f+T_1(f, g),$$

where T and T_1 are operators from $L(\Pi_{s-2}(\Omega)\times\Gamma_{s-3/2}(\Gamma), \Pi_{s-1}(\Omega))$.

At the points of the boundary Γ

$$lR(f, g)=\sum l\varphi_i R_i(\psi_i f, \psi_i g)=\sum\varphi_i lR_i(\psi_i f, \psi_i g)+T'(f, g)=$$

$$\sum\varphi_i g+T'_1(f, g)=g+T'_1(f, g),$$

where T' and T'_1 are operators from $L(\Pi_{s-2}(\Omega)\times\Gamma_{s-3/2}(\Gamma), \Gamma_{s-1/2}(\Gamma))$. Here, we make use of the fact that $l\varphi_i = 0$ in a neighborhood of Γ_0 so that the support of function T'(f, g) lies outside this neighborhood.

Thus, if a solution of our problem is found in the form $u = R(\tilde f, \tilde g)$, then the problem reduces to solving a system of equations

$$\tilde f+T_1(\tilde f, \tilde g)=f, \quad \tilde g+T'_1(\tilde f, \tilde g)=g,$$

to which the Fredholm theorem is applicable.

We shall now construct the operator R_i for $i = 1, \ldots, N$. For simplicity we drop the index i and assume the functions f and g to be different from 0 only in a small neighborhood U of a point P lying on Γ_0.

Let R_D be a parametrix for the Dirichlet problem in the region U. Then

$$LR_D(f, g)=f+T_D(f, g) \text{ in } U,$$

$$R_D(f, g)=g+T'_D(f, g) \text{ on } \partial U.$$

Here

$$T_D\times T'_D\in L(H_{s-2}(U)\times H_{s-1/2}(\partial U), H_{s-1}(U)\times H_{s+1/2}(\partial U)).$$

If u is a solution of the problem (1.1)-(1.2), then

$$Llu = lf + Au \text{ in } U, \quad lu = g \text{ on } \Gamma.$$

Therefore,

$$u = SR_D(lf + Au, g).$$

According to Lemma 3.2, which is also applicable here, we can represent the operator $u \mapsto SR_D(Au, 0)$ in the form $A_1 + A_2$, where

$$A_1 \in L(H_s(U), H_s(U)) \text{ and } \|A_1\| < 1/2, \text{ but } A_2 \in L(H_s(U), H_{s+1}(U)).$$

Thus, $(I - A_1 - A_2)u = SR_D(lf, g)$. Define the operator Φ such that if $w \in H_s(U)$ is represented in the form $w = (I - A_1 - A_2)z$ for some $z \in H_s(U)$, then Φw coincides with the element z of minimum norm. If w is not represented in this form, then we assume $\Phi w = \Phi w'$, where w' is the projection of w onto

$$\text{Im}(I - A_1 - A_2) = \{(I - A_1 - I_2)z;\ z \in H_s(U)\}.$$

We now assume that

$$R(\psi f, \psi g) = \Phi(SR_D(l\psi f, \psi g))$$

and verify that R is the desired parametrix.

Let $v = R(\psi f, \psi g)$. Clearly $v \in H_{s-1}(U)$. We shall show that $v \in \Pi_s(U)$. We have

$$(I - A_1 - A_2)v = SR_D(l\psi f, \psi g) + T_0(\psi f, \psi g),$$

where $T_0(\psi f, \psi g) \in \Pi_{s+2}(U)$. From this and from the definition of A_1 and A_2 it follows that

$$v = SR_D(l\psi f + Av, \psi g) + T_0(\psi f, \psi g),$$

so that

$$lv = R_D(l\psi f + Av, \psi g) + lT_0(\psi f, \psi g).$$

Clearly $lv \in H_{s-1}(U)$ and

$$L(lv) = l\psi f + Av + T_1(\psi f, \psi g),$$

that is,

$$l(Lv - \psi f) = T_1(\psi f, \psi g) \in H_s(U).$$

It now follows that

$$Lv - \psi f = ST_1(\psi f, \psi g) \in H_s(U), \tag{5.7}$$

$$lv - \psi g = T'(\psi f, \psi g) \in H_{s+1/2}(\partial U). \tag{5.8}$$

So far we have made use of the fact that $f \in H_{s-2}(\Omega)$ and $g \in H_{s-3/2}(\Gamma)$. But it is also known that

$$l(hf) \in H_{s-2}(\Omega), \quad hg \in H_{s-1/2}(\Gamma).$$

It is obvious from (5.7) and (5.8) that $v \in H_s$ outside a d/2-neighborhood of manifold Γ_0. Therefore,

$$l(hv) - hg \in H_{s-1/2}(\Gamma \cap \partial U).$$

Since $Lhv - hf \in H_{s-1}(U)$, we have

$$Lhv \in \Pi_{s-2}(U), \quad lhv|_\Gamma \in \Gamma_{s-3/2}(\Gamma).$$

Repeating the proof of Theorem 5.2 we find that $v \in \Pi_s(U)$.

Thus, the proof of Theorem 5.3 is complete.

5.5. Note that Example 4.1 with odd p illustrates that the *a priori* estimates of the solution cannot be improved in the case of a class II manifold either.

CHAPTER V. WAVE FRONT SETS. FOURIER INTEGRAL OPERATORS

§1. Wave Front Sets. Definitions and Examples

1.1. The fundamental concept of a wave front set was introduced by A. Sato (for the analytic case) and L. Hörmander. This concept provides a more exact description of singularities of solutions to certain classes of differential equations. This has made it possible to more accurately define the process of propagation of singularities of solutions and to simplify the study of their smoothness properties. This concept is also very natural and useful for the theory of multiple trigonometric series and multidimensional Fourier integrals.

The basis of the definition of a wave front set is the simple idea that a singularity of a distribution in $\mathbb{R}^n$, for $n > 1$, is characterized not only by its location in this space, but also by its direction in the dual space. If, for example, a distribution u is a plane wave, that is, a function of the type $f(\omega, x)$, where $\omega \in \mathbb{R}_n \setminus 0$, then this function is smooth in any direction orthogonal to the direction of ω. Since any function, according to the Radon theorem, can be represented as a sum of plane waves, its "singular" directions at every point can be determined as a sum of directions singular for these plane waves. The applications of the concept of a wave front set to the theory of differential equations are based on the fact that the wave front set of a distribution which is a solution of an equation with a smooth right-hand side possesses at each point a bicharacteristic through the point of the principal symbol corresponding to the differential operator.

1.2. Definition 1.1. A point $(x_0, \xi^0), \in T^*\Omega \setminus 0$, does not belong to the *wave front set* WF of a distribution $u \in \mathcal{D}'(\Omega)$ if there exist two functions φ and ψ such that

1) $\varphi \in C_0^\infty(\Omega)$, $\varphi = 1$ in a neighborhood of x_0;

2) $\psi \in C^\infty(\mathbb{R}_n)$, $\psi(t\xi) = \psi(\xi)$ when $t \geqslant 1$, $|\xi| \geqslant 1$, $\psi(\xi^0) \neq 0$;

3) $\psi(D)\varphi u(x) \in C^\infty(\mathbb{R}^n)$.

For applications it is often convenient to replace condition 3) by the assumption of a decreasing function $\psi(\xi)\widetilde{\varphi u}(\xi)$ such that

$$|\widetilde{\varphi u}(\xi)| \leqslant C_N (1 + |\xi|)^{-N} \tag{1.1}$$

for any N and for $\xi \in \Gamma$, where $\Gamma = \operatorname{supp} \psi$.

Proposition 1.1. A point (x_0, ξ^0) does not belong to WF(u) if and only if there exists a function $\varphi \in C_0^\infty(\Omega)$ such that $\varphi(x_0) \neq 0$ and there is a cone Γ in $\mathbb{R}_n$ with vertex at the origin which contains in its interior a ray $\{\xi;\ \xi = t\xi^0,\ t > 0\}$ such that (1.1) is satisfied for $\xi \in \Gamma$.

Proof. Let inequality (1.1) be satisfied and let there be a function $\psi \in C^\infty$ such that $\operatorname{supp} \psi \subset \Gamma$, $\psi(\xi^0) = 1$ and $\psi(t\xi) = \psi(\xi)$ when $t \geqslant 1$, $|\xi| \geqslant |\xi^0|$. Then, obviously, $\psi(D)\varphi u(x) \in C^\infty$ and hence $(x_0, \xi^0) \notin \mathrm{WF}(u)$.

Conversely, suppose $(x_0, \xi^0) \in \mathrm{WF}(u)$ and φ and ψ satisfy the conditions of Definition 1.1. We shall show that $\psi(\xi)\widetilde{\varphi u}(\xi) \in \mathcal{S}$ and, in particular, $\widetilde{\varphi u}(\xi)$ rapidly decreases in a conic neighborhood of vector ξ^0, that is, inequality (1.1) is satisfied.

If $h \in C_0^\infty$, then the function $h\psi\varphi u$ also is in C_0^∞ and it suffices to prove that the inequalities

$$|D^\alpha (1-h)\psi(D)v(x)| \leqslant C_{\alpha, N}(1+|x|)^{-N}$$

are valid for any α and N, where $v = \varphi u$. Suppose $\varphi(x) = 0$ when $|x - x_0| > \rho$ and $h = 1$ in a neighborhood of the support of φ.

Note that $v \in \mathscr{E}'(\mathbb{R}^n)$ and by Theorem 3.6 of Chapter I, $v = \Delta^k W$, where k is some integer and W is a continuous function. Therefore

$$\psi(D)v(x) = (2\pi)^{-n}\int\int e^{i(x-y)\xi}\psi(\xi)v(y)\,dy\,d\xi =$$

$$(2\pi)^{-n}\int\int e^{i(x-y)\xi}|\xi|^{2k}\psi(\xi)W(y)\,dy\,d\xi.$$

Using the equality

$$(-\Delta_\xi)^N e^{i(x-y)\xi} = |x-y|^{2N}e^{i(x-y)\xi}$$

and integrating by parts, we write

$$\psi(D)v(x) = (2\pi)^{-n}\int\int e^{i(x-y)\xi}(-\Delta_\xi)^N[|\xi|^{2k}\psi(\xi)]w(y)|x-y|^{-2N}\,dy\,d\xi$$

if $x \in \operatorname{supp}(1-h)$ and $y \in \operatorname{supp} w$, since in this case

$$|x-y| \geqslant c_0(1+|x|), \qquad c_0 = \text{const} > 0.$$

If $2N > 2k + n + 1$ and $2N > n + 1$, then the last integral converges absolutely and the inequality

$$|\psi(D)v(x)| \leqslant C(1+|x|)^{-2N+n+1}$$

is valid.

Similarly, the inequalities

$$|D^\alpha\psi(D)v(x)| \leqslant C^1_{\alpha, N}(1+|x|)^{-2N+n+1}$$

are verified, which completes the proof.

1.3. Example 1.1. Let the distribution u be a plane wave, that is $u(x) = f(x^1)$, where $f \in \mathscr{D}'(\mathbb{R})$. Then any direction nonparallel to the x^1-axis is nonsingular and every point (x_0, ξ^0) in $T^*\Omega\setminus 0$ is not in WF(u) if $\xi^0_j \neq 0$ for some $j \neq 1$.

Indeed, if $\varphi \in C_0^\infty$, then

$$\widetilde{\varphi u}(\xi) = \tilde{\varphi}(\xi) * \tilde{u}(\xi) = \tilde{\varphi}(\xi) * \tilde{f}(\xi_1)\delta(\xi') = \int \tilde{\varphi}(\xi_1 - \eta_1, \xi')\tilde{f}(\eta_1)\,d\eta_1.$$

Suppose a cone Γ with its vertex at the origin contains a ray defined by the vector ξ^0 and does not contain the ξ_1-axis and the directions near to it, so that

$$|\xi_1| \leqslant c|\xi'|$$

for $\xi \in \Gamma$. Then

$$\varphi\tilde{u}(\xi) = \int \tilde{\varphi}(\xi_1 - \eta_1, \xi')\tilde{f}(\eta_1)\,d\eta_1.$$

If $\xi \in \Gamma$, then

$$|\tilde{\varphi}(\xi_1-\eta_1, \xi')| \leqslant C_{N,N'}(1+|\xi_1-\eta_1|)^{-N}(1+|\xi'|)^{-N'} \leqslant$$

$$C_{N,N'}(1+|\eta_1|)^{-N_1}(1+|\xi_1|)^N(1+|\xi'|)^{-N'} \leqslant C'_{N,N'}(1+|\eta_1|)^{-N}(1+|\xi|)^{N-N'}.$$

Without loss of generality, we suppose $f(x^1) = 0$ for $|x^1 - x^1_0| \geqslant \delta > 0$. Therefore, there exist C and k such that

$$|\tilde{f}(\eta_1)| \leqslant C(1+|\eta_1|)^k.$$

Choosing $N_1 = k + 2$ we find that for any l the estimate

$$|\widetilde{\varphi u}(\xi)| \leqslant C_l(1+|\xi|)^{-l}$$

is valid, indicating that $(x_0, \xi^0) \notin WF(u)$. This completes the proof.

1.4. Example 1.2. If $\varphi \in C_0^\infty$, $\tilde{\varphi} \geqslant 0$, $\tilde{\varphi}(0) = 1$, $\eta \neq 0$ then the function

$$u(x) = \sum_{k=1}^{\infty} k^{-2l}\varphi(kx)\, e^{ik^2(x,\eta)}$$

is in $C^{l-1}(\mathbb{R}^n)$ for $l \geqslant 0$ and $WF(u) = \{(0, t\eta), t > 0\}$.

Proof. It is clear that at every point $x \neq 0$ only a finite number of terms differ from zero and therefore $u \in C^\infty(\mathbb{R}^n \setminus 0)$. Thus the points (x, ξ) are not in $WF(u)$ when $x \neq 0$. If $\tilde{\varphi}$ is the Fourier transform of the function φ, then

$$\tilde{u}(\xi) = \sum_{k=1}^{\infty} k^{-2l-n}\tilde{\varphi}\left(\frac{\xi - k^2\eta}{k}\right),$$

so $\tilde{u}(k^2\eta) \geqslant k^{-2l-n}$; that is, $(0, \eta) \in WF(u)$.

If ξ^0 is not collinear to η, then we find $c > 0$ such that the inequality

$$|\xi - \eta| \geqslant c(|\xi| + |\eta|)$$

is satisfied in some conic neighborhood Γ of the vector ξ^0.

The estimates

$$\left|\frac{\xi - k^2\eta}{k}\right| \geqslant c\frac{|\xi| + k^2|\eta|}{k} \geqslant c_1\sqrt{|\xi|}, \quad c_1 > 0,$$

are therefore valid for all $k \geqslant 1$. We see that for all N and $\xi \in \Gamma$

$$|\tilde{u}(\xi)| \leqslant \sum_{k=1}^{\infty} k^{-2l-n}C_N(1+\sqrt{|\xi|})^{-N} \leqslant C_N'(1+|\xi|)^{-N/2},$$

that is, $(0, \xi^0) \notin WF(u)$. This completes the proof.

§2. Basic Properties of a Wave Front Set

2.1. We shall first prove the following statement.

Lemma 2.1. If $h \in C_0^\infty(\mathbb{R}^n)$ and $u \in \mathscr{D}'(\mathbb{R}^n)$, then

$$WF(hu) \subset WF(u).$$

Proof. Let $v = hu$ and $(x_0, \xi^0) \notin WF(u)$. By Proposition 1.1, there exists a cone Γ with vertex at the origin and a function $\varphi \in C_0^\infty$ such that $\varphi = 1$ in a neighborhood of the point x_0 and if $\xi \in \Gamma$, then $|\widetilde{\varphi u}(\xi)| \leqslant C_N(1+|\xi|)^{-N}$ for every N. Therefore, for $\xi \in \Gamma'$, where Γ' is a cone which lies inside Γ and contains the point ξ^0, we have

$$\widetilde{\varphi hu}(\xi) = \int_\Gamma \tilde{h}(\xi - \eta)\widetilde{\varphi u}(\eta)\, d\eta + \int_{R_n \setminus \Gamma} \tilde{h}(\xi - \eta)\widetilde{\varphi u}(\eta)\, d\eta.$$

To estimate these integrals we need the following inequality:

$$1 + |\xi| \leqslant (1+|\eta|)(1+|\xi - \eta|).$$

In the first integral the integrand is estimated by

$$C_N'(1+|\xi - \eta|)^{-N}(1+|\eta|)^{-N-n-1} \leqslant C_N''(1+|\xi|)^{-N}(1+|\eta|)^{-n-1}.$$

As to the second integral, the inequality $|\xi - \eta| \geqslant c_0|\xi|$ with some $c_0 > 0$ is satisfied for $\xi \in \Gamma'$ and therefore the integrand is estimated by

$$B_N(1+|\xi-\eta|)^{-N-2k-n-1}(1+|\eta|)^k \leqslant B'_N(1+|\xi|)^{-N}(1+|\eta-\xi|)^{-n-1},$$

since $|\widetilde{\varphi u}(\eta)| \leqslant C(1+|\eta|)^k$, according to Theorem 2.3 of Chapter I, and

$$1+|\eta| \leqslant (1+|\xi|)(1+|\xi-\eta|) \leqslant C(1+|\xi-\eta|)^2.$$

Thus, we see that for $\xi \in \Gamma$,

$$|\widetilde{h\varphi u}(\xi)| \leqslant C_N^0(1+|\xi|)^{-N}$$

for all N, that is $(x_0, \xi^0) \notin WF(u)$. This completes the proof.

2.2. The relation of a wave front set to the singular support of a distribution is given in the following statement.

Proposition 2.1. If π: $T^*\Omega \to \Omega$ is a natural projection, then $\pi WF(u)$ = sing supp u.

Proof. If $x_0 \in$ sing supp u, then there is a function $\varphi \in C_0^\infty(\Omega)$ equal to 1 in a neighborhood of the point x_0 and $\varphi u \in C_0^\infty(\Omega)$. Therefore, $|\widetilde{\varphi u}(\xi)| \leqslant C_N(1+|\xi|)^{-N}$ for every N and all $\xi \in \mathbb{R}_n$ so that $(x_0, \xi) \notin WF(u)$ for an arbitrary vector $\xi \in \mathbb{R}_n \setminus 0$.

Conversely, if $(x_0, \xi) \notin WF(u)$ for all vectors $\xi \in \mathbb{R}_n$ when $|\xi| = 1$, then for every point $\xi^0 \in S^{n-1}$ we can find a cone Γ with vertex at the origin and containing a point ξ^0 such that $|\widetilde{\varphi u}(\xi)| \leqslant C_N(1+|\xi|)^{-N}$ for all N and for some function $\varphi \in C_0^\infty(\Omega)$ with $\varphi(x) = 1$ in a neighborhood of the point x_0. Let us denote by ω_{ξ_0} a neighborhood of the point x_0 on the sphere S^{n-1} in the interior of Γ. From a covering of S^{n-1} by such neighborhoods we choose a finite subcovering of neighborhoods $\omega_1, \ldots, \omega_k$ corresponding to points $\xi^1, \ldots, \xi^k$. To each of these points corresponds a function $\varphi_j(x)$ and a cone Γ_j, φ_j equal to 1 in a neighborhood of x_0 and $\bigcup \Gamma_j = \mathbb{R}_n$. If we assume that $\varphi_0(x) = \varphi_1(x) \ldots \varphi_k(x)$, then $\varphi_0(x) = 1$ in some neighborhood of x_0, $\varphi_0 \in C_0^\infty$ and by Lemma 2.1,

$$|\widetilde{\varphi_0 u}(\xi)| \leqslant C_N(1+|\xi|)^{-N}$$

for every N and for any vector $\xi \in \mathbb{R}_n \setminus 0$. This implies that $\varphi_0 u \in C_0^\infty(\Omega)$ and therefore $x_0 \notin$ sing supp u. This completes the proof.

2.3. Proposition 2.2. Let $u \in \mathscr{E}'(\Omega)$. If $(x_0, \xi^0) \notin WF(u)$, then $(x_0, \xi^0) \notin WF(Au)$ for any pseudo-differential operator, that is, $WF(Au) \subset WF(u)$.

Proof. By hypothesis there exists a function $\varphi \in C_0^\infty$ equal to 1 in a neighborhood of x_0 and a nonsingular cone Γ with vertex at the origin and axis collinear to ξ^0 such that

$$|\widetilde{\varphi u}(\xi)| \leqslant C_N(1+|\xi|)^{-N}$$

for all N and all $\xi \in \Gamma$.

Let $\psi \in C^\infty$ be a function such that supp $\psi \subset \Gamma$ and $\psi(\xi) = 1$ in a conic neighborhood of ξ^0. Denote by Q the pseudo-differential operator $\psi(D)\varphi(x)$. Clearly the symbol of this operator is equal to 1 in a neighborhood ω of (x_0, ξ^0) and $Qu \in C^\infty$.

Let B be a pseudo-differential operator with symbol $\psi_1(\xi)\varphi_1(x)$, the support of which lies in ω, homogeneous of degree zero for $|\xi| > |\xi^0|$ and equal to unity in a small conic neighborhood ω' of the point (x_0, ξ^0). We shall show that $BAu \in C^\infty$, from which it will follow that $(x_0, \xi^0) \notin WF(Au)$.

We have

$$BAu = ABu + [B, A]u = ABQu + AB(I-Q)u + [B, A]Qu + [B, A](I-Q)u.$$

All terms on the right-hand side are infinitely differentiable functions, since $Qu \in C^\infty$ and the symbol of B is equal to zero in a neighborhood of the support of the symbol of $Q - I$.

The proposition is proven.

2.4. Since the principal symbols of pseudo-differential operators are invariant under canonical transformations on the cotangent space, it follows from the following proposition that the wave front set of a distribution is also invariant under these transformations.

Proposition 2.3. A point (x_0, ξ^0) is not in the wave front set of $u \in \mathscr{E}'(\Omega)$ if and only if there is a pseudo-differential operator A of order zero with principal symbol $a(x, \xi)$ such that $Au \in C^\infty$ and $a(x_0, \xi^0) \neq 0$.

Proof. If $(x_0, \xi^0) \notin WF(u)$, then by Proposition 1.1 the operator $A = \psi(D)\varphi(x)$ satisfies the hypotheses of the proposition.

Conversely, if A is an operator with given properties, then arguing as in the proof of Theorem 3.1 of Chapter II we may find an operator B such that the symbol of BA is identically unity when $|\xi| > 1$, in a conic neighborhood ω of (x_0, ξ^0). Let the function $\psi(\xi)\varphi(x)$ be infinitely differentiable, have its support in ω, and be equal to 1 in a smaller conic neighborhood of (x_0, ξ^0), $\psi(t\xi) = \psi(\xi)$ when $|\xi| \geqslant 1$ and $t \geqslant 1$. It suffices to prove that $\psi(D)\varphi u \in C^\infty$.

We have

$$\psi(D)\varphi u = BA\psi(D)\varphi u + (I - BA)\psi(D)\varphi u =$$

$$\psi(D)\varphi BAu + [BA, \psi(D)\varphi]u + (I - BA)\psi(D)\varphi u.$$

The first term is infinitely differentiable, since $Au \in C^\infty$. Since the symbols of $[BA, \psi(D)\varphi]$ and $(I - BA)\psi(D)$ are identically zero, the remaining terms are also infinitely differentiable.

The proof is complete.

§3. Another Approach to the Study of Wave Front Sets

3.1. In this section we shall consider another method of studying the wave front set of a distribution, which involves decomposition of the distribution into plane waves.

Let X and Y be smooth manifolds and let $f\colon X \to Y$ be a smooth mapping. Recall that f is called *proper* if the preimage of every compact set in Y is a compact set in X. If f is proper, then the mapping $f^*\colon C_0^\infty(Y) \to C_0^\infty(X)$ is defined and associates to every function $\varphi \in C_0^\infty(Y)$ the function

$$f^*\varphi(x) = \varphi(f(x)),$$

which obviously belongs to $C_0^\infty(X)$.

Thus, if $u \in \mathscr{D}'(X)$, then we may define f_*u — the *push-forward* of u — as a distribution in $\mathscr{D}'(Y)$ for which

$$f_*u(\varphi) = u(f^*\varphi).$$

In particular, the push-forward f_*u is defined for every distribution u in $\mathscr{E}'(X)$.

If f is a diffeomorphism, then one can define a mapping $f^*\colon \mathscr{D}'(Y) \to \mathscr{D}'(X)$ which is an extension of the mapping $C_0^\infty(Y) \to C_0^\infty(X)$. The distribution f^*u in $\mathscr{D}'(X)$ is in this case called the *pull-back* of the distribution $u \in \mathscr{D}'(Y)$.

A pull-back of a distribution can also be defined when f is a *submersion*, that is when the preimage of every point y is a smooth submanifold diffeomorphic to a fixed manifold Z of dimension k. In this case, having integrated along the fibers one can associate a smooth function $f_*\varphi \in C_0^\infty(Y)$ with every function $\varphi \in C_0^\infty(X)$. Thereafter, every distribution $u \in \mathcal{D}'(Y)$ may be associated with its pull-back f_*u by setting

$$f^*u(\varphi) = u(f_*\varphi).$$

Example 3.1. Let $\pi: \mathbb{R}^n \to \mathbb{R}^1$ be a projection onto the x^1-axis. In this case, both π^*u and π_*v are defined for $u \in \mathcal{E}'(R^1)$ and $v \in \mathcal{E}'(\mathbb{R}^n)$ respectively, and

$$\pi^*u(x) = u(x^1) \otimes 1_{x^2, \ldots, x^n}; \qquad \pi_*v(x^1) = \int v(x^1, x^2, \ldots, x^n)\, dx^2 \ldots dx^n.$$

3.2. Let us now define the wave front set of a distribution using concepts described earlier.

Theorem 3.1. Let Ω be a domain in $\mathbb{R}^n$ and Ω_1 a domain in $\mathbb{R}^m$; x_0 and y_0 are points in Ω and Ω_1. Let $u \in C^\infty(\Omega_1; \mathcal{D}'(\Omega))$ and $(x_0, \xi^0) \in T^*\Omega \setminus 0$. The point $(x_0, \xi^0) \notin WF(u(y_0))$ if and only if there exists a function $\varphi \in C_0^\infty(\Omega)$ such that $\varphi(x_0) \neq 0$ and for every smooth function $f: \operatorname{supp}\varphi \to \mathbb{R}$, for which $\operatorname{grad} f(x_0) = \xi^0$ the function $f_*(\varphi u)(t)$ is smooth in t and in y when $(t, y) \in \mathbb{R} \times \omega$ and ω is a neighborhood of the point y_0 in Ω_1.

To prove Theorem 3.1 we need the following lemmas.

Lemma 3.1. Theorem 3.1 holds for distributions u which are plane waves.

Lemma 3.2. (Radon theorem). If $u \in \mathcal{E}'(\mathbb{R}^n)$, then

$$u = (2\pi)^{-n+1} \int_{S^{n-1}} g_\omega^* I^{n-1} (g_\omega)_* u\, d\omega,$$

where $g_\omega: \mathbb{R}^n \to \mathbb{R}$ is a mapping transforming x into $x \cdot \omega$ for $\omega \in S^{n-1}$ and

$$I^{n-1}v(\tau) = \frac{1}{2\pi} \int_0^\infty t^{n-1} \tilde{v}(t)\, e^{it\tau}\, dt.$$

Lemma 3.3. If $u \in \mathcal{E}'(\mathbb{R}^n)$, then $\widetilde{(g_\omega)_* u}(t) = \tilde{u}(t\omega)$.

Further we shall show how Theorem 3.1 follows from these lemmas.

3.3. Proof of Theorem 3.1. Let $(x_0, \omega^0) \notin WF(u)$. Then the function $\tilde{u}(t\omega)$ rapidly decreases as $t \to +\infty$ for all ω near ω^0. The smoothness of function $(g_\omega)_*u$ follows from Lemma 3.3, and hence $g_\omega^* I^{n-1}(g_\omega)_*u \in C_0^\infty$ for ω near ω^0. Therefore, if $u = u_1 + u_2$, where u_1 stands for the integral as in Lemma 3.2, but in a neighborhood of ω^0 on S^{n-1}, then $u_1 \in C^\infty$ and $f_*(\varphi u_1) \in C^\infty$.

If the direction of vector $df(x_0)$ is near ω^0, then the function $f_*(\varphi u_2)$ is smooth, by Lemma 3.1. Thus, $f_*(\varphi u) = f_*(\varphi u_1) + f_*(\varphi u_2)$ is infinitely differentiable. It is not hard to see that in this case the statement concerning dependence of smoothness on parameters also holds.

Conversely, if the assumptions of the theorem are valid for any function f, then assuming $f = g_\omega$ we see that $(g_\omega)_*\varphi u$ is a smooth function for ω near ω^0. By Lemma 3.3, it now follows that $\widetilde{\varphi u}(t\omega)$ rapidly decreases as $t \to +\infty$ if the direction of ω is near the direction of ω^0. This implies that $(x_0, \omega^0) \notin WF(u)$, which proves the theorem.

3.4. Proof of Lemma 3.1. Let $u = f(x^1)$, where $f \in \mathcal{D}'(\mathbb{R})$ and let $g: \Omega \to \mathbb{R}$ be a smooth function whose vector grad g is noncollinear to the x^1-axis. Using a smooth transformation of coordinates it may be assumed that $g = x^2$. Then

$$(g_*\varphi f)(x^2) = \int \left(\int f(x^1)\, \varphi(x^1, x^2, \ldots, x^n)\, dx^1 \right) dx^3 \ldots dx^n$$

is an infinitely differentiable function.

Since, as in Example 1.1,

$$\mathrm{WF}\, u = \{(x, \xi);\ x^1 \in \operatorname{sing\,supp} f,\ \xi_1 \neq 0,\ \xi_2 = \ldots = \xi_n = 0\},$$

the lemma is proven.

Proof of Lemma 3.2. Note that

$$u(x) = (2\pi)^{-n} \int \tilde{u}(\xi) e^{ix\xi} d\xi = (2\pi)^{-n} \int_0^\infty dr \int_{S^{n-1}} \tilde{u}(r\omega) r^{n-1} e^{ir\omega\cdot x} d\omega.$$

Since $\tilde{u}(r\omega) = \widetilde{(g_\omega)_* u}(r)$, it follows that

$$u(x) = (2\pi)^{-n+1} \int_{S^{n-1}} I^{n-1} (g_\omega)_* u(\omega \cdot x) d\omega,$$

where

$$I^{n-1} v(\tau) = \frac{1}{2\pi} \int_0^\infty r^{n-1} \tilde{v}(r) e^{ir\tau} dr,$$

that is,

$$u(x) = (2\pi)^{-n+1} \int_{S^{n-1}} (g_\omega)^* [I^{n-1}((g_\omega)_* u)] d\omega,$$

which proves the lemma.

Proof of Lemma 3.3. If $u \in \mathscr{E}'(\mathbb{R}^n)$, then

$$\tilde{u}(t\omega) = \int u(y) e^{-it\omega y} dy.$$

Replacing ωy by z^1 we obtain

$$\tilde{u}(t\omega) = \int u(z) e^{-itz^1} dz^1 \ldots dz^n = \int (g_\omega)_* u e^{-itz^1} dz^1 = \widetilde{(g_\omega)_* u}(t),$$

which proves the lemma.

§4. Wave Front Sets and Mappings

4.1. We first consider the wave front set of a push-forward map of a distribution.

Theorem 4.1. Let the mapping f: $X \to Y$ be a submersion and let $u \in \mathscr{E}'(X)$. Then

$$\mathrm{WF}(f_* u) \subset \{(f(x), \eta):\ x \in X,\ (x, {}^t f'_x \eta) \in \mathrm{WF}(u) \text{ or } {}^t f'_x \eta = 0\},$$

where f'_x is the Jacobian of f and ${}^t f'_x$ is the transposed matrix.

Proof. Let $x_0 \in X$, $y_0 = f(x_0) \in Y$ and the vector η^0 be such that ${}^t f'_{x_0} \eta^0 \neq 0$. Assuming that $(x_0, {}^t f'_{x_0} \eta^0) \notin \mathrm{WF}(u)$, we shall prove that $(y_0, \eta^0) \notin \mathrm{WF}(f_* u)$.

Let g: $Y \to \mathbb{R}$ be a smooth mapping and $g'(y_0) = \eta^0$. By Theorem 3.1, in order that $(y_0, \eta^0) \notin \mathrm{WF}(f, u)$ it is sufficient that $g_* f_* u \in C^\infty(\mathbb{R})$. Let us verify this. To do so we consider a mapping $g \cdot f$: $X \to \mathbb{R}$. By the differentiation rule for composite functions, $(g \cdot f)'_{x_0} = {}^t f'_{x_0} \eta^0$. But $(x_0, {}^t f_{x_0} \eta^0) \notin \mathrm{WF}(u)$ and, by Theorem 3.1, $(g \cdot f)_* u \in C^\infty(\mathbb{R})$. Since $(g \cdot f)_* \omega = g_*(f_* u)$ we obtain, by the definition of a push-forward map, that $g_*(f_* u) \in C^\infty(\mathbb{R})$.

The theorem is proven.

Example 4.1. Let f: $\mathbb{R}^2 \to \mathbb{R}$ be the projection $(f(x^1, x^2) = x^1$. Then

$f'_x=(1,\ 0)$, ${}^tf'_x\eta=\binom{\eta}{0}$ for $\eta\in\mathbb{R}$. It is clear from Proposition 4.1 that $\mathrm{WF}\left(\int u(x^1,\ x^2)\,dx^2\right)\subset\{(x^1,\ \eta):\ (x^1,\ x^2,\ \eta,\ 0)\in\mathrm{WF}(u)$ for some $x^2\}$, that is, $\mathrm{WF}(f_*u)$ consists of projections of those points of the wave front set of u for which the singular directions are parallel to the x^1-axis.

4.2. Theorem 4.1 enables us to to prove the following interesting result:

Theorem 4.2. (D. Ludwig). Let X and Y be smooth manifolds $Z = X \times Y$, π: $Z \to X$ be a natural projection. Let $f\colon Z\to\mathbb{R}$ be a smooth mapping. For every $y \in Y$ we shall consider in X a surface

$$S(y)=\{x\in X,\ f(x,\ y)=0\}.$$

If $u\in\mathscr{E}'(\mathbb{R})$, $\operatorname{sing\,supp} u=\{0\}$, $h\in C_0^\infty(Z)$, then

$$\mathrm{WF}(\pi_*hf^*u)\subset\Lambda=\{(x,\ \xi);\ \exists\,(y,\ t)\in Y\times\mathbb{R}_+,\ f(x,\ y)=0,$$

$$f'_y(x,\ y)=0,\ \xi=tf'_x(x,\ y)\}.$$

Example 4.2. Let $X=\mathbb{R}^n$, Y be a smooth hypersurface in $\mathbb{R}^n$, $f(x,\ y)=|x-y|^2-a^2$. Then $S(y)=\{x\in\mathbb{R}^n,\ |x-y|^2=a^2\}$ is a sphere of radius a centered at the point y, and Λ is a set of points normal to the envelope of this family of spheres.

Let $u=\delta(t)$, $t\in\mathbb{R}$. It can be easily verified that

$$f^*u=\delta(|x-y|^2-a^2)\in\mathscr{D}'(X\times Y),\qquad \pi_*hf^*u=\int\limits_Y h(x,\ y)\,\delta(|x-y|^2-a^2)\,dy.$$

Theorem 4.2 reveals that when distributions interfere with singularities on spheres of constant radius and with centers on a smooth surface the singularities vanish at points belonging to several spheres; singularities remain only at their envelope. In mathematical physics this is known as Huygens' principle.

Proof of Theorem 4.2. By Lemma 3.1, if $df(x,\ y) \neq 0$ then

$$\mathrm{WF}(f^*u)=\{(x,\ y,\ tf_x,\ tf_y):\ f=0,\ t\neq 0\}.$$

By Lemma 2.1, $\mathrm{WF}(hf^*u) \subset \mathrm{WF}(f^*u)$. The mapping $\pi'_{x,y}$ is defined by the matrix $\|I,\ 0\|$, where I and 0 stand, respectively, for the unit and zero matrices of order $n \times n$. Therefore, if $\xi\in\mathbb{R}_n$, then ${}^t\pi'_{x,y}\xi = (\xi,\ 0)$. From Theorem 4.1 it follows that

$$\mathrm{WF}(\pi_*hf^*u)\subset\{(x,\ \xi):\ (x,\ y,\ \xi,\ 0)\in\mathrm{WF}(hf^*u)\},$$

that is,

$$\mathrm{WF}(\pi_*hf^*u)\subset\Lambda,$$

which proves the theorem.

4.3. Now we shall study the *wave front set of the pull-back* of a distribution.

Let X and Y be smooth manifolds and F: $X \to Y$ be a smooth mapping. Let $u\in\mathscr{D}'(Y)$. When can we define a distribution $F^*u\in\mathscr{D}'(X)$ such that for $u \in C^\infty(Y)$

$$(F^*u)(x)=u(F(x))\in C^\infty(X)?$$

The answer to this question is given in the following theorem.

Define the set

$$N_F=\{(y,\ \eta);\ \exists x\in X,\ y=F(x),\ {}^tF'(x)\eta=0\}\subset T^*Y.$$

Theorem 4.3. If $\mathrm{WF}(u) \cap N_F = \emptyset$, then the distribution $F^*u \in \mathcal{D}'(X)$ is defined and

$$\mathrm{WF}(F^*u) \subset \{(x, \xi);\ \exists (y, \eta) \in \mathrm{WF}(u);\ F(x) = y,\ \xi = {}^tF'(x)\eta\}.$$

Remark. We have already discussed a particular case of this theorem in Example 1.1, where $Y = \mathbb{R}$ and

$$N_F = \{(y, \eta);\ \exists x \in X,\ y = F(x),\ F'(x) = 0,\ \eta \in \mathbb{R}\}.$$

For this case the theorem asserts that 1) the distribution $F^*u \in \mathcal{D}'(X)$ is defined if $F'(x) \neq 0$ in sing supp u; 2) for this condition

$$\mathrm{WF}(F^*u) \subset \{(x, \xi);\ \exists (y, \eta) \in \mathrm{WF}(u),\ y = F(x),\ \xi = {}^tF'(x)\eta\},$$

that is, sing supp F^*u is a set of level lines $F(x) = y$, where $y \in$ sing supp u, and the vectors ξ are normals to these level lines.

4.4. Proof of Theorem 4.3. Since the theorem is local in nature, we can assume that $X \subset \mathbb{R}^n$, $Y \subset \mathbb{R}^m$, $u \in \mathcal{E}'(Y)$, Y is sufficiently small in a neighborhood of the point y_0, and $u(y_0) \neq 0$.

Let first $u \in C_0^\infty(Y)$. Then

$$(F^*u)(x) = u(F(x)) = (2\pi)^{-n} \int \tilde{u}(\eta)\, e^{iF(x)\cdot\eta}\, d\eta.$$

Assume that $\chi \in C_0^\infty(X)$, $\chi(x_0) \neq 0$ and $F(x_0) = y_0$. Then

$$F^*u(\chi) = (2\pi)^{-n} \int I_\chi(\eta)\, \tilde{u}(\eta)\, d\eta, \tag{4.1}$$

where

$$I_\chi(\eta) = \int \chi(x)\, e^{iF(x)\eta}\, dx. \tag{4.2}$$

We shall prove that if the conditions of the theorem are satisfied, then formula (4.1) holds for every u in $\mathcal{E}'(Y)$ and χ in $C_0^\infty(X)$ when the support of χ is contained in a sufficiently small neighborhood U of x_0. Thus, the first statement of the theorem on the existence of a distribution F^*u will be proven.

Let $V_0 = \{\eta \in \mathbb{R}^m;\ {}^tF'(x_0)\eta = 0\}$ so that if $\eta \in V_0$, then

$$\sum_{j=1}^{m} \frac{\partial F^j(x_0)}{\partial x^k}\eta_j = 0, \qquad k = 1, \ldots, n.$$

Let V be a conic neighborhood of the set V_0. Then for $x \in U$, $\eta \notin V$ the inequality

$$\sum_{k=1}^{n} \left| \sum_{j=1}^{m} \frac{\partial F^j(x)}{\partial x^k}\eta_j \right|^2 \geq c|\eta|^2, \qquad c = \mathrm{const} > 0, \tag{4.3}$$

is satisfied. By hypothesis those directions of η along which the function $\tilde{u}(\eta)$ does not decay faster than any negative degree lie outside V. We shall prove that for any V, $I_\chi(\eta) = O(|\eta|^{-N})$ outside V as $\eta \to \infty$. This implies that (4.1) converges for every function χ in $C_0^\infty(U)$, so the distribution F^*u is defined.

Note that

$$\frac{\partial^2}{\partial (x^k)^2} e^{iF(x)\eta} = \left[i \sum_{j=1}^{m} \frac{\partial^2 F^j}{(\partial x^k)^2}\eta_j - \left(\sum_{j=1}^{m} \frac{\partial F^j}{\partial x^k}\eta_j \right)^2 \right] e^{iF(x)\eta}, \qquad k = 1, \ldots, n.$$

Summing up over k we obtain

$$e^{iF(x)\cdot\eta}=\left[i\sum_{k=1}^{n}\sum_{j=1}^{m}\frac{\partial^2 F_j^j}{(\partial x^k)^2}\eta_j-\sum_{k=1}^{n}\left(\sum_{j=1}^{m}\frac{\partial F_j}{\partial x^k}\eta_j\right)^2\right]^{-1}\Delta_x e^{iF(x)\eta}=A(x,\ \eta)\,\Delta_x e^{iF(x)\eta}$$

and the estimate $|A(x, \eta)| \leqslant C_1|\eta|^{-2}$, when $|\eta| \geqslant 1$, holds by (4.3). Inserting this inequality in (4.2) and integrating by parts, we find

$$I_\chi(\eta)=\int\Delta[\chi(x)\,A(x,\ \eta)]\,e^{iF(x)\eta}\,dx$$

so that the inequality

$$|I_\chi(\eta)|\leqslant C_2|\eta|^{-2}$$

is satisfied, when $\eta \notin V$, by (4.3).

Repeating the argument N times one can show that the estimate

$$|I_\chi(\eta)|\leqslant C_{2N}|\eta|^{-2N}$$

holds when $\eta \notin V$. This proves the theorem.

We now consider $WF(F^*u)$. Let $\chi \in C_0^\infty(U)$ and U be a small neighborhood of x_0. The integral

$$\widetilde{\chi F^*u}(\xi)=(2\pi)^{-n}\int\tilde{u}(\eta)\,d\eta\int\chi(x)\,e^{iF(x)\eta-ix\xi}\,dx \tag{4.4}$$

is written in such a way that the gradient in x of the exponent vanishes on the set

$$V_1=\{(\xi,\ \eta):\ {}^tF'(x)\,\eta=\xi\}.$$

Outside a conic neighborhood V_2 of the set V_1 the inequality

$$|\xi-{}^tF'(x)\,\eta|\geqslant\varepsilon(|\xi|+|\eta|),\qquad \varepsilon=\mathrm{const}>0,$$

is satisfied. Repeating the above steps and integrating by parts, one easily sees that

$$\left|\int e^{iF(x)\eta-ix\xi}\chi(x)\,dx\right|\leqslant C_N(1+|\xi|+|\eta|)^{-N}$$

when $(\xi, \eta) \notin V_2$. Since the distribution u is of finite order, that is, $|\tilde{u}(\eta)| \leqslant C_3(1 + |\eta|)^k$, we see that

$$\left|\int_{(\xi,\ \eta)\notin V_2}\tilde{u}(\eta)\,d\eta\int e^{iF(x)\eta-ix\xi}\chi(x)\,dx\right|\leqslant C_N'(1+|\xi|)^{-N}.$$

On the other hand, if η is outside a conic neighborhood of the set

$$\{{}^tF'(x_0)\,\xi;\ (F(x_0),\ \xi)\in \mathrm{WF}(u)\},$$

then $\tilde{u}(\eta) = O(|\eta|^{-N})$ for all N when the neighborhood U of x_0 is sufficiently small. Besides, if $(\xi, \eta) \in V_2$, then $|\xi| \leqslant C_4|\eta|$. Therefore

$$\left|\int_{(\xi,\ \eta)\in V_2}\tilde{u}(\eta)\,d\eta\int e^{iF(x)\eta-ix\xi}\chi(x)\,dx\right|\leqslant C_N''(1+|\xi|)^{-N},$$

provided supp χ and supp u are sufficiently small.

Thus Theorem 4.3 is proven.

4.5. <u>Example 4.3.</u> Let M be a smooth k-dimensional submanifold in $\mathbb{R}^n$ and $\rho(x)$ be a smooth density given on M. Let $u=\rho\otimes\delta(M)$ be a single-layer potential on M. Then

$$\mathrm{WF}(u)\subset N(M)\setminus 0,$$

where N(M) is the conormal bundle, $N(M) \subset T^*(\mathbb{R}^n)$. Besides, if $\rho(x_0) \neq 0$, then $(x_0, \xi) \in WF(u)$ if and only if the vector ξ is directed along the normal to M.

Indeed, by Theorem 4.3 it is sufficient to consider the case when M is defined by the equation $x'' = 0$, where $x' = (x^1, \ldots, x^k)$, $x'' = (x^{k+1}, \ldots, x^n)$ and $u = \rho(x')\delta(x'')$, $\rho \in C^\infty(\mathbb{R}^k)$. If ρ has compact support, then the function $\tilde{u}(\xi) = \tilde{\rho}(\xi')$ rapidly decreases in every direction of ξ for which $\xi' \neq 0$, but does not decrease in directions $\xi = (0, \xi'')$ provided $\rho \not\equiv 0$.

4.6. Now we shall list some corollaries of Theorem 4.3.

Corollary 4.1. Let M be the smooth k-dimensional submanifold in $\mathbb{R}^n$ defined by the equation $x'' = 0$, where $x' = (x^1, \ldots, x^k)$, $x'' = (x^{k+1}, \ldots, x^n)$ and $u \in \mathscr{D}'(\mathbb{R}^n)$. If $WF(u) \cap N(M) = \emptyset$, then a natural restriction

$$u|_M \in \mathscr{D}'(M)$$

is defined and

$$WF(u|_M) \subset \{(x', \xi'),\ \xi' \neq 0,\ \exists \xi'',\ (x', 0, \xi', \xi'') \in WF(u)\}.$$

Proof. Consider a natural embedding $i: M \to \mathbb{R}^n$ such that $i(x') = (x', 0)$. Then

$$i'(x') = (I, 0), \quad (\xi, \eta)\, {}^t i'(x') = \xi.$$

It is not hard to see that the set N_F defined in Theorem 4.3 agrees with N(M) for F = 1. Therefore, if $WF(u) \cap N(M) = \emptyset$, then $i^* u \in \mathscr{D}'(M)$ is defined and

$$WF(i^* u) \subset \{(x', \xi');\ \xi' \neq 0,\ \exists \xi'':\ (x', 0, \xi', \xi'') \in WF(u)\},$$

which proves the corollary.

4.7. Corollary 4.2. If a distribution $u \in \mathscr{D}'(\mathbb{R}^{n+1})$ satisfies the differential equation

$$P(t, x, D_t, D_x)\, u = f \in C^\infty(\mathbb{R}^{n+1}), \quad t \in \mathbb{R}^1,\ x \in \mathbb{R}^n,$$

and the plane t = 0 is a noncharacteristic, that is, $p^0(0, x, 1, 0) \neq 0$, then the traces $D_t^k u(0, x) \in \mathscr{D}'(\mathbb{R}^n)$ are defined for all k and $u(t, x) \in C^\infty(\mathbb{R}, \mathscr{D}'(\mathbb{R}^n))$.

Proof. From the condition $p^0(0, x, 1, 0) \neq 0$ it follows that the points $(0, x, \tau, 0) \notin WF(u)$ so that $WF(u) \cap N(M_0) = \emptyset$, where M_0 stands for the plane t = 0.

The remaining statements are proved analogously.

4.8. Corollary 4.3. If $u_1, u_2 \in \mathscr{D}'(X)$, where X is a smooth n-dimensional manifold and

$$WF(u_1) + WF(u_2) \subset T^*X \setminus 0,$$

then the distribution $u_1 u_2$ is defined and

$$WF(u_1 u_2) \subset \{(x, \xi + \eta);\quad (x, \xi) \in WF(u_1) \quad \text{or} \quad \xi = 0,$$

$$(x, \eta) \in WF(u_2) \ \text{or}\ \eta = 0;\ \xi + \eta \neq 0\}.$$

Proof. The distribution $u_1 \otimes u_2$ is well defined in $\mathscr{D}'(X \times X)$. In this case, if u_1 and u_2 have compact support in some coordinate neighborhood, then

$$\widetilde{u_1 \otimes u_2}(\xi, \eta) = \tilde{u}_1(\xi)\, \tilde{u}_2(\eta), \quad \xi \in \mathbb{R}_n,\ \eta \in \mathbb{R}_n.$$

The function $\tilde{u}_1(t\xi)\tilde{u}_2(t\eta)$ rapidly decreases as $t \to +\infty$ if only $(\xi, \eta) \notin WF(u_1) \otimes WF(u_2)$ and $\xi \neq 0$, $\eta \neq 0$.

Let us consider a diagonal mapping Δ: $X \to X \times X$ so that $\Delta(x) = (x, x)$ and Theorem 4.3 is applicable. This may be done if

$$\mathrm{WF}(u_1 \otimes u_2) \cap N_\Delta = \emptyset.$$

Note that $N_\Delta = \{(x, x, \xi, -\xi),\ \xi \in \mathbb{R}_n\}$ and $\Delta^*(u_1 \otimes u_2) = u_1 u_2$. Therefore, Theorem 4.3 is applicable provided

$$\mathrm{WF}(u_1) + \mathrm{WF}(u_2) \subset T^*X \setminus 0.$$

By this theorem

$$\mathrm{WF}(u_1\ u_2) \subset \{(x,\ \xi + \eta);\ (x,\ x,\ \xi,\ \eta) \in \mathrm{WF}(u_1 \otimes u_2)\}.$$

The corollary follows immediately.

§5. Wave Front Sets and Fourier Integral Operators

5.1. Let us first consider the behavior of the wave front set of a distribution when an integral operator acts on it. Suppose that $K \in \mathscr{D}'(\Omega_1 \times \Omega_2)$, $\Omega_1 \subset \mathbb{R}^n$, $\Omega_2 \subset \mathbb{R}^m$ and the mapping $K\colon C_0^\infty(\Omega_2) \to \mathscr{D}'(\Omega_1)$ is defined by the equality

$$Ku(\varphi) = K(u \otimes \varphi), \quad u \in C_0^\infty(\Omega_2), \quad \varphi \in C_0^\infty(\Omega_1).$$

Theorem 5.1. If $u \in C_0^\infty(\Omega_2)$, then

$$\mathrm{WF}(Ku) \subset \{(x,\ \xi) \in T^*\Omega_1 \setminus 0;\ \exists y \in \Omega_2,\ (x,\ y,\ \xi,\ 0) \in \mathrm{WF}(K)\}.$$

Proof. Without loss of generality, K may be assumed to have compact support in $\Omega_1 \times \Omega_2$ and $\mathrm{WF}(K) \subset \Omega_1 \times \Omega_2 \times \Gamma$, where Γ is a closed cone in $\mathbb{R}^{n+m}$ with vertex at the origin. It is sufficient to prove that $\mathrm{WF}(Ku) \subset \Omega_1 \times \Gamma_0$, where $\Gamma_0 = \{\xi;\ (\xi,\ 0) \in \Gamma\}$.

Note that

$$\widetilde{Ku}(\xi) = (2\pi)^{-m} \int \tilde{K}(\xi,\ -\eta)\, \tilde{u}(\eta)\, d\eta.$$

If a closed cone Γ_1 in $\mathbb{R}^n$ with vertex at the origin does not intersect Γ_0, then there is an $\varepsilon > 0$ such that

$$|\tilde{K}(\xi,\ -\eta)| \leqslant C_N (1 + |\xi|)^{-N}, \quad \text{if} \quad |\eta| < \varepsilon |\xi|, \quad \xi \in \Gamma_1.$$

As $|\tilde{K}(\xi,\ -\eta)| \leqslant C(1 + |\xi| + |\eta|)^{+k}$ for some k, the inequality

$$|\widetilde{Ku}(\xi)| \leqslant C_N (1 + |\xi|)^{-N} \int_{|\eta| < \varepsilon|\xi|} |\tilde{u}(\eta)|\, d\eta +$$

$$c \int_{|\eta| > \varepsilon|\xi|} (1 + |\eta|)^k |\tilde{u}(\eta)|\, d\eta \leqslant c'_N (1 + |\xi|)^{-N}$$

is satisfied for $\xi \in \Gamma_1$ and $\widetilde{Ku}(\xi)$ rapidly decreases if $\xi \in \Gamma_1$, $\xi \to \infty$. Thus, $\mathrm{WF}(Ku) \subset \Omega_1 \times \Gamma_0$, which proves the theorem.

5.2. We now consider a distribution Ku, where $u \in \mathscr{E}'(\Omega_2)$.

Theorem 5.2. Let $u \in \mathscr{E}'(\Omega_2)$, $K \in \mathscr{D}'(\Omega_1 \times \Omega_2)$ and

$$M_1 \equiv \{(y,\ \eta);\ (x,\ y,\ 0, -\eta) \in \mathrm{WF}(K)\} \cap \mathrm{WF}(u) = \emptyset,$$

$$M_2 \equiv \{(x,\ \xi);\ (x,\ y,\ \xi,\ 0) \in \mathrm{WF}(K)\} = \emptyset.$$

Then the distribution

$$Ku(x) = \int K(x,\ y)\, u(y)\, dy \in \mathscr{D}'(\Omega_1)$$

is defined and

$$\mathrm{WF}(Ku) = \{(x,\ \xi);\ (x,\ y,\ \xi,\ -\eta) \in \mathrm{WF}(K),\ (y,\ \eta) \in \mathrm{WF}(u)\}.$$

Remark 5.1. We shall consider the formal adjoint K^* which acts according to the formula

$$K^*v(y) = \int \overline{K(x, y)}\, v(x)\, dx.$$

By Theorem 5.1 K^* takes $C_0^\infty(\Omega_1)$ into $C^\infty(\Omega_2)$ if

$$M_3 \equiv \{(y, \eta),\ (x, y, 0, \eta) \in \mathrm{WF}(K)\} = \emptyset.$$

Therefore, by duality, K takes $\mathscr{E}'(\Omega_2)$ into $\mathscr{D}'(\Omega_1)$.

Remark 5.2. Let $1(x)$ be a function which is identically one on Ω_1. By Corollary 4.3, the product

$$K(x, y)\, u(y) = K(x, y)\ [u(y) \otimes 1(x)]$$

is defined if the wave front sets of the distributions do not contain opposite directions. But

$$\mathrm{WF}[u(y) \otimes 1(x)] = \{(x, y, 0, \eta):\ (y, \eta) \in \mathrm{WF}(u)\}.$$

Therefore, in order that the product $K(x, y)u(y)$ be defined it is sufficient that the set WF(K) does not contain points of the form $(x, y, 0, -\eta)$ where $(y, \eta) \in \mathrm{WF}(u)$, that is, $M_1 = \emptyset$.

Remark 5.3. By Theorem 5.1, the condition $M_2 = \emptyset$ implies that $K: C_0^\infty(\Omega_2) \to C^\infty(\Omega_1)$, that is, the images of smooth functions do not have singularities.

Proof of Theorem 5.2. Using a partition of unity the proof is easily reduced to the case where K has compact support, $\mathrm{WF}(K) \subset \Omega_1 \times \Omega_2 \times \Gamma$, $\mathrm{WF}(u) \subset \Omega_2 \times \Gamma'$, where Γ and Γ' are closed cones with vertices at the origin of the spaces $\mathbb{R}_{n+m}$ and $\mathbb{R}_m$, respectively. Suppose these cones are minimal so that, by the assumptions of the theorem,

$$(\xi, \eta) \in \Gamma \Rightarrow 1) \text{ if } \xi = 0, \text{ then } -\eta \notin \Gamma';\ 2)\ \eta \neq 0. \tag{5.1}$$

It is necessary to prove that

$$\mathrm{WF}(Ku) \subset \Omega_1 \times \{\xi;\ \exists \eta \in \Gamma',\ (\xi, -\eta) \in \Gamma\}. \tag{5.2}$$

To do so we introduce functions $\psi(\xi, \eta)$ and $\varphi(\eta)$ homogeneous of degree zero, which are equal to unity in a neighborhood of the cones Γ and Γ', respectively. Suppose the supports of ψ and φ are so close to the cones Γ and Γ' that the condition (5.1) remains valid on replacing Γ by supp ψ and Γ' by supp φ.

We now write the Fourier transformation

$$\widetilde{Ku}(\xi) = (2\pi)^{-n} \int \tilde{K}(\xi, -\eta)\, \tilde{u}(\eta)\, d\eta$$

in the form

$$\widetilde{Ku}(\xi) = (2\pi)^{-n} (I_1 + I_2 + I_3 + I_4),$$

where

$$I_1 = \int_{|\eta| < \varepsilon |\xi|} [1 - \varphi(\eta)]\, \tilde{K}(\xi, -\eta)\, \tilde{u}(\eta)\, d\eta;$$

$$I_2 = \int_{|\eta| > \varepsilon |\xi|} [1 - \varphi(\eta)]\, \tilde{K}(\xi, -\eta)\, \tilde{u}(\eta)\, d\eta;$$

$$I_3 = \int \varphi(\eta)\, [1 - \psi(\xi, -\eta)]\, \tilde{K}(\xi, -\eta)\, \tilde{u}(\eta)\, d\eta;$$

$$I_4 = \int \varphi(\eta)\, \psi(\xi, -\eta)\, \tilde{K}(\xi, -\eta)\, \tilde{u}(\eta)\, d\eta.$$

By construction $[1 - \psi(\xi, -\eta)]\tilde{K}(\xi, -\eta) = 0$ when $(\xi, -\eta) \in \Gamma$, so that $|(1 - \psi)\tilde{K}| \leqslant C_l(1 + |\xi| + |\eta|)^{-l}$ for all (ξ, η) and $|I_3| \leqslant C_l'(1 + |\xi|)^{-l}$ for all l.

The integrands in I_1 and I_2 differ from 0 only if $\eta \notin \Gamma'$. In view of condition (5.1), in integral I_1 the point $(\xi, -\eta)$ lies outside Γ so that $|I_1| \leqslant C_1''(1 + |\xi|)^{-1}$. Integral I_2 is estimated using the inequality $|\xi|^N|\eta|^{-1} \leqslant C_1'|\xi|^{N-1/2}|\eta|^{-1/2}$ so that $|I_2| \leqslant C_1'''(1 + |\xi|)^{-1}$.

In integral I_4, the expression under the integral sign differs from 0 only when $(\xi, -\eta) \in \operatorname{supp} \psi$, $\eta \in \operatorname{supp} \varphi$. Therefore

$$\operatorname{WF}(Ku) \subset \Omega_1 \times \{(\xi, -\eta) \in \operatorname{supp} \psi,\ \eta \in \operatorname{supp} \varphi\}$$

which proves the theorem.

5.3. Let now P be a pseudo-differential operator with symbol $p(x, \xi)$ and let $u \in \mathscr{E}'(\Omega)$ so that

$$Pu(x) = (2\pi)^{-n} \iint p(x, \xi)\, u(y)\, e^{i(x-y)\xi}\, dy\, d\xi.$$

Clearly P may be regarded as an integral operator with kernel

$$K(x, y) = (2\pi)^{-n} \int p(x, \zeta)\, e^{i(x-y)\zeta}\, d\zeta, \quad x \in \Omega,\ y \in \Omega. \tag{5.3}$$

Lemma 5.1. $\operatorname{WF}(K) \subset \{(x, x, \xi, -\xi),\ x \in \Omega,\ \xi \in \mathbb{R}^n \setminus 0\}$.

Proof. We have already seen in Chapter II that sing supp $K \subset \{(x, x), x \in \Omega\}$. It may be assumed that $p(x, \xi) = 0$ when $|x| \geqslant R$. Let us consider the Fourier transformation

$$\tilde{K}(\xi, \eta) = (2\pi)^{-n} \iiint p(x, \zeta)\, e^{i(x-y)\zeta - ix\xi - iy\eta}\, d\zeta\, dx\, dy.$$

Since

$$(2\pi)^{-n} \int e^{-iy(\zeta+\eta)}\, dy = \delta(\zeta + \eta)$$

we obtain after integration with respect to y and ζ that

$$\tilde{K}(\xi, \eta) = \int p(x, -\eta)\, e^{-ix(\xi+\eta)}\, dx.$$

In the cone $\{(\xi, \eta);\ |\xi + \eta| > \varepsilon(|\xi| + |\eta|)\}$ we use the identity

$$e^{-ix(\xi+\eta)} = (1 + |\xi + \eta|^2)^{-N} (1 - \Delta_x)^N e^{-ix(\xi+\eta)},$$

from which, after integration by parts, it follows that

$$\tilde{K}(\xi, \eta) = \int (1 + |\xi + \eta|^2)^{-N} (1 - \Delta_x)^N p(x, -\eta)\, e^{-ix(\xi+\eta)}\, dx.$$

It now follows that

$$|\tilde{K}(\xi, \eta)| \leqslant C_N (1 + |\eta|)^m (1 + |\xi + \eta|)^{-2N} \leqslant$$

$$C_N' (1 + |\eta|)^m (1 + \varepsilon|\xi| + \varepsilon|\eta|)^{-2N},$$

that is, K decays faster than any power of $|\xi| + |\eta|$ outside a conic neighborhood of the set $\xi + \eta = 0$. The lemma is proven.

5.4. We shall now prove that a pseudo-differential operator P is pseudolocal from the viewpoint of wave front sets — a more precise statement than Theorem 2.5 of Chapter II.

Theorem 5.3. If P is a pseudo-differential operator and $u \in \mathscr{E}'(\Omega)$, then $\operatorname{WF}(Pu) \subset \operatorname{WF}(u)$.

Proof. We shall use Theorem 5.2. In the given case $M_1 = \emptyset$, $M_2 = \emptyset$ so that $\operatorname{WF}(Pu) \subset \{(x, \xi);\ (x, y, \xi, -\eta) \in \operatorname{WF}(K),\ (y, \eta) \in \operatorname{WF}(u)\}$. But if $(x, y, \xi, -\eta) \in \operatorname{WF}(K)$, then, by Lemma 5.1, $y = x$, $\eta = \xi$. Hence $\operatorname{WF}(Pu) \subset \operatorname{WF}(u)$, which proves the theorem.

5.5. We shall now look at the determination of WF(u) in terms of WF(Pu). To do so we introduce the set

$$\operatorname{Char} P = \{(x, \xi) \in T^*\Omega \setminus 0;\ p^0(x, \xi) = 0\}.$$

Theorem 5.4. $WF(u) \subset WF(Pu) \cup \operatorname{Char} P$.

Proof. Let $(x_0, \xi^0) \notin WF(Pu)$ and $p^0(x_0, \xi^0) \neq 0$. We shall prove that in this case $(x_0, \xi^0) \notin WF(u)$.

By the definition of $WF(Pu)$, there exist functions $\varphi \in C_0^\infty(\mathbb{R}^n)$, $\psi \in C^\infty(\mathbb{R}_n \setminus 0)$, $\varphi(x) = 1$ in a neighborhood of the point x_0, $\psi(t\xi) = \psi(\xi)$ for $t > 0$, $\psi(\xi^0) \neq 0$ such that $\psi(\xi)\widetilde{\varphi Pu}(\xi)$ decays faster than any power of $(1 + |\xi|)^{-1}$. Let $P' = \psi(D)\varphi P$. Then $P'u \in C^\infty$ and $p'^0(x_0, \xi^0) = \psi(\xi^0)p^0(x_0, \xi^0) \neq 0$.

We now construct a symbol $q(x, \xi)$ of a parametrix for which

$$\sum \frac{1}{\alpha!}(iD_\xi)^\alpha q(x, \xi)\, D_x^\alpha p'(x, \xi) \cong 1$$

for points (x, ξ) in a sufficiently small neighborhood of (x_0, ξ^0). Let Q be an operator with symbol $q(x, \xi)$; let $R = QP$, and let $r(x, \xi)$ be the symbol of R. Then $Re \in C^\infty$ and the symbol of the operator is equal to 1 in some conic neighborhood of (x_0, ξ^0). We now choose functions φ' and ψ' so that $\psi'(t\xi) = \psi'(\xi)$ when $t > 1$, $|\xi| > 1$, $\varphi' = 1$ in a neighborhood of x_0, $\psi' = 1$ in a neighborhood of ξ^0, and

$$\varphi'(x)\psi'(\xi)r(x, \xi) = \varphi'(x)\psi'(\xi).$$

Then $\psi'(D)\varphi'(x)Ru - \psi'(D)\varphi'(x)u \in C^\infty$ and therefore $\psi'(D)\varphi'(x)u \in C^\infty$. Hence, $(x_0, \xi^0) \notin WF(u)$, which proves the theorem.

Corollary 5.1. If P is an elliptic operator, then

$$WF(Pu) = WF(u).$$

Remark. Recall that an operator P is called hypoelliptic if

$$\operatorname{sing\,supp} Pu \supset \operatorname{sing\,supp} u.$$

If $WF(Pu) \supset WF(u)$, then the operator P is called *microlocally hypoelliptic*.

5.6. We shall now consider the variation of the wave front set of a distribution under the action of a Fourier integral operator. Let

$$Au(x) = (2\pi)^{-n} \int a(x, \xi)\, \tilde{u}(\xi)\, e^{iS(x, \xi)}\, d\xi, \tag{5.4}$$

where $a \in S^0(\mathbb{R}^n)$ and the function S satisfies the conditions of Definition 4.4 of Chapter II.

The kernel of this operator is of the form

$$K(x, y) = (2\pi)^{-n} \int a(x, \zeta)\, e^{iS(x, \zeta) - i(y, \zeta)}\, d\zeta.$$

In order to apply Theorem 5.2 we need to describe $WF(K)$.

Lemma 5.2.

$$WF(K) \subset \left\{(x, y, \xi, \eta);\ y = \frac{\partial S(x, \eta)}{\partial \eta},\quad \xi = \frac{\partial S(x, \eta)}{\partial x}\right\}.$$

Using a partition of unity in x and y, the theorem can be reduced to the case where $K(x, y) = 0$ outside a small neighborhood of (x_0, y_0). Let us consider a cone

$$\Gamma = \left\{(\xi, \eta);\ \left|y - \frac{\partial S(x, \eta)}{\partial \eta}\right| + \left|\xi - \frac{\partial S(x, \eta)}{\partial x}\right|(|\xi| + |\eta|)^{-1} \geqslant \varepsilon,\ (x, y) \in U\right\}.$$

We shall show that the Fourier transform $\tilde{K}(\xi, -\eta)$ decreases faster than any power of $(|\xi| + |\eta|)^{-1}$ as $|\xi| + |\eta| \to \infty$ outside Γ. Note that

$$\tilde{K}(\xi, -\eta)=(2\pi)^{-n}\iiint h(x, y)a(x, \zeta)\, e^{iS(x,\zeta)-i(y,\zeta)-i(x,\xi)+i(y,\eta)}\, d\zeta\, dx\, dy,$$

where $h\in C_0^\infty(\mathbb{R}^{2n})$, $h=1$ in U. Using the identities

$$(1-\Delta_y)\, e^{iy(\eta-\zeta)}=(1+|\eta-\zeta|^2)\, e^{-iy(\eta-\zeta)},$$

$$(1-\Delta_x)\, e^{iS(x,\zeta)-i(x,\xi)}=(1+|S_x(x, \zeta)-\xi|^2-i\Delta_x S(x, \zeta))\, e^{iS(x,\zeta)-i(x,\xi)},$$

$$(1-(1+|\zeta|)^2\Delta_\zeta)\, e^{iS(x,\zeta)-i(y,\zeta)}=$$

$$1+(1+|\zeta|)^2|y-S_\zeta(x, \zeta)|^2-i(1+|\zeta|^2)\,\Delta_\zeta S(x, \zeta))\, e^{iS(x,\zeta)-i(y,\zeta)}$$

and integrating by parts, $\tilde{K}$ may be represented as the integral of a function which is estimated by

$$C_N(1+|\zeta|)^{-n-1}(1+|\xi|+|\eta|)^{-N}$$

for any N. The lemma follows.

5.7. The lemma proved above and Theorem 5.2 make it possible to describe WF(Au).

Theorem 5.5. Let $u\in\mathscr{E}'(\Omega)$ and A be an operator of the form (5.4). Suppose that

$$\text{if } \frac{\partial S}{\partial x}(x, \xi)=0, \text{ then } \left(\frac{\partial S(x, \xi)}{\partial \xi}, \xi\right)\notin \mathrm{WF}(u).$$

Then

$$\mathrm{WF}(Au)\subset\left\{\left(x, \frac{\partial S(x, \eta)}{\partial x}\right);\ \left(\frac{\partial S(x, \eta)}{\partial \eta}, \eta\right)\in \mathrm{WF}(u)\right\}.$$

Thus, if S is the generating function of a canonical transformation, then WF(Au) is a map of WF(u), $(y, \eta)\longmapsto(x, \xi)$, generated by the function $S(x, \eta)$. Since a canonical transformation is invertible, the inclusion sign in Theorem 5.5 may be replaced by an equality sign.

§6. Propagation of Singularities. Solvability of Equations of Principal Type with a Real-Valued Principal Symbol

6.1. In this section we shall prove Hörmander's theorem on the propagation of singularities of solutions of the equation Pu = f, where P is a pseudo-differential operator with a real-valued principal symbol $p^0(x, \xi)$. We assume that P is an operator of principal type; that is, Definition 4.3 of Chapter II is satisfied.

Let us consider the Hamiltonian system

$$\frac{dx^j}{dt}=\frac{\partial p^0}{\partial \xi_j}(x, \xi), \quad \frac{d\xi_j}{dt}=-\frac{\partial p^0(x, \xi)}{\partial x^j}, \qquad j=1, \ldots, n. \tag{6.1}$$

The phase curves $x = x(t)$, $\xi = \xi(t)$ of this system are called the bicharacteristics of P. It is obvious that along a bicharacteristic $dp^0(x(t), \xi(t))/dt = 0$, so $p^0(x(t), \xi(t)) = \text{const}$. Therefore, if I is a connected portion of the bicharacteristic, then either $I \subset \mathrm{Char}\, P$ or $I\cap \mathrm{Char}\, P=\emptyset$.

We now introduce the differential operator

$$H_{p^0}=\sum_{j=1}^{n}\left(\frac{\partial p^0(x, \xi)}{\partial \xi_j}\frac{\partial}{\partial x^j}-\frac{\partial p^0(x, \xi)}{\partial x^j}\frac{\partial}{\partial \xi_j}\right)$$

in the space $T^*\Omega$. Clearly

$$H_{p^0}f(x(t), \xi(t))=\frac{d}{dt}f(x(t), \xi(t))$$

for any function f in $C^1(T^*\Omega)$.

6.2. Theorem 6.1. If $u \in \mathscr{D}'(\Omega)$, I is a connected piece of the bicharacteristic, and $I \cap \mathrm{WF}(Pu) = \emptyset$, then either $I \subset \mathrm{WF}(u)$ or $I \cap \mathrm{WF}(u) = \emptyset$.

In other words, in the complement of WF(Pu) the set WF(u) is invariant under shifts along the trajectories of the Hamiltonian system (6.1).

Proof. We shall first consider the special case of the operator $P = D_1$. Then $p^0(x, \xi) = \xi_1$ and the bicharacteristics are of the form $x^1 = x_0^1 + t$, $x^j = x_0^j$, $\xi_k = \xi_k^0$, $j = 2, \ldots, n$; $k = 1, \ldots, n$. In other words

$$I = \{(x_0^1 + t,\ x_0',\ \xi^0),\ t \in \mathbb{R}\}.$$

At points on the bicharacteristic, vectors (x, ξ) with $\xi_1 \neq 0$ are noncharacteristic and therefore are not in WF(u). If $\xi = 0$, then by Corollary 4.1, if $(x, \xi) \in \mathrm{WF}(u)$, then $(x, \xi') \subset \mathrm{WF}(u|_{x^1 = x_0^1})$. Therefore, the function $\psi(D')\varphi(x')u|_{x^1=x_0^1} \in C^\infty$, if $\psi(D')\varphi(x')u \in C^\infty$ for $x^1 = a$, and the portion of the bicharacteristic $[a, x_0^1]$ does not intersect $\mathrm{WF}(D_1 u)$.

In the general case one may assume m = 1 since the theorem is invariant under the multiplication of P by an elliptic operator. By Proposition 4.2 of Chapter II there exists an integral operator A of order zero such that

$$P = A^* D_1 A + T_{-\infty},$$

where $T_{-\infty}$ is a smoothing operator; therefore, if we assume Au = v, then $D_1 v = Af + T'_{-\infty}$.

Thus the proof reduces to a consideration of the above case. Since under the mapping A straight lines parallel to the x^1-axis correspond to the bicharacteristics of P, the theorem is proven for the general case.

6.3. For applications it is convenient to refine the concept of singularity by calling singular those points at which function u is not in the Sobolev space H_s.

Definition 6.1. We say that $u \in H_s(x_0, \xi^0)$ or $(x_0, \xi^0) \notin \mathrm{WF}_s(u)$ if u can be represented as $u = u_1 + u_2$, where $u_1 \in H_s$ and $(x_0, \xi^0) \notin \mathrm{WF}(u_2)$.

It is not difficult to see that Theorem 6.1 may be restated in terms of $\mathrm{WF}_s(u)$ as follows.

Theorem 6.2. If $u \in \mathscr{D}'(\Omega)$, I is a connected portion of a bicharacteristic and $I \cap \mathrm{WF}_s(Pu) = \emptyset$, where P stands for a pseudo-differential operator of order m, then either $I \subset \mathrm{WF}_{s+m-1}(u)$ or $I \cap \mathrm{WF}_{s+m-1}(u) = \emptyset$.

In this case the proof of Theorem 6.2 does not essentially differ from that of Theorem 6.1. We need only to make use of the estimate of Chapter III, Section 4.3, asserting that the operator A and its parametrix are bounded in H_s.

6.4. Theorems 6.1 and 6.2 make it possible to prove the existence of solution of equations of type Pu = f, where P is a pseudo-differential operator of principal type with a real-valued principal symbol.

Theorem 6.3. For every point $x_0 \in \Omega$ there is a neighborhood U such that if $f \in H_s(\Omega)$, then there exists a function $u \in \mathring{H}_{s+m-1}(\Omega)$ satisfying the equation Pu = f in U and $\|u\|_{s+m-1} \leq C\|f\|_s$ with constant C depending only on s.

Proof. Let U be a neighborhood of x_0 such that every null bicharacteristic of p^0 passing through a point in $T^*U \setminus 0$ has points in $T^*U/\partial U$. Since P is an operator of principal type such a neighborhood always exists.

Note that operator P^* is also an operator of principal type and if

$P^*\varphi \in H_{s-m+1}(U)$, $\varphi \in \mathscr{E}'(U)$, then $\varphi \in H_s(U)$. Indeed, by Theorem 6.2, the set of points at which $\varphi \in H_s$ is invariant under the Hamiltonian flow H_{p^0}, and since the distribution φ vanishes in a neighborhood of the boundary of U, every null bicharacteristic has points where $\varphi \in C^\infty$. Thus, $\varphi \in H_s$ at every point in T^*U. Using a partition of unity we can prove that $\varphi \in H_s(U)$.

By the Banach theorem on open mappings, there exists a constant C depending only on s such that

$$\|\varphi\|_s \leqslant C\|P^*\varphi\|_{s-m+1}, \quad \varphi \in C_0^\infty(U).$$

From this inequality it follows that when $f \in H_s(\Omega)$ the integral $\int f\overline{\varphi}\,dx$ is a bounded linear functional of $\overline{P^*\varphi} \in H_{-s-m+1}$. By the Hahn-Banach theorem there exists a function $u \in H_{s+m-1}$ for which

$$\int f\overline{\varphi}\,dx = \int u\,\overline{P^*\varphi}\,dx, \quad \varphi \in C_0^\infty(U),$$

and

$$\|u\|_{s+m-1} \leqslant C\|f\|_s.$$

By definition, u is a generalized solution of the equation $Pu = f$, and the theorem is proven.

6.5. The theorem on the propagation of singularities makes it also possible to obtain more general results on semiglobal solvability, that is, on solvability in any compact subset. The following example illustrates that these results are nontrivial.

Example 6.1. Let Ω be an annulus in the plane defined by the inequalities $1 \leqslant r \leqslant 2$, where r stands for the polar radius. The operator $D_\varphi = \frac{1}{i}\frac{\partial}{\partial\varphi}$ is an operator of principal type, but for solvability of the equation $D_\varphi u = f$ in Ω it is necessary that

$$\int_0^{2\pi} f(r, \varphi)\,d\varphi = 0$$

for all $r \in [1, 2]$. Thus, for the equation to be solvable an infinite number of orthogonality conditions must be satisfied.

The above example establishes the following:

(A) Every null bicharacteristic passing through a point in $T^*\Omega \setminus 0|_K$, where K is compact in Ω, has points whose projection on Ω lies outside K.

As usual, the equation $Pu = f$ has a solution only if f is orthogonal to the kernel of the adjoint operator. Let us consider this kernel in detail.

Lemma 6.1. Let $P^*\varphi = 0$, $\varphi \in \mathscr{E}'(K)$ and let condition (A) be satisfied. Then $\varphi \in C_0^\infty(K)$. The set of all such φ such that $\|\varphi\|_0 \leqslant 1$ is compact in $C_0^\infty(K)$ and has finite dimension.

Proof. The smoothness of φ follows from Theorem 6.1 since by hypothesis every null bicharacteristic has points which do not belong to $WF(\varphi)$. Considering the set

$$N(K) = \{\varphi;\ \varphi \in C_0^\infty(K),\ P^*\varphi = 0\} \tag{6.2}$$

and the ball $\{\varphi;\ \|\varphi\|_0 \leqslant 1\}$ contained in it, we find by Theorem 6.2 that all such functions φ are in $H_1(K)$ and $\|\varphi\|_1 \leqslant C_0$. By Theorem 3.11 of Chapter I, this set is compact and therefore N(K) has finite dimension. The lemma is proven.

We can now formulate a theorem on solvability.

Theorem 6.4. Let a pseudo-differential operator P of order m and of principal type have a real-valued principal symbol p^0 in a domain $\Omega \subset \mathbb{R}^n$. Suppose condition (A) is satisfied for a compact subset $K \subset \Omega$. Then the set N(K) has finite dimension and for every function $f \subset H_s(\Omega)$ orthogonal to N(K) one can find a function $u \subset H_{s+m-1}(\Omega)$ satisfying the equation Pu = f in K. In this case there exists a constant C independent of f such that

$$\|u\|_{s+m-1} \leqslant C\|f\|_s.$$

Proof. In Lemma 6.1 it was proven that N(K) has finite dimension. Obviously, if Pu = f, then

$$u(P^*\varphi) = \int f\bar{\varphi}\,dx$$

for $\varphi \in N(K)$. Therefore, the condition that function f be orthogonal to N(K) is necessary for solvability of the equation.

We shall now show that the inequality

$$\|\varphi\|_{-s} \leqslant C_1\|P^*\varphi\|_{-s-m+1} \tag{6.3}$$

holds for functions $\varphi \in C_0^\infty(K)$ orthogonal to N(K). Suppose this is not true. Then one can find a sequence of functions $\{\varphi_k\} \in C_0^\infty(K)$ for which

$$\|\varphi_k\|_{-s} = 1, \quad \|P^*\varphi_k\|_{-s-m+1} \leqslant \frac{1}{k} \text{ and } \int \varphi_k\bar{\psi}\,dx = 0 \text{ for } \psi \in N(K).$$

By Theorem 6.2 and the closed graph theorem (see K. Yosida[1]), there exists an estimate

$$\|\varphi_k\|_{-s} \leqslant C_1(\|P^*\varphi_k\|_{-s-m+1} + \|\varphi_k\|_{-s-1}). \tag{6.4}$$

Since the sequence $\{\varphi_k\}$ is compact in H_{-s-1}, one may assume, passing to a subsequence, that $\{\varphi_k\}$ converges in H_{-s-1}. But

$$\|\varphi_k - \varphi_{k'}\|_{-s} \leqslant C_1(\|P^*\varphi_k\|_{-s-m+1} + \|P^*\varphi_{k'}\|_{-s-m+1} + \|\varphi_k - \varphi_{k'}\|_{-s-1});$$

hence $\{\varphi_k\}$ converges in $H_{-s}(K)$ to some function $\varphi \in \overset{0}{H}_{-s}(K)$. In this case, $P^*\varphi = 0$ and $\|\varphi\|_{-s} = 1$; that is, $\varphi \in N(K)$, which contradicts the condition $\int \varphi\bar{\psi}\,dx = 0$ for $\psi \in N(K)$. Thus, inequality (6.3) is proven. It follows from this inequality that the functional $\int f\bar{\varphi}\,dx$ on $P^*\varphi$ is bounded in H_{-s-m+1} so

$$\left|\int f\bar{\varphi}\,dx\right| \leqslant \|f\|_s\|\varphi\|_{-s} \leqslant C_1\|f\|_s\|P^*\varphi\|_{-s-m+1}.$$

By the Riesz representation theorem, there exists an element $u \in H_{s+m-1}$ for which

$$\int u\overline{P^*\varphi}\,dx = \int f\bar{\varphi}\,dx,$$

and $\|u\|_{s+m-1} \leqslant C_1\|f\|_s$. The function u is the desired solution. The theorem is proven.

CHAPTER VI. NECESSARY CONDITIONS FOR LOCAL SOLVABILITY

§1. Examples

1.1. In this section we shall consider some equations which are not solvable, even locally, in the class of distributions. An equation of this type was first found by H. Lewy in 1956. He proved that the equation

$$\frac{\partial u}{\partial x^1}+i\frac{\partial u}{\partial x^2}+i(x^1+ix^2)\frac{\partial u}{\partial x^3}=f(x^1, x^2, x^3), \tag{1.1}$$

where f is a certain function in $C^\infty(\mathbb{R}^3)$, is not solvable in any open subset of the space $\mathbb{R}^3$.

Example 1.1. The simplest equation which is not solvable in any neighborhood of the origin was constructed by V. V. Grushin, using an idea of P. Garabedian.

This equation is written

$$\frac{\partial u}{\partial x}+ix\frac{\partial u}{\partial y}=f(x, y), \tag{1.2}$$

where $f\in C_0^\infty(\mathbb{R}^2)$, $f\geqslant 0$, $f(x, y)=f(-x, y)$ and for $x \geqslant 0$ the function f equals zero everywhere outside the set $\bigcup_{n=1}^{\infty} D_n$, where $\{D_n\}$ is a sequence of disjoint circles which lie in the half-plane $x > 0$ and converges to zero as $n \to \infty$. Clearly, the space of such functions has infinite dimension. Suppose $f > 0$ in D_n and denote by D the support of f.

We shall now show that every equation of this type has no distribution solution in any neighborhood of the origin.

Indeed, let u be such a solution. Suppose that

$$v(x, y)=\frac{u(x, y)+u(-x, y)}{2}, \quad w(x, y)=\frac{u(x, y)-u(-x, y)}{2}.$$

Comparing equation (1.1) with

$$\frac{\partial u(-x, y)}{\partial x}+ix\frac{\partial u(-x, y)}{\partial y}=-f(-x, y),$$

we obtain

$$\frac{\partial v}{\partial x}+ix\frac{\partial v}{\partial y}=0, \quad \frac{\partial w}{\partial x}+ix\frac{\partial w}{\partial y}=f(x, y).$$

The first of these equations changes after substituting $s = x^2/2$ into the Cauchy-Riemann equation

$$\frac{\partial v}{\partial s}+i\frac{\partial v}{\partial y}=0,$$

which implies that v is an analytic function of $s = iy$, and $v = F(s + iy)$. The smoothness of v follows from Theorem 3.4 of Chapter II. The function w is also an analytic function of $x^2/2 + iy$ everywhere where $x > 0$ outside the set D. Since $\mathbb{R}^2\setminus D$ is connected and $w = 0$ when $x = 0$, we obtain by construction that w = 0 in the region $\mathbb{R}^2 \setminus D$. By Theorem 3.4 of Chapter II, w is infinitely differentiable when $x > 0$. Therefore w = 0 at the boundary of every circle D_n. But using Green's formula,

$$\iint_{D_n} f\,dx\,dy = \iint_{D_n} (u_x + ixu_y)\,dx\,dy = \oint_{\partial D_n} u\,dy - ixu\,dx = 0,$$

and we reach a contradiction.

It is easy to see that the above argument applies to the more general equation

$$\frac{\partial u}{\partial x} + ix^{2k+1}\frac{\partial u}{\partial y} = f(x, y). \tag{1.3}$$

1.2. Another example of an unsolvable equation, this time of a pseudo-differential equation, can be obtained from Example 1.2 of Chapter IV.

Indeed, the procedure described in Section 3, Chapter III, enables the boundary problem

$$\Delta u = 0 \text{ in } \Omega, \quad \frac{\partial u}{\partial x^1} = g \text{ on } \Gamma \tag{1.4}$$

to be reduced to an equivalent pseudo-differential equation

$$Pu = g \text{ on } \Gamma. \tag{1.5}$$

However, the boundary problem (1.4) is solvable only when g satisfies an infinite number of orthogonality conditions of the form: $\int \Phi_i g\,dx = 0$, where Φ_i, i = 1, 2,..., are functions in $C^\infty(\Gamma)$ which form a linearly independent system. Therefore, (1.5) is solvable only if these necessary conditions are fulfilled. Moreover, it can be shown that even for the existence of a solution of equation (1.5) in an arbitrarily small neighborhood of an arbitrary point in Γ_0 it is necessary that g satisfies an infinite number of orthogonality conditions in this neighborhood.

§2. Hörmander's Theorem

2.1. In this section a theorem of Hörmander on necessary conditions for local solvability is proven. It is precisely this theorem which first explained the unsolvability of G. Lewy's equation. To prove this theorem Hörmander developed certain methods which are used in all further works dedicated to this problem.

Theorem 2.1 (L. Hörmander). Let P(x, D) be a pseudo-differential operator of order m and the point $(x_0, \xi^0) \in T^*\Omega$ be such that $\xi \neq 0$, $p^0(x_0, \xi^0) = 0$ and $c^0(x_0, \xi^0) > 0$, where

$$c^0(x, \xi) = 2\,\mathrm{Im} \sum_{j=1}^{n} \frac{\partial p^0(x, \xi)}{\partial x^j} \frac{\overline{\partial p^0(x, \xi)}}{\partial \xi_j} \tag{2.1}$$

is the principal symbol of the commutator $C = P^*P - PP^*$. Then for every neighborhood ω of point x_0 one can find a function $f \in C_0^\infty(\omega)$ such that there is no distribution $u \in \mathscr{E}'(\Omega)$ for which Pu = f in ω.

2.2. Before giving the proof of Theorem 2.1 we shall list some examples and certain corollaries of the theorem.

Example 2.1. For Lewy's operator with symbol

$$p^0(x, \xi) = i\xi_1 - \xi_2 + i(x^1 + ix^2)\xi_3$$

the $c^0 = 4\xi_3$. Therefore, the conditions of Theorem 3.1 are satisfied at every point (x, ξ) where $\xi_3 > 0$, $\xi_1 = -x^2\xi_3$, $\xi_2 = -x^1\xi_3$. Thus, the equation Pu = f is unsolvable in any open set $\omega \subset \mathbb{R}^3$.

For operator (1.2) with symbol

$$p^0(x, \xi) = i\xi_1 - x^1\xi_2$$

the $c^0 = 2\xi_2$. The conditions of Theorem 3.1 are satisfied at points (x, ξ), where $x^1 = 0$, $\xi_1 = 0$, and $\xi_2 > 0$. Therefore, (1.1) is solvable in every open set in the plane (x, y) having a nonempty intersection with the y-axis.

Example 2.2 (G. Lewy). If equation (1.1), where $f \in C^\infty(\mathbb{R}^3)$, has no solutions in the class of distributions in any open domain of $\mathbb{R}^3$, then the homogeneous equation

$$P(x, D)u - fu = 0$$

cannot have a nontrivial solution in $C^1(\omega)$ in any open subset of $\mathbb{R}^3$.

Indeed, if u is such a solution, then $u \neq 0$ in some neighborhood ω and the function $v = \ln u$ satisfies the equation $P(x, D)v = f$ in ω, which contradicts the assumptions.

The following statement shows that the set of smooth functions $f \in C_0^\infty(\omega)$ for which the equation $Pu = f$ has a distribution solution is a first category set according to Baire's classification (that is, it is the union of a countable number of closed nowhere nondense sets).

Corollary 2.1. Let P(x, D) be a pseudo-differential operator of order m. Let Ω be a domain in $\mathbb{R}^n$ and suppose that for every subdomain $\omega \subset \Omega$ there exists a function $\varphi \subset C_0^\infty(\omega)$ for which the equation $Pu = \varphi$ has no solution in ω in $\mathscr{E}'(\Omega)$. Then there is a function $f \in \mathscr{S}(\Omega)$ for which the equation $Pu = f$ cannot be solved in any subdomain $\omega \subset \Omega$. The set of these functions f is a second category set in $\mathscr{S}(\Omega)$.

Recall that $\mathscr{S}(\Omega)$ is defined as the complement of the space $C_0^\infty(\Omega)$ in the topology of the space $\mathscr{S}$.

Proof. First let ω be a fixed subdomain in Ω and s be some integer. We consider the set

$$L(\omega, s) = \{f \in \mathscr{S}(\Omega);\ u \in H_s,\ Pu = f \text{ in } \omega\}.$$

By hypothesis $L(\omega, s) \neq \mathscr{S}(\Omega)$.

This set is the projection onto $\mathscr{S}(\Omega)$ of the set

$$K(\omega, s) = \{(u, f) \in H_s \times \mathscr{S}(\Omega);\ Pu = f \text{ in } \omega\},$$

which is obviously closed in $H_s \times \mathscr{S}(\Omega)$. It follows that $L(\omega, s)$ is a closed set, nowhere nondense in $\mathscr{S}(\Omega)$, i.e., it is a first category set.

Let now $\Omega = \bigcup_j \omega_j$, where ω_j are open subsets in Ω which form a denumerable basis for the open subsets of Ω. For example, as $\{\omega_j\}$ we can take a sequence of spheres contained in Ω of rational radii having centers with rational coordinates. Then the set $\bigcup_j L(\omega_j, s)$ is a first category set — the union of a countable number of such sets. Hence its complement in $\mathscr{S}$ is a second category set and if f belongs to this complement, then the equation $Pu = f$ has no distribution solution in any open subset $\omega \subset \Omega$. This proves the corollary.

Corollary 2.2. If P(x, D) is a differential operator of order m, then Theorem 3.1 holds provided at a point $(x_0, \xi^0) \in T^*\Omega \setminus 0$

$$p^0(x_0, \xi^0) = 0, \quad c^0(x_0, \xi^0) \neq 0.$$

Proof. If P is a differential operator of order m, then C(x, D) is also a differential operator and its order is equal to $2m - 1$. Therefore, if $c^0(x_0, \xi^0) \neq 0$, then one of the two numbers $c^0(x_0, \xi^0)$ and $c^0(x_0, -\xi^0)$ is positive and therefore Theorem 3.1 is applicable.

2.3. The proof of Theorem 2.1 is based on the following lemmas.

Lemma 2.1. Let the equation $Pu = f$ be solvable in the domain ω; that is, to every function $f \in C_0^\infty(\omega)$ suppose one can find a distribution $u \in \mathcal{E}'(\mathbb{R}^n)$ for which $Pu = f$ in ω. Then there exist constants C, s, and t such that

$$\|\varphi\|_s \leqslant C\|P^*\varphi\|_t, \quad \varphi \in C_0^\infty(\omega'), \tag{2.2}$$

where $\omega' \Subset \omega$.

Proof. Without loss of generality we can assume that the set ω is compact and that the support of u lies in a compact set Ω, since the symbol $p(x, \xi)$ can be varied outside a set compact in x without changing the values of Pu in ω.

The functional $\int f\bar{v}\,dx$ is continuous when f is in $C_0^\infty(\omega)$ with a topology defined by the denumerable set of norms $\|\cdot\|_s$ for integers s, and v is in $C_0^\infty(\omega')$, where $\omega' \Subset \omega$, in which the topology is defined by the denumerable set of norms $\|P^*v\|_t$ for integers t. Indeed, the continuity in f for the fixed function $v \in C_0^\infty(\omega)$ is obvious, and the continuity in v follows from the identity

$$\int f\bar{v}\,dx = \int u\overline{P^*v}\,dx,$$

where u is a solution of the equation $Pu = f$ in ω. By the Banach-Steinhaus theorem (see K. Yosida[1]), from the continuity of the functional $\int f\bar{v}\,dx$ separately in f and v there follows its continuity; that is, there follows the existence of C, s, and t such that

$$\left|\int f\bar{v}\,dx\right| \leqslant C\|f\|_{-s}\|P^*v\|_t.$$

This inequality holds for any function $f \in C_0^\infty(\Omega)$, since $f\bar{v} = hf\bar{v}$, where $h \in C_0^\infty(\omega)$, $h = 1$ in ω', and so

$$\left|\int f\bar{v}\,dx\right| \leqslant C\|hf\|_{-s}\|P^*v\|_t \leqslant C'\|f\|_{-s}\|P^*v\|_t.$$

Therefore

$$\|v\|_s \leqslant C\|P^*v\|_t, \quad v \in C_0^\infty(\omega'),$$

which proves the lemma.

2.4. By Lemma 2.1, Theorem 2.1 follows from the following theorem, which will be proven below.

Theorem 2.2. Let $P(x, D)$ be a pseudo-differential operator of order m and let

$$p^0(x_0, \xi^0) = 0, \quad c^0(x_0, \xi^0) > 0,$$

at a point $(x_0, \xi^0) \in T^*\Omega \setminus 0$. Then there is no neighborhood ω of x_0 where the constants C, s, and t exist such that

$$\|\varphi\|_s \leqslant C\|P^*\varphi\|_t, \quad \varphi \in C_0^\infty(\omega). \tag{2.3}$$

The following lemma enables the proof of Theorem 2.2 to be reduced to a special case.

Lemma 2.2. To prove Theorem 2.2 it is sufficient to consider the case when $m = 1$, $t = 0$.

Proof. Suppose inequality (2.3) is valid for some m, s, t, ω, and C for all $u \in C_0^\infty(\omega)$. Let $Q(x, D)$ be a pseudo-differential operator of order 1 whose symbol, when $|\xi| \geqslant 1$, is $|\xi|^{1-m}p^0(x, \xi)$ and let $A(x, D)$ be a pseudo-differential operator whose symbol, when $|\xi| \geqslant 1$, is $|\xi|^{1-m-t}h(x)$, where $h \in C_0^\infty(\omega)$, $h(x) = 1$ in $\omega' \subset \omega$.

Since A is elliptic in ω', assuming $u = Av$ we find that for $v \in C_0^\infty(\omega')$ we have

$$\|v\|_r \leqslant C(\|u\|_s + \|v\|_{r-1}),$$

where $r = 1 - m - t + s$. Using (2.3) we see that

$$\|v\|_r \leqslant C_1(\|Pu\|_t + \|v\|_{r-1}) \leqslant C_2(\|Qv\|_0 + \|v\|_{r-1})$$

for all $v \in C_0^\infty(\omega')$; that is, inequality (2.3) is satisfied for the operator Q in the domain ω' for $t = 0$.

We now show that the conditions of Theorem 2.2 are satisfied simultaneously for operators P and Q. By B we denote a pseudo-differential operator with symbol $|\xi|^{1-m}$ for $|\xi| > 1$. Then up to lower order terms $Q = BP + \ldots$ and $B^* = B$. Therefore,

$$[Q^*, Q] = P^*B^*BP - BPP^*B + \ldots = B^2C + A_1 + \ldots,$$

where $a_1^0(x, \xi) = 0$, provided $p^0(x, \xi) = 0$. Thus, the function $b_0^2c^0 = |\xi|^{2(1-m)}c^0$ plays the role of the function c^0 for operator Q. Therefore, the conditions of Theorem 2.2 are satisfied for P if and only if they are satisfied for Q. The lemma is proven.

2.5. The following lemma is due to L. Hörmander.

Lemma 2.3. If for a first-order operator P with symbol having compact support the inequality

$$\|u\|_s \leqslant C(K)(\|Pu\|_0 + \|u\|_{s-1}), \quad u \in C_0^\infty(K),$$

is satisfied, where K stands for the compact subset of points in Ω, then for any integer $N > 0$ and any compact set $M \subset K \times \{\mathbb{R}_n \setminus 0\}$ there exists a constant C such that

$$\|\varphi\|_0 \leqslant C\lambda^{1-s}\left(\left\| \sum_{2j+|\alpha+\beta|<N} \frac{1}{\alpha!\,\beta!} \frac{\partial^{\alpha+\beta}p^j(x, \xi)}{\partial\xi^\alpha\,\partial x^\beta} y^\beta D^\alpha \varphi \lambda^{-j-\frac{|\alpha+\beta|}{2}} \right\|_0 + \lambda^{-\frac{N}{2}} \sum_{|\alpha+\beta|\leqslant N} \|y^\beta D^\alpha \varphi\|_0 \right) \tag{2.4}$$

for all $\varphi \in C_0^\infty(\mathbb{R}^n)$, $\lambda \geqslant 1$, $(x, \xi) \in M$, $p^j(x, \lambda\xi) = \lambda^{1-j}p^j(x, \xi)$.

Proof. It is sufficient to prove inequality (2.4) for large λ. Let $(x, \xi) \in M$ and

$$u(y) = h(y)\varphi((y-x)\sqrt{\lambda})e^{i\lambda(y,\xi)},$$

where $h \in C_0^\infty(K)$ and $h = 1$ in a neighborhood of the compact set of K'. Here K' is the projection of M onto $\mathbb{R}_x^n$. Let

$$v(y) = \varphi((y-x)\sqrt{\lambda})e^{i\lambda(y,\xi)}.$$

Then

$$\tilde{v}(\eta) = \lambda^{-n/2}\tilde{\varphi}((\eta - \lambda\xi)\lambda^{-1/2})e^{i(x, \lambda\xi - \eta)}.$$

Note that if $h = 1$ at points which are at a distance less than $\delta > 0$ from K', then for $\sqrt{\lambda} > \delta^{-1}$

$$(1-h)v = 0.$$

Thus $\|u\|_s = \|v\|_s$, so

$$\|u\|_s^2 = \lambda^{-n}\int (1+|\eta|^2)^s |\tilde{\varphi}((\eta-\lambda\xi)\lambda^{-1/2})|^2 d\eta = \lambda^{-n/2}\int (1+|\lambda\xi + \zeta\lambda^{1/2}|^2)^s |\tilde{\varphi}(\zeta)|^2 d\zeta.$$

For $|\xi| < \lambda^{1/2}/2$, $1 + |\lambda\xi + \xi\lambda^{1/2}|^2$ can be estimated from above and below by $4^{\pm 1}\lambda^2$. Since

$$\int_{|\zeta|>\lambda^{1/2}/2} |\tilde{\varphi}(\zeta)|^2 d\zeta \leqslant 2^{2N}\lambda^{-N}\int |\zeta|^{2N} |\tilde{\varphi}(\zeta)|^2 d\zeta,$$

we obtain

$$\|u\|_s^2 \geq c\lambda^{2s-n/2}(\|\varphi\|_0^2 - \lambda^{-N}\|\varphi\|_N^2).$$

Similarly

$$\|u\|_{s-1}^2 \leq c_1\lambda^{2(s-1)-n/2}(\|\varphi\|_0^2 + \lambda^{-N}\|\varphi\|_N^2).$$

Furthermore $Pu = Qv$, where $Q = P(x, D)h(x)$. Clearly

$$Pu(y) = (2\pi)^{-n}\int q(y, \eta)\tilde{v}(\eta)e^{i(y, \eta)}d\eta =$$

$$(2\pi)^{-n}\int q(y, \eta)\lambda^{-n/2}\tilde{\varphi}((\eta - \lambda\xi)\lambda^{-1/2})e^{i(x, \lambda\xi - \eta) + i(y, \eta)}d\eta =$$

$$(2\pi)^{-n}\int q(y, \lambda\xi + \zeta\lambda^{1/2})\tilde{\varphi}(\zeta)e^{-i(x, \zeta)\lambda^{1/2}}e^{i(y, \lambda\xi + \zeta\lambda^{1/2})}d\zeta.$$

Expanding $q(y, \lambda\xi + \zeta\lambda^{1/2})$ in ζ by Taylor's formula, we obtain

$$q(y, \lambda\xi + \zeta\lambda^{1/2}) = \sum_{|\alpha| < N-1}\frac{1}{\alpha!}q^{(\alpha)}(y, \lambda\xi)(\zeta\lambda^{1/2})^\alpha + r(y, \zeta, \lambda). \tag{2.5}$$

In this case

$$r(y, \zeta, \lambda) = \sum_{|\alpha| = N}\frac{N}{\alpha!}\int_0^1(1-t)^{N-1}q^{(\alpha)}(y, \lambda\xi + t\zeta\lambda^{1/2})(\zeta\lambda^{1/2})^\alpha dt$$

and $q^{(\alpha)}(x, \xi) = (iD\xi)^\alpha q(x, \xi)$. When $|\zeta| < \lambda^{1/2}/2$ the remainder term $r(y, \zeta, \lambda)$ does not exceed $|\zeta|^N\lambda^{1-N/2}$, and for $|\zeta| > \lambda^{1/2}/2$ estimating the remaining terms in (2.5), we obtain

$$|r(y, \zeta, \lambda)| \leq C_2\sum_{|\alpha| < N}\lambda^{1-|\alpha|}|(\zeta\lambda^{1/2})^\alpha| \leq C_3|\zeta|^N\lambda^{1-N/2}.$$

If $w \in C_0^\infty(\Omega)$ is an arbitrary function and R is an operator with symbol $r(y, \zeta, \lambda)$, then

$$\langle Ru, w\rangle = \iint r(y, \zeta, \lambda)\tilde{\varphi}(\zeta)e^{-i(x, \zeta)\lambda^{1/2}}e^{i(y, \lambda\xi + \zeta\lambda^{1/2})}d\zeta\, w(y)\, dy =$$

$$\iint \tilde{r}(\eta - \zeta\lambda^{1/2}, \zeta, \lambda)\tilde{\varphi}(\zeta)e^{-i(x, \zeta)\lambda^{1/2}}\tilde{w}(\lambda\xi - \eta)\, d\zeta\, d\eta,$$

where $\tilde{r}(\eta, \zeta, \lambda)$ is the Fourier transform of $r(y, \zeta, \lambda)$ in the first argument. Hence

$$|\langle Ru, w\rangle| \leq C_4\iint|\zeta|^N(1 + |\eta - \zeta\lambda^{1/2}|)^{-n-1}\lambda^{1-N/2}|\tilde{\varphi}(\zeta)||\tilde{w}(\lambda\xi - \eta)|\, d\zeta\, d\eta \leq$$

$$C_5\lambda^{1-N/2}\left(\int|\zeta|^{2N}|\tilde{\varphi}(\zeta)|^2 d\zeta\right)^{1/2}\left(\int|\tilde{w}(\lambda\xi - \eta)|^2 d\eta\right)^{1/2} \leq C_5\lambda^{1-N/2}\|\varphi\|_N\|w\|_0.$$

Therefore

$$\|Ru\|_0 \leq C_5\lambda^{1-N/2}\|\varphi\|_N.$$

Thus, with permissible accuracy, Pu agrees with

$$\int\sum_{|\alpha| < N}\frac{1}{\alpha!}q^{(\alpha)}(y, \lambda\xi)(\zeta\lambda^{1/2})^\alpha\tilde{\varphi}(\zeta)e^{i(y-x)\zeta\lambda^{1/2}}e^{i\lambda(y, \xi)}d\zeta =$$

$$\sum_{|\alpha| < N}\frac{1}{\alpha!}q^{(\alpha)}(y, \lambda\xi)D_y^\alpha\varphi((y - x)\lambda^{1/2})e^{i\lambda(y, \xi)}.$$

Now by Taylor's formula we expand the derivatives $q^{(\alpha)}(y, \xi)$ with respect to y at a point x:

$$q^{(\alpha)}(y, \lambda\xi) = \sum_{|\alpha+\beta| < N}\frac{1}{\beta!}q_{(\beta)}^{(\alpha)}(x, \lambda\xi)(y - x)^\beta + \sum_{|\alpha+\gamma| = N}r_\gamma^\alpha(x, y, \xi)(y - x)^\gamma,$$

where $q_{(\beta)}^{(\alpha)}(x, \xi) = (iD_x)^\beta(iD_\xi)^\alpha q(x, \xi)$. We have

$$\left\| \sum_{|\alpha+\gamma|=N} \frac{1}{\alpha!} r_{\gamma}^{\alpha}(x, y, \xi)(y-x)^{\gamma} \overline{D_y^{\alpha}}\varphi((y-x)\lambda^{1/2}) e^{i\lambda\langle y, \xi\rangle} \right\|_0^2 \leqslant$$

$$C_6 \sum_{|\alpha+\gamma|=N} \lambda^{-n/2} \| \lambda^{1-|\alpha+\gamma|/2} z^{\gamma} D_z^{\alpha}\varphi(z) \|_0^2 = C_6 \lambda^{2-n/2-N} \sum_{|\alpha+\gamma|=N} \| z^{\gamma} D^{\alpha}\varphi \|_0^2.$$

Thus we obtain

$$\lambda^{2s-n/2} \| \varphi \|_0^2 \leqslant C \left\{ \lambda^{-n/2} \left\| \sum_{|\alpha+\beta|<N} \frac{1}{\alpha!\beta!} q_{(\beta)}^{(\alpha)}(x, \lambda\xi) y^{\beta} D^{\alpha}\varphi \lambda^{(|\alpha|-|\beta|)/2} \right\|_0^2 + \right.$$

$$\left. \lambda^{2-n/2-N} \sum_{|\alpha+\beta|=N} \| y^{\beta} D^{\alpha}\varphi \|_0^2 \right\}.$$

Recalling that $p(x, \xi) = \sum_{j<N/2} p^j(x, \xi) + r(x, \xi)$, where r stands for the symbol of an operator of order $1 - N/2$, we arrive at (2.4) and the lemma is proven.

2.6. The following two lemmas are proven in Chapter II.

Lemma 2.4. Let T be an arbitrary canonical transformation mapping the set $M \subset K \times \{\mathbb{R}^n \setminus 0\}$ into a compact set M'. Inequality (2.4) is satisfied for all $(x, \xi) \in M$ if and only if an analogous inequality, where $p^j(x, \xi)$ is replaced by functions $q_j(x', \xi')$, where $q^0(x', \xi') = p^0(x, \xi)$ and $q^j(x', \lambda\xi') = \lambda^{1-j} q^j(x', \xi')$, is satisfied for all $(x', \xi') \subset M'$.

Lemma 2.5. Let p^0 be a real function in $C^{\infty}(U \times \{\mathbb{R}^n \setminus 0\})$ and $p^0(x, t\xi) = tp^0(x, \xi)$ for $t \geqslant 1$ and $|\xi| \geqslant 1$. Suppose $p^0(x_0, \xi^0) = 0$, but $\mathrm{grad}_{x,\xi} p^0(x_0, \xi^0)$ is not collinear to $(\xi^0, 0)$. Then there exists a canonical transformation under which $p^0(x, \xi)$ transforms to a function equal to ξ_1 in some neighborhood of the point (x_0, ξ^0).

2.7. From the condition $c^0(x_0, \xi^0) > 0$ it follows that $\mathrm{grad}_{\xi} p^0(x_0, \xi^0) \neq 0$. Therefore, using Lemmas 2.2 and 2.5 we may assume that $p^0(x, \xi) = i\xi_1 + q(x, \xi)$ in a neighborhood of the point $(0, \xi^0)$, that $\xi_1^0 = 0$, $q(0, \xi^0) = 0$, that

$$c^0(0, \xi^0) = -\frac{\partial q(0, \xi^0)}{\partial x^1} > 0,$$

and that q is a real-valued function independent of ξ_1.

Lemma 2.6. To any $N > 0$ in a neighborhood of the origin there corresponds a function $w(x)$ such that $w(0) = 0$, $\mathrm{grad}\, w(0) = 0$,

$$\mathrm{Im}\, w(x) \geqslant c_0 |x|^2, \quad c_0 = \mathrm{const} > 0,$$

and $p^0(x, \xi^0 + \mathrm{grad}\, w(x)) = O(|x|^N)$ as $x \to 0$.

Proof. Since $\mathrm{grad}\, w(0) = 0$ it suffices to solve the equation

$$\sum_{|\alpha|<N/2} \frac{1}{\alpha!} \overline{p^{0(\alpha)}}(x, \xi^0)[\mathrm{grad}\, w(x)]^{\alpha} = O(|x|^N) \text{ as } x \to 0$$

or

$$\frac{1}{i}\frac{\partial w}{\partial x^1} + \sum_{|\alpha|<N/2} \frac{1}{\alpha!} q^{0(\alpha)}(x, \xi^0)[\mathrm{grad}\, w(x)]^{\alpha} = O(|x|^N) \text{ as } x \to 0. \tag{2.6}$$

Furthermore, replacing the coefficients $q^{0(\alpha)}(x, \xi^0)$ by the partial sums of the Taylor series to order $N - 2|\alpha|$ we obtain a first-order equation with analytic coefficients (if zero is substituted for the right-hand side). In this case, equation (2.6) reveals that

$$\frac{1}{i}\frac{\partial w_0}{\partial x^1}+\sum_{j=1}^{n}\frac{\partial q(0,\xi^0)}{\partial x^j}x^j+\sum_{j=2}^{n}\frac{\partial q(0,\xi^0)}{\partial \xi_j}\frac{\partial w_0}{\partial x^j}=O(|x|^2)$$

as $x \to 0$, $w = w_0 + w_1$, and $w_1 = O(|x|^3)$.

When $\operatorname{grad}_\xi q(0, \xi^0) = 0$ we may assume that

$$w_0=-iq_{(1)}(0,\xi^0)(x^1)^2/2+iM((x^2)^2+\ldots+(x^n)^2)-i\sum_{j=2}^{n}q_{(j)}(0,\xi^0)x^1x^j,$$

where $q_{(1)}(0, \xi^0) < 0$ by hypothesis and, if $M > 0$ is sufficiently large, then $\operatorname{Im} w_9(x) \geqslant c_0|x|^2$.

If $\operatorname{grad}_\xi q(0, \xi^0) \neq 0$, then we may assume that

$$w_0=-i\Big[q_{(1)}(0,\xi^0)+\sum q^{(j)}(0,\xi^0)^2\Big]\frac{(x^1)^2}{2}+i((x^2)^2+\ldots+(x^n)^2)/2+v(x),$$

where $v(x)$ is a quadratic form with real coefficients satisfying the equation

$$\frac{1}{i}\frac{\partial v}{\partial x^1}+\sum_{j=1}^{n}b_jx^j+\sum_{j=2}^{n}q^{(j)}(0,\xi^0)\frac{\partial v}{\partial x^j}=0,$$

and also

$$b_1=-\sum_{j=2}^{n}q^{(j)}(0,\xi^0)^2,\quad b_j=q_{(j)}(0,\xi^0)+iq^{(j)}(0,\xi^0),\qquad j=2,\ldots,n.$$

If $v=\sum_{j,l=1}^{n}A_{jl}x^jx^l$, then it is easy to see that

$$A_{1j}=-\operatorname{Im}b_j,\quad \sum_{l=1}^{n}A_{lj}q^{(l)}(0,\xi^0)=-\operatorname{Re}b_j,\qquad j=1,\ldots,n.$$

This set of equations relative to A_{1j} is compatible since

$$\sum_{j=1}^{n}\operatorname{Im}b_jq^{(j)}(0,\xi^0)=+\operatorname{Re}b_1.$$

Therefore the problem reduces to the construction of a symmetric matrix A of order $n - 1$ for which $A\alpha = \beta$, where $\alpha = (q^2(0, \xi^0),\ldots, q^{(n)}(0, \xi^0))$ is not zero by hypothesis. Let $e = (1, 0,\ldots, 0)$ and T be a nonsingular matrix of order $n - 1$ such that $\alpha = Te$. If we assume $T^*AT = B$, then the equation takes the form $Be = T^*\beta$. The first column of B is determined uniquely by this equation; we determine the remaining elements arbitrarily, but in such a way that matrix B is symmetric. Then let $A = T^{*-1}BT^{-1}$.

The lemma is proven.

2.8. Proof of Theorem 2.2. By Lemma 2.2 we assume the order of operator P to be equal to unity. Suppose

$$\|\varphi\|_s\leqslant C(\|P^*\varphi\|_0+\|\varphi\|_{s-1}),\qquad \varphi\in C_0^\infty(\omega) \tag{2.7}$$

for some constants C and s. We shall show that this violates the conditions of Theorem 2.2.

According to Lemma 2.3, it follows from (2.7) that

$$\|\psi\|_0\leqslant C\lambda^{1-s}\Bigg\{\Bigg\|\sum_{2j+|\alpha+\beta|<N}r^{j\,(\alpha)}_{(\beta)}(x,\xi)\frac{1}{\alpha!\beta!}y^\beta D^\alpha\psi(y)\lambda^{-j-\frac{|\alpha+\beta|}{2}}\Bigg\|_0+\lambda^{-N/2}\sum_{|\alpha+\beta|\leqslant N}\|y^\beta D^\alpha\psi\|_0\Bigg\},\quad \psi\in C_0^\infty(\mathbb{R}^n), \tag{2.8}$$

for all $\lambda \geqslant 1$, $(x, \xi) \in M$, where M is a compact neighborhood of (x_0, ξ^0), $r^0 = p^0$, and $\sum r^j$ is the symbol of the operator P^*.

Applying Lemmas 2.4 and 2.5 we can restrict our discussion to the case in which

$$p^0(x, \xi) = i\xi_1 + q(x, \xi),$$

where $q(x_0, \xi^0) = 0$, $q_{(1)}(x_0, \xi^0) < 0$, $\xi_1^0 = 0$, and q is a real-valued function independent of ξ_1. Subsequently, we shall assume for brevity that $x_0 = 0$. Let $\psi_1(y) = \psi(y\lambda^{-1/2})$. Inserting function ψ_1 into (2.8) we obtain

$$\|\psi\|_0 \leqslant C\lambda^{1-s}\left\{\left\| \sum_{|\alpha+\beta|<N} q^{(\alpha)}_{(\beta)}(0, \xi^0)\frac{1}{\alpha!\beta!}y^\beta D^\alpha\psi \cdot \lambda^{-|\alpha|} - \lambda^{-1}\frac{\partial\psi}{\partial y^1} + \sum_{\substack{j>0\\ 2j+|\alpha+\beta|<N}} r^{j(\alpha)}_{(\beta)}(0, \xi^0)\frac{1}{\alpha!\beta!}y^\beta D^\alpha\psi\lambda^{-j-|\alpha|}\right\|_0 + \lambda^{-N/2}\sum_{|\alpha+\beta|\leqslant N}\|y^\beta D^\alpha\psi\|_0\lambda^{\frac{|\beta|-|\alpha|}{2}}\right\}, \tag{2.9}$$

This inequality is satisfied by hypothesis for all $\lambda \geqslant 1$ and any $\psi \in C_0^\infty(\mathbb{R}^n)$. The number N will be indicated later.

Let ω be a small neighborhood of y = 0 in which a function w is defined for which

$$A(y) \equiv \frac{1}{i}\frac{\partial w}{\partial y^1} + \sum_{|\alpha+\beta|<N} q^{(\alpha)}_{(\beta)}(0, \xi^0)\frac{1}{\alpha!\beta!}y^\beta(\operatorname{grad} w)^\alpha = O(|y|^k)$$

and

$$\operatorname{Im} w(y) \geqslant c_0((y^1)^2 + \ldots + (y^n)^2), \quad c_0 = \text{const} > 0.$$

The existence of such a function w and such a neighborhood ω is proved in Lemma 2.6 (the number k will be chosen later).

We shall now find a function of the form

$$\psi(y) = \sum_{j=0}^{l}\varphi_j(y)\lambda^{-j}e^{i\lambda w(y)}, \quad \varphi_j \in C_0^\infty(\omega),$$

for which (2.9) cannot be satisfied for large λ. Note that

$$\sum_{2j+|\alpha+\beta|<N} r^{j(\alpha)}_{(\beta)}(0, \xi^0)\frac{1}{\alpha!\beta!}y^\beta D^\alpha\psi\lambda^{-j-|\alpha|} = e^{i\lambda w(y)}\sum_{j=0}^{M} a_j(y)\lambda^{-j},$$

where $M = l - N + 1$,

$$a_0(y) = \varphi_0(y)A(y), \quad a_1(y) = \varphi_1(y)A(y) + i\frac{\partial\varphi_0}{\partial y^1} + \sum_{j=1}^{n}A_j(y)\frac{\partial\varphi_0}{\partial y^j} + B(y)\varphi_0,$$

$$a_m(y) = \varphi_m(y)A(y) + i\frac{\partial\varphi_{m-1}}{\partial y^1} + \sum_{j=1}^{n}A_j(y)\frac{\partial\varphi_{m-1}}{\partial y^j} + B(y)\varphi_{m-1} + C_m(y),$$

and C_m stands for a linear combination of functions $\varphi_0, \varphi_1, \ldots, \varphi_{m-2}$ and their derivatives. Since $A(y) = O(|y|^k)$ we can find a function $\varphi_0 \not\equiv 0$ such that $a(y) = O(|y|^{k-1})$. To do so it is sufficient to solve the equation

$$i\frac{\partial\varphi_0}{\partial y^1} + \sum_{j=1}^{n}A_j(y)\frac{\partial\varphi_0}{\partial y^j} + B(y)\varphi_0 = O(|y|^{k-1}) \quad \text{as} \quad y \to 0,$$

so that $\varphi_0(0) = 1$. According to the Cauchy-Kovalevsky theorem, we can find a function φ_0 such that $\varphi_0 = 1$ when $y^1 = 0$. The analytic solution obtained should then be multiplied by a cutoff function equal to 1 in the domain $\omega' \subset \omega$.

Similarly the functions $\varphi_1, \varphi_2, \ldots, \varphi_M \in C_0^\infty(0)$ can be determined so that $a_m(y) = O(|y|^{k-m})$ as $y \to 0$.

Note that

$$\|\psi\|_0^2 = \int \left| \sum_{j=0}^{l} \varphi_j(y)\lambda^{-j} \right|^2 e^{-2\lambda \operatorname{Im} w(y)}\, dy.$$

Let $y = z\lambda^{-1/2}$. Then as $\lambda \to \infty$

$$\|\psi\|_0^2 \geqslant \lambda^\varkappa \left[\int |\varphi_0(0)|^2 e^{-2C((z^1)^2+\ldots+(z^n)^2)}\, dz + O(1)\right] \geqslant c_1\lambda^\varkappa, \tag{2.10}$$

where $c_1 = \text{const} > 0$ and $\varkappa = -n/2$. On the other hand,

$$\left\| e^{i\lambda w(y)} \sum_{j=0}^{M} a_j(y)\lambda^{-j} \right\|_0^2 \leqslant C_2 \sum_{j=0}^{M} \int |y|^{2(k-j)} e^{-2\lambda \operatorname{Im} w(y)} \lambda^{-2j}\, dy \leqslant$$

$$C_3\lambda^\varkappa \sum_{j=0}^{M} \lambda^{-j-k} \int |z|^{2(k-j)} e^{-2a_0((z^1)^2+\ldots+(z^n)^2)}\, dz,$$

where $a_0 > 0$ and $k > M$. Hence

$$\lambda^{1-s} \left\| \sum_{2j+|\alpha+\beta|<N} r_{j(\beta)}^{(\alpha)}(0, \xi^0) \frac{1}{\alpha!\beta!} y^\beta D^\alpha \psi \lambda^{-j-|\alpha|} \right\| \leqslant C_4\lambda^{1-s+(\varkappa-k)/2}. \tag{2.11}$$

Finally, since $|\operatorname{grad} w(y)| \leqslant C|y|$ we have

$$\int |y^\beta D^\alpha \psi(y)|^2 dy\, \lambda^{|\beta|-|\alpha|} \leqslant C_5 \sum_{j=0}^{|\alpha|} \int |y|^{2|\beta|} \lambda^{2(|\alpha|-j)} |y|^{2(|\alpha|-j)} e^{-2\lambda \operatorname{Im} w(y)} \lambda^{|\beta|-|\alpha|}\, dy \leqslant$$

$$C_6\lambda^\varkappa \sum_{j=0}^{|\alpha|} \int |z|^{2(|\beta|+|\alpha|-j)} e^{-2c_0((z^1)^2+\ldots+(z^n)^2)}\, dz\, \lambda^{-j} \leqslant C_7\lambda^\varkappa,$$

so that

$$\lambda^{1-s-N/2} \sum_{|\alpha+\beta| \leqslant N} \|y^\beta D^\alpha \psi\|_0 \lambda^{(|\beta|-|\alpha|)/2} \leqslant C_8\lambda^{1-s-N/2+\varkappa/2}. \tag{2.12}$$

Now let $k \geqslant 2(1-s)$ and N be so large that $2(1-s) - N/2 \leqslant 0$; that is, $N \geqslant 4(1-s)$. (Note that in (2.7) the value of s is less than unity.) The estimates (2.10)-(2.12) reveal that from (2.9) the inequality

$$\sqrt{c_1}\, \lambda^{\varkappa/2} \leqslant C\,(C_4\lambda^{\varkappa/2+1-s-k/2} + C_8\lambda^{\varkappa/2+1-s-N/2})$$

follows; this inequality cannot be satisfied with the same constant C for all $\lambda \geqslant 1$, that is, we arrive at a contradiction. The proof of Theorem 2.1 is complete.

§3. Finite-Order Zero Theorem

3.1. In this section we shall prove a theorem giving the necessary conditions for local solvability in the case when the Lie algebra generated by the operators P and P^* contains an operator elliptic at the given point. Such a theorem was proved simultaneously and independently by L. Nirenberg and F. Trèves[2] and the author[11] (for somewhat broader conditions).

In the above-mentioned work of L. Nirenberg and F. Trèves the (Ψ) condition was formulated for the first time and it was suggested that this condition was necessary and sufficient for local solvability of pseudo-differential equations. The sufficiency of this condition will be proven in the following chapter.

Let us formulate this condition. Let P be a pseudo-differential operator with principal symbol p^0 and let $a = \operatorname{Re} zp^0$ and $b = \operatorname{Im} zp^0$, where z is a complex number different from zero. Let $(x_0, \xi^0) \in T^*\Omega \setminus 0$ be a characteristic point. By γ we shall denote the bicharacteristic of b passing through the point (x_0, ξ^0) so that

$$\gamma = \Big\{(x(t), \xi(t));\ t \in \mathbb{R},\ \dot{x}(t) = \frac{\partial b}{\partial \xi}(x(t), \xi(t)),$$

$$\dot{\xi}(t) = -\frac{\partial b(x(t), \xi(t))}{\partial x},\ x(0) = x_0,\ \xi(0) = \xi^0\Big\}.$$

Suppose $h(t) = a(x(t), \xi(t))$.

<u>(Ψ) condition:</u> For every $z \in \mathbb{C} \setminus 0$, it follows from the condition $h(t_1) > 0$ that $h(t) \geqslant 0$ for all $t \geqslant t_1$.

Clearly, the (Ψ) condition is considered only on that portion of the curve γ where the vector tangent to it has no "radial" direction, that is, the tangent vector is not collinear to the vector $(0, \xi)$.

In L. Nirenberg and F. Trèves[2] it is shown that it suffices in practice to verify the (Ψ) condition for one value of $z \neq 0$.

3.2. Let $\operatorname{Im} p^0 = a_1$ and $\operatorname{Re} p^0 = a_2$. By H_i we denote the operator of differentiation along the bicharacteristic of a_i:

$$H_i = \sum_{j=1}^{n} \left(\frac{\partial a_i}{\partial x^j}\frac{\partial}{\partial \xi_j} - \frac{\partial a_i}{\partial \xi_j}\frac{\partial}{\partial x^j}\right), \qquad i = 1, 2.$$

If $\alpha = (\alpha_1, \dots, \alpha_r)$, $\beta = (\beta_1, \dots, \beta_r)$, where α_j and β_j are non-negative integers, then by $H_1^\alpha H_2^\beta$ we denote the operator $H_1^{\alpha_1} H_2^{\beta_1} \dots H_1^{\alpha_r} H_2^{\beta_r}$.

Denote by $k_0(x, \xi)$ the least number k for which $H_1^k a_2(x, \xi) \neq 0$, and by $k_1(x, \xi)$, the least k for which $H_1^\alpha H_2^\beta a_2(x, \xi) \neq 0$ and $|\alpha + \beta| = k$. If $a_1(x, \xi) + ia_2(x, \xi) \neq 0$, then we assume $k_0(x, \xi) = k_1(x, \xi) = 0$. Thus, the functions k_0 and k_1 have values different from zero only at characteristic points.

In this section we shall prove

<u>Theorem 3.1.</u> Let P be a pseudo-differential operator of order m, let $(x_0, \xi^0) \in T^*\Omega \setminus 0$ be a characteristic point, and let $k(x_0, \xi^0) = k < \infty$. Suppose for some $z \in \mathbb{C} \setminus 0$ the (Ψ) condition is violated at (x_0, ξ^0). Then for every neighborhood ω of x_0 one can find a function $f \in C_0^\infty(\omega)$ such that the equation $Pu = f$ has no solutions in ω belonging to $\mathscr{E}'(\Omega)$.

Note that Theorem 2.1 is a special case of Theorem 3.1, which corresponds to the value $k = 1$.

3.3. In the following section we shall prove

<u>Theorem 4.1.</u> Let $k_1(x_0, \xi^0) = k$, where $0 < k < \infty$. Let $c^0(x, \xi) \leqslant 0$ at every characteristic point of some neighborhood of (x_0, ξ^0). Then for all real numbers ρ except perhaps a finite number of them (which is not greater than k) the operator $(1 + i\rho)P$ is such that $k_1(x_0, \xi^0) = k_0(x_0, \xi^0) = k$.

By the assumption of Theorem 3.1, (Ψ) is violated for some $z \in \mathbb{C} \setminus 0$. Thus along the bicharacteristic $x = x(t)$, $\xi = \xi(t)$ of $\operatorname{Im} zp^0$ the function $h(t) = \operatorname{Re} zp^0(x, (t), \xi(t))$ is such that it assumes negative values on some sequence $t_j \to 0$, $t_j > 0$, and positive values on another sequence $t_j' \to 0$, $t_j' < 0$. By continuity it follows that (Ψ) is also violated for nearby values of z. Since the (Ψ) condition is invariant under multiplication by real numbers, one may assume these values of z to be of the form

$z = 1 + i\rho$ for real ρ. By Theorem 4.1, we assume $k_1 = k_0 = k$ for the operator $(1 + i\rho)P$, k being odd, and if the principal symbol of this operator is equal to $a_1 + ia_2$, then $H^k_{a_2} a_1(x_0, \xi^0) < 0$. In particular, from this inequality it follows that $\text{grad}_{x,\xi}\, a_2(x_0, \xi^0)$ is collinear to $(\xi^0, 0)$. Therefore, applying a canonical transformation $a_2 = \xi_1$ so that $p^0(x, \xi) = i\xi_1 + q(x, \xi)$ in some neighborhood of (x_0, ξ^0), k being an odd number, $\xi^0_1 = 0$ and $\partial^j_1 q(x_0, \xi^0)$ when $j < k$, $\partial^k_1 q(x_0, \xi^0) < 0$.

3.4. Lemma 3.1. Let $F \in C^\infty(\mathbb{R}^2)$ and

$$F(x, y) = ax^k + byx^p + \varphi(x, y),$$

where $2p \leqslant k - 1$, $b \neq 0$, $a \neq 0$. If k is odd, then $a < 0$ and

$$\varphi(x, y) = O(|y|^2 + |yx^{p+1}| + |x|^{k+1}), \quad \frac{\partial\varphi(x, y)}{\partial x} = O(|y|^2 + |yx^p| + |x|^k)$$

as $(x, y) \to (0, 0)$. Then in every neighborhood of the origin there is a point (x_0, y_0) for which $F(x_0, y_0) = 0$, $\dfrac{\partial F(x_0, y_0)}{\partial x} < 0$.

Proof. Let $y = f(x)$ be a function such that

$$F(x, f(x)) = 0, \quad f(0) = 0.$$

The existence of such a function f in a small neighborhood of $x = 0$ follows from the following considerations. Let $z = yx^{p-k}$. Then

$$F(x, y) = x^k[a + bz + O(|z|^2|x|^{k-2p} + |z||x| + |x|)]$$

as $(x, y) \to (0, 0)$. As $b \neq 0$ one may find $z = z(x)$ from the equation $F(x, y) = 0$ by the Implicit Function Theorem. Obviously

$$f(x) = z(x)x^{k-p} = \left[-\frac{a}{b} + O(|x|)\right]x^{k-p}$$

as $x \to 0$. We now consider the function $\partial F(x, y)/\partial x$ when $y = f(x)$. We have

$$\left.\frac{\partial F(x, y)}{\partial x}\right|_{y=f(x)} = akx^{k-1} + bpyx^{p-1} + O(|y|^2 + |yx^p| + |x|^k) =$$

$$ax^{k-1}(k - p) + O(|x|^k + |x|^{2(k-p)}) \quad \text{as} \quad x \to 0.$$

But $2k - 2p > k$ by assumption. Hence

$$\left.\frac{\partial F(x, y)}{\partial x}\right|_{y=f(x)} = x^{k-1}[a(k - p) + o(1)] \quad \text{as} \quad x \to 0.$$

The lemma is proven.

3.5. Thus, Lemma 3.1 reveals that if

$$\text{grad}_{x',\xi'}\, q(x_0, \xi^0) \neq 0 \quad \text{or} \quad \text{grad}_{x',\xi'} \frac{\partial^j q(x_0, \xi^0)}{\partial (x^1)^j} \neq 0 \qquad (3.1)$$

for $2j \leqslant k - 1$, then in every neighborhood of (x_0, ξ^0) we can find a point (x, ξ) where the conditions of Theorem 2.1 are satisfied. Therefore, in proving Theorem 3.1 we can assume that

$$\text{grad}_{x',\xi'} \frac{\partial^j q(x_0, \xi^0)}{\partial (x^1)^j} = 0 \text{ for } j = 0, 1, \ldots, \left[\frac{k-1}{2}\right].$$

Further we shall use Lemmas 2.1, 2.2, and 2.3 without alteration. Instead of Lemma 2.6, we shall prove the following statement.

Lemma 3.2. Let $p^0 \in C^\infty(T^*\Omega)$ and $p^0(x, \xi) = i\xi_1 + q(x, \xi)$ in a neighborhood of $(0, \xi^0)$, where $|\xi^0| \geqslant 1$, $\xi_1^0 = 0$. Suppose $q(0, \xi^0) = 0$, and $\partial_1^j q(0, \xi^0) = 0$ when $j < k = 2l - 1$, but $\partial_1^k q(0, \xi^0) < 0$ and

$$\frac{\partial^{j+1} q(0, \xi^0)}{(\partial x^1)^j \partial x^s} = \frac{\partial^{j+1} q(0, \xi^0)}{(\partial x^1)^j \partial \xi_t} = 0 \text{ when } j < l,\ s = 2, \ldots, n,$$

$t = 1, \ldots, n$, and q is independent of ξ_1. Then for every $\sigma > 0$ in a neighborhood of the origin one can find a function $w(x)$ having a second-order zero at $x = 0$ and such that $\operatorname{Im} w(x) \geqslant c_0((x^1)^{2l} + (x^2)^2 + \ldots + (x^n)^2$ and

$$\overline{p^0}(x, \xi^0 + \operatorname{grad} w(x)) = O(|x|^\sigma) \text{ when } x \to 0.$$

Proof. Since grad $W(0) = 0$ it suffices to solve the equation

$$\sum_{|\alpha| < \sigma/2} \frac{1}{\alpha!} \frac{\partial^\alpha p^0(x, \xi^0)}{\partial \xi^\alpha} [\operatorname{grad} w(x)]^\alpha = O(|x|^\sigma) \text{ when } x \to 0$$

or

$$-i\frac{\partial w}{\partial x^1} + \sum_{|\alpha| < \sigma/2} \frac{1}{\alpha!} \frac{\partial^\alpha q(x, \xi^0)}{\partial \xi^\alpha} [\operatorname{grad} w(x)]^\alpha = O(|x|^\sigma) \text{ when } x \to 0.$$

Furthermore, replacing the coefficients $\partial^\alpha q(x, \xi^0)/\partial\xi^\alpha$ by the terms of their Taylor series up to order $\sigma - 2|\alpha|$ we obtain an equation with analytic coefficients (if the right-hand side is replaced by zero). For this equation there exists a unique solution to the Cauchy problem with initial conditions:

$$w = i((x^2)^2 + \ldots + (x^n)^2) \text{ when } x^1 = 0, \quad \operatorname{grad} w(0) = 0.$$

We now show that

$$\operatorname{Im} w(x) = [b(x^1)^{2l} + (x^2)^2 + \ldots + (x^n)^2](1 + o(1)) \text{ when } x \to 0,$$

where $b > 0$. Note that our equation can be rewritten in the form

$$\frac{1}{i}\frac{\partial w}{\partial x^1} + q^0(x, \xi^0) + \sum_{s=2}^{n} \frac{\partial q(x, \xi^0)}{\partial \xi_s} \frac{\partial w}{\partial x^s} + \frac{1}{2} \sum_{s,t=2}^{n} \frac{\partial^2 q(x, \xi^0)}{\partial \xi_s \partial \xi_t} \frac{\partial w}{\partial x^s} \frac{\partial w}{\partial x^t} + \ldots = 0.$$

From this equation it is seen that $\partial w/\partial x^1 = 0$ at $x = 0$. Differentiating with respect to x^s when $2 \leqslant s \leqslant n$, we obtain $\partial^2 w(0)/\partial x^1 \partial x^s = 0$. Now suppose $j < 1$ and

$$\frac{\partial^j w(0)}{(\partial x^1)^{j-1} \partial x^s} = \frac{\partial^{j-1} w(0)}{(\partial x^1)^{j-2} \partial x^s} = \ldots = \frac{\partial^2 w(0)}{\partial x^1 \partial x^s} = 0.$$

Then we shall show that $\partial^{j+1} w(0)/(\partial x^1)^j \partial x^s = 0$. We differentiate the equation once with respect to x^s and $j - 1$ times with respect to x^1. By hypothesis $\partial^j q(0, \xi^0)/\partial x^s (\partial x^1)^{j-1} = 0$. It is not hard to see that the derivative $\frac{\partial^j}{\partial x^s (\partial x^1)^{j-1}} \left(\frac{\partial q}{\partial \xi_s} \frac{\partial w}{\partial x^s}\right)$ is also equal to 0 at $x = 0$, since differentiating, we obtain a finite sum whose every term contains a factor of form $\frac{\partial^{i+1} q}{\partial \xi_s (\partial x^1)^i}$ or $\frac{\partial^{i+1} w}{\partial x^s (\partial x^1)^i}$ when $i \leqslant j - 1$, and these factors are zero at $x = 0$. All the following terms also have a factor $\frac{\partial^{i+1} w}{\partial x^s (\partial x^1)^i}$ when $i \leqslant j - 1$. Thus, we have shown that $\frac{\partial^{j+1} w}{\partial x^s (\partial x^1)^j} = 0$ when $j = 0, 1, \ldots, l$.

Now suppose $j < 2l - 1$ and $\frac{\partial w(0)}{\partial x^1} = \ldots = \frac{\partial^j w(0)}{(\partial x^1)^j} = 0$. We differentiate the equation j times with respect to x^1. By hypothesis $\partial_1^j q(0, \xi^0) = 0$. The derivative of $\frac{\partial q(x, \xi^0)}{\partial \xi_s} \frac{\partial w}{\partial x^s}$ is also equal to 0 at x = 0, since differentiating, we obtain a finite sum whose terms have a factor of the form $\frac{\partial^{r+1} q}{\partial \xi_s (\partial x_1)^r}$ or $\frac{\partial^{r+1} w}{\partial x^s (\partial x^1)^r}$ when $r \leqslant l - 1$; these factors are equal to 0 at x = 0 by the above arguments. Differentiating the remaining terms on the left-hand side we obtain a finite sum, in which every term has a factor equal to the derivative $\frac{\partial^{r+1} w}{\partial x^s (\partial x^1)^r}$ when $r \leqslant l - 1$ or $\partial^i w/(\partial x^1)^i$ when $i \leqslant j$ and $2 \leqslant r \leqslant n$.

Hence, $\partial_1^{j+1} w(0) = 0$ and we have proven that $\partial_1^p w(0) = 0$ when $p \leqslant 2l - 2$. We shall now differentiate the equation $2l - 1$ times with respect to x^1. By hypothesis $-\partial_1^{2l-1} q(0, \xi^0) = a > 0$. Again the derivative of $\frac{\partial q}{\partial \xi_s} \frac{\partial w}{\partial x^s}$ vanishes, since differentiating we obtain a finite sum in which every term has a factor of the form $\partial_1^r \partial \xi_s q$ or $\partial_1^r \partial_s w$. This also holds for the remaining terms on the left-hand side of the equation. Hence $\partial_1^{2l} w(0) = ia$, where $a > 0$.

Thus, we have proven that

$$w = \left[(x^2)^2 + \ldots + (x^n)^2 + \frac{a}{(2l)!}(x^1)^{2l}\right](1 + o(1)) \quad \text{as} \quad x \to 0,$$

since $|(x^1)^j x^s| = O((x^2)^2 + \ldots + (x^n)^2 + (x^1)^{2l})$ when $x \to 0$, if $j \geqslant l + 1$ and $2 \leqslant s \leqslant n$. Lemma 3.2 is proven.

3.6. Proof of Theorem 3.1. By Lemma 2.2 we may assume the order of the operator P to be equal to unity. Suppose

$$\|\varphi\|_s \leqslant C(\|P^*\varphi\|_0 + \|\varphi\|_{s-1}), \quad \varphi \in C_0^\infty(\omega) \tag{3.2}$$

with some real coefficients C and s, $C > 0$, $s < 1$. We shall prove that this contradicts the hypothesis of the theorem.

By Lemma 2.3, it follows from (3.2) that

$$\|\psi\|_0 \leqslant C\lambda^{1-s}\left\{\left\|\sum_{2j+|\alpha+\beta|<N} r^{j\,(\alpha)}_{(\beta)}(x, \xi)\frac{1}{\alpha!\,\beta!} y^\beta D^\alpha \psi \lambda^{-j-|\alpha+\beta|/2}\right\|_0 + \lambda^{-N/2}\sum_{|\alpha+\beta|\leqslant N}\|y^\beta D^\alpha \psi\|_0\right\} \tag{3.3}$$

for all $\psi \in C_0^\infty(\mathbb{R}^n)$, $\lambda \geqslant 1$, $(x, \xi) \in M$, where M is a compact neighborhood of $(0, \xi^0)$, $r^0 = \overline{p^0}$, and $\sum r^j$ denotes the symbol of the operator P^*.

By Theorem 4.1 we can replace P by the operator $(1 + i\rho)P$, whose principal symbol is $a_1 + ia_2$, where $k_1(0, \xi^0) = k_0(0, \xi^0) = k$ is an odd number, $k = 2l - 1$, and $H_{a_2}^k a_1(0, \xi^0) < 0$.

Applying Lemmas 2.4 and 2.5 we can restrict our discussion to the case when

$$p^0(x, \xi) = i\xi_1 + q(x, \xi),$$

where $q(0, \xi^0) = 0$, $\partial_1^k q(0, \xi^0) > 0$, and q is a real-valued function independent of ξ_1. Furthermore, for brevity we shall assume $\xi_1^0 = 0$. Let $\psi_1(y) = \psi(y\lambda^{-1/2})$. Inserting ψ_1 into (3.3) we obtain

$$\|\psi\|_0 \leqslant C\lambda^{1-s}\left\{\left\|\sum_{\alpha+\beta<N} q^{(\alpha)}_{(\beta)}(0, \xi^0)\frac{1}{\alpha!\,\beta!} y^\beta D^\alpha \psi \lambda^{-|\alpha|} - \lambda^{-1}\frac{\partial\psi}{\partial y^1} + \sum_{\substack{j>0 \\ 2j+|\alpha+\beta|<N}} r^{j\,(\alpha)}_{(\beta)}(0, \xi^0)\frac{1}{\alpha!\,\beta!} y^\beta D^\alpha \psi \lambda^{-j-|\alpha|}\right\|_0 + \right.$$

$$\lambda^{-N/2}\sum_{|\alpha+\beta|\leqslant N}\|y^\beta D^\alpha\psi\|_0\,\lambda^{(|\beta|-|\alpha|)/2}\Big\}. \qquad (3.4)$$

This equality is satisfied by hypothesis for all $\lambda \geqslant 1$ and any $\psi \in C_0^\infty(\mathbb{R}^n)$. The number N will be chosen below.

We assume condition (3.1) to be satisfied from Lemma 3.1 as, otherwise, one could find a point in any neighborhood of $(0, \xi^0)$ at which the function $C^0 = -\partial_1 q$ assumes positive values; therefore Theorem 2.1 is applicable.

Let ω be a sufficiently small neighborhood of the point $y = 0$ so that there is a function w defined on it for which

$$A(y)\equiv\frac{1}{i}\frac{\partial w}{\partial y^1}+\sum_{|\alpha+\beta|<N}q_{(\beta)}^{(\alpha)}(0,\xi^0)\frac{1}{\alpha!\,\beta!}y^\beta(\operatorname{grad}w)^\alpha=O(|y|^\sigma)$$

as $y \to 0$ and

$$\operatorname{Im}w(y)\geqslant c_0((y^1)^{2l}+(y^2)^2+\ldots+(y^n)^2).$$

The existence of such a function w and of a neighborhood ω follows from Lemma 3.2 (the number σ will be chosen below).

We shall now find a function $\psi(y)=\sum_{j=0}^{L}\varphi_j(y)\lambda^{-j}e^{i\lambda w(y)}$, where $\varphi_j \in C_0^\infty(\omega)$, for which inequality (3.4) cannot be satisfied for large λ. Note that

$$\sum_{2j+|\alpha+\beta|<N}r_{(\beta)}^{j\,(\alpha)}(0,\xi^0)\frac{1}{\alpha!\,\beta!}y^\beta D^\alpha\psi\cdot\lambda^{-j-|\alpha|}=e^{i\lambda w(y)}\sum_{j=0}^{k}a_j(y)\lambda^{-j},$$

where $K = L - N + 1$,

$$a_0(y)=\varphi_0(y)A(y),\qquad a_1(y)=\varphi_1(y)A(y)+\frac{1}{i}\frac{\partial\varphi_0}{\partial y^1}+\sum_{j=1}^{n}A_j(y)\frac{\partial\varphi_0}{\partial y^j}+B(y)\varphi_0,$$

$$a_s(y)=\varphi_s(y)A(y)+\frac{1}{i}\frac{\partial\varphi_{s-1}}{\partial y^1}+\sum_{j=1}^{n}A_j(y)\frac{\partial\varphi_{s-1}}{\partial y^j}+B(y)\varphi_{s-1}+C_s(y),$$

C_s being a linear combination of functions $\varphi_0, \varphi_1, \ldots, \varphi_{s-2}$ and their derivatives, and $A_j(0) = 0$. Since $A(y) = O(|y|^\sigma)$ as $y \to 0$ we can find a function $\varphi_0 \neq 0$ such that $a_1(y) = O(|y|^{\sigma-1})$. To do so it is sufficient to solve the equation

$$\frac{1}{i}\frac{\partial\varphi_0}{\partial y^1}+\sum_{j=1}^{n}A_j(y)\frac{\partial\varphi_0}{\partial y^j}+B(y)\varphi_0=O(|y|^{\sigma-1})\ \text{when}\ y\to 0$$

so that $\varphi_0(0) = 1$. The existence of such a function is guaranteed by the Cauchy-Kovalevsky theorem, provided the Cauchy conditions are imposed at $y^1 = 0$. (For example, it may be assumed that $\varphi_0 = 1$ at $y^1 = 0$.) Then the obtained analytic solution, defined in some neighborhood of the origin, must be multiplied by a smooth cutoff function equal to unity in the domain $\omega' \subset \omega$.

Similarly we can define functions $\varphi_1, \varphi_2, \ldots, \varphi_L$ in $C_0^\infty(\omega)$ for which $a_s(y) = O(|y|^{\sigma-s})$ when $y \to 0$.

Note that

$$\|\psi\|_0^2=\int\Big|\sum_{j=0}^{L}\varphi_j(y)\lambda^{-j}\Big|^2e^{-2\lambda\operatorname{Im}w(y)}dy.$$

Let $y^1=z^1\lambda^{-1/2l}$, $y^2=z^2\lambda^{-1/2}$, ..., $y^n=z^n\lambda^{-1/2}$. Then when $\lambda \to \infty$

$$\|\psi\|_0^2 \geqslant \lambda^{\varkappa}\left[\int |\varphi_0(0)|^2 e^{-2C((z^1)^{2l}+(z^2)^2+\ldots+(z^n)^2)}\,dz + o(1)\right] \geqslant C_1\lambda^{\varkappa}, \tag{3.5}$$

where $\varkappa = -\frac{1}{2l} - \frac{n-1}{2}$, $C > 0$, $C_1 > 0$. On the other hand,

$$\left\| e^{i\lambda w(y)} \sum_{j=0}^{K} a_j(y)\lambda^{-j} \right\|_0^2 \leqslant C_2 \sum_{j=0}^{K} \int |y|^{2(\sigma-j)}\lambda^{-2j} e^{-2\lambda\,\mathrm{Im}\,w(y)}\,dy \leqslant$$

$$C_3\lambda^{\varkappa} \sum_{j=0}^{K} \lambda^{-2j-(\sigma-j)/l} \int |z|^{2(\sigma-j)} e^{-2a_0((z^1)^{2l}+(z^2)^2+\ldots+(z^n)^2)}\,dz,$$

where $a_0 > 0$ and $\sigma > K$. Hence

$$\lambda^{1-s} \left\| \sum_{2j+|\alpha+\beta|<N} r_{(\beta)}^{j(\alpha)}(0,\xi^0)\frac{1}{\alpha!\beta!} y^{\beta} D^{\alpha}\psi \cdot \lambda^{-j-|\alpha|} \right\|_0 \leqslant C_4 \lambda^{\frac{\varkappa}{2}+1-s-\frac{\sigma}{2l}}. \tag{3.6}$$

Finally, since $|\mathrm{grad}\, w(y)| \leqslant c|y|$ we have

$$\int |y^{\beta} D^{\alpha}\psi(y)|^2\, dy \lambda^{|\beta|-|\alpha|} \leqslant$$

$$C_5 \sum_{j=0}^{|\alpha|} \int |y|^{2|\beta|} \lambda^{2(|\alpha|-j)} |y|^{2(|\alpha|-j)} e^{-2\lambda\,\mathrm{Im}\,w(y)}\, dy \lambda^{|\beta|-|\alpha|} \leqslant$$

$$C_6\lambda^{\varkappa} \sum_{j=0}^{|\alpha|} \int |z|^{2(|\beta+\alpha|-j)} e^{-2a_0[(z^1)^{2l}+(z^2)^2+\ldots+(z^n)^2]}\, dz\, \lambda^{|\alpha+\beta|-2j-\frac{|\alpha+\beta|-j}{2l}} \leqslant C_7 \lambda^{\varkappa+|\alpha+\beta|(1-1/2l)},$$

so

$$\lambda^{1-s-N/2} \sum_{|\alpha+\beta|\leqslant N} \|y^{\beta} D^{\alpha}\psi\| \lambda^{(|\beta-\alpha|)/2} \leqslant C_8 \lambda^{\varkappa/2+1-s-N/2+N(1-1/l)/2}. \tag{3.7}$$

Now let $\sigma > 2(1 - s)l$ and let N be so large that $2(1 - s) < N/2l$, that is, $N > 4l(1 - s)$. (Recall that $s < 1$.) The estimates (3.5)-(3.7) then reveal that inequality (3.4) may not be satisfied with the same constant C for all $\lambda \geqslant 1$; that is, we arrive at a contradiction. The proof of Theorem 3.1 is complete.

§4. Structure of the Symbol

4.1. Let us recall the theorem on the principal symbol structure.

Theorem 4.1. Let $k_1(x_0, \xi^0) = k$, where $0 < k < \infty$. Let $c^0(x, \xi) \leqslant 0$ at every characteristic point in some neighborhood ω of (x_0, ξ^0). Then for all real numbers ρ except, perhaps, a finite number of them (not exceeding k), the operator $(1 + i\rho)P$ is such that $k_1(x_0, \xi^0) = k_0(x_0, \xi^0) = k$.

In proving this theorem we shall obtain in passing the following

Theorem 4.2. The conditions of Theorem 4.1 are invariant under multiplication of operator P by an arbitrary elliptic operator.

4.2. It is easy to see that all the assertions of Theorem 4.1 are invariant under general canonical transformations of the phase space of (x, ξ). Without loss of generality we assume in proving this theorem $p^0(x, \xi)$ is a polynomial of degree k.

As, by hypothesis $k < \infty$, $\mathrm{grad}_{x,\xi}\, p^0(x_0, \xi^0)$ is collinear to $(\xi^0, 0)$. Otherwise for any function r homogeneous in ξ

$$H_{p^0 r}(x_0, \xi^0) = \sum_{j=1}^{n} \left(\frac{\partial p^0}{\partial x^j} \frac{\partial r}{\partial \xi_j} - \frac{\partial p^0}{\partial \xi_j} \frac{\partial r}{\partial x^j} \right)(x_0, \xi^0) =$$

$$\varkappa \sum_{j=1}^{n} \xi_j^0 \frac{\partial r(x_0, \xi^0)}{\partial \xi_j} = \varkappa m r(x_0, \xi^0),$$

where m is the order of homogeneity of r and $\varkappa$ is the proportionality factor: $\mathrm{grad}_{x,\xi}\, p^0(x_0, \xi^0) = (\varkappa\xi^0, 0)$. Thus, if $c_I(x_0, \xi^0) = 0$, then $H_{p^0 c_I}(x_0, \xi^0) = 0$, so $k(x_0, \xi^0) = \infty$.

We assume $\mathrm{Im}\, \mathrm{grad}_{x,\xi}\, p^0(x_0, \xi^0)$ to be collinear to $(\xi^0, 0)$. If this condition is not satisfied, then we multiply P by i and interchange the position of the functions a_1 and a_2. The only element that is not invariant under such a rearrangement is the definition of the function k_0. However, as will be evident from the proof, the number $k_0(x, \xi)$ in the statement of the theorem can be replaced by a number $k_2(x, \xi)$ equal to the minimal value of k for which $H_1^k a_2(x, \xi) \neq 0$. Therefore, for definiteness $\mathrm{Im}\, \mathrm{grad}_{x,\xi}\, p^0(x_0, \xi^0)$ may be assumed to be noncollinear to $(\xi^0, 0)$ and, using a canonical transformation, the equality

$$a_2(x, \xi) = \xi_1, \quad \xi_1^0 = 0 \tag{4.1}$$

may be satisfied in a neighborhood of (x_0, ξ^0). Obviously, we shall then have $H_2^j a_1(x, \xi) = \partial_1^j a_1(x, \xi)$ and the coefficients $(x^1 - x_0^1)^j$ in the monomial $a_1(x, \xi)$ must be vanish if $j < k_0$.

For simplicity of notation we take (x_0, ξ^0) to be the origin.

4.3. Lemma 4.1. Let Λ be the linear space over the field of complex numbers generated by monomials $\xi^\alpha x^\beta$ for $1 \leqslant |\alpha + \beta| \leqslant k$, except for monomials of the form $(x^1)^j$, when $j \leqslant k$, and $\xi_l (x^1)^j$, $x^l (x^1)^j$ when $j < k/2$, $l = 2, \ldots, n$. Let $\Lambda_s = (\mathrm{ad}\, \Lambda)^s \Lambda$. Then all the elements of Λ_s are 0 for $x = \xi = 0$, $s \leqslant k$.

Recall that $k = k_1(0, 0)$ and $(\mathrm{ad}\, f) g(x, \xi) = \{f, g\}(x, \xi)$.

Proof. We shall assume that some element of Λ_s differs from zero at $x = \xi = 0$, $s \leqslant k$. Every element in Λ_s is a polynomial, and if the polynomial at $x = \xi = 0$ assumes values other than zero, then its free term is different from zero. Let us see how this term comes about. Since in Λ we have only one monomial with a nonzero gradient, namely, ξ_1, a nonzero constant may appear only upon applying the operator $H_2 = \partial_1$ to x^1. In the general case, some earlier steps would have involved the application of operator H_2 and we must find out how $(x^1)^r$ with some r, $r \leqslant k$, appears. Since such monomials were absent among the generators of Λ, $(x^1)^r$ could appear only on computing Poisson brackets of the type $\{(x^1)^a \xi_s, (x^1)^b x^s\}$, where $a + b = r$, and s is one of the values $2, \ldots, n$. Note that in this case either $(x^1)^a \xi_s$ or $(x^1)^a x^s$ must be in Λ. Since ξ_s and x^s are equivalent, we assume that $(x^1)^a \xi_s \in \Lambda$, so that $a \geqslant k/2$.

If $(x^1)^b x^s$ also belongs to Λ, then $b \geqslant k/2$ and in order to find a nonzero constant we should compute $\geqslant k + 1$ Poisson brackets, which contradicts our assumptions. On the other hand, $(x^1)^b x^s$ may be obtained only by computing the following Poisson brackets:

$$\{\xi_1, (x^1)^{b+1} x^s\}, \ \{(x^1)^c x^s \xi_l, (x^1)^d x^l\}, \ \{(x^1)^c \xi_l, (x^1)^d x^l x^s\},$$

where $c + d = b$. We assume that every element of Λ is always in the first position, in this case, ξ_l and x^l can of course be interchanged. We see that in the first two cases we again return to finding the origin of a monomial of type $(x^1)^i x^l$, and only the third method comes to our rescue. However, in this case $c \geqslant k/2$ and, hence, $b \geqslant k/2$. Since $a \geqslant k/2$, then $r \geqslant k$, and we again arrive at a contradiction. Lemma 4.1 is proved.

4.4. Thus, if $p^0(0, 0) = 0$, $\mathrm{grad}_{x', \xi'} a_1(0, 0) = 0$ and $\mathrm{grad}_{x', \xi'} \partial_1^j \times a_1(0, 0) = 0$ for $j < k/2$, then the unique method of finding a value $H_1^\alpha H_2^\beta \times a_1(0, 0)$ different from 0 for $|\alpha + \beta| \leqslant k$ consists in assuming $|\alpha| = k$, $|\beta| = 0$. Thus, we immediately see that $k_1(0, 0)$ and $k_1(0, 0)$ agree by definition; this means that Theorem 4.1 is nontrivial only for $k(0, 0) \geqslant 2$. Further we assume that $\partial_1 a_1(0, 0) = 0$.

4.5. Now suppose either $\mathrm{grad}_{x', \xi'} a_1(0, 0) \neq 0$ or $\mathrm{grad}_{x', \xi'} \partial_1^j a_1(0, 0) \neq 0$ for some j, $1 \leqslant j < k/2$. Since the variables x' and ξ' are equivalent, we suppose that

$$\partial_{\xi_2} \partial_1^s a_1(0, 0) \neq 0 \tag{4.2}$$

for some s, $0 \leqslant s < k/2$. Now if $s > 0$, we take $\partial_{\xi_2} \partial_1^j a_1(0, 0) = 0$ for $j < s$. Thus, the coefficient of $(x^1)^s \xi_2$ in a_1 is assumed to be different from 0. Applying a canonical transformation in the space of variables x' and ξ', we can make all the coefficients of monomials of the type $(x^1)^s (x')^{\beta'} \xi^\alpha$, except $(x^1)^s \xi_2$, equal to 0 for $\alpha_1 = 0$. (To do so we replace $\xi_2 + \sum_{\beta_1 = s} a_{\alpha\beta} (x')^{\beta'} \xi^\alpha$ by ξ_2.) In particular, if $s = 0$, then after such a transformation we obtain

$$\mathrm{grad}_{x', \xi'} a_1(0, 0) = \mathrm{grad}_{x', \xi'} \xi_2.$$

4.6. <u>Lemma 4.2.</u> If conditions (4.1) and (4.2) are satisfied and $c^0(x, \xi) \leqslant 0$ at all characteristic points of some neighborhood ω of (x_0, ξ^0), then

$$\partial_{\xi_l} \partial_1^j a_1(0, 0) = \partial_m \partial_1^j a_1(0, 0) = 0 \quad \text{when} \quad j < k/2, \; l = 3, \ldots, n, \; m = 1, \ldots, n.$$

<u>Proof.</u> Suppose one of these derivatives, say, $\partial_{\xi_3} \partial_1^j a_1(0, 0) \neq 0$ for some j, $j < k/2$. Then we prove that c^0 assumes positive values at some characteristic point in ω.

Assume $\xi_1 = 0$, $\xi_4 = \ldots = \xi_n = 0$, $x^2 = \ldots = x^n = 0$ and consider function a_1 on this three-dimensional space. We have

$$a_1(x, \xi) = A (x^1)^k + B (x^1)^s \xi_2 + C (x^1)^j \xi_3 + \varphi(x, \xi),$$

where $\varphi(x, \xi) = O(|x^1|^{k+1} + |x^1|^{s+1} |\xi_2| + |x^1|^{j+1} |\xi_3| + \xi_2^2 + \xi_3^2)$ as $(x, \xi) \to 0$. In this case $B \neq 0$ and $C \neq 0$. Let $x^1 = t$, $\xi_2 = y t^{k-s}$ and $\xi_3 = z t^{k-j}$. Then $a_1(x, \xi) = t^k [A + By + Cz + O(|t| + |t||y| + |t||z| + |t| y^2 + |t| z^2)]$, since $k > 2s$ and $k > 2j$. By the Implicit Function Theorem, there exist functions $y(t)$ and $z(t)$ such that $a_1(x, \xi) = 0$ when $y = y(t)$, $z = z(t)$, and as $t \to 0$, $y(t) = y_0 + O(|t|)$ and $z(t) = z_0 = O(|t|)$ for arbitrary y_0 and z_0 such that $A + By_0 + Cz_0 = 0$. On the other hand,

$$H_2 a_1(x, \xi) = kA (x^1)^{k-1} + sB (x^1)^{s-1} \xi_2 + jC (x^1)^{j-1} \xi_3 + \partial_1 \varphi$$

and on our curve

$$H_2 a_1(x, \xi) = t^{k-1} [kA + sBy_0 + jCz_0] + O(|t|^k).$$

Varying y_0 and z_0 on the plane $A + By_0 + Cz_0 = 0$ we can change the sign of $kA + sBy_0 + jCz_0$ (recall that $j \neq s$) and hence the sign of $H_2 a_1(x, \xi)$, which contradicts our assumptions. Lemma 4.2 is proved.

<u>Lemma 4.3.</u> Suppose P satisfies conditions (4.1) and (4.2) and $c^0(x, \xi) \leqslant 0$ at all the characteristic points in some neighborhood ω of $(x_0, \xi^0) = (0, 0)$. Then in $a_1(x, \xi)$ the coefficients are 0 of $(x^1)^a (x^2)^b \xi_l$, $(x^1)^a (x^2)^b x^m$, where $l = 3, \ldots, n$; $m = 1, \ldots, n$; $a + b(s+1) < K/2$. Here $K = \min[a + b(s + 1)]$ and the minimum is taken over all a and b for which the coefficients of $(x^1)^a (x^2)^b$ in a_1 are not zero.

<u>Proof.</u> Suppose the coefficient of $(x^1)^a (x^2)^b \xi_3$ does not vanish for some a and b such that $a + b(s + 1) = p < K/2$. (ξ_3 can be replaced by ξ_l when $l = 3, \ldots, n$ or x^m when $m = 1, \ldots, n$.)

As above, we assume $x^3 = \dots = x^n = 0$, $\xi_4 = \dots = \xi_n = 0$. Then

$$a_1(x, \xi) = \sum_{i+j(s+1)=K} A_{ij}(x^1)^i(x^2)^j + B(x^1)^s\xi_2 + \sum_{a+b(s+1)=p} C_{ab}(x^1)^a(x^2)^b\xi_3 + \varphi_1(x, \xi),$$

where

$$\varphi_1(x, \xi) = O\Big(\sum_{i+j(s+1)=K} |x^1|^i|x^2|^j(|x^1|+|x^2|) + \xi_2^2 + \xi_3^2 + |x^2\xi_2| + |(x^1)^{s+1}\xi_2| + \sum_{a+b(s+1)=p} |(x^1)^a(x^2)^b||\xi_3|(|x^1|+|x^2|)\Big).$$

Substituting $x^1 = xt$, $x^2 = yt^{3+1}$, $\xi_2 = zt^{K-s}$, $\xi_3 = ut^{K-p}$, we obtain

$$a_1(x, \xi) = t^K\Big[\sum_{i+j(s+1)=K} A_{ij}x^iy^j + Bzx^s + \sum_{a+b(s+1)=p} C_{ab}x^ay^bu + O\Big(|t|\Big[\sum_{i+j(s+1)=K} |x^iy^j|(|x|+|y|) + |x^{s+1}z| + |yz| + z^2 + u^2 + \sum_{a+b(s+1)=p} |x|^a|y|^b|z|(|x|+|y|)\Big]\Big),$$

since $K > 2s$, $K > 2p$. By the Implicit Function Theorem there exist functions $x = x(t)$, $y = y(t)$, $z = z(t)$ and $u = u(t)$ such that $a_1(x, \xi) = 0$ on the curve, and when $t \to 0$

$$x(t) = x_0 + O(t), \quad y(t) = y_0 + O(t), \quad z(t) = z_0 + O(t), \quad u(t) = u_0 + O(t),$$

for any x_0, y_0, z_0, and u_0 such that

$$\sum_{i+j(s+1)=K} A_{ij}x_0^iy_0^j + Bz_0x_0^s + \sum_{a+b(s+1)=p} C_{ab}x_0^ay_0^bu_0 = 0. \tag{4.3}$$

In this case

$$H_2a_1(x, \xi) = \sum_{i+j(s+1)=K} A_{ij}i(x^1)^{i-1}(x^2)^j + Bs(x^1)^{s-1}\xi_2 + \sum_{a+b(s+1)=p} C_{ab}a(x^1)^{a-1}(x^2)^b\xi_3 + \partial_1\varphi_1 = t^{K-1}\Big[\sum_{i+j(s+1)=K} A_{ij}ix_0^{i-1}y_0^j + Bsx_0^{s-1}z_0 + \sum_{a+b(s+1)=p} C_{ab}ax_0^{a-1}y_0^bu_0 + O(|t|)\Big].$$

Now we have to prove that (4.3) contains points (x_0, y_0, z_0, u_0) such that

$$\sigma = \sum A_{ij}ix_0^{i-1}y_0^j + Bsx_0^{s-1}z_0 + \sum C_{ab}ax_0^{a-1}y_0^bu_0$$

takes values with different signs. Note that if x_0, y_0 and u_0 are any fixed numbers and $x_0 \neq 0$, then z_0 is uniquely determined by (4.3). Multiplying the left-hand side in (4.3) by s and subtracting from $x_0\sigma$, we obtain

$$\sigma_1 = \sum A_{ij}(i-s)x_0^iy_0^j + \sum C_{ab}x_0^ay_0^bu_0(a-s).$$

Since not all the coefficients C_{ab} vanish for $a + b(s+1) = p$ and $a \neq s$, we can find (x_0, y_0) such that $x_0 > 0$ and $\sum C_{ab}(a-s)x_0^ay_0^b \neq 0$. (Recall that $C_{a,b} = 0$ when $a = s$.) Since the value of u_0 can be chosen arbitrarily (once (x_0, y_0) have been fixed), we find a value of $u_0 = u_0^{(1)}$ for which

σ_1, and hence σ, will be >0, and a value of $u_0 = u_0^{(2)}$ for which σ and σ_1 will be <0. This contradicts the conditions of Lemma 4.3.

4.7. The next step is analogous to Lemma 4.1. We shall prove that the coefficients of $(x^1)^a(x^2)^b$ cannot vanish for all a and b for which $a + b(s + 1) \leqslant k_1$.

Lemma 4.4. Let $l = \min(k_1, K)$. Let Λ be a linear space over the field of complex numbers generated by the monomials ξ_1 and $\xi^\alpha x^\beta$ when $2 \leqslant |\alpha + \beta| \leqslant l$, except for monomials of type $(x^1)^a(x^2)^b$, for $a + b(s + 1) \leqslant l$, $(x^1)^c\xi_2$, for $c < s$, and of types $(x^1)^d x^s$, $(x^1)^d \xi_t$, $(x^1)^e (x^2)^f x^s$, $(x^1)^e (x^2)^f \xi_t$ when $d < l/2$, $e + f(s+1) < l/2$, $s = 2, \ldots, n$; $t = 3, \ldots, n$. Then all elements of the set $\Lambda_j = (\mathrm{ad}\ \Lambda)^s\Lambda$ vanish at $x = \xi = 0$, $j \leqslant l$.

Proof. As in the proof of Lemma 4.1, we must indicate the method of finding a nonzero constant in no more than l steps. Since in all the generating monomials of Λ, except for ξ_1, the gradient is 0 at $x = \xi = 0$, the last step involves the use of operator ad ξ_1 and we should therefore indicate the method of finding a monomial of the type $(x^1)^r$ for some r: $1 \leqslant r \leqslant l$.

If $(x^1)^r$ is obtained by computing the Poisson brackets $\{(x^1)^a\xi_j, (x^1)^b x^j\}$, when $j \neq 2$, then $a \geqslant l/2$ and one has to know about the origin of $(x^1)^b x^j$. As seen in the proof of Lemma 4.1, it is not possible to find this monomial in $\leqslant l/2$ steps without using $(x^1)^s\xi_2$. If use is made of this monomial, then one must indicate the method of obtaining in $\leqslant l/2$ steps a monomial of the form $(x^1)^a(x^2)^b x^j$.

If $(x^1)^r$ is obtained by computing the Poisson brackets $\{(x^1)^s\xi_2, (x^1)^b x^2\}$, then the origin of $(x^1)^b x^2$ should be indicated. It could appear in the following forms:

$$\{\xi_1,\ (x^1)^{b+1}x^2\},\quad \{(x^1)^s\xi_2,\ (x^1)^{b-s}(x^2)^2\},$$

$$\{(x^1)^a x^2\xi_2,\ (x^1)^{b-a}x^2\},\quad \{(x^1)^a x^2\xi_j,\ (x^1)^{b-a}x^j\},\quad \{(x^1)^a\xi_j,\ (x^1)^{b-a}x^2x^j\},\quad j \neq 2.$$

(As always, x^j and ξ_j can be interchanged when $j \geqslant 3$.) In the first three forms the origin remains unknown. In the fourth form $a > l/2 - s - 1$; in the fifth form $a \geqslant l/2$, by the conditions of the lemma. In these cases one should elucidate the origin of the monomial of the form $(x^1)^b(x^2)^c x^j$ (as in the previous paragraph).

Such a monomial can be found in the following forms:

$$\{\xi_1,\ (x^1)^{b+1}(x^2)^c x^j\},\quad \{(x^1)^s\xi_2,\ (x^1)^{b-s}(x^2)^{c+1}x^j\},\quad \{(x^1)^{b_1}(x^2)^{c_1}x^j\xi_i,\ (x^1)^{b-b_1}(x^2)^{c-c_1}x^i\},$$

$$\{(x^1)^{b_1}(x^2)^{c_1}(x^j)^2,\ (x^1)^{b-b_1}(x^2)^{c-c_1}\xi_j\},\quad \{(x^1)^{b_1}(x^2)^{c_1}\xi_i,\ (x^1)^{b-b_1}(x^2)^{c-c_1}x^i x^j\}.$$

Here, the first four forms do not indicate the origin of the monomial; therefore either (i) use should be made of the fifth case, or (ii) the monomial of the form $(x^1)^b(x^2)^c x^j$ is found among the generating monomials of Λ.

In (i) $b_1 + c_1(s + 1) \geqslant l/2$, so $b + c(s + 1) \geqslant l/2$. In (ii) we also have $b + c(s + 1) \geqslant l/2$ (by the conditions of the lemma). Going in the reverse order we compute the number of steps necessary for obtaining a nonzero constant. In accordance with what has been proven above, we should apply the operators ad ξ_1 and $\mathrm{ad}(x^1)^s\xi_2$ to $(x^1)^b(x^2)^c x^j$ p and q times, respectively. Then a monomial of the form $(x^1)^{b+p+sq}(x^2)^{c-q}x^j$ is obtained. Now, if in the next step we wish to eliminate x^j, then, as seen above, it is necessary that $q = c$ or $q = c = 1$.

If $q = c$, then $b - p + sq = t - a$ so that $t = a + b - p + sq$ and $r = s + t = s + a + b - p + q$ and the total number of steps performed in obtaining a nonzero constant will not be less than $r + 1 + p + q = s + a +$

$b + sq + 1 + q > l/2 + b + (s + 1)q$ since $a > l/2 - s - 1$. But $b + q(s + 1) \geqslant l/2$ and therefore $N > l$.

If $q = c - 1$, then $b - p + sq = t - a \geqslant 0$, where $a \geqslant l/2$. Here again $r = s + t = s + a + b - p + sq$ and the number of steps involved in finding a nonzero constant will be less than $r + 1 + p + q = s + a + b + 1 + (s + 1)q > s + l/2 + b + 1 + (s + 1)(c - 1)$ as $a \geqslant l/2$. But $b + c(s + 1) \geqslant l/2$ so that $N > s + l/2 + [b + c(s + 1)] + 1 - (s + 1) \geqslant l$, that is, in this case $N > l$. Thus, we have proved that there does not exist any method of finding a nonzero constant in not more than l steps. Lemma 4.4 is proved.

4.8. Lemma 4.5. Suppose the conditions of Lemma 4.4 are fulfilled. Then

$$k_1(0,\ 0) = K = \min[a + b(s+1)],$$

where the minimum is taken over those a and b for which $\partial_1^a \partial_2^b a_1(0, 0) \neq 0$.

Proof. Lemma 4.4 reveals that $K \leqslant k_1$. Therefore it suffices to verify that if $K = a_0 + b_0(s + 1)$, then

$$H_2^{b_0 s} H_1^{b_0} H_2^{a_0} a_1(0,\ 0) \neq 0, \tag{4.4}$$

since it immediately follows that $k_1 \leqslant K$ and therefore $k_1 = K$.

Inequality (4.4) follows at once from the proof of Lemma 4.2: We see that even the most simple method of finding a nonzero constant from independent monomials of Λ involves $>K$ steps if, in the process, we use a monomial containing a factor x^j for $j \geqslant 3$. Therefore, if we add monomials $(x^1)^a(x^2)^b$, when $a + b(s + 1) = K$, to the space Λ described in Lemma 4.4, then for computing the left-hand side of (4.4) we can omit in $a_1(x, \xi)$ and H_1 all monomials except those which correspond to $(x_1)^s\xi_2$ and $(x^1)^s\xi_2$. After this only simple calculations are required. Lemma 4.5 is proved.

4.9. We shall now show that if the number $k = k_1(0, 0)$ is odd, then condition (4.2) is incompatible with the condition $c^0(x, \xi) \geqslant 0$ at the characteristic points. Therefore, if the last condition is fulfilled and k is even, then we deal with the situation considered in Lemma 4.1.

Lemma 4.6. If k is even and the function $c^0(x, \xi)$ is non-negative at characteristic points, then

$$\operatorname{grad}_{x',\ \xi'} \partial_1^j a_1(0,\ 0) = 0 \tag{4.5}$$

for $j < k/2$.

Proof. Suppose this is not the case and condition (4.2) is fulfilled. Then, by virtue of Lemmas 4.3 and 4.5, we have for $\xi_1 = 0$, $\xi_3 = \dots = \xi_n = 0$, $x^3 = \dots = x^n = 0$

$$a_1(x,\ \xi) = \sum_{a + b(s+1) = k} A_{ab}(x^1)^a (x^2)^b + B(x^1)^s \xi_2 + \varphi(x,\ \xi),$$

where

$$\varphi(x,\ \xi) = O(|x^1|^{k+1} + |x^2|^{(k+1)/(s+1)} + |x^1|^{s+1}|\xi_2| + \xi_2^2 + |x^2\xi_2|).$$

Let $x^1 = t$, $x^2 = yt^{s+1}$, and $\xi_2 = zt^{k-s}$. Then

$$a_1(x,\ \xi) = t^k[\sum A_{ab}y^b + Bz] + O(|t|^{k+1})$$

since $2(k - s) \geqslant k + 1$ and $k \geqslant 2(s + 1)$.

As usual, by the Implicit Function Theorem, we conclude that there are functions $y(t)$ and $z(t)$ such that $a_1(x, \xi) = 0$ when $x = t$, $y = y(t)$, $z = z(t)$, and $y(t) = y_0 + O(t)$, $z(t) = z_0 + O(t)$ as $t \to 0$ for any y_0, z_0 for which

$$\sum A_{ab}y_0^b + Bz_0 = 0.$$

On the other hand $H_2a_1(x, \xi) = \sum aA_{ab}(x^1)^{a-1}(x^2)^b + sB(x^1)^{s-1}\xi_2 + \partial_1\varphi(x, \xi)$ and on our curve

$$H_2a_1(x, \xi) = t^{k-1}\left(\sum aA_{a,b}y_0^b + sBz_0\right) + O(t^k) \quad \text{as} \quad t \to 0.$$

The number $k - 1$ is odd and we have to prove only the existence of such y_0 and z_0 for which

$$\sum A_{ab}y_0^b + Bz_0 = 0, \quad \sum aA_{a,b}y_0^b + sBz_0 \neq 0.$$

Since, by the assumption, $B \neq 0$ it suffices to prove the existence of y_0 for which $\sum (a-s) A_{ab}y_0^b \neq 0$. The latter is obvious since $\sum_{a \neq s} |A_{ab}| \neq 0$ by Lemma 4.5. (Recall that we had earlier agreed that $A_{ab} = 0$ when $s = 0$, see Section 4.4.) Lemma 4.6 is proved.

Remark 4.1. If k is odd, then from the proof of Lemma 2.6 it is evident that the polynomial $\sum (a-s) A_{ab}y_0^b$ does not change sign.

4.10. Proof of Theorem 4.1. Let ρ be a real number. The leading symbol of the operator $(1 + i\rho)P$ equals $a_1 - \rho a_2 + i(a_2 + \rho a_1)$. Note that

$$[\mathrm{ad}\,(a_2 + \rho a_1)]^k (a_1 - \rho a_2) = (1 + \rho^2)[\mathrm{ad}\,(a_2 + \rho a_1)]^{k-1}(-c_1) =$$

$$(1 + \rho^2) \sum_{|\alpha + \beta| = k-1} (\mathrm{ad}\,a_1)^\alpha (\mathrm{ad}\,a_2)^\beta \rho^{|\alpha|}(-c_1),$$

where

$$c_1 = -H_2a_1(x, \xi), \quad (\mathrm{ad}\,f)g = \{f, g\},$$

$$(\mathrm{ad}\,a_1)^\alpha (\mathrm{ad}\,a_2)^\beta = (\mathrm{ad}\,a_1)^{\alpha_1} \cdot (\mathrm{ad}\,a_2)^{\beta_1} \ldots (\mathrm{ad}\,a_1)^{\alpha_r}(\mathrm{ad}\,a_2)^{\beta_r}.$$

The obtained polynomial in ρ has no more than l roots; here, l stands for the maximum value of α_0 for which

$$\sum_{\substack{|\alpha + \beta| = k-1 \\ |\alpha| = \alpha_0}} (\mathrm{ad}\,a_1)^\alpha (\mathrm{ad}\,a_2)^\beta c_1(x, \xi) \neq 0. \tag{4.6}$$

As we have seen in the proof of Lemma 4.5, the value of l agrees with the maximum value of a for which $a + b(s + 1) = k_1$ and the derivative $\partial_1^a\partial_2^b a_1(0, 0) \neq 0$. It is not hard to verify that all the terms in (4.6) have the same sign if only one of them is nonzero (see Lemma 4.5). Therefore, for all ρ, except perhaps l values $\rho_1, \ldots, \rho_l$, we have

$$[\mathrm{ad}\,(a_2 + \rho a_1)]^k (a_1 - \rho a_2)(0, 0) \neq 0,$$

that is, the value $k_0(0, 0)$ agrees with $k = k_1(0, 0)$. This completes the proof.

4.11. Proof of Theorem 4.2. Let the principal symbol of operator Q be $q_1 + iq_2$, and let the principal symbol of operator P be $a_1 + ia_2$. Then the principal symbol of the operator QP is

$$a_1' + ia_2' = a_1q_1 - a_2q_2 + i(a_2q_1 + a_1q_2).$$

Note that $a_1'(x, \xi) + ia_2'(x, \xi) = 0$ if and only if $a_1(x, \xi) + ia_2(x, \xi) = 0$. At these points

$$c_1'(x, \xi) = -\mathrm{ad}\,(a_2q_1 + a_1q_2) \cdot (a_1q_1 - a_2q_2) = -(q_1^2 + q_2^2)\,\mathrm{ad}\,(a_2) \cdot a_1 = (q_1^2 + q_2^2)\,c_1(x, \xi),$$

whence one immediately sees the invariance of the condition: $c_1 \leqslant 0$ at characteristic points.

In order to verify the invariance of the condition $k_1(x, \xi) < \infty$ we consider two cases corresponding to conditions (4.5) and (4.2). (As above, we assume that condition (4.1) is fulfilled.) If condition (4.5) is fulfilled, then $\operatorname{grad} a_1(0, 0) = \varkappa \operatorname{grad} \xi_1$ for some $\varkappa$ and $\partial_1^j a_1(0, 0) = 0$ for $j < k$, but $\partial_1^k a_1(0, 0) \neq 0$. We first assume that $q_1(0, 0) + \varkappa q_2(0, 0) \neq 0$. This implies that $\operatorname{grad} a_2'(0, 0) = [\varkappa q_2(0, 0) + q_1(0, 0)] \operatorname{grad} \xi_1 \neq 0$. Clearly $\partial_1^j a_i'(0, 0) = 0$, $\partial_1^l \operatorname{grad}_{x', \xi'} a_i'(0, 0) = 0$ when $j < k$, $l < k/2$, $i = 1, 2$, so that

$$[\operatorname{ad}(a_2 q_1 + a_1 q_2)]^k \cdot (a_1 q_1 - a_2 q_2)(0, 0) = [\varkappa q_2(0, 0) + q_1(0, 0)]^{k-1} (q_1^{\circ} + q_2^{\circ}) (\operatorname{ad} a_2)^k a_1(0, 0),$$

and since $(\operatorname{ad} a_2)^k a_1(0, 0) = 0$ and $(\operatorname{ad} a_2')^k a_1'(0, 0) \neq 0$ so that $k_1(0, 0) = k_0(0, 0) = k_1'(0, 0) = k_0'(0, 0)$.

If $q_1(0, 0) + \varkappa q_2(0, 0) = 0$, then $q_2(0, 0) - \varkappa q_1(0, 0) \neq 0$, that is $\operatorname{grad} a_1'(0, 0) = [\varkappa q_1(0, 0) - q_2(0, 0)] \operatorname{grad} \xi_1 \neq 0$. In this case

$$[\operatorname{ad}(a_1 q_1 - a_2 q_2)]^k (a_2 q_1 + a_1 q_2)(0, 0) = [\varkappa q_1(0, 0) - q_2(0, 0)]^{k-1} (q_1^{\circ} + q_2^{\circ}) (\operatorname{ad} a_1)^k a_2(0, 0)$$

and again

$$k_1'(0, 0) = k_2'(0, 0) = k_2(0, 0) = k_1(0, 0).$$

Similarly, if condition (4.2) is fulfilled, then

$$\partial_1^j \partial_{\xi_2} a_i'(0, 0) = 0 \text{ when } j < s, \quad i = 1, 2,$$

$$\partial_1^s \partial_{\xi_2} a_1'(0, 0) = q_1(0, 0) \partial_1^s \partial_{\xi_2} a_1(0, 0), \quad \partial_1^s \partial_{\xi_2} a_2'(0, 0) = q_2(0, 0) \partial_1^s \partial_{\xi_2} a_1(0, 0),$$

and

$$\partial_1^j \partial_{\xi_l} a_i'(0, 0) = \partial_1^j \partial_m a_i'(0, 0) = 0$$

$$i = 1, 2; \quad j < k/2; \quad l = 3, \ldots, n; \quad m = 1, \ldots, n,$$

by Lemma 4.2. By Lemmas 4.3 and 4.5, we have

$$\partial_1^a \partial_2^b \partial_{\xi_l} a_i'(0, 0) = \partial_1^a \partial_2^b \partial_m a_i'(0, 0) = 0$$

for $i = 1, 2$; $a + b(s+1) < k/2$; $l = 3, \ldots, n$; $m = 1, \ldots, n$.

Besides, we see that

$$\partial_1^a \partial_2^b a_i'(0, 0) = 0 \text{ when } a + b(s+1) < k; \quad i = 1, 2,$$

and if $a + b(s + 1) = k$, then

$$\partial_1^a \partial_2^b a_1'(0, 0) = q_1(0, 0) \partial_1^a \partial_2^b a_1(0, 0), \quad \partial_1^a \partial_2^b a_2'(0, 0) = q_2(0, 0) \partial_1^a \partial_2^b a_1(0, 0).$$

First consider the case when

$$q_1(0, 0) + \varkappa q_2(0, 0) \neq 0,$$

$\operatorname{grad} a_2'(0, 0) \neq 0$. As in the proof of Lemma 4.5, it can be verified that if $k_1 = a_0 + b_0(s+1)$, $\partial_1^{a_0} \partial_2^{b_0} a_1(0, 0) \neq 0$ and $q_1(0, 0) \neq 0$, then

$$H_2'^{sb_0} H_1'^{b_0} H_2'^{a_0} a_1'(0, 0) \neq 0,$$

that is, $k_1'(0, 0) = k_1(0, 0)$. (Here, $H_i' = \operatorname{ad} a_i'$ stands for the operator of differentiation along the bicharacteristic curve of the function $a_i'(x, \xi)$, $i = 1, 2$.) If $q_1(0, 0) = 0$, then $q_2(0, 0) \neq 0$ and $q_1(0, 0) + \varkappa q_2(0, 0) \neq 0$ so that

$$H_1'^{sb_0} H_2'^{b_0} H_1'^{a_0} a_2'(0, 0) \neq 0,$$

that is, again $k_1'(0, 0) = k_1(0, 0)$.

Similarly, if $q_1(0, 0) + \varkappa q_2(0, 0) = 0$, then $q_2(0, 0) - \varkappa q_1(0, 0) \neq 0$. It follows that $q_2(0, 0) \neq 0$, and by repeating the proof of Lemma 4.5 it may be verified that

$$H_1'^{b_0 s} H_2'^{b_0} H_1'^{a_0} a_2'(0,\ 0) \neq 0, \quad \text{if} \quad a_0 + b_0(s+1) = k_0, \quad \partial_1^{b_0(s+1)} \partial_2^{a_0} a_1(0,\ 0) \neq 0.$$

The proof of Theorem 4.2 is complete.

§5. Infinite-Order Zero Theorem

5.1. In this section we shall prove a theorem on necessary conditions for local solvability without assuming that a = Re p^0 has a finite-order zero along the bicharacteristics of b = Im·p^0. It contains as particular cases Theorems 2.1 and 3.1 proved earlier.

Of significance in Theorems 2.1 and 3.1 is the satisfaction of condition (A) given below. In order to formulate it we shall consider a characteristic point $(x_0,\ \xi^0)$ and a bicharacteristic curve of the function b passing through it so that $\dot{x}(t) = \mathrm{grad}_\xi b(x(t),\ \xi(t)),\quad \dot{\xi}(t) = -\mathrm{grad}_x \times b(x(t),\ \xi(t))$ and $(x(0),\ \xi(0)) = (x_0,\ \xi^0)$ along γ. As above, we assume $\mathrm{grad}_{x,\xi} \times b(x_0,\ \xi^0)$ to be collinear to vector $(\xi^0,\ 0)$. By γ_+ and γ_-, respectively, we denote a portion of the curve γ corresponding to positive and negative values of parameter t.

Suppose the following conditions

$$a(x(t),\ \xi(t)) \geqslant 0 \text{ on } \gamma_-, \quad a(x(t),\ \xi(t)) \leqslant 0 \text{ on } \gamma_+$$

(A) are satisfied and there exist sequences of points $(x_{(k)},\ \xi_{(k)}) \in \gamma_-$ and $(\tilde{x}_{(k)},\ \tilde{\xi}_{(k)}) \in \gamma_+$ converging to $(x_0,\ \xi^0)$ as $k \to \infty$ such that function a has nonzero values at these points.

Thus, we assume that function a does not vanish on any segment of the curve which contains $(x_0,\ \xi^0)$. Let us denote by I(t) the operator of integration along the curve γ so that

$$I(t) f = \int_0^t f(x(s),\ \xi(s))\, ds.$$

By condition (A), function $I(t)a$ assumes negative values for all $t \neq 0$.

Also suppose that condition (B), which generalizes conditions (4.5), is satisfied:

Let l be a smooth vector field in $T^*\Omega \setminus 0$, defined at the points of γ, and tangent to the surface $b(x,\ \xi) = 0$ containing this curve. Then

$$\text{(B)} \qquad I(t)\, la \leqslant c_0 \sqrt{-I(t)\, a}$$

when $|t| \leqslant \varepsilon$; $\varepsilon > 0$, c_0 = const > 0, constant c_0 being small enough (depending on $\sum_{|\alpha| \leqslant \beta} |D^\alpha q(x_0,\ \xi^0)|$) and independent of l.

Let us now formulate the main result of this section.

Theorem 5.1. Let conditions (A) and (B) be fulfilled at the characteristic point $(x_0,\ \xi^0)$ of operator P. Then to any neighborhood ω of point x_0 one can find a function $f \in C_0^\infty(\omega)$ such that the equation Pu = f has no solution in ω belonging to the class $\mathscr{E}'(\Omega)$.

As above, we may use Lemmas 2.1, 2.2, and 2.5 and assume that m = 1 and

$$p^0(x,\ \xi) = i\xi_1 + q(x,\ \xi').$$

Besides, we assume the function $p^* - \bar{p}^0$ to be independent of ξ_1; here p^* stands for the symbol of operator P^*.

It is easy to see that the conditions of Theorem 5.1 are invariant under canonical transformations and conditions (A) and (B) can be reformulated as:

(A) Function $q(x^1, x_0', \xi^0)(x^1 - x_0^1)$ is non-negative when $|x^1 - x_0^1| < \varepsilon$ and is not identically zero on any segment containing x_0^1.

There exist positive constants c_0 and ε such that

(B) $$\left|\int_0^{x^1} \operatorname{grad}_{x',\xi'} q(s, x_0', \xi^0)\, ds\right|^2 \leqslant -c_0^2 \int_0^{x^1} q(s, x_0', \xi^0)\, ds$$

when $|x^1 - x_0^1| < \varepsilon$, and constant c_0 is sufficiently small.

In Egorov and Popivanov[1] a more rough condition

$$|\operatorname{grad}_{x',\xi'} q(x^1, x_0', \xi^0)|^2 \leqslant c^2 |q(x^1, x_0', \xi^0)|$$

is considered instead of (B). They also give different conditions sufficient for the inequality to be fulfilled.

5.2. The following statement is analogous to Lemma 2.3 and is proved similarly.

Lemma 5.1. Let P be a first-order operator with principal symbol $p^0(x, \xi) = i\xi_1 + q(x, \xi)$, where q is a smooth real-valued function independent of ξ_1. Suppose there are real constants C and s for which the inequality

$$\|u\|_s \leqslant C(K)(\|P^*u\|_0 + \|u\|_{s-1}), \quad u \in C_0^\infty(K), \tag{5.1}$$

holds. Here, K is a compact subset of points in Ω. Then for every integer $N > 0$ and any compact set $M \subset K \times \{\mathbb{R}^n \setminus 0\}$ there exists a constant $C > 0$ such that

$$\lambda^s \|\psi\|_0 \leqslant C\left\{\left\| -\frac{\partial\psi}{\partial y^1} + \sum_{|\alpha+\beta|<N} \frac{1}{\alpha!\beta!} q_{(\beta)}^{(\alpha)}(x^1, x_0', \xi^0)\, y^\beta D_y^\alpha \psi(x^1, y) \times \right.\right.$$
$$\lambda^{1-|\alpha+\beta|/2} + \sum_{\substack{j>0\\ 2j+|\alpha+\beta|<N}} \frac{1}{\alpha!\beta!} p_{(\beta)}^{j(\alpha)}(x^1, x_0', \xi^0)\, y^\beta D_y^\alpha \psi(x^1, y) \times$$
$$\left.\left.\lambda^{1-j-|\alpha+\beta|/2}\right\|_0 + \lambda^{1-N/2} \sum_{|\alpha+\beta|\leqslant N} \|y^\beta D^\alpha \psi(x^1, y)\|_0\right\} \tag{5.2}$$

for all $\lambda \geqslant 1$ and all functions $\psi \in C_0^\infty((-\varepsilon, \varepsilon) \times \mathbb{R}^{n-1})$, provided $\varepsilon > 0$ is small enough.

To abbreviate notations we transfer the origin to x^0. By analogy with Lemma 3.2 we shall prove the following statement.

Lemma 5.2. Let the principal symbol of operator P be equal to $i\xi + q(x, \xi')$ in a neighborhood of the point $(0, \xi^0)$ and satisfy conditions (A) and (B). Then for every natural number $\sigma > 0$ in a neighborhood ω of the origin in $\mathbb{R}^n$ one can define a smooth function $w(x^1, y)$ such that

$$\frac{1}{i}\frac{\partial w}{\partial x^1} + \sum_{|\alpha+\beta|<N} \frac{1}{\alpha!\beta!} q_{(\beta)}^{(\alpha)}(x^1, x_0', \xi^{0\prime})\, y^\beta (\operatorname{grad}_y w)^\alpha = O(|y|^\sigma), \tag{5.3}$$

$$\operatorname{Im} w(x^1, y) \geqslant a_0\rho^2, \quad |\operatorname{grad}_y w(x^1, y)| \leqslant a_1\rho, \tag{5.4}$$

where $a_0 = \text{const} > 0$; $a_1 = \text{const} > 0$,

$$\rho^2(x^1, y) = |y|^2 - \int_0^{x^1} q(t, x_0', \xi^{0\prime})\, dt.$$

Proof. We shall try to find $w(x^1, y)$ in the form

$$w(x^1, y) = i|y|^2 + \sum_{|\alpha|<\sigma} w_\alpha(x^1) y^\alpha,$$

where $w_\alpha \in C^\infty$, $w_\alpha(0) = 0$.

Inserting this expression in (5.3) and equating the coefficients of y^α, $|\alpha| < \sigma$, to zero, we obtain for determining functions w_α a system of ordinary differential equations of the form

$$w'_\alpha(x^1) = \Phi_\alpha(x^1, w_\beta(x^1)), \quad |\alpha| < \sigma, \quad |\beta| < \sigma,$$

with initial conditions: $w_\alpha(0) = 0$. Functions Φ_α are, in general, complex-valued, but this system can be readily transformed into a system of equations with real-valued right-hand sides by separately equating the real and imaginary parts so that the number of equations in the system doubles. The system obtained has a unique solution for $|x^1| \leqslant \varepsilon$ if ε is a sufficiently small positive number.

Note that

$$w'_0(x^1) + iq(x^1, 0, \xi^{0'}) + i\sum_{j=1}^{n-1} q^{(j)}(x^1, 0, \xi^{0'}) w_j(x^1) + \sum_{j,l=1}^{n-1} c^{jl}(x^1) w_j(x^1) w_l(x^1) = 0, \tag{5.5}$$

$$w'_j(x^1) + iq_{(j)}(x^1, 0, \xi^{0'}) - 2q^{(j)}(x^1, 0, \xi^{0'}) + x^1\sum_{l=1}^{n-1} a_{jl}(x^1) q^{(l)}(x^1, 0, \xi^{0'}) + x^1\sum_{l=1}^{n-1} b_j^l(x^1) w_l(x^1) = 0, \tag{5.6}$$

where a_{jl}, b_j^l, and c^{jl} are smooth functions whose maxima are estimated by a value depending on $\sum_{\alpha \leqslant 2} |D^\alpha q(x_0, \xi^0)|$, and also $x^1 a_{jl} = x^1 a_{lj} = \frac{1}{2}\frac{\partial^2 w(0)}{\partial y^j \partial y^l}$.

Let us introduce the notation

$$Q(x^1) = \int_0^{x^1} q(t, 0, \xi^{0'})dt, \quad Q_j(x^1) = \int_0^{x^1} q_{(j)}(t, 0, \xi^{0'})dt, \quad Q^j(x^1) = \int_0^{x^1} q^{(j)}(t, 0, \xi^{0'})dt.$$

Recall that the inequality

$$\sum_{j=1}^{n-1} \{|Q_j(x^1)|^2 + |Q^j(x^1)|^2\} \leqslant -2c_0^2(n-1)Q(x^1) \tag{5.7}$$

is fulfilled in view of condition (B), and the function $Q(x^1)$ is monotonic at $x^1 \neq 0$ by virtue of condition (A).

From equations (5.6) it follows that

$$|w_j(x^1)| \leqslant |Q_j(x^1)| + 2|Q^j(x^1)| + C_1\varepsilon\sum_{l=1}^{n-1}|Q^l(x^1)| + C_2\int_0^{x^1}\sum_{l=1}^{n-1}|w_l(t)|dt.$$

Adding these inequalities over j and applying inequality (5.7) we obtain

$$\sum_{j=1}^{n-1}|w_j(x^1)| \leqslant C_3\int_0^{x^1}\sum_{j=1}^{n-1}|w_j(t)|dt + C_4 c_0\sqrt{|Q(x^1)|},$$

whence, by the Gronwall inequality it follows that

$$\sum_{j=1}^{n-1}|w_j(x^1)|^2 \leqslant C_5 c_0^2|Q(x^1)|. \tag{5.8}$$

From (5.5) it is seen that, by (5.8),

$$\left| w_0(x^1) + i \sum_{j=1}^{n-1} \int_0^{x^1} q^{(j)}(t,\ 0,\ \xi^{0'})\, w_j(t)\, dt + iQ(x^1) \right| \leqslant C_6 c_0^2 \varepsilon\, |Q(x^1)|.$$

Integrating by parts gives

$$\int_0^{x^1} q^{(j)}(t,\ 0,\ \xi^{0'})\, w_j(t)\, dt = Q^j(x^1)\, w_j(x^1) - \int_0^{x^1} Q^j(t)\, w_j'(t)\, dt.$$

From (5.7) and (5.8) it is seen that

$$|Q^j(x^1)\, w_j(x^1)| \leqslant C_7 c_0^2 |Q(x^1)|.$$

Using again equation (5.6) we obtain

$$-\int_0^{x^1} Q^j(t)\, w_j'(t)\, dt = \int_0^{x^1} Q^j(t) \Big[iq_{(j)}(t,\ 0,\ \xi^{0'}) - 2q^{(j)}(t,\ 0,\ \xi^{0'}) +$$

$$t \sum_{l=1}^{n-1} a_{jl}(t)\, q^{(l)}(t,\ 0,\ \xi^{0'}) + \sum_{l=1}^{n-1} b_j^l(t)\, w_l(t) \Big]\, dt.$$

Now we consider the four integrals appearing on the right-hand side. The first of them is imaginary and therefore is of no interest. The second integral equals

$$-2\int_0^{x^1} Q^j(t)\, q^{(j)}(t,\ 0,\ \xi^{0'})\, dt = -[Q^j(x^1)]^2.$$

Furthermore, integration by parts gives

$$\int_0^{x^1} Q^j(t)\, t a_{jl}(t)\, q^{(l)}(t,\ 0,\ \xi^{0'})\, dt = Q^j(x^1)\, Q^l(x^1)\, x^1 a_{jl}(x^1) -$$

$$\int_0^{x^1} Q^l(t)\, Q^j(t)\, (t a_{jl})'\, dt - \int_0^{x^1} Q^l(t)\, q^{(j)}(t,\ 0,\ \xi^{0'})\, t a_{jl}(t)\, dt.$$

It is seen that

$$\sum_{j,\, l=1}^{n-1} \int_0^{x^1} t a_{jl}(t)\, q^{(l)}(t,\ 0,\ \xi^{0'})\, Q^j(t)\, dt =$$

$$\frac{1}{2} \sum_{j,\, l=1}^{n-1} x^1 a_{jl}(x^1)\, Q^j(x^1)\, Q^l(x^1) - \frac{1}{2} \sum_{j,\, l=1}^{n-1} \int_0^{x^1} Q^j(t)\, Q^l(t)\, [t a_{jl}(t)]'\, dt$$

and therefore

$$\left| \sum_{j,\, l=1}^{n-1} \int_0^{x^1} t a_{jl}(t)\, q^{(l)}(t,\ 0,\ \xi^{0'})\, Q^j(t)\, dt \right| \leqslant C_8 \varepsilon c_0^2 |Q(x^1)|.$$

Finally, we consider the last integral

$$\int_0^{x^1} Q^j(t) \sum_{l=1}^{n-1} b_j^l(t)\, w_l(t)\, dt.$$

From (5.7) and (5.8) it follows that this integral in absolute value does not exceed $C_9 \varepsilon c_0^2 |Q(x^1)|$.

Summing up the estimates, we obtain

$$|\operatorname{Im} w_0(x^1) + Q(x^1)| \leqslant c_{10} c_0^2 |Q(x^1)|.$$

Thus,

$$\operatorname{Im} w(x^1,\ y) \geqslant -Q(x^1) + |y|^2 - |y| \sum_{j=1}^{n-1} |w_j(x^1)| - c_{10} c_0^2 |Q(x^1)| \geqslant$$

$$- Q(x^1)\left[1 - c_{10}c_0^2 - \frac{1}{2} C_5 c_0^2\right] + \frac{1}{2} |y|^2,$$

and if c_0 is so small that

$$1 - c_{10}c_0^2 - \frac{1}{2} c_5 c_0^2 > \frac{1}{2},$$

then $\operatorname{Im} w(x^1, y) > \frac{1}{2}(|y|^2 - Q(x^1))$, which proves the lemma.

5.3. To prove Theorem 5.1 we shall need another lemma:

Lemma 5.3. Let w(x) be a function of Lemma 5.2. Then there exist constants C_α such that

$$|D^\alpha e^{i\lambda w}| \leqslant C_\alpha \lambda^{|\alpha|} e^{-\lambda \operatorname{Im} w \cdot 2^{-|\alpha|}} \tag{5.9}$$

for all $\lambda \geqslant 1$ and $x \in \omega$.

Proof. When $\alpha = 0$ this inequality is obvious and $C_0 = 1$. If $|\alpha| = 1$, then

$$|D^\alpha e^{i\lambda w}|^2 = |\lambda D^\alpha w e^{i\lambda w}|^2 = \lambda^2 |D^\alpha w|^2 e^{-2\lambda \operatorname{Im} w}.$$

From (5.4) it follows that

$$|D^\alpha w|^2 e^{-\lambda \operatorname{Im} w} \leqslant a_1^2 \rho^2 e^{-\lambda a_0 \rho^2} \leqslant C\lambda^{-1}.$$

Therefore

$$|D^\alpha e^{i\lambda w}|^2 \leqslant C_1 \lambda e^{-\lambda \operatorname{Im} w}.$$

Suppose inequality (5.9) is proved for $|\alpha| < l$, where $l > 1$. Then for $|\alpha| = l$ we have

$$|D_j D^\alpha e^{i\lambda w}|^2 = |D^\alpha (D_j w \cdot \lambda e^{i\lambda w})|^2 \leqslant$$

$$C\lambda^2 \sum_\beta |D_j D^\beta w \cdot D^{\alpha-\beta} e^{i\lambda w}|^2 \leqslant C_1 \lambda^2 \sum_{|\beta| \geqslant 1} \lambda^{|\alpha-\beta|} e^{-\lambda \operatorname{Im} w \cdot 2^{-|\alpha-\beta|}} +$$

$${}_1C\lambda^2 |D_j w \cdot D^\alpha e^{i\lambda w}|^2 \leqslant C_\alpha \lambda^{|\alpha|+1} e^{-\lambda \operatorname{Im} w \cdot 2^{-|\alpha|-1}},$$

which proves the lemma.

5.4. Proof of Theorem 5.1. Using Lemmas 2.1, 2.2, and 2.5 we may assume that m = 1 and

$$p^0(x, \xi) = i\xi_1 + q(x, \xi)$$

in some neighborhood of the point (x_0, ξ^0), and the inequality

$$\|\varphi\|_s \leqslant C \|P^* \varphi\|_t, \quad \varphi \in C_0^\infty(\omega),$$

is valid. Then (5.2) is fulfilled in view of Lemma 5.1. Let ω be a so small neighborhood of the origin of coordinates in $\mathbb{R}^n$ that Lemma 5.2 is valid.

In inequality (5.2) we replace y by $y\lambda^{1/2}$. Then it changes to inequality

$$\lambda^s \|\psi\|_0 \leqslant C_1 \Big\{ \Big\| -\partial_1 \psi + \sum_{|\alpha+\beta|<N} \frac{1}{\alpha!\,\beta!} q_{(\beta)}^{(\alpha)}(x^1, x_0', \xi^{0\prime}) y^\beta D_y^\alpha \psi(x^1, y) \lambda^{1-|\alpha|} +$$

$$\sum_{j>0} \sum_{|\alpha+\beta|<N-2j} \frac{1}{\alpha!\,\beta!} p_{(\beta)}^{j(\alpha)}(x^1, x_0', \xi^{0\prime}) y^\beta D_y^\alpha \psi(x^1, y) \lambda^{1-j-|\alpha|} \Big\|_0 +$$

$$\lambda^{1-N/2} \sum_{\substack{|\alpha+\beta| \leqslant N \\ \alpha_1 = 0}} \lambda^{\frac{|\beta|-|\alpha|}{2}} \|y^\beta D^\alpha \psi(x^1, y)\|_0 \Big\}. \tag{5.10}$$

By $A(x^1, y)$ we denote the expression on the left-hand side of (5.3). The numbers σ and N will be found below.

Let

$$\psi(x^1, y) = \sum_{j=0}^{L} \varphi_j(x^1, y)\lambda^{-j} e^{i\lambda w(x^1, y)},$$

where $\varphi_j \in C_0^\infty(\omega)$. We have

$$-\frac{\partial\psi}{\partial x^1} + \sum_{|\alpha+\beta| \leqslant N} \frac{1}{\alpha!\beta!} q_{(\beta)}^{(\alpha)}(x^1, x_0', \xi^{0'}) y^\beta D_y^\alpha \psi(x^1, y)\lambda^{1-|\alpha|} +$$

$$\sum_{j>0} \sum_{|\alpha+\beta|<N-2j} \frac{1}{\alpha!\beta!} p_{(\beta)}^{j\,(\alpha)}(x^1, x_0', \xi^{0'}) y^\beta D_y^\alpha \psi(x^1, y)\lambda^{1-j-|\alpha|} =$$

$$e^{i\lambda w(x^1, y)} \sum_{j=0}^{L-N+1} a_j(x^1, y)\lambda^{-j+1},$$

where $a_j \in C_0^\infty(\omega)$. We wish to find functions φ_j for which $a_j = 0$. It is not hard to verify that

$$a_0(x^1, y) = A(x^1, y)\varphi_0(x^1, y) = O(|y|^\sigma) \text{ when } y \to 0,$$

$$a_1(x^1, y) = A(x^1, y)\varphi_1(x^1, y) - \partial_1\varphi_0 + \frac{1}{i}\sum_{j=1}^{n-1} A^j(x^1, y)\partial_j\varphi_0 + B(x^1, y)\varphi_0,$$

$$a_m(x^1, y) = A(x^1, y)\varphi_m(x^1, y) - \partial_1\varphi_{m-1} +$$

$$\frac{1}{j}\sum_{j=1}^{n-1} A^j(x^1, y)\partial_j\varphi_{m-1} + B(x^1, y)\varphi_{m-1} + C_m(x^1, y),$$

where C_m stands for a linear combination of the functions $\varphi_0, \varphi_1, \ldots, \varphi_{m-2}$ and their derivatives, and $A^j(0, 0) = 0$.

We now define φ_0 as a solution of the equation

$$-\partial_1\varphi_0 + \sum_{j=1}^{n-1} A_j(x^1, y) D_j\varphi_0 + B(x^1, y)\varphi_0 = O(|y|^{\sigma-1}) \text{ when } y \to 0,$$

with the condition that $\varphi_0(0, y) = 1$. Just as for w, we find φ_0 with the use of Kovalevskaya's theorem. Similarly, the functions $\varphi_{m-1}(x, y)$, when $2 \leqslant m \leqslant L$, are found from the conditions: $-\partial_1\varphi_{m-1} + \sum_{j=1}^{n-1} A_j(x^1, y) D_j\varphi_{m-1} + B(x^1, y)\varphi_{m-1} + C_m(x^1, y) = O(|y|^{\sigma-m+1})$ and $\varphi_{m-1}(0, y) = 0$. It is clear that in this case the functions φ_m may be assumed to vanish in a neighborhood of the boundary of domain ω. Note that since $\varphi_0(0, 0) = 1$, we have

$$\|\psi(x^1, y)\|^2 = \iint \left|\sum_{j=0}^{L} \varphi_j(x^1, y)\lambda^{-j}\right|^2 e^{-2\lambda \operatorname{Im} w(x^1, y)} dx^1 dy \geqslant$$

$$\iint \left|\sum_{j=0}^{L} \varphi_j(x^1, y)\lambda^{-j}\right|^2 e^{-2\lambda C_1((x^1)^2 + |y|^2)} dx^1 dy =$$

$$\lambda^{-n/2} \iint \left|\sum_{j=0}^{L} \varphi_j\left(\frac{x^1}{\sqrt{\lambda}}, \frac{y}{\sqrt{\lambda}}\right)\lambda^{-j}\right|^2 e^{-2C_1((x^1)^2 + |y|^2)} dx^1 dy =$$

$$\lambda^{-n/2}\left[\iint |\varphi_0(0, 0)|^2 e^{-2C_1((x^1)^2 + |y|^2)} dx^1 dy + O\left(\frac{1}{\lambda}\right)\right] = \lambda^{-n/2}\left[c_0 + O\left(\frac{1}{\lambda}\right)\right] \geqslant \frac{c_0}{2}\lambda^{-n/2},$$

where $c_0 > 0$ and λ is sufficiently large. On the other hand,

$$\left\| e^{i\lambda w(x^1, y)} \sum_{j=0}^{L-N+1} a_j(x^1, y)\lambda^{-j} \right\|^2 \leqslant C_2 \sum_{j=0}^{L-N+1} \lambda^{-2j} \iint_\omega |y|^{2(\sigma-j)} e^{-2\lambda C_1 \rho^2(x^1, y)} dx^1 dy \leqslant$$

$$C_2 \sum_{j=0}^{L-N+1} \lambda^{-\sigma-j} \iint_{\omega} (\lambda\rho^2)^{\sigma-j} e^{-2\lambda C_1 \rho^2 (x^1, y)} \, dx^1 \, dy \leqslant C_3 \lambda^{-\sigma}.$$

Finally, by Lemma 5.3,

$$\iint |y^{\beta} D^{\alpha} \psi(x^1, y)|^2 \, dx^1 \, dy \leqslant C_4 \iint |y|^{2|\beta|} \lambda^{|\alpha|} e^{-\lambda c_0 \rho^2 (x^1, y)} dx^1 \, dy \leqslant$$

$$C_5 \lambda^{|\alpha|} \iint |y|^{2|\beta|} e^{-\lambda c_0 |y|^2} \, dx^1 \, dy \leqslant C_6 \lambda^{|\alpha|-\beta-(n-1)/2},$$

when $|\alpha + \beta| \leqslant N$, so that

$$\lambda^{1-N/2} \sum_{|\alpha+\beta| \leqslant N} \lambda^{(|\beta|-|\alpha|)/2} \| y^{\beta} D^{\alpha} \psi \| \leqslant C_7 \lambda^{1-N/2-(n-1)/4}.$$

Substituting the estimates obtained in (5.10) we obtain

$$\frac{c_0}{2} \lambda^{s-n/4} \leqslant C_3 \lambda^{-\sigma/2+1} + C_7 \lambda^{1-N/2-(n-1)/4},$$

where the constants c_0, C_3, and C_7 are independent of λ and $c_0 > 0$. Choosing N and σ so large that $\sigma/2 > -s + n/4 + 1$ and $N > -2s + 5/2$ we obtain an inequality which cannot be fulfilled for all $\lambda \geqslant \lambda_0$. This contradiction proves the validity of Theorem 5.1.

5.5. Here we shall list certain conditions under which the conditions of Theorem 5.1 are fulfilled.

Proposition 5.1. Let condition (A) be fulfilled for operator P at (x_0, ξ^0) and function $c_1 = \{a, b\}$ assume nonpositive values at every characteristic point of some neighborhood of (x_0, ξ^0). Then each of the following conditions suffices for (B) to hold.

(B_1). Function a is monotone along γ in some neighborhood of the point (x_0, ξ^0).

(B_2). There exists a constant $C > 0$ such that $|a(x, \xi)| \leqslant C|a(y, \eta)|$ provided $(x, \xi) \in \gamma$, $(y, \eta) \in \gamma$ and (x, ξ) lies closer to (x_0, ξ^0) than (y, η).

(B_3). Along γ, function a has a finite-order zero at (x_0, ξ^0).

(B_4). If $a(x, \xi) = 0$, then $\text{grad}_{x;\xi'} a(x, \xi) = 0$.

(B_5). In some neighborhood of (x_0, ξ^0) there exists a smooth manifold S transverse to the field $(\text{grad}_\xi b, -\text{grad}_x b)$ and having the following properties: at every point of the surface S at which $b(x, \xi) = 0$ function a changes its sign from minus to plus in moving in the positive direction along the null bicharacteristic of function b which passes through this point.

Proof. Assume as we may that $m = 1$ and $p^0(x, \xi) = i\xi_1 + q(x, \xi')$ so that the inequality $\partial_1 q(x, \xi') \geqslant 0$ is fulfilled at those points where $q(x, \xi') = 0$. We shall prove that the inequality

$$|\text{grad}_{x', \xi'} q(x^1, x_0', \xi^{0\prime})|^2 \leqslant C |q(x^1, x_0', \xi^{0\prime})|$$

follows from conditions (B_1)—(B_5).

It is not hard to see that from this inequality follows

$$\left| \int_0^{x^1} \text{grad}\, q(t, x_0', \xi^{0\prime}) \, dt \right|^2 \leqslant C |x^1| \left| \int_0^{x^1} q(t, x_0', \xi^{0\prime}) \, dt \right|$$

and, therefore, condition (B) holds.

(B_1). The proof of this case and also of (B_2) is contained in L. Nirenberg and F. Trèves.[2] It involves proofs of two lemmas.

Lemma 5.4. Let $\varphi \in C^2(-r, r)$ and $|\varphi''(y)| \leqslant M$, where φ has real values and $\varphi(0) < M(\frac{1}{2}Cr)^2$. Here C is some constant, $C > \sqrt{2}$. Then

a) if $|\varphi'(0)| > C\sqrt{M|\varphi(0)|}$, then on the segment $|y| \leqslant \frac{2}{C}\sqrt{\frac{|\varphi(0)|}{M}}$ there is contained exactly one point y_0 at which φ vanishes and $\varphi'(y_0) \neq 0$;

b) conversely, if $\varphi(y_0) = 0$ for $|y_0| < \frac{1}{C}\sqrt{\frac{|\varphi(0)|}{M}}$ then $|\varphi'(0)| \geqslant \frac{3}{4}C\sqrt{M|\varphi(0)|}$.

Lemma 5.5. Let $R = \{(t, y) \in \mathbb{R}^2, |y| < r_0, 0 \leqslant t \leqslant \delta_0\}$ and $\varphi \in C^2(R)$, where

1) $\varphi(0, 0) = 0$, $\varphi(0, t)$ monotonically increases;

2) if $\varphi(t, y) = 0$, then $\varphi_t(t, y) \leqslant 0$;

3) $|\varphi_{yy}(t, y)| \leqslant M$ in R.

Then the inequality

$$|\varphi_y(t, 0)|^2 \leqslant 16M|\varphi(t, 0)|, \quad 0 \leqslant t < \delta,$$

is fulfilled for a sufficiently small $\delta > 0$.

Proof of Lemma 5.4. a) The transformation

$$y \to Ty = -\frac{1}{\varphi'(0)}\left[\varphi(0) + \int_0^y (y - \eta)\varphi''(\eta)\,d\eta\right]$$

continuously maps the segment $|y| \leqslant \rho$ into itself. Therefore, there exists a point y_0 for which $y_0 = Ty_0$. Since $|\varphi'(y)| > 0$ for $|y| \leqslant \rho$, we find this point to be unique and $\varphi(y_0) \neq 0$.

b) Let $\varphi(y_0) = 0$ and $|y_0| \leqslant \rho$. We assume that $y_0 > 0$ and $\varphi(0) > 0$. Then

$$-y_0\varphi'(0) \geqslant \varphi(0) - \frac{1}{2}My^2 \geqslant \varphi(0)\left(1 - \frac{1}{2C^2}\right) \geqslant \frac{3}{4}\varphi(0)$$

and

$$|\varphi'(0)| = -\varphi'(0) \geqslant \frac{3}{4}\frac{\varphi(0)}{y_0} \geqslant \frac{3C}{4}\sqrt{M|\varphi(0)|}.$$

Proof of Lemma 5.5. Let r_0 be so small that $\varphi(\delta_0, y) > 0$ when $|y| < r_0$. Suppose that

$$|\varphi_y(t_0, 0)|^2 > 16M|\varphi(t_0, 0)|, \quad 0 < t_0 < \delta_0.$$

Here it may be assumed that $|\varphi(t_0, 0)| < Mr^2$. By Lemma 5.4, for C = 4 there exists, in the interval $|y| < \frac{1}{2}\varphi(t_0, y_0) = 0$, a unique point y_0, where $\varphi_y(t_0, y_0) = 0$. In this case, $\varphi_y(t_0, y_0) \neq 0$. For definiteness we assume that $y_0 > 0$. Since $\varphi(t_0, y) > 0$ for $0 \leqslant y \leqslant y_0$, we see that $\varphi_y(t_0, y_0) < 0$. By the Implicit Function Theorem, the null set of functions φ forms, in a neighborhood of the point (t_0, y_0), a smooth curve Γ defined by the equation $y = f(t)$. On this curve $\varphi_y < 0$. On the other hand, $\varphi_y f_t + \varphi_t = 0$. It follows that $f_t \leqslant 0$ on Γ, that is, f is decreasing. Since $\varphi(t, 0)$ is increasing by hypothesis, the inequalities

$$0 < f(t) \leqslant f(t_0) \leqslant \frac{1}{2}|\varphi(t, 0)/M|^{1/2}$$

hold when $t \geqslant t_0$.

By Lemma 5.4 it follows that

$$|\varphi_y(t, 0)| \geqslant \frac{3}{2}\sqrt{M|\varphi(t, 0)|}, \quad t \geqslant t_0.$$

Using Lemma 5.4 with constant C = 3/2, we find that f is defined at all t for which $t_0 \leqslant t \leqslant \delta_0$, where $f(\delta_0) < r_0$, so that $\varphi(\delta_0, f(\delta_0)) = 0$. This contradicts the assumption that $\varphi(\delta_0, y) > 0$ when $|y| < r_0$. The lemma is proved.

(B_2) To prove this case it is sufficient to apply the earlier arguments to the function

$$\psi(t, y) = \frac{\varphi(t, y)}{\varphi(t, 0)} \max_{s \leqslant t} \varphi(s, 0).$$

In this case

$$|\varphi_y(t, 0)| \leqslant 4\sqrt{CM|\varphi(t, 0)|}.$$

(B_3) See Lemmas 4.2 and 4.6.

(B_4) The assertion readily follows from the following lemma.

Lemma 5.6. Let $f \in C^2(\mathbb{R})$ and $f'(x) = 0$ at every point x where $f(x) = 0$. Let $M = \max|f''(x)|$. Then

$$f'^2(x) \leqslant 2M|f(x)|.$$

Proof. Let α and β be two neighboring roots of the function f. Suppose $f > 0$ on (α, β) (otherwise, f may be replaced by $-f$). We find the maximum of the function $g = f'^2 f^{-1}$ on $[\alpha, \beta]$. If it is at an internal point x_0 of this segment, then $g'(x_0) = 0$ and therefore

$$2f''(x_0) f(x_0) - f'^2(x_0) = 0,$$

so that $g(x_0) = 2f''(x_0) \leqslant 2M$. If $g(x) = g_{max}$ at the boundary point x of $[\alpha, \beta]$, for example, when $x = \alpha$, then by by l'Hôpital's rule

$$g_{\max} = \lim_{x \to \alpha} \frac{f'^2(x)}{f(x)} = \lim_{x \to \alpha} \frac{2f'(x) f''(x)}{f'(x)} = 2f''(\alpha).$$

Thus, we have proved the inequality $f'^2(x) \leqslant 2Mf(x)$. This proves the assertion.

(B_5) By hypothesis $\text{grad}_{x; \xi} q(x, \xi) = 0$ at $(x, \xi) \in S$. Since at zeros not belonging to S the sign of q does not vary, condition (B_4) is fulfilled.

CHAPTER VII. SUFFICIENT CONDITIONS FOR LOCAL SOLVABILITY

§1. Review of Results

1.1. Not too long ago the theory of partial differential equations had no theorems on solvability of differential equations except for the classes of operators with analytic coefficients and the basic classes of equations — elliptic, parabolic, and hyperbolic — of mathematical physics.

The first general results in this direction were obtained by L. Hörmander[1] in 1955 when he proved the following

Theorem 1.1. For local solvability of a differential operator $P(x, D)$ it is sufficient that two conditions be fulfilled:

1) $\mathrm{grad}_\xi\, p^0(x, \xi) \neq 0$;

2) $c_1^0(x, \xi) = 2\,\mathrm{Im} \sum_{j=1}^{n} \frac{\partial p^0(x, \xi)}{\partial \xi_j} \frac{\overline{\partial p^0(x, \xi)}}{\partial x^j} = 0, \quad (x, \xi) \in T^*\Omega \setminus 0.$

Condition (2) is fulfilled, for example, when p^0 is a real-valued function or when p^0 is independent of x.

Recall that operator P is called solvable in $x_0 \in \Omega$, if there is a neighborhood ω of this point such that

$$P(\mathscr{D}'(\omega)) \supset C_0^\infty(\omega).$$

For nonlocal pseudo-differential operators P, solvability at the point x^0 means the existence of two neighborhoods ω and ω' of this point such that $\omega' \Supset \omega$ and

$$\gamma_\omega P(\mathscr{E}'(\omega')) \supset C_0^\infty(\omega),$$

where $\gamma_\omega f$ is the restriction of function f to the domain ω.

This theorem was generalized in Hörmander[3] where he defined a new class of *principally normal* operators characterized by the condition: there exists a differential operator $Q(x, D)$ of order $m - 1$ with coefficients from $C^1(\overline{\Omega})$ such that

$$c_1^0(x, \xi) = 2\,\mathrm{Re}\, p^0(x, \xi)\, q(x, \xi).$$

Theorem 1.2. Principally normal operators satisfying condition (1) are locally solvable.

A more complete analysis allowing for higher order derivatives in characteristic points is given in L. Nirenberg and F. Trèves[1] for the case of first-order differential operators

$$P(x, D) = \sum_{j=1}^{n} a^j(x) D_j + a(x).$$

They have given necessary and sufficient conditions for solvability in the case of operators with analytic coefficients a_j. Let $\{,\}$ be the Poisson bracket

$$\{f, g\} = \sum_{j=1}^{n} \left(\frac{\partial f}{\partial \xi_j} \frac{\partial g}{\partial x^j} - \frac{\partial f}{\partial x^j} \frac{\partial g}{\partial \xi_j} \right).$$

Then the number of the first nonvanishing function

$$p^0, \; c_1 = \{\overline{p^0}, \; p^0\}, \quad c_2 = \{\overline{p^0}, \; c_1\}, \; \ldots$$

must be finite and even or infinite for all $\xi \neq 0$ and $x \in \omega$, where ω is some neighborhood of the origin.

1.2. Further advances in the solvability theory are associated with the development of the theory of pseudo-differential operators.

On the one hand, we come naturally to pseudo-differential equations in studying boundary problems for differential equations. In this case, the solvability of the pseudo-differential operator obtained is equivalent to that of the boundary problem. On the other hand, the theory of pseudo-differential operators proves to be very useful in studying the properties of differential equations. For example, to study the solvability of equations of principal type it suffices to consider the case of first-order differential equations, which significantly alleviates the study.

The most important step in further development of the theory was taken by L. Hörmander (see Hörmander[6]). He proved the validity of

Theorem 1.3. For the solvability of equation $Pu = f$ it is sufficient that the condition

$$p^0(x, \; \xi) = 0, \quad (x, \; \xi) \in T^*\Omega \setminus 0 \Rightarrow c_1(x, \; \xi) > 0,$$

is fulfilled. In this case, for $f \in H_s(\omega)$ and any s, there exists a local solution u of the class $H_{s+m-1/2}(\Omega)$.

Hörmander[6] has proved Theorem 1.2 for arbitrary pseudo-differential operators.

1.3. The earlier mentioned results of L. Nirenberg and F. Trèves and Theorem 1.3 were generalized by the author in his work[4] on subelliptic operators which will be discussed in Chapter VIII of the present book.

Theorems on sufficient conditions for local solvability are proved in F. Trèves,[3,4] L. Nirenberg and F. Trèves,[1-3] and Yu. Egorov.[7,9]

For differential operators of principal type with analytic coefficients in the principal part, the local solvability is proved in L. Nirenberg and F. Trèves[2] under this condition $(\mathscr{P})$: *On every bicharacteristic of the function* Re zp^0, Im zp^0 *does not change its sign* $(z \in \mathbb{C} \setminus 0)$.

The sufficiency of this condition for operators with smooth (not necessarily analytic) coefficients was proved in R. Beals and C. Fefferman.[1] Their further improvement of the proof led to the development of a very wide theory of pseudo-differential operators, which covers new classes of operators (see R. Beals[2]).

1.4. In this section we give the proof of a theorem on sufficient conditions for local solvability which does not overlap those mentioned in the earlier works, since condition $(\mathscr{P})$ is assumed to be fulfilled in it. The methods used in this section are comparatively elementary; they rely on a discussion of quadratic forms and the use of Gårding's inequality.

Although our results are not the best possible and do not contain the theorems of L. Nirenberg—F. Trèves and R. Beals—C. Fefferman, they are fairly obviously related to the necessary conditions for solvability in Chapter VI.

Suppose that $P(x, D)$ is a pseudo-differential operator of order m that satisfies condition (ψ):

(Ψ) If γ is a bicharacteristic of function $\operatorname{Im} zp^0$ and passes through the characteristic point (x, ξ^0) and $h(t) = \operatorname{Re} zp^0(x(t), \xi(t))$, where $x = x(t)$, $\xi = \xi(t)$ are equations that determine γ, then from condition $h(t_1) > 0$ it follows that $h(t) \geqslant 0$ for all $t \geqslant t_1$, provided $x(t)$ belongs to Ω for all $z \in \mathbb{C} \setminus 0$.

Besides, we assume that the following conditions are fulfilled for every characteristic point (x_0, ξ^0):

(A) A smooth function $\alpha(x, \xi)$ is defined in a neighborhood of (x_0, ξ^0) such that $\alpha(x, t\xi) = \alpha(x, \xi)$ for $t \geqslant 1$, $|\xi| \geqslant 1$ and

$$c_1^0(x, \xi) + \operatorname{Re} \alpha(x, \xi) p^0(x, \xi) + N |p^0(x, \xi)|^2 |\xi|^{-1} \geqslant 0$$

with some constant $N \geqslant 0$. Here, as usual,

$$c_1^0(x, \xi) = 2 \operatorname{Im} \sum_{j=1}^{n} p^{0(j)}(x, \xi) \overline{p^0_{(j)}(x, \xi)}.$$

(B) Let $a = \operatorname{Re} p^0$, $b = \operatorname{Im} p^0$, and a be the symbol of an operator of principal type. Let $\nabla' b$ be the derivative of function b in the direction of the vector field tangent to the surface $a(x, \xi) = 0$ and transverse to the field

$$H_a = \sum_{j=1}^{n} \left(\frac{\partial a}{\partial \xi_j} \frac{\partial}{\partial x^j} - \frac{\partial a}{\partial x^j} \frac{\partial}{\partial \xi_j} \right).$$

Then there is a constant K such that

$$|\nabla' b(x, \xi)|^2 \leqslant K |b(x, \xi)|$$

for all (x, ξ) from some neighborhood of (x_0, ξ^0) and lying on the surface $a(x, \xi) = 0$, $|\xi| = 1$.

The main result of this section is

Theorem 1.4. Let $P(x, D)$ be a pseudo-differential operator of principal type of order m in Ω that satisfies condition (Ψ). Suppose that condition (A) or (B) is fulfilled at every characteristic point $(x_0, \xi^0) \in T^*\Omega \setminus 0$. Then for every point $x_0 \in \Omega$ and every real-valued s, $s \leqslant n/2 + m - 1$ we can find a neighborhood $\omega \subset \Omega$ such that in ω the equation $Pu = f$, where $f \in \mathring{H}_{s-m+1}(\Omega)$, has a solution $u \in \mathring{H}_s(\Omega)$, and

$$\|u\|_s \leqslant C\varepsilon \|f\|_{s-m+1}, \quad \varepsilon = \operatorname{diam} \omega,$$

where the constant C is independent of f.

Let us indicate some conditions that are sufficient for the validity of conditions (A) and (B).

Clearly, condition (A) is fulfilled when $c_1^0(x_0, \xi^0) > 0$. In this case we can assume that $\alpha = 0$ and $N = 0$.

The validity of this condition is less obvious when the vectors $\operatorname{grad}_{x,\xi} \operatorname{Re} p^0$, $\operatorname{grad}_{x,\xi} \operatorname{Im} p^0$, and $(\xi^0, 0)$ are linearly independent. If $\operatorname{Re} p^0 = a$ and $\operatorname{Im} p^0 = b$, then a and b are symbols of operators of principal type with linearly independent gradients. Therefore, the characteristic set $M = \{(x, \xi);\ p^0(x, \xi) = 0\}$ in a neighborhood of (x_0, ξ^0) is a smooth $(2n - 2)$-dimensional manifold and the vectors grad a and grad b form a basis in every plane normal to M. It follows that there exists a smooth function α satisfying the condition

$$\alpha(x, t\xi) = \alpha(x, \xi) \text{ for } t \geqslant 1,\ |\xi| \geqslant 1,$$

for which the function

$$h(x, \xi) = c_1^0(x, \xi) + \operatorname{Re} \alpha(x, \xi)\, p^0(x, \xi)$$

has at the points of M only zeros of second order. Thus, there exists a constant $N \geqslant 0$ such that

$$h(x, \xi) + N\, |p^0(x, \xi)|^2 \times |\xi|^{-1} \geqslant 0$$

in a neighborhood of (x_0, ξ^0), that is, condition (A) is fulfilled.

Condition (B) is fulfilled, for example, when the function b does not change sign on the surface $a(x, \xi) = 0$. This follows, for example, from Lemma 5.6 of Chapter VI. This is true also when b changes sign, but has a second or higher-order zero so that $\nabla' b(x, \xi) = 0$ at every characteristic point.

Further, condition (B) is fulfilled when function b changes sign at points of the smooth manifold S, transverse to the bicharacteristic of function a. In this case the proof is also based on the use of Lemma 5.6 of Chapter VI.

1.5. Of course, another approach to the problem concerning solvability is also possible: the class of generalized solutions considered may be widened by going beyond the traditional limits of distribution theory. Evidently, any equation of principal type is solvable in a class of functionals on the space of analytic functions. A class of hyperfunctions which may be regarded as boundary values in a real space of holomorphic functions, was introduced in M. Sato.[1] However, P. Shapira[1] has proved that the equation $D_1u + ix^1D_2u = f$ is unsolvable in a class of hyperfunctions in every domain $\Omega \subset \mathbb{R}^2$ containing the origin of coordinates for some function $f \in C^\infty(\mathbb{R}^2)$.

F. Trèves[2] has considered the class K^s of ultradistributions, which are images for Fourier transforms of the functions $\tilde{v}(\xi)$ such that

$$\int |\tilde{v}(\xi)|^2 e^{-2s|\xi|}\, d\xi < \infty.$$

For this class, he proved the local solvability of the equation $P(x, D)u = f(x)$ (and even of the Cauchy problem for such equations) for very wide assumptions.

In a number of papers of M. I. Vishik and G. I. Eskin, M. I. Vishik and V. V. Grushin, and V. G. Mazja and B. P. Paneyakh use has been made, in the absence of solvability, of "coboundary" conditions which save the situation. This technique comes about through the replacement of the equation $Pu = f$ by an equation of the type $Pu = f + \delta(\Gamma) \otimes \rho$, where Γ is a characteristic set and ρ stands for an unknown function which is sought simultaneously with u.

L. Nirenberg and F. Trèves[2] have shown how one can obtain from Theorem 1.4 a theorem on local solvability of a system of equations $Pu = f$ assuming that $\det P(x, \xi)$ satisfies condition (Ψ).

§2. Reduction to First-Order Operators

In this section we shall show how the proof of Theorem 1.4 can be reduced to the consideration of a special case.

Lemma 2.1. It is sufficient to prove Theorem 1.4 for the case $s = 0$, $m = 1$.

Proof. Suppose that Theorem 1.4 holds for $m = 1$, $s = 0$. Let $Q = P\Lambda^{1-m}$ be a first order operator with symbol $q(x, \xi) = P(x, \xi)(1 + |\xi|^2)^{(1-m)/2}$ and let $A = h\Lambda^{-s}$ be an operator with symbol $h(x)(1 + |\xi|^2)^{-s/2}$, where $h \in C_0^\infty(\omega)$ and $h(x) = 1$ in some smaller neighborhood ω' of the point x_0 which may be presumed to be the origin of coordinates. In so doing we can, of course, assume that $\operatorname{diam} \omega' = \varepsilon/2$.

Let us verify that the conditions of the theorem are fulfilled simultaneously for operators P, Q. Clearly, if $p^0(x, \xi) = 0$, then

$$\operatorname{grad}(p^0(x, \xi)|\xi|^{1-m}) = |\xi|^{1-m} \operatorname{grad} p^0(x, \xi),$$

so that Q is an operator of principal type. If $zp^0 = a + ib$, where $z \in \mathbb{C} \setminus 0$, then $zq_0 = |\xi|^{1-m}(a + ib)$ so that

$$\operatorname{sign} \operatorname{Re} zp^0 = \operatorname{sign} \operatorname{Re} zq^0,$$

and the bicharacteristics of functions $\operatorname{Re} zp^0$ and $\operatorname{Re} zq^0$, along which $\operatorname{Re} zp^0$ coincide differing only in parametrization.

By assumption, the equation $Qu = f$ is solvable, that is, for functions $f \in L_2(\Omega)$ with compact support there exists a function $u \in L_2(\Omega)$ for which

$$\|u\|_0 \leqslant C\varepsilon\|f\|_0, \quad (\varphi, f) = (Q^*\varphi, u), \quad \varphi \in C_0^\infty(\omega).$$

It is seen that

$$\|\varphi\|_0 = \sup_{\|f\|_0 = 1} |(f, \varphi)| \leqslant \sup_{\|f_0\| = 1} \|u\|_0 \|Q^*\varphi\|_0 \leqslant C\varepsilon\|Q^*\varphi\|_0,$$

that is,

$$\|\varphi\|_0 \leqslant C\varepsilon\|Q^*\varphi\|_0, \quad \varphi \in C_0^\infty(\omega). \tag{2.1}$$

Let now $\psi \in C_0^\infty(\omega')$ so that $h\psi = \psi$. Substituting $\varphi = A\psi = h\Lambda^{-s}\psi$ in (2.1) and considering that $Q^* = \Lambda^{1-m}P^*$, we obtain

$$\|h\Lambda^{-s}\psi\|_0 \leqslant C\varepsilon\|\Lambda^{1-m}P^*h\Lambda^{-s}\psi\|_0 = C\varepsilon\|P^*h\Lambda^{-s}\psi\|_{1-m}.$$

It follows that

$$\|\psi\|_{-s} \leqslant \|h\Lambda^{-s}\psi\|_0 + C_1\|\psi\|_{-s-1} \leqslant C\varepsilon\|P^*h\Lambda^{-s}\psi\|_{1-m} + C_1\|\psi\|_{-s-1} \leqslant$$
$$C\varepsilon(\|P^*\psi\|_{1-m-s} + C_2\|\psi\|_{-s}) + C_1\|\psi\|_{-s-1}.$$

If $CC_2\varepsilon < \frac{1}{2}$, then from this follows the inequality

$$\|\psi\|_{-s} \leqslant 2C\varepsilon\|P^*\psi\|_{1-m-s} + 2C_1\|\psi\|_{-s-1}.$$

Since $s \leqslant n/2$, in view of Theorem 3.11 of Chapter I

$$\|\psi\|_{-s-1} \leqslant C_3\varepsilon\|\psi\|_{-s}.$$

If $4C_1C_3\varepsilon < 1$, then we obtain

$$\|\psi\|_{-s} \leqslant 4C\varepsilon\|P^*\psi\|_{1-m-s}.$$

From this estimate it is seen that if $g \in H_s(\Omega)$, then the functional (g, ψ) of $P^*\psi \in H_{-s-m+1}$ is bounded and therefore there exists an element $v \in H_{s+m-1}$ for which

$$(P^*\psi, v) = (\psi, g), \quad \psi \in C_0^\infty(\omega'), \qquad \|v\|_{s+m-1} \leqslant 4C\varepsilon\|g\|_s.$$

Thus, the lemma is proved.

2.2. In proving inequality (2.1) we shall need a partition of unity on a unit sphere $\{\xi \in \mathbb{R}_n, |\xi| = 1\}$. Let g_j, $j = 1, \ldots, r$ be r non-negative C^∞ functions of ξ when $|\xi| = 1$ and $\sum g_j = 1$. Extending them to C^∞ functions defined for all ξ and homogeneous of zero degree for $|\xi| \geqslant 1/2$, we define a pseudo-differential operator $g_j(D)$ of zero order after assuming

$$g_j(D)u(x) = (2\pi)^{-n}\int e^{ix\xi}g_j(\xi)\tilde{u}(\xi)\,d\xi,$$

where $\tilde{u}$ stands for the Fourier transform of function u. Operator $\sum g_j(D) - I$ will be infinitely smooth and therefore

$$\|u\|_0 \leqslant \sum\|g_j(D)u\|_0.$$

2.3. We assume that x^0 is the origin. If $p^0(0, \xi) \neq 0$ and $|\xi^0| = 1$, then

$$|p(x, \xi)| \geq c_0, \quad c_0 = \text{const} > 0$$

for all (x, ξ) in some neighborhood of $(0, \xi^0)$.

If the support of $g(\xi)h(x)$ is in this neighborhood, then

$$\|h(x)g(D)u\|_1 \leq C_1(\|P^*(x, D)hgu\|_0 + \|hgu\|_0), \quad u \in C_0^\infty(\omega').$$

Here we assume that diam supp $u = \varepsilon$ and $h(x) = 1$ in a neighborhood of the support of u so that $(1 - h)u = 0$. Clearly

$$\|hgu\|_0 \leq C_2\varepsilon\|hgu\|_1.$$

For small ε, from these inequalities the estimate

$$\|hgu\|_0 \leq C_3\varepsilon\|P^*hgu\|_0 \tag{2.2}$$

follows.

2.4. Let now $p^0(0, \xi^0) = 0$ and $|\xi^0| = 1$. Since P is an operator of principal type, there exists a homogeneous canonical transformation φ for which $\text{grad}_\xi\, \tilde{p}^0(0, \xi^0) \neq 0$, for example, $\partial_{\xi_1}\widetilde{p^0}(0, \xi^0) \neq 0$, where $\widetilde{p^0} = \varphi \circ p^0$. Furthermore, there is smooth function $\lambda(x, \xi')$ in some neighborhood of $(0, \xi^0)$ such that

$$\widetilde{p^0}(x, \xi) = [\xi_1 - \lambda(x, \xi')]a(x, \xi),$$

where $a \in C^\infty$ and $|a| \geq C_1 > 0$. Let $\lambda = \alpha + i\beta$, where α and β are smooth real-valued functions of x and ξ'. Then there exists a homogeneous canonical transformation φ_1 in the space of variables (x', ξ') depending on parameter x^1, under which $\xi_1 - \alpha(x, \xi')$ changes to ξ'. In this case, function $\beta(x, \xi')$ transforms to $q(x, \xi')$.

Let $H(x, \xi')$ and $H_1(x, \xi')$ be smooth functions on $\mathbb{R}^n \times S^{n-2}$ with supports in that neighborhood of the point $(0, \xi^0)$ where the function λ and transformation φ_1 are defined, equal to 1 in some smaller neighborhood ω of $(0, \xi^0)$, and such that $HH_1 = H_1$. We extend them to smooth functions which are homogeneous of degree zero when $|\xi| \geq 1/2$. Let a be extended to a smooth function, homogeneous of degree zero when $|\xi| \geq 1/2$, such that $|a(x, \xi)| \geq c_1 > 0$ and let $A(x, D)$ be an operator with symbol $a(x, \xi)$. Let Φ and Φ_1 be Fourier integral operators corresponding to homogeneous canonical transformations φ and φ_1. Finally, by $Q(x, D')$ and $H_1(x, D')$ we denote the operators with symbols $q(x, \xi')H(x, \xi')$ and $H_1(x, \xi')$, respectively. Then the order of the operator

$$H_1(x, D')[P^* - \Phi^* A\Phi_1^*(iD_1 + Q)\Phi_1\Phi]$$

equals 0, where A is an elliptic operator of first order and the symbol $Q(x, \xi')$ of operator Q does not change sign from plus to minus as x^1 increases.

Now we shall prove the following

Theorem 2.1. If operator P satisfies condition (Ψ), then there exists a constant C such that

$$\|\varphi\|_0 \leq C\varepsilon\|(iD_1 + Q)\varphi\|_0, \quad \varphi \in C_0^\infty(\omega), \tag{2.3}$$

where ε = diam ω.

Substituting $\varphi = \Phi_1\Phi hgu$, where $h(x)$ and $g(\xi)$ are defined in the same way as in Section 2.3, into (2.3) we obtain

$$\|hgu\|_0 = \|\Phi_1\Phi hgu\|_0 = \|\varphi\|_0 \leq C\varepsilon\|(iD_1 + Q)\Phi_1\Phi hgu\|_0 = C\varepsilon\|\Phi_1^*(iD_1 + Q)\Phi_1\Phi hgu\|_0 \leq$$

$$CC_1\varepsilon(\|A\Phi_1^*(iD_1+Q)\Phi_1\Phi hgu\|_0+\|hgu\|_0)=$$

$$CC_1\varepsilon(\|\Phi^*A\Phi_1^*(iD_1+Q)\Phi_1\Phi hgu\|_0+\|hgu\|_0)\leqslant CC_1\varepsilon(\|P^*hgu\|_0+C_2\|hgu\|_0).$$

If $CC_1C_2\varepsilon < 1/2$, then we obtain

$$\|hgu\|_0\leqslant C_3\varepsilon\|P^*hgu\|_0.$$

This inequality is similar to (2.2).

2.5. Thus, to every point $\xi \in S^{n-1}$ we can associate a function $g(\xi) \times h(x)$ as is done in Sections 2.3 and 2.4. The supports of functions g cover the unit sphere. From this covering we choose a finite subcovering and construct a smooth partition of unity $\sum_{j=1}^{r} g_j(\xi)=1$ corresponding to this covering. To every function g_j corresponds a neighborhood ω_j of the origin of coordinates in $\mathbb{R}^n$ by the constructions of Sections 2.3 and 2.4. By ω we denote the intersection of these neighborhoods. Then we have the estimates

$$\|h(x)g_j(D)u\|_0\leqslant C\varepsilon\|P^*h(x)g_j(D)u\|_0,\quad u\in C_0^\infty(\omega), \tag{2.4}$$

when $j = 1, \ldots, r$.

Let now $h(x) = 1$ be a neighborhood ω' of the origin and let $u \in C_0^\infty(\omega'')$, where $\omega''\Subset\omega'$, so that $hu = u$. Note that

$$\|[h, g_j]u\|_1\leqslant C_\varepsilon\|u\|_{-1},\quad u\in C_0^\infty(\omega''), \tag{2.5}$$

as the symbol of operator $[h, g_j]$ equals 0 in the neighborhood ω' of the support u.

The relationship between C_ε and ε is of significance for us. Suppose here diam $\omega = \varepsilon$, diam $\omega' \leqslant \varepsilon/2$, diam $\omega'' \leqslant \varepsilon/4$. Note that passing to new y coordinates by the formulas $x = \varepsilon y$, we obtain from (2.5) a similar inequality where the symbol of operator $[h, g_j]$ is independent of ε. Therefore,

$$\|[h, g_j]u\|_1\leqslant C_1\|u\|_{-1}.$$

Reverting to original variables x, we see that

$$\|[h, g_j]u\|_0\leqslant C_1\varepsilon^{-1}\|u\|_{-1},\quad \|\partial_l[h, g_j]u\|_0\leqslant C_1\varepsilon^{-2}\|u\|_{-1}.$$

Therefore, from (2.4) it follows that

$$\|g_j(D)u\|_0\leqslant C\varepsilon\|P^*g_j(D)u\|_0+CC_1(1+\varepsilon^{-1})\|u\|_{-1},\quad j=1,\ \ldots,\ r.$$

Summing these inequalities over j, we obtain

$$\|u\|_0\leqslant\sum_{j=1}^{r}\|g_j(D)u\|_0\leqslant C\varepsilon\sum\|P^*g_j(D)u\|_0+C_2\varepsilon^{-1}\|u\|_{-1}\leqslant$$

$$C\varepsilon\sum\|g_j(D)P^*u\|_0+C_3\varepsilon\|u\|_0+C_2\varepsilon^{-1}\|u\|_{-1}\leqslant C_4\varepsilon\|P^*u\|_0+C_3\varepsilon\|u\|_0+C_2\varepsilon^{-1}\|u\|_{-1}.$$

Let $C_3\varepsilon < 1/2$. Then we see that

$$\|u\|_0\leqslant 2C_4\varepsilon\|P^*u\|_0+2C_2\varepsilon^{-1}\|u\|_{-1}.$$

Let now $u \in C_0^\infty(\tilde{\omega})$ and diam $\tilde{\omega} \leqslant \delta\varepsilon$, where $\delta > 0$ is some small number. Then

$$\|u\|_{-1}\leqslant C_5\delta\varepsilon\|u\|_0,$$

where C_5 is independent of δ and ε. If δ is so small that

$$4C_2C_5\delta<1$$

(note that δ is independent of ε), then we obtain

$$\|u\|_0 \leq C_6 \varepsilon \|P^* u\|_0, \qquad u \in C_0^\infty(\tilde{\omega}),$$

where C_6 is independent of ε.

Thus, we have reduced the proof of Theorem 1.4 to the proof of Theorem 2.1.

§3. Proof of Theorem 2.1

3.1. Thus, we suppose that operator P is of the form

$$P(x, D) = -iD_1 + q(x, D'),$$

where the real-valued function q is independent of ξ_1 and $q(x, t\xi') = tq(x, \xi')$ for $t \geqslant 1$, $|\xi'| \geqslant 1$. Here, as usual $\xi' = (\xi_2, \ldots, \xi_n)$.

By condition (Ψ), q does not change sign from minus to plus as x^1 increases, provided the values of other arguments are fixed.

As mentioned in Sections 2.4 and 2.5, it is sufficient to prove (2.3) locally, that is, for functions of the type $u = h(x)g(D)\varphi$, $\varphi \in C_0^\infty(\omega)$.

3.2. If $p^0(x_0, \xi^0) \neq 0$, then the desired inequality is obtained in Section 2.3. Let (x_0, ξ^0) be a characteristic point, $\xi^0 \neq 0$. First we shall consider the case when condition (A) is fulfilled at this point.

We extend the function $\alpha(x, \xi)$ to a smooth function on $T^*(\omega)$ so that $\alpha(x, t\xi) = \alpha(x, \xi)$ for $|\xi| \geqslant 1$ and $t \geqslant 1$. Note that

$$\|Pu\|^2 = \|P^*u\|^2 - \operatorname{Re}(C_1 u, u) \tag{3.1}$$

and the principal symbol of operator C_1 is equal to c_1^0. Suppose that A is an operator with symbol $\alpha(x, \xi)$ and $S = C_1 + \frac{1}{2}(AP + P^*A^*) + N\Lambda^{-1}P^*P$ so that S is a first-order operator and its principal symbol $s^0(x, \xi) = c_1^0(x, \xi) + \operatorname{Re}\alpha(x, \xi)p^0(x, \xi) + N|p^0(x, \xi)|^2|\xi|^{-1}$ is non-negative in a conic neighborhood of (x_0, ξ^0). Let $H(x, \xi)$ be an auxiliary function with support in this neighborhood, non-negative, equal to 1 in supp $h(x)g(\xi)$ and such that $H(x, t\xi) = H(x, \xi)$ for $t \geqslant 1$, $|\xi| \geqslant 1$. Let $H(x, D)$ be a pseudo-differential operator with symbol $H(x, \xi)$.

Since

$$\operatorname{Re}(C_1 u, u) = \operatorname{Re}(SHu, u) + \operatorname{Re}(S(I - H)u, u) + \operatorname{Re}(Pu, A^*u) + N(\Lambda^{-1}P^*Pu, u),$$

we obtain, using the Gårding inequality,

$$\operatorname{Re}(C_1 u, u) \geqslant -c_1\|u\|_0^2 - \frac{1}{2}\|Pu\|_0^2.$$

Using this estimate, we infer from (3.1) that

$$\frac{1}{2}\|Pu\|_0^2 \leqslant \|P^*u\|_0^2 + c_1\|u\|_0^2.$$

But $P^* - P = 2iD_1 + A_0$, where A_0 is a zero-order operator. Therefore

$$\|D_1 u\|_0 \leqslant \frac{1}{2}\|(P^* - P)u\|_0 + c_2\|u\|_0 \leqslant \frac{3}{2}\|P^*u\|_0 + c_3\|u\|_0.$$

From this estimate and the inequality

$$\|u\|_0 \leqslant \varepsilon\|D_1 u\|_0, \quad u \in C_0^\infty(\omega),$$

it follows that for $c_3\varepsilon < 1/2$ the following estimate holds:

$$\|u\|_0 \leqslant 3\varepsilon\|P^*u\|_0.$$

Thus, we have obtained inequality (2.3).

3.3. Let now condition (A) be fulfilled at (x_0, ξ^0). If $P = -iD_1 + q(x, D')$, then this condition can be written as

$$\sum_{l=2}^{n} (|q^{(l)}(x, \xi')|^2 |\xi'| + |q_{(l)}(x, \xi')|^2 |\xi'|^{-1}) \leqslant K |q(x, \xi)|. \tag{3.2}$$

Let $\theta(x')$ be a function such that $\theta \in C_0^\infty(\mathbb{R}^{n-1})$, $\theta(x) \geqslant 0$ everywhere; $\theta(x') = 0$ when $|x^j| \geqslant 3/4$ for any $j = 2, \ldots, n$; $\theta(x') = 1$ if $|x^j| \leqslant 1/2$ for all $j = 2, \ldots, n$. Let $g_1, g_2, \ldots$ be points with integer coordinates in $\mathbb{R}^{n-1}$ and

$$\varphi_j(x') = \theta(x' - g_j)\left(\sum_{i=1}^{n} \theta(x' - g_i)^2\right)^{-1/2},$$

so that $\varphi_j \in C_0^\infty(\mathbb{R}^{n-1})$ and $\sum \varphi_j^2(x') = 1$. Note that

$$\sum_{i=1}^{\infty} |D^\alpha \varphi_j(x')| \leqslant C_\alpha$$

for all α and if $\varphi_j(x') \neq 0$, $\varphi_j(y') \neq 0$, then $|x' - y'| \leqslant 2\sqrt{n-1}$.

Let $\psi_j(D') = \varphi_j(D'|D'|^{-1/2})$. Then

$$\|P^* u\|^2 = \sum_j \|\psi_j(D')[iD_1 + Q(x, D')]u\|^2 =$$

$$\sum_j \|[iD_1 + Q(x, D')]\psi_j(D')u + [\psi_j, Q]u\|^2 \geqslant$$

$$\frac{1}{2}\sum_j \|[iD_1 + Q(x, D')]\psi_j(D')u\|^2 - \sum_j \|[\psi_j, Q]u\|^2. \tag{3.3}$$

Further, if we assume that

$$P_j^* = iD_1 + Q(x, \xi'^j) + \sum_{l=2}^{n} Q^{(l)}(x, \xi'^j)(D_l - \xi_l'^j),$$

where ξ'^j is any point from supp ψ_j, then

$$\sum_i \|[P_i^* - P]\psi_j(D')u\|^2 \leqslant C_1 \sum_j \sum_{l=2}^{n} |\xi'^j|^{-1} \left\|(\xi_l - \xi_l'^j)^2 \psi_j(\xi')\tilde{u}(\xi')\right\|^2 \leqslant C_2 \|u\|^2,$$

so that $|\xi' - \xi_j'| \leqslant c_0 |\xi'^j|^{1/2}$, provided $\xi' \in$ supp ψ_j. Therefore

$$\sum_j \|P^* \psi_j(D')u\|^2 \leqslant \frac{1}{2}\sum_j \|P_j^* \psi_j(D')u\|^2 - C_2\|u\|^2. \tag{3.4}$$

Further

$$\|P_j^* \psi_j(D')u\|^2 = \sum_k \|\varphi_k(x'|\xi^j|^{1/2}) P_j^* \psi_j(D')u\|^2 \geqslant$$

$$\frac{1}{2}\sum_k \|P_j^*(x, D)\varphi_k(x'|\xi^j|^{1/2})\psi_j(D')u\|^2 -$$

$$\sum_k \sum_{l=2}^{n} \|Q^{(l)}(x, \xi'^j)|\xi'^j|^{1/2} \varphi_{k(l)}(x'|\xi^j|^{1/2})\psi_j(D')u\|^2. \tag{3.5}$$

And again if we assume that $\varphi_k(x'|\xi^j|^{1/2})\psi_j(D')u = v_{jk}(x)$ and

$$P_{jk}^* = iD_1 + Q(x^1, x_{jk}', \xi'^j) + \sum_{l=2}^{n} Q^{(l)}(x^1, x_{jk}', \xi'^j)(D_l - \xi_l'^j) +$$

$$\sum_{l=2}^{n} Q_{(l)}(x^1, x_{jk}', \xi'^j)(x^l - x_{jk}^l),$$

where x_{jk} is any point from supp $\varphi_k(x'|\xi^j|^{1/2})$, then we obtain

$$\sum_k \| (P_j^* - P_{jk}^*) v_{jk} \|^2 \leqslant 2 \sum_k \| [Q(x, \xi'^j) - Q(x^1, x'_{jk}, \xi'^j) -$$

$$\sum_{l=2}^n Q_{(l)}(x^1, x'_{jk}, \xi'_j)(x^l - x^l_{jk})] v_{jk}(x) \|^2 + 2 \sum_k \left\| \sum_{l=2}^n [Q^{(l)}(x, \xi'^j) - \right.$$

$$\left. Q^{(l)}(x^1, x'_{jk}, \xi'^j)] (D_l - \xi'^j_l) v_{jk} \right\|^2 \leqslant C_3 \| \psi_j u \|^2. \tag{3.6}$$

P^*_{jk} is a first-order differential operator. Let α_{jk} be a value of variable x^1 from $[-\varepsilon, +\varepsilon]$ such that

$$\operatorname{sgn}(x^1 - \alpha_{jk}) Q(x^1, x'_{jk}, \xi'^j) \leqslant 0$$

for all x^1 of this interval. If function $Q(x^1, x'_{jk}, \xi'^j)$ does not change sign, then we assume α_{jk} to be equal to ε or $-\varepsilon$. Let $\max_{x^1} \int |v_{jk}|^2 dx' = \int |v_{jk}(\beta_{jk}, x')|^2 dx'$. If $\beta_{jk} \leqslant \alpha_{jk}$, then integrating over $-\varepsilon \leqslant x^1 \leqslant \beta_{jk}$ gives the equality

$$\operatorname{Re} \int P^*_{jk} v_{jk} \bar{v}_{jk}\, dx = \frac{1}{2} \int_{x^1 = \beta_{jk}} |v_{jk}|^2 dx' + \int | Q(x^1, x'_{jk}, \xi'^j) | \, |v_{jk}|^2 dx +$$

$$\operatorname{Re} \sum_{l=2}^n \int [Q^{(l)}(x^1, x'_{jk}, \xi'^j)(D_l - \xi^j_l) + Q_{(l)}(x^1, x'_{jk}, \xi'^j)(x^l - x^l_{jk})] v_{jk} \bar{v}_{jk}\, dx.$$

If $\beta_{jk} > \alpha_{jk}$, then an analogous equality is obtained after integrating with respect to x^1 between ε and β_{jk}. Further, if we integrate the function $\operatorname{sgn}(x^1 - \alpha_{ij}) p^*_{jk} v_{jk} \bar{v}_{jk}$ over the entire strip $|x^1| < \varepsilon$, then we obtain a similar equality containing integrals of $|Q(x^1, x_{jk}, \xi'^j| \, |v_{jk}|^2$ taken over the entire strip.

Thus, we obtain the inequality

$$2 \| P^*_{jk} v_{jk} \| \| v_{jk} \| \geqslant \frac{1}{2} \max_{x^1} \int |v_{jk}|^2 dx' + \int | Q(x^1, x'_{jk}, \xi'^j) | \, |v_{jk}|^2 dx -$$

$$2 \sum_{l=2}^n \int [| Q^{(l)}(x^1, x'_{jk}, \xi'^j)(D_l - \xi^j_l) v_{jk} | +$$

$$| Q_{(l)}(x^1, x'_{jk}, \xi'^j)(x^l - x^l_{jk}) v_{jk} |] \, | \overline{v_{jk}} | \, dx. \tag{3.7}$$

Note that

$$\max_{x^1} \int |v_{jk}|^2 dx' \geqslant \frac{1}{2\varepsilon} \| v_{jk} \|^2.$$

It is not hard to see that

$$\int | Q^{(l)}(x^1, x'_{jk}, \xi'^j)(D_l - \xi^j_l) v_{jk} | \, | \overline{v_{jk}} | \, dx \leqslant$$

$$\delta \int | Q^{(l)}(x^1, x'_{jk}, \xi'^j) |^2 | \xi'^j | \, | v_{jk} |^2 dx + \frac{4}{\delta} \int | v_{jk} |^2 dx,$$

where $\delta > 0$. By (3.2)

$$\sum_{l=2}^n \int | Q^{(l)}(x^1, x'_{jk}, \xi'^j) |^2 | \xi'^j | \, | v_{jk} |^2 dx \leqslant K \int | Q(x^1, x'_{jk}, \xi'^j) | \, | v_{jk} |^2 dx.$$

Similarly,

$$\sum_{l=2}^n \int | Q_{(l)}(x^1, x'_{jk}, \xi'^j)(x^l - x^l_{jk}) | \, | v_{jk} |^2 dx \leqslant$$

$$\delta \sum_{l=2}^n \int | Q_{(l)}(x^1, x'_{jk}, \xi'^j) |^2 | \xi'^j |^{-1} | v_{jk} |^2 dx + \frac{4}{\delta}(n-1) \int | v_{jk} |^2 dx \leqslant$$

$$K\delta \int |Q(x^1, x'_{jk}, \xi'^j)|\,|v_{jk}|^2\,dx + \frac{4}{\delta}(n-1)\int |v_{jk}|^2\,dx.$$

Let δ be such that $2K\delta = 1/4$ and let ε be so small that $16\varepsilon(n-1) < \delta$. Then from (3.7) it follows that

$$\frac{1}{8\varepsilon}\|v_{jk}\|^2 + \frac{1}{4}\int |Q(x^1, x'_{jk}, \xi'^j)|\,|v_{jk}|^2\,dx \leqslant \|P^*_{jk}v_{jk}\|\cdot\|v_{jk}\|.$$

Let us now evaluate the remaining terms in (3.3) and (3.5). We have

$$\sum_j \|[\psi_j, Q]u\|^2 \leqslant 2\sum_j \sum_{l=2}^n \|\psi_j^{(l)}(D')Q_{(l)}(x, D')u\|^2 + C_4\|u\|^2.$$

Since

$$\sum_j \|\psi_j^{(l)}(D')g\|^2 \leqslant C_5 \sum_j |\xi^j|^{-1}\|\psi_j(D')g\|^2, \quad g \in L_2(\mathbb{R}^n),$$

we have

$$\sum_j \|[\psi_j, Q]u\|^2 \leqslant 2C_5 \sum_{j,k}\sum_{l=2}^n |\xi^j|^{-1}\|\varphi_k(x'|\xi^j|^{1/2})\psi_j(D')Q_{(l)}(x, D')u\|^2 +$$

$$C_4\|u\|^2 \leqslant 2C_5\sum_{j,k}\sum_{l=2}^n |\xi^j|^{-1}\|Q_{(l)}(x^1, x'_{jk}, \xi'^j)v_{jk}\|^2 + C_6\|u\|^2.$$

Similarly

$$\sum_{j,k}\sum_{l=2}^n \|Q^{(l)}(x, \xi^{(j)})|\xi'^j|^{1/2}\varphi_{k(l)}(x'|\xi^j|^{1/2})\psi_j(D')u\|^2 \leqslant$$

$$C_7\sum_{j,k}\sum_{l=2}^n \|Q^{(l)}(x^1, x'_{jk}, \xi'^j)|\xi'^j|^{1/2}v_{jk}\|^2 + C_8\|u\|^2.$$

As above, from (3.2) it is deduced that

$$\sum_{l=2}^n (|\xi^j|^{-1}\|Q_{(l)}(x^1, x'_{jk}, \xi'^j)v_{jk}\|^2 + |\xi'^j|\,\|Q^{(l)}(x^1, x'_{jk}, \xi'^j)v_{jk}\|^2) \leqslant$$

$$K\int |Q(x^1, x'_{jk}, \xi'^j)|\cdot|v_{jk}|^2\,dx.$$

Gathering all the earlier inequalities, starting from (3.3), we obtain

$$\|P^*u\|^2 \geqslant \frac{1}{4}\sum_j \|P_j^*\psi_j(D')u\|^2 - \frac{1}{2}C_2\|u\|^2 - \sum_j \|[\psi_j, Q]u\|^2 \geqslant \frac{1}{8}\sum_{j,k}\|P_j^*v_{jk}\|^2 -$$

$$\frac{1}{4}\sum_{j,k}\sum_{l=2}^n \|Q^{(l)}(x, \xi'^j)|\xi'^j|^{1/2}\varphi_{k(l)}(x'|\xi'^j|^{1/2})\psi_j(D')u\|^2 -$$

$$\frac{1}{2}C_2\|u\|^2 - 2C_5\sum_j\sum_{l=2}^n |\xi'^j|^{-1}\|Q_{(l)}(x^1, x'_{jk}, \xi'^j)v_{jk}\|^2 -$$

$$C_6\|u\|^2 \geqslant \frac{1}{16}\sum_{j,k}\|P_{jk}^*v_{jk}\|^2 - \frac{1}{8}C_3\|u\|^2 -$$

$$\frac{1}{4}C_7\sum_{j,k}\sum_{l=2}^n \|Q^{(l)}(x^1, x'_{jk}, \xi'^j)|\xi'^j|^{1/2}v_{jk}\|^2 - \frac{1}{4}C_8\|u\|^2 -$$

$$\frac{1}{2}C_2\|u\|^2 - 2C_5\sum_{j,k}\sum_{l=2}^n |\xi^j|^{-1}\|Q_{(l)}(x^1, x'_{jk}, \xi'^j)v_{jk}\|^2 - C_6\|u\|^2 \geqslant$$

$$\frac{1}{16}\sum_{j,k}\|P_{jk}^*v_{jk}\|^2 - K\left(\frac{1}{4}C_7 + 2C_5\right)\int |Q(x^1, x'_{jk}, \xi'^j)|\,|v_{jk}|^2\,dx - C_9\|u\|^2. \quad (3.9)$$

From (3.8) we deduce that

$$\frac{1}{8\varepsilon}\|v_{jk}\|^2+\frac{1}{4}\int|Q(x^1,\ x'_{jk},\ \xi'')||v_{jk}|^2dx\leqslant\frac{1}{16\varepsilon}\|v_{jk}\|^2+4\varepsilon\|P^*_{jk}v_{jk}\|^2$$

and therefore

$$\frac{1}{\varepsilon^2}\|v_{jk}\|^2+\frac{4}{\varepsilon}\int|Q(x^1,\ x'_{jk},\ \xi'')||v_{jk}|^2\,dx\leqslant 64\|P^*_{jk}v_{jk}\|^2.$$

From this and (3.9) it follows that

$$\|P^*u\|^2\geqslant\frac{1}{16\cdot 64\cdot\varepsilon^2}\sum\|v_{jk}\|^2-C_9\|u\|^2\geqslant\frac{1}{32\cdot 64\cdot\varepsilon^2}\|u\|^2,$$

if only ε is so small that

$$\varepsilon K\left(\frac{1}{4}C_7+2C_5\right)<4 \quad\text{and}\quad 32\cdot 64\cdot\varepsilon^2\cdot C_9<1.$$

Thus, inequality (2.3) has also been obtained for case (B). As mentioned above, this suffices for completing the proof of Theorem 2.1.

CHAPTER VIII. SUBELLIPTIC OPERATORS

§1. Definition and Basic Properties

1.1. Definition 1.1. A pseudo-differential operator $P(x, D)$ of order m is called *subelliptic* in Ω if for every compact subset $K \in \Omega$ there exist real constants C, s, and δ such that

$$\|u\|_s \leqslant C(\|Pu\|_{s-m+\delta} + \|u\|_{s-1}), \qquad u \in C_0^\infty(K), \tag{1.1}$$

where $0 \leqslant \delta < 1$.

If $\delta = 0$, then operator P is elliptic and the estimate (1.1) is equivalent to the condition: $p^0(x, \xi) \neq 0$ in $T^*\Omega \setminus 0$.

The following theorem is proved in Hörmander.[6]

Theorem 1.1. Estimate (1.1) with $\delta = 1/2$ is fulfilled if and only if

$$p^0(x, \xi) = 0 \Rightarrow c_1(x, \xi) > 0,$$

where $(x, \xi) \in T^*\Omega \setminus 0$ for $c_1(x, \xi) = 2\,\mathrm{Im} \sum_{j=1}^{n} \frac{\partial p^0(x, \xi)}{\partial x^j} \frac{\overline{\partial p^0(x, \xi)}}{\partial \xi_j}$.

In Ref. 4 Egorov formulated

Theorem 1.2. Estimate (1.1) with $\delta = k/(k + 1)$ is fulfilled if and only if conditions (Ψ') and ($\mathscr{B}$) are fulfilled. If $p^0(x_0, \xi^0) = 0$, $(x_0, \xi^0) \in T^*\Omega \setminus 0$ and all the j-fold Poisson brackets composed of functions p^0 and and $\bar{p}^0$ vanish at (x_0, ξ^0) when $j < l$, then (1.1) cannot be fulfilled for $\delta < l(l + 1)^{-1}$, $x_0 \in K$.

Condition (Ψ) is given in Chapters VI and VII. If $p^0 = a_1 + ia_2$, where a_1 and a_2 are real-valued functions, and grad $a_2(x_0, \xi^0) \neq 0$, $p^0(x_0, \xi^0) = 0$, then we consider the integral curve $x = x(t)$, $\xi = \xi(t)$ of the system of equations

$$\dot{x}(t) = \partial_\xi a_2(x(t), \xi(t)), \quad \dot{\xi}(t) = -\partial_x a_2(x(t), \xi(t)),$$

which passes through (x_0, ξ^0), and the function $h(t) = a_1(x(t), \xi(t))$. Our condition consists of the following:

(Ψ'): if $h(t_0) < 0$, $h(t) \leqslant 0$ when $t > t_0$.

To formulate condition ($\mathscr{B}$) we consider the operators

$$H_i = \sum_{j=1}^{n} \left| \partial_j a_i(x, \xi) \frac{\partial}{\partial \xi_j} - \partial_{\xi_j} a_i(x, \xi) \frac{\partial}{\partial x^j} \right|, \qquad i = 1, 2.$$

If $\alpha = (\alpha_1, \ldots, \alpha_r)$, $\beta = (\beta_1, \ldots, \beta_r)$, then by $H_1^\alpha H_2^\beta$ we denote the operator $H_1^{\alpha_1} H_2^{\beta_1} \ldots H_1^{\alpha_r} H_2^{\beta_r}$. By $k_0(x, \xi)$ we shall denote the least number k for which $H_2^k a_1(x, \xi) \neq 0$, and by $k_1(x, \xi)$ the least k for which $H_1^\alpha H_2^\beta a_1(x, \xi) \neq 0$, $|\alpha + \beta| = k$. If $a_1(x, \xi) + ia(x, \xi) = 0$, we assume $k_1(x, \xi) = k_0(x, \xi) = 0$. Thus, k_0 and k_1 have nonzero values only at the characteristic points.

The second condition of Theorem 2.1 takes the form

$$(\mathscr{B})\ \sup k_1(x, \xi) \leqslant k, \quad (x, \xi) \in T^*(\Omega), \quad x \in K, \quad |\xi| = 1.$$

1.2. A complete proof of Theorem 1.2 was published in Egorov[14,15] (see also Ref. 18). In our earlier works we had examined several particular cases, for example, in Ref. 3 we studied the case with $\delta \leqslant 2/3$; Ref. 8 is devoted to the study of operators whose symbol has a gradient which is not proportional to a real vector; other special classes are considered in Ref. 7.

H. Suzuki[1] has considered necessary conditions for subellipticity of a differential operator in two independent variables. His proofs are simple enough, but the results are not quite accurate.

A complete proof of Theorem 1.2 for differential operators with analytic coefficients is given in F. Trèves.[7] Note that for differential operators condition (Ψ') is replaced by a simpler condition ($\mathscr{P}$): function h does not change its sign.

Special-type subelliptic operators which are obtained upon uniform degeneracy of elliptic operators on some smooth submanifold are also discussed in G. I. Eskin.[3]

Finally, subelliptic operators are considered in L. Hörmander[18] who presents the results of Egorov[14,15] and recommends some improvement concerning selection of a special partition of unity in the proof for the sufficiency of conditions of Theorem 1.2. We have made use of this recommendation in proving Theorem 1.2.

1.3. In this chapter we shall incidentally prove the following

Theorem 1.3. If the conditions of Theorem 1.2 are fulfilled, then for every real number s and any compact subset $K \subset \Omega$ there exists a constant $C = C(s, K)$ such that

$$\|P^*u\|_s \leqslant C(\|Pu\|_s + \|u\|_{s+m-1}), \qquad u \in C_0^\infty(K).$$

Also note that from Theorem 1.2 and Theorem 4.1 of Chapter VI follows

Theorem 1.4. Let P be a pseudo-differential operator satisfying condition ($\mathscr{B}$). For the operator P to be hypoelliptic it is necessary and sufficient that condition (Ψ') be fulfilled.

When P is a differential operator, Theorem 1.2 takes the following form

Theorem 1.5. For the differential operator P to be subelliptic it is necessary and sufficient that the function $k_1(x, \xi)$ should assume at every point even values not exceeding some integer k. In this case, $\delta = k/(k + 1)$. If the function $k_1(x, \xi)$ takes an odd value in some point (x_0, ξ^0), then (1.1) cannot be fulfilled in any compact K containing x_0 for any $\delta \in \mathbb{R}$.

Note that for a subelliptic differential operator $k_0(x, \xi) = k_1(x, \xi)$.

1.4. The problem concerning conditions for which (1.1) is fulfilled for matrix operators P is a very difficult one and has not been solved even for the most simple case when $\delta = 1/2$. Here, we shall list the conditions sufficient for such an estimate to hold when $0 \leqslant \delta < 1$, which can be expressed in terms of the characteristic form of an operator. In so doing, by the norm $\|u\|_s$ of vector $u = (u_1, \ldots, u_N)$ we mean $(\|u_1\|_s^2 + \ldots + \|u_N\|_s^2)^{1/2}$.

Theorem 1.6. Operator P is subelliptic and (1.1) holds with $\delta = k/(k + 1)$, provided the conditions of Theorem 1.2 are fulfilled for a scalar vector $R_0(x, D)$, where R_0 is the principal part of the operator $R(x, D) = \det \| P_{ij}(x, D) \|$, $i, j = 1, \ldots, N$.

Proof. We denote by $Q_{ij}(x, \xi)$ the cofactor of the element $P_{ij}(x, \xi)$ in matrix $\|P_{ij}(x, \xi)\|$. Function Q_{ij} is positively homogeneous in ξ for $|\xi| \geqslant 1$, of degree $m(N-1)$. Operator R is of order mN. Let $Pu = f$, where $u = u(u_1, \ldots, u_N)$, $f = (f_1, \ldots, f_N)$. Then every component u_j satisfies the scalar equation

$$R(x, D) u_j = \sum_{l=1}^{N} Q_{lj}(x, D) f_l + \sum_{l=1}^{N} T_{lj}(x, D) u_l(x),$$

where the order of operators T_{ij} does not exceed $mN - 1$. By Theorem 1.2

$$\| u_j \|_s \leqslant C \left(\sum_{l=1}^{N} \| Q_{lj}(x, D) f_l \|_{s-mN+\delta} + \sum_{l=1}^{N} \| u_l \|_{s+\delta-1} \right).$$

Hence it follows that

$$\| u \|_s \leqslant C (\| f \|_{s-m+\delta} + \| u \|_{s+\delta-1}),$$

which proves the theorem.

1.5. We shall now prove some simple theorems on the properties of subelliptic operators.

Proposition 1.1. Every subelliptic operator is hypoelliptic.

This proposition follows from the following

Proposition 1.2. Let P be a subelliptic operator and $k_1(x, \xi) \leqslant k$ for all $(x, \xi) \in T^*\Omega$. Let ω be a subdomain in Ω. If $u \in \mathscr{D}'(\Omega)$, $Pu \in H_l(\omega)$, then $u \in H_{l+m-\delta}(\omega)$, where $\delta = k/(k+1)$.

Proof. Without loss of generality we may assume the closure of domain ω to be compact. Then there exists a real number t such that $u \in H_t(\omega)$. Let $\psi \in C_0^\infty(\omega)$ and let ω_h be a function $\in C_0^\infty(\mathbb{R}^n)$ such that $\omega_h \geqslant 0$, $\omega_h(x) = 0$ when $|x| \geqslant h$, $\int \omega_h(x)\, dx = 1$. Suppose that $u_h(x) = \int \omega_h(x-y) \psi(y) \times u(y)\, dy$. As seen in Chapter I, $u_h \in C_0^\infty(\mathbb{R}^n)$ and

$$\| u_h - \psi u \|_t \to 0 \quad \text{when} \quad h \to 0. \tag{1.2}$$

Let $t \leqslant l + m - 1$. We shall then prove that $\psi u \in H_{t+1-\delta}(\omega)$. If $t + 1 - \delta \leqslant l + m - 1$, then repetition of the same argument gives

$$u \in H_{t+2(1-\delta)}(\omega'),$$

where $\omega' \subset \omega$, and so on. Finally we find that $u \in H_{l+m-\delta}(\omega_0)$, where ω_0 is an arbitrary subdomain in ω.

Note that

$$Pu_h = P(\omega_h * \psi u) = \omega_h * P\psi u + T_h u,$$

where T_h is an operator of order $m - 1$ and $\|T_h u\|_s \leqslant C\|u\|_{s+m-1}$ for $u \in H_{s+m-1}(\omega)$ and the constant C is independent of h. Inserting $u = u_h$, $s = t + 1 - \delta$, and $K = \omega$ into (1.1) we obtain

$$\| u_h \|_{t+1-\delta} \leqslant C (\| Pu_h \|_{t-m+1} + \| u_h \|_{t-\delta}).$$

It follows that

$$\| u_h \|_{t+1-\delta} \leqslant C (\| \psi Pu \|_{t-m+1} + \| u \|_t + \| \psi u \|_t) \leqslant C_1 (\| Pu \|_{t-m+1} + \| u \|_t),$$

where the constant C_1 is independent of h. Thus, the norms $\|u_h\|_{t+1-\delta}$ are uniformly bounded in h. From this and (1.2) it follows that $\psi u \in H_{t+1-\delta}(\omega)$, which proves the proposition.

1.6. As the following statement implies, the equations containing subelliptic operators are solvable on a compact manifold without a boundary.

Proposition 1.3. Let Ω be a smooth closed manifold and P be a subelliptic operator on Ω. There exists a finite set of functions $\psi_1, \ldots, \psi_d \in C^\infty(\Omega)$ such that if $f \in H_s(\Omega)$, $\int f\psi_j\, d\Omega = 0$, $j = 1, \ldots, d$, then the equation $P^*u = f$ has a solution u in the class $H_{s+m-\delta}(\Omega)$.

Proof. Let N be the kernel of operator P, that is, the collection of functions $v \in \mathscr{D}'(\Omega)$ for which $Pv = 0$. By virtue of Proposition 1.2, these functions are infinitely differentiable and, since the operator P is subelliptic, it follows from (1.1) that

$$\| v \|_s \leqslant C \| v \|_{s-1}, \quad v \in N.$$

Hence it follows that a unit ball of the linear space $N \cap H_{s-1}(\Omega)$ is compact and, in view of Kolmogorov's theorem, finite-dimensional. Let $\bar{\psi}_1, \ldots, \bar{\psi}_d$ be the base of space N. We shall prove that if $\int u\psi_j\, d\Omega = 0$, $j = 1, \ldots, d$, $u \in C^\infty(\Omega)$, then

$$\| u \|_s \leqslant C \| Pu \|_{s-m+\delta}, \tag{1.3}$$

where the constant $C = C(s)$ is independent of u. Indeed, we assume this inequality to be false. We can then find a sequence of functions $u_\nu \in C^\infty(\Omega)$ for which

$$\int u_\nu \psi_j\, d\Omega = 0, \quad j = 1, \ldots, d; \quad \| u_\nu \|_s = 1, \quad \| Pu_\nu \|_{s-m+\delta} \to 0.$$

Since the sequence $\{u_\nu\}$ is bounded in $H_s(\Omega)$, it is weakly compact in this space. Let $u_\nu \to u$ weakly in $H_s(\Omega)$ and strongly in $H_{s-1}(\Omega)$. Then from (1.1) it follows that $\| u_\nu - u \|_s \to 0$ when $\nu \to \infty$. Therefore, $u \neq 0$ and $Pu = 0$, that is $u \in N$. But from the equalities $\int u_\nu \psi_j\, d\Omega = 0$ it follows that $\int u\psi_j\, d\Omega = 0$, that is, function u is orthogonal to N. This contradiction proves the validity of inequality (1.3).

Let now $f \in H_s(\Omega)$, $f \neq 0$ and $\int f\psi_j\, d\Omega = 0$ for $j = 1, \ldots, d$. If $u \in C^\infty(\Omega)$, then $u = u_1 + u_2$, where $u_1 \in N$ and $\int u_2\psi_j\, d\Omega = 0$ for $j = 1, \ldots, d$. We have

$$|(f, u)| = |(f, u_2)| \leqslant C \| f \|_s \| u_2 \|_{-s}.$$

But by (1.3), $\| u_2 \|_{-s} \leqslant C_1 \| Pu_2 \|_{-s-m+\delta} \leqslant C_1 \| Pu \|_{-s-m+\delta}$. Thus $|(f, u)| \leqslant \| Pu \|_{-s-m+\delta} \times \| f \|_s$ for all $u \in C^\infty(\Omega)$. This implies that (f, u) is a linear continuous functional of Pu and, by the Hahn-Banach theorem, may be extended to a continuous linear functional on $H_{-s-m+\delta}(\Omega)$, that is, there exists a function $v \in H_{s+m-\delta}(\Omega)$ such that $\| v \|_{s+m-\delta} \leqslant CC_1 \| f \|_s$ and $(f, u) = (v, Pu)$ for all $u \in C^\infty(\Omega)$. The last equality implies that $P^*v = f$.

Proposition 1.3 is proved.

1.7. The following assertion enables the proof of Theorem 1.2 to be reduced to a particular case.

Proposition 1.4. To prove Theorem 1.2 it is sufficient to examine the special case when $m = 1$, $s = 1 - \delta$.

Proof. Let P(x, D) be an operator of order m. Denote by P_1 the operator with symbol $p(x, \xi)(1 + |\xi|^2)^{(1-m)/2}$ and by A the operator with symbol $(1 + |\xi|^2)^{(1-\delta-s)/2} h(x)$, where $h \in C_0^\infty(K)$ and $h(x) = 1$ in $K' \subset K$. Suppose that (1.1) is fulfilled for the operator P when $\delta \in [0, 1)$. Since A is an elliptic operator in K, assuming $u = Av$, where $v \in C_0^\infty(K')$, we obtain

$$\| v \|_{1-\delta} \leqslant C (\| u \|_s + \| v \|_{-\delta}).$$

Using (1.1) we find that

$$\| v \|_{1-\delta} \leqslant C(K) (\| Pu \|_{s-m+\delta} + \| u \|_{s-1} + \| v \|_{-\delta}) \leqslant C_1 (\| PAv \|_{s-m+\delta} + \| v \|_{-\delta}) \leqslant$$

$$C_2 (\| APv \|_{s-m+\delta} + \| v \|_0) \leqslant C_3 (\| P_1 v \|_0 + \| v \|_0)$$

for all $v \in C_0^\infty(K')$.

Conversely, assume the estimate

$$\|v\|_{1-\delta} \leqslant C_1(K)(\|P_1 v\|_0 + \|v\|_0), \quad v \in C_0^\infty(K), \tag{1.4}$$

to be valid for operator P_1. Denote by B the operator with symbol $(1 + |\xi|^2)^{(s-1+\delta)/2} h(x)$, where h is the same function as above. Since B is an elliptic operator in K', assuming v = Bu, we obtain that

$$\|u\|_s \leqslant C(\|v\|_{1-\delta} + \|u\|_{s-1})$$

for $u \in C_0^\infty(K')$.

Using (1.4) we write

$$u\|_s \leqslant CC_1(\|P_1 v\|_0 + \|v\|_0) + C\|u\|_{s-1} \leqslant C_2(\|P_1 Bu\|_0 + \|u\|_{s-1+\delta}) \leqslant$$

$$C_3(\|BP_1 u\|_0 + \|u\|_{s-1+\delta}) \leqslant C_4(\|Pu\|_{s-m+\delta} + \|u\|_{s-1+\delta}),$$

and inequality (1.1) follows immediately. Thus, we have proved the equivalence of inequalities (1.1) and (1.4).

We proceed to show that the conditions of the theorem are also invariant under the passage from P to P_1. Note that $p^0(x_0, \xi^0) = 0$ if and only if $p_1^0(x_0, \xi^0) = |\xi^0|^{1-m} p^0(x_0, \xi^0) = 0$ so that the sets of characteristic points of operators P and P_1 agree. The null bicharacteristics of functions Im p_1^0 are defined as integral curves of the system of equations

$$\dot{x}^j = \frac{\partial}{\partial \xi_j} \operatorname{Im} p_1^0(x, \xi) = |\xi|^{1-m} \frac{\partial}{\partial \xi_j} \operatorname{Im} p^0(x, \xi);$$

$$\dot{\xi}_j = -|\xi|^{1-m} \frac{\partial}{\partial x^j} \operatorname{Im} p^0(x, \xi),$$

along which Im $p^0(x, \xi) = 0$. It is easily seen that the vector on the right-hand side of this system of equations is proportional (with a continuous coefficient) at every point (x, ξ) to the vector (grad_ξ Im $p^0(x, \xi)$, $-\operatorname{grad}_x$ Im $p^0(x, \xi)$), that is, the null bicharacteristics corresponding to functions Im p^0 and Im p_1^0 agree. On the other hand, Re $p^0(x, \xi) = |\xi|^{m-1}$ Re $p_1^0(x, \xi)$ and, therefore, condition (Ψ') is fulfilled for P if and only if it is fulfilled for P_1.

To prove the invariance of condition ($\mathscr{B}$) we verify that $k_1(x, \xi) = k_1'(x, \xi)$, where k_1' is a function defined for P_1 by the same rule as k_1 for P. It is sufficient to prove that if

$$p^0(x_0, \xi^0) = 0, \quad H_1^\alpha H_2^\beta a_1(x_0, \xi^0) = 0 \text{ when } |\alpha + \beta| < j,$$

then

$$p_1^0(x_0, \xi^0) = 0, \quad H_1'^\alpha H_2'^\beta a_1'(x_0, \xi^0) = 0 \text{ when } |\alpha + \beta| < j$$

and

$$H_1'^\alpha H_2'^\beta a_1'(x_0, \xi^0) = |\xi|^{(1-m)(j+1)} H_1^\alpha H_2^\beta a_1(x_0, \xi^0) \text{ when } |\alpha + \beta| = j.$$

Here $a_1'(x, \xi) = a_1(x, \xi)|\xi|^{1-m}$, $a_2'(x, \xi) = a_2(x, \xi)|\xi|^{1-m}$, H_1' and H_2' are differential operators along the bicharacteristics of functions a_1' and a_2', respectively.

Note that for $|\alpha + \beta| = l$, where $l \leqslant j$,

$$H_1'^\alpha H_2'^\beta a_1'(x_0, \xi^0) = H_1'^{\alpha_1} H_2'^{\beta_1} \ldots H_1'^{\alpha_r} H_2'^{\beta_r} a_1'(x_0, \xi^0) =$$

$$(|\xi|^{1-m} H_1)^{\alpha_1} (|\xi|^{1-m} H_2)^{\beta_1} \ldots (|\xi|^{1-m} H_1)^{\alpha_r} (|\xi|^{1-m} H_2)^{\beta_r} (|\xi|^{1-m} a_1(x_0, \xi^0)).$$

This expression is a sum of a large number of terms of the form $c_{\rho,\sigma} H_1^\rho H_2^\sigma a_1(x_0, \xi^0)$ for $|\rho + \sigma| \leqslant l$. However, if $|\rho + \sigma| < l$, that is, even if we once differentiate the factor $|\xi|^{1-m}$, then these terms are equal to 0 by hypothesis. Therefore, when $|\alpha + \beta| = j$,

$$H_1'^{\alpha} H_2'^{\beta} a_1' (x_0, \xi^0) = |\xi|^{(1-m)(j+1)} H_1^{\alpha} H_2^{\beta} a_1 (x_0, \xi^0).$$

This completes the proof of Proposition 1.4.

Proposition 1.5. To prove Theorem 1.3 it is sufficient to examine the case when $m = 1$, $s = 0$.

The following assertion is almost obvious.

Proposition 1.6. In proving Theorems 1.1-1.5 we assume that $p(x, \xi) = p^0(x, \xi)$ when $|\xi| \geqslant 1/2$.

Proof. Let P^0 be an operator with a smooth symbol equal to $p^0(x, \xi)$ when $|\xi| \geqslant 1/2$. Then $P - P^0$ is an operator of order $m - 1$ and $|\|Pu\|_{s-m+\delta} - \|P^0 u\|_{s-m+\delta}| \leqslant C\|u\|_{s-1+\delta}$. Since to every $\varepsilon > 0$ one can find a constant C_ε for which

$$\|u\|_{s+\delta-1} \leqslant \varepsilon\|u\|_s + C_\varepsilon \|u\|_{s-1}, \qquad u \in C_0^\infty(K),$$

the proposition is proved.

1.8. Also note that a subelliptic operator is an operator of principal type, which is evident from the following assertions.

Proposition 1.7. If P is a subelliptic operator, then vector $\operatorname{grad}_{x,\xi} p^0(x, \xi)$ does not vanish when $|\xi| \geqslant 1/2$.

Proof. In accordance with Propositions 1.4 and 1.6 we assume that $m = 1$ and $P = P^0$. For $0 \leqslant \delta < 1$, from (1.1) it follows, by Lemma 2.3 of of Chapter VI, that

$$\|\psi\|_0 \leqslant C\left\{\lambda^\delta \left\| \sum_{|\alpha+\beta| \leqslant 1} p_{(\beta)}^{0(\alpha)}(x, \xi)\, y^\beta D^\alpha \psi \lambda^{-|\alpha+\beta|/2} \right\|_0 + \lambda^{\delta-1} \sum_{|\alpha+\beta| \leqslant 2} \|y^\beta D^\alpha \psi\|_0 \right\} \tag{1.5}$$

for all $\psi \in C_0^\infty(\mathbb{R}^n)$ and $\lambda \geqslant 1$. Here, as usual $\|\cdot\|_0$ is the norm in $L_2(\mathbb{R}^n)$. If the vector $\operatorname{grad}_{x,\xi} p^0(x_0, \xi^0) = 0$ and $|\xi^0| \geqslant 1/2$, then $p^0(x_0, \xi^0) = \sum_{j=1}^n \xi_j^0 \partial_{\xi_j} p^0(x_0, \xi^0) = 0$ and inequality (1.5) takes the form

$$\|\psi\|_0 \leqslant C\lambda^{\delta-1} \sum_{|\alpha+\beta| \leqslant 2} \|y^\beta D^\alpha \psi\|_0,$$

where C is a constant. Since $\delta < 1$, it is clear that this inequality cannot be fulfilled for all $\lambda > 1$.

Proposition 1.8. If operator P satisfies condition $(\mathscr{B})$, then the vector $\operatorname{grad}_{x,\xi} p^0(x_0, \xi^0)$ is not collinear to the vector $(\xi^0, 0)$ when $|\xi| \geqslant 1/2$.

Proof. Suppose that $\operatorname{grad}_\xi p^0(x_0, \xi^0) = 0$ and $\operatorname{grad}_x p^0(x_0, \xi^0) = \varkappa \xi^0$ for $|\xi^0| < 1/2$. Then $p^0(x_0, \xi^0) = \sum_{j=1}^n \xi_j^0 \partial_{\xi_j} p^0(x_0, \xi^0)$ and

$$H_{p^0} = \sum_{j=1}^n \left(\partial_j p^0(x_0, \xi^0)\, \partial_{\xi_j} - \partial_{\xi_j} p^0(x_0, \xi^0)\, \partial_j\right) = \varkappa \sum_{j=1}^n \xi_j^0 \partial_{\xi_j}.$$

Therefore, if the function $r(x, \xi)$ is homogeneous of degree m for $|\xi| \geqslant 1/2$, then

$$H_{p^0} r(x_0, \xi^0) = \varkappa m r(x_0, \xi^0), \quad H_{\overline{p^0}} r(x_0, \xi^0) = \bar{\varkappa} m r(x_0, \xi^0).$$

It is clear that $k_1(x_0, \xi^0)$ can be defined as the least value of k_0 for which

$$H_{p^0}^\alpha H_{\overline{p^0}}^\beta p^0(x_0, \xi^0) \neq 0 \text{ when } |\alpha+\beta| = k.$$

But if $C_I(x_0, \xi^0) = 0$, then $H_{p^0}C_I(x_0, \xi^0) = 0$ so that $H_{\bar{p}^0}C_I(x_0, \xi^0) = 0$. Proposition 1.8 is proved.

§2. Localization of Estimates

2.1. In proving Theorems 1.2 and 1.3 we shall find it convenient to make use of local estimates similar to estimates (1.5).

Theorem 2.1. If the first-order operator P is subelliptic and $0 < \delta < 1$, then to every integer $k > 0$ and every compact set $K \subset \Omega$ one may find a constant $C = C(K)$ such that for all $x \in K$, $\xi \in \mathbb{R}^n$ ($|\xi| = 1$), $\lambda \geqslant 1$ and all $\psi \in C_0^\infty(\mathbb{R}^n)$ the following inequality is fulfilled:

$$\|\psi(y)\| \leqslant C\Big\{\lambda^\delta \| T_k^{(x,\xi)} p^0(x + y\lambda^{\delta-1}, \xi + D\lambda^{-\delta})\psi(y)\|_0 + \lambda^{\delta-(k+1)(1-\delta)} \sum_{|\alpha+\beta|\leqslant k+1} \|y^\beta D^\alpha \psi\|_0 \lambda^{|\alpha|(1-2\delta)}\Big\}. \tag{2.1}$$

Here we have used the notation

$$T_k^{(x,\xi)} f(y, \eta) = \sum_{|\alpha+\beta|\leqslant k} \frac{1}{\alpha!\,\beta!} f_{(\beta)}^{(\alpha)}(x, \xi)(y - x)^\beta (\eta - \xi)^\alpha,$$

where $f_{(\beta)}^{(\alpha)}(x, \xi) = (iD_\xi)^\alpha (iD_x)^\beta f(x, \xi)$ for $f \in C^\infty(T^*\Omega)$.

Theorem 2.2. If to every compact set $K \subset \Omega$ there exist constants $C = C(K)$, $N - N(K)$, and ε_0 such that for all $x \in K$, $|\xi| \geqslant 1/2$, $\lambda \geqslant 1$ for all $\psi(y) \in C_0^\infty(\mathbb{R}^n)$ the inequality

$$\|\psi\|_0 + \|\bar{P}_\lambda \psi\|_0 \leqslant C\Big\{\|P_\lambda \psi\|_0 + \lambda^{-\varepsilon_0} \sum_{\alpha+\beta|\leqslant N} \lambda^{-|\alpha|(k-1)} (\|y^\beta D^\alpha \psi\|_0 \lambda^{-|\alpha|(k-1)} + \|y^\beta D^\alpha P_\lambda \psi\|_0 + \|y^\beta D^\alpha \bar{P}_\lambda \psi\|_0) \tag{2.2}$$

is valid, where $P_\lambda(y, D)\psi = \lambda^k T_k^{(x,\xi)} p^0(x + y\lambda^{-1}, \xi + D\lambda^{-k})\psi$ and $p^0(x, t\xi) = tp^0(x, \xi)$ for $t \geqslant 1$, $|\xi| \geqslant 1/2$, and $p^0 \in C^\infty(T^*\Omega)$, then the estimate

$$\|P^* u\|_0 + \|u\|_{1/(k+1)} \leqslant C_K(\|Pu\|_0 + \|u\|_0), \quad u \in C_0^\infty(K), \tag{2.3}$$

is fulfilled for operator P with principal symbol p^0.

2.2. Proof of Theorem 2.1. By Lemma 2.3 of Chapter VI, the following estimate follows from (1.1):

$$C\Big(\lambda^\delta \Big\|\sum_{j=0}^{k/2} T_k^{(x,\xi)} p^j(x + y\lambda^{-1/2}, \xi + D\lambda^{-1/2})\psi(y)\Big\|_0 + \lambda^{\delta-(k+1)/2} \sum_{|\alpha+\beta|\leqslant k+1} \|y^\beta D^\alpha \psi\|_0\Big), \quad \psi \in C_0^\infty(\mathbb{R}^n).$$

It is easily seen that (2.1) follows from this estimate after we have replaced $y = z\lambda^{\delta-1/2}$.

2.3. Proof of Theorem 2.2. Let $u \in C_0^\infty(K)$. Without loss of generality, we may assume that $p(x, \xi) = 0$ when $x \in \mathbb{R}^n \setminus K'$, where K' is some compact set containing K.

Let $\theta \in C_0^\infty(\mathbb{R}^n)$, where $\theta(x) = 1$ when $|x| \leqslant 1$ and $\theta(x) = 0$ when $|x| \geqslant 2$. Let $g_0, g_1 \ldots$ be points in $\mathbb{R}^n$ with integral coordinates so that $g_0 = 0$ and $\varphi_i(x) = \theta(x - g_i)[\sum \theta(x - g_j)^2]^{-1/2}$. Then $\varphi_i \in C_0^\infty(\mathbb{R}^n)$ and $\sum \varphi_i^2(x) = 1$. Suppose that

$$\psi_i(\xi) = \varphi_i(\varepsilon\xi|\xi|^{-k/(k+1)}),$$

where ε is some small positive number which will be chosen later. We have $\|Pu\|^2 = \sum \|\psi_j(D)Pu\|^2$. Note that $\sum_j |\psi_j(\xi) - \psi_j(\eta)|^2 \leqslant C\varepsilon^2|\xi - \eta|^2|\xi|^{-2k/(k+1)} + 16|\xi -$

$\eta|^2|\xi|^{-2}$. The first term on the right-hand side is obtained in estimating the left-hand side in the domain where $2|\xi - \eta| < |\xi|$ and the second, in estimating in the complement to this domain. Let $\{v_j\}$ be an arbitrary sequence of functions $C_0^\infty(\mathbb{R}^n)$ for which the series $\sum\|v_j\|^2$ converges. We have

$$\sum_j \left|\int[\Psi_j(D)Pu - P\Psi_j(D)u]v_j(y)\,dy\right| = \sum_j \left|\iint \tilde{p}(\eta-\xi,\ \xi)[\psi_j(\xi)-\psi_j(\eta)]\tilde{u}(\xi)\tilde{v}_j(-\eta)\,d\xi\,d\eta\right| \leqslant$$

$$C_l\iint(1+|\eta-\xi|)^{-l}|\xi||\tilde{u}(\xi)|\sqrt{\sum|\psi_j(\xi)-\psi_j(\eta)|^2}\times$$

$$\sqrt{\sum|\tilde{v}_j(-\eta)|^2}\,d\xi\,d\eta \leqslant C_l\iint(1+|\eta-\xi|)^{-l}|\xi|^{1/(k+1)}\times$$

$$|\tilde{u}(\xi)|(C\varepsilon^2+16|\xi|^{-2/(k+1)})^{1/2}|\xi-\eta|\sqrt{\sum_j|\tilde{v}_j(-\eta)|^2}\,d\xi\,d\eta,$$

where $l \geqslant 0$ is an arbitrary integer. Hence for $l = n + 2$ it follows that

$$\sum_j\left|\int[\Psi_j(D)Pu - P\Psi_j(D)u]v_j(y)\,dy\right| \leqslant C_0[\varepsilon\|u\|_{1/(k+1)} + \|u\|_0]\|v\|_0,$$

where $\|v\|_0 = \sqrt{\sum_j\|v_j\|_0^2}$. Therefore

$$\sum_j\|\Psi_j(D)Pu - P\Psi_j(D)u\|^2 \leqslant 2C_0[\varepsilon^2\|u\|^2_{1/(k+1)} + \|u\|_0^2].$$

We have

$$P^0(x,\ D)\Psi_j(D)u = (2\pi)^{-n}\int p^0(x,\ \xi)\psi_j(\xi)\tilde{u}(\xi)e^{ix\xi}\,d\xi.$$

Fixing an arbitrary point $\xi^j \in \operatorname{supp}\psi_j$ we expand the function $p^0(x,\ \xi)$ into a Taylor series in the variable ξ:

$$p^0(x,\ \xi) = \sum_{|\alpha|\leqslant k}\frac{1}{\alpha!}p^{0(\alpha)}(x,\ \xi^j)(\xi-\xi^j)^\alpha + r_j(x,\ \xi).$$

Since $|\xi - \xi^j| \leqslant C_1\varepsilon^{-1}|\xi|^{k/(k+1)}$ when $j \geqslant j_0$ for $\xi \in \operatorname{supp}\psi_j$, we see that $|r_j(x,\ \xi)| \leqslant C_2$, where C_2 is independent of ξ and j, but depends on ε. Repeating the reasoning above with v_j we obtain

$$\sum_j\|r_j(x,\ D)\Psi_j(D)u\|^2 \leqslant C_3\|u\|^2.$$

Thus, we have

$$\left|\|Pu\|^2 - \sum_j\left\|\sum_{|\alpha|\leqslant k}\frac{1}{\alpha!}p^{0(\alpha)}(x,\ \xi^j)(D-\xi^j)^\alpha\psi_j(D)u\right\|^2\right| \leqslant C\varepsilon^2\|u\|^2_{1/(k+1)} + C_4\|u\|^2,$$

where the constant C is independent of ε. Let us now examine the expression

$$\sum_{i,j}\left\|\sum_{|\alpha|\leqslant k}\frac{1}{\alpha!}p^{0(\alpha)}(x,\ \xi^j)[\varphi_i(\varepsilon x|\xi^j|^{1/(k+1)})(D-\xi^j)^\alpha -\right.$$

$$\left.(D-\xi^j)^\alpha\varphi_i(\varepsilon x|\xi^j|^{1/(k+1)})]\Psi_j(D)u\right\|^2.$$

It is easy to verify that it does not exceed

$$C_5\sum_j\sum_{|\alpha|=1}^{k}\sum_{\nu\leqslant\alpha}|\xi^j|^{2(1-|\alpha|)}\varepsilon^{2|\nu|}|\xi^j|^{2|\nu|/(k+1)}\|(D-\xi^j)^{\alpha-\nu}\Psi_j(D)u\|^2 \leqslant$$

$$C_5\sum_j\sum_{|\alpha|=1}^{k}\sum_{|\nu|\geqslant 1}|\xi^j|^{2\left(1-\frac{|\alpha|}{k+1}-|\nu|\frac{k-1}{k+1}\right)}\|\Psi_j(D)u\|^2\varepsilon^{4|\nu|-2|\alpha|} \leqslant C_6\varepsilon^2\|u\|^2_{1/(k+1)} + C_6\|u\|^2$$

and therefore

$$\left| \| Pu \|^2 - \sum_{l,j} \left\| \sum_{|\alpha| \leqslant k} \frac{1}{\alpha!} p^{0(\alpha)}(x, \xi^j)(D - \xi^j)^\alpha \varphi_l(\varepsilon x |\xi^j|^{1/(k+1)}) \Psi_j(D) u \right\|^2 \right| \leqslant$$

$$C_7 \varepsilon^2 \| u \|^2_{1/(k+1)} + C_\varepsilon \| u \|^2_0.$$

Let $z_{lj}(x) = e^{-i(x,\xi^j)} \varphi_l(\varepsilon x |\xi^j|^{1/(k+1)}) \Psi_j(D) u(x)$. Expand $p^{0(\alpha)}(x, \xi^j)$ in a Taylor series in x at any point $\in$ supp z_{lj}:

$$p^{0(\alpha)}(x, \xi^j) = \sum_{|\alpha+\beta| \leqslant k} p^{0(\alpha)}_{(\beta)}(x^{lj}, \xi^j) \frac{1}{\beta!}(x - x^{lj})^\beta + r_{\alpha lj}(x).$$

It is easily seen that when $j \geqslant j_0$

$$|r_{\alpha lj}(x)| \leqslant C_8 |\xi^j|^{1 - |\alpha| - \frac{k+1-|\alpha|}{k+1}} \varepsilon^{|\alpha| - k - 1}.$$

Therefore

$$\sum_l \sum_{j \geqslant j_0} \left\| \sum_{|\alpha| \leqslant k} r_{\alpha l_j}(x) \frac{1}{\alpha!} D^\alpha z_{lj}(x) \right\|^2_0 \leqslant C_9 \varepsilon^{-2(k+1)} \| u \|^2_0$$

and hence

$$\left| \| Pu \|^2_0 - \sum_{l,j} \left\| \sum_{|\alpha+\beta| \leqslant k} p^{0(\alpha)}_{(\beta)}(x^{lj}, \xi^j) \frac{1}{\alpha!\,\beta!}(x - x^{lj})^\beta D^\alpha z_{lj}(x) \right\|^2_0 \right| \leqslant$$

$$C_7 \varepsilon^2 \| u \|^2_{1/(k+1)} + C'_\varepsilon \| u \|^2_0. \tag{2.4}$$

A similar estimate can of course be written for $\|P^*u\|^2_0$ if we replace $p^{0(\alpha)}_{(\beta)}(x, \xi)$ by $\overline{p^{0(\alpha)}_{(\beta)}(x, \xi)}$. On the other hand,

$$\sum_{l,j} \| z_{lj}(x) \|^2_0 |\xi^j|^{2/(k+1)} = \sum_j |\xi^j|^{2/(k+1)} \| \Psi_j(D) u \|^2_0 \geqslant \frac{1}{2} \| u \|^2_{1/(k+1)} - C_{10} \| u \|^2_0.$$

Using estimate (2.2) with $\lambda = |\xi^j|^{1/(k+1)}$ and $y = (x - x^{lj}) |\xi^j|^{1/(k+1)}$, $\xi = \xi^j / |\xi^j|$ we obtain

$$\frac{1}{2} \| u \|^2_{1/(k+1)} + \| P^* u \|^2_0 \leqslant C \{ \| Pu \|^2_0 + C_{11} \varepsilon^2 \| u \|^2_{1/(k+1)} + C_\varepsilon \| u \|^2_0 +$$

$$\sum_{l,j} |\xi^j|^{-\varepsilon_0/(k+1)} \sum_{|\alpha+\beta| \leqslant k+1} |\xi^j|^{[2 - 2|\alpha|(k-1)]/(k+1)} \times$$

$$[\| y^\beta D^\alpha z_{lj} \|^2_0 + \| y^\beta D^\alpha P_\lambda z_{lj} \|^2_0 + \| y^\beta D^\alpha \bar{P}_\lambda z_{lj} \|^2_0]\}.$$

It is not hard to see that

$$\sum_l \sum_{j \geqslant j_0} \sum_{|\alpha+\beta| \leqslant k+1} |\xi^j|^{\frac{2 - 2|\alpha|(k-1)}{k+1}} \int |y^\beta D^\alpha z_{lj}|^2 dx \leqslant$$

$$C_{12} \sum_{j \geqslant j_0} \sum_{|\alpha+\beta| \leqslant k+1} |\xi^j|^{\frac{2 - 2|\alpha|k}{k+1}} \int \left| D^\alpha_x \left[e^{-i(x,\xi^j)} \Psi_j(D) u(x) \right] \right|^2 dx =$$

$$C_{12} \sum_{j \geqslant j_0} \sum_{|\alpha+\beta| \leqslant k+1} |\xi^j|^{\frac{2 - 2|\alpha|k}{k+1}} \int |\zeta - \xi^j|^{2|\alpha|} |\psi_j(\zeta) \tilde{u}(\zeta)|^2 d\zeta \leqslant$$

$$C_{13} \sum_{j \geqslant j_0} \int |\zeta|^{2/(k+1)} |\psi_j(\zeta) \tilde{u}(\zeta)|^2 d\zeta \leqslant C_{14} \| u \|^2_{1/(k+1)}.$$

Finally,

$$\sum_l \sum_{j \leqslant j_0} \sum_{|\alpha+\beta| \leqslant k+1} |\xi^j|^{\frac{2 - 2|\alpha|(k-1)}{k+1}} \int |y^\beta D^\alpha z_{lj}|^2 dx \leqslant C_{15} \| u \|^2_0.$$

Similarly,

$$\sum_{l,\,j}\ \sum_{|\alpha+\beta|\leqslant k+1} |\xi^j|^{\frac{2-2|\alpha|(k-1)}{k+1}} (\|y^\beta D^\alpha P_\lambda z_{lj}\|_0^2 + \|y^\beta D^\alpha \bar{P}_\lambda z_{lj}\|_0^2) \leqslant$$

$$C_{16}(\|Pu\|_0^2 + \|P^*u\|_0^2 + 2C_7\varepsilon^2\|u\|_{1/(k+1)}^2 + 2C'_\varepsilon\|u\|_0^2),$$

where the last inequality follows from (2.4). Finally we obtain that, if $|\xi^j| \geqslant \lambda_0$ when $j \geqslant j_0$, then

$$\frac{1}{2}\|u\|_{1/(k+1)}^2 + \|P^*u\|_0^2 \leqslant C[\|Pu\|_0^2 + C_7\varepsilon^2\|u\|_{1/(k+1)}^2 +$$

$$C_\varepsilon\|u\|_0^2 + \lambda^{-2\varepsilon_0}(C_{14}\|u\|_{1/(k+1)}^2 + C_{15}\|u\|_0^2 +$$

$$C_{16}(\|Pu\|_0^2 + \|P^*u\|_0^2 + 2C_7\varepsilon^2\|u\|_{1/(k+1)}^2 + 2C'_\varepsilon\|u\|_0^2))].$$

If ε and λ_0 are such that

$$CC_7\varepsilon^2 + \lambda^{-2\varepsilon_0}C_{14} + 2\lambda^{-2\varepsilon_0}C_7C_{16}\varepsilon^2 < 1/4, \quad C_{16}\lambda^{-2\varepsilon_0} < 1/2,$$

then the inequality

$$\frac{1}{4}\|u\|_{1/(k+1)}^2 + \frac{1}{2}\|P^*u\|_0^2 \leqslant \frac{3}{2}C\|Pu\|_0^2 + C_{17}\|u\|_0^2,$$

equivalent to (2.3), follows. Theorem 2.2 is proved.

§3. Necessary Conditions for Subellipticity

3.1. In this section we shall prove the necessity of the conditions of Theorem 1.2.

Theorem 3.1. Suppose that the condition

$$l \leqslant k_1(x_0,\ \xi^0) \leqslant \infty$$

and $l < \infty$ is fulfilled for the first order operator P. Then the estimate

$$\|u\|_{1-\delta} \leqslant C(\|Pu\|_0 + \|u\|_0), \qquad u \in C_0^\infty(K) \tag{3.1}$$

cannot hold for $\delta < l/(l+1)$ if the compact set K contains the point x^0.

Proof. Suppose that (3.1) is valid. Note that the assertion of the lemma is trivial when $l = 0$. Therefore, we assume $l \geqslant 1$.

By Proposition 1.8, the zero vector $\mathrm{grad}_{x,\xi}p^0(x_0,\ \xi^0)$ is not proportional to the vector $(\xi^0,\ 0)$. By Theorem 2.1 of Chapter VI, if $C_1(x_1,\ \xi^1)$ and $p^0(x_1,\ \xi^1) = 0$, then (3.1) cannot be fulfilled for any $\delta \in \mathbb{R}$ in any neighborhood of x_1. Therefore, $C_1(x_1,\ \xi^1) \leqslant 0$ and $C_1(x,\ \xi) \leqslant 0$ at all of the characteristic points of a neighborhood of $(x_0,\ \xi^0)$.

According to Theorem 2.1, the inequality

$$\|\psi\|_0 \leqslant c\Big\{\lambda^\delta\|T_{l-1}^{(x,\ \xi)}p^0(x_0 + y\lambda^{\delta-1},\ \xi^0 + D\lambda^{-\delta})\psi(y)\|_0 +$$

$$\lambda^{\delta-l(1-\delta)}\sum_{|\alpha+\beta|\leqslant l}\|y^\beta D^\alpha\psi\|_0\lambda^{|\alpha|(1-2\delta)}\Big\} \tag{3.2}$$

follows from (3.1) for all $\psi \in C_0^\infty(\mathbb{R}^n)$ and $\lambda \geqslant 1$.

We shall now show that the latter is not possible. Note immediately that $\delta - l(1-\delta) < 0$ since $\delta < l/(l+1)$. Without loss of generality we assume that $(l-1)/l \leqslant \delta$. Since operator P can be replaced by operator iP we may without loss of generality assume that $\mathrm{Im}\ \mathrm{grad}\ p^0(x_0,\ \xi^0) \neq 0$. By Theorem 4.2 of Chapter II, estimate (3.2) is invariant under ordinary canonical transformations. Therefore, we assume that $\mathrm{Im}\ p^0(x,\ \xi) = \xi_1$ in some neighborhood of the considered point so that $\xi_1^0 = 0$.

If $l = 1$, then $\delta < 1/2$ and inequality (3.2) takes the form

$$\|\psi\|_0 \leqslant C\left\{\lambda^{\delta}\|p^0(x_0, \xi^0)\psi(y)\|_0 + \sum_{|\alpha+\beta|\leqslant 1}\|y^{\beta}D^{\alpha}\psi\|_0\lambda^{(-1+|\alpha|)(1-2\delta)}\right\}.$$

But $p^0(x_0, \xi^0) = 0$ and $2\delta - 1 < 0$. And this inequality is not fulfilled as $\lambda \to \infty$ even for one fixed function $\psi \in C_0^{\infty}(\mathbb{R}^n)$, provided $\psi \neq 0$.

Therefore, we assume that $l \geqslant 2$. In accord with the analysis made in Section 4, Chapter VI, two cases will be distinguished in the further discussion.

A. Let function $a_1(x, \xi) = \operatorname{Re} p^0(x, \xi)$ be such that

$$\partial_1^j \operatorname{grad}_{x', \xi'} a_1(x_0, \xi^0) = 0 \text{ when } 0 \leqslant j < l/2. \tag{3.3}$$

Here $\operatorname{grad} a_1(x_0, \xi^0) = \varkappa \operatorname{grad} \xi_1$. After we have replaced $y^1 = z^1$, $y' = z'\lambda^{1/2-\delta}$ the inequality (3.2) will change to

$$\|\psi\|_0 \leqslant C\left\{\left\|\partial_1\psi(z) + \sum_{|\alpha+\beta|\leqslant l-1}\frac{1}{\alpha!\,\beta!}a_{1(\beta)}^{(\alpha)}(x_0, \xi^0)z^{\beta}D^{\alpha}\psi(z)\times\right.\right.$$
$$\left.\lambda^{\delta-\delta\alpha_1-(1-\delta)\beta_1-|\alpha'+\beta'|/2}\right\|_0 + \lambda^{\delta-l(1-\delta)}\times$$
$$\left.\sum_{|\alpha+\beta|\leqslant l}\|z^{\beta}D^{\alpha}\psi\|_0\lambda^{\alpha_1(1-2\delta)+\left(\frac{1}{2}-\delta\right)|\alpha'+\beta'|}\right\} \tag{3.4}$$

for all $\psi \in C_0^{\infty}(\mathbb{R}^n)$. The exponent

$$\delta - \delta\alpha_1 - (1-\delta)\beta_1 - |\alpha'+\beta'|/2$$

will be negative in the following cases:

1) $|\alpha' + \beta'| \geqslant 2$;

2) $\alpha_1 \geqslant 2$;

3) $\alpha_1 = 1$, $|\alpha'| + |\beta| \neq 0$;

4) $\alpha_1 = 0$, $|\alpha' + \beta'| = 1$, $\delta - 1/2 - (1-\delta)\beta_1 < 0$.

For $2 \leqslant |\alpha + \beta| \leqslant l - 1$, there are still combinations (α, β) for which $\alpha_1 = 0$, $|\alpha' + \beta'| = 1$, $\beta_1(1-\delta) \leqslant \delta - 1/2$ or $\alpha = 0$, $\beta' = 0$, $\beta_1 < l$.

But $(\delta - 1/2)(1-\delta)^{-1} < (l-1)/2$ since $\delta < l/(l+1)$. By (3.3) the derivatives $a_{1(\beta)}^{(\alpha)}(x_0, \xi^0)$ are equal to 0 when $\alpha_1 = 0$, $|\alpha' + \beta'| = 1$ and $\beta_1(1 - \delta) \leqslant \delta - 1/2$.

On the other hand, the condition $k_1(x_0, \xi^0) \geqslant l$ implies, in particular, that $D_1^j a_1(x_0, \xi^0) = 0$ when $j < l$. Hence, the inequality (3.4) takes the form

$$\|\psi\|_0 \leqslant C|1 - i\varkappa|\,\|\partial_1\psi\|_0, \quad \psi \in C_0^{\infty}(\mathbb{R}^n).$$

However, the last inequality cannot be fulfilled for all $\psi \in C_0^{\infty}(\mathbb{R}^n)$ with the same constant C. If, for instance, $\psi_m(z) = \psi_0(z/m)$, where $\psi_0 \neq 0$, then

$$\|\psi_m\|_0 \geqslant c_0 m^{n/2}, \quad \|\partial_1\psi\|_0 \leqslant c_1 m^{n/2-1},$$

where $c_0 > 0$, and when $m \to \infty$ we arrive at a contradiction.

B. If condition (3.3) is not fulfilled, that is,

$$\partial_1^s \partial_{\xi_2} a_1(x_0, \xi^0) \neq 0$$

for some s, $0 \leqslant s < l/2$, then by applying a canonical transformation we achieve

$$a_{1(\beta)}^{(\alpha)}(x_0, \xi^0) = 0 \text{ for } \alpha_1 = 0, \ \beta_1 = s, \ |\alpha + \beta| \leqslant l - 1,$$

with the exception of the derivative $\partial_1^s \partial \xi_2 a_1(x_0, \xi^0)$. Besides, we assume that

$$\partial_1^j \partial_{\xi_2} a_1(x_0, \xi^0) = 0 \quad \text{for} \quad j < s. \tag{3.5}$$

By Lemma 4.2 of Chapter VI we have

$$\partial_1^a \partial_2^b \partial_{\xi_j} a_1(x_0, \xi^0) = \partial_1^a \partial_2^b \partial_m a_1(x_0, \xi^0) = 0$$

when $a = b(s + 1) < 1/2$; $j = 3, \ldots, n$; $m = 1, \ldots, n$;

$$\partial_1^a \partial_2^b a_1(x_0, \xi^0) = 0 \quad \text{for} \quad a + b(s+1) < l. \tag{3.6}$$

Let us replace the variables under the integral sign in (3.2)

$$z^1 = y^1, \quad z^2 = y^2 \lambda^{s(1-\delta)}, \quad z'' = y'' \lambda^{\delta - 1/2},$$

where $y'' = (y^3, \ldots, y^n)$. Now inequality (3.2) changes to

$$\|\psi\|_0 \leqslant C \left\{ \left\| \partial_1 \psi(z) + \sum_{|\alpha+\beta| \leqslant l-1} \frac{1}{\alpha! \beta!} a_{1(\beta)}^{(\alpha)}(x_0, \xi^0) z^\beta D^\alpha \psi(z) \lambda^{N_{\alpha,\beta}} \right\|_0 + \right.$$

$$\left. \lambda^{\delta - l(1-\delta)} \sum_{|\alpha+\beta| \leqslant l} \| z^\beta D^\alpha \psi(z) \|_0 \lambda^{M_{\alpha,\beta}} \right\}, \tag{3.7}$$

where

$$N_{\alpha,\beta} = \delta - \delta\alpha_1 - (1-\delta)\beta_1 - \delta\alpha_2 + \alpha_2 s(1-\delta) - \beta_2(1-\delta)(s+1) - |\alpha'' + \beta''|/2,$$

$$M_{\alpha,\beta} = (\alpha_1 + \alpha_2)(1 - 2\delta) + (\alpha_2 - \beta_2)s(1-\delta) + \left(\frac{1}{2} - \delta\right)|\alpha'' + \beta''|.$$

Note that the exponent λ in the last sum on the right-hand side of (3.7) will always be negative, since

$$\delta - l(1-\delta) < 0, \quad 2\delta > 1,$$

and, as we shall now show (the coefficient of α_2),

$$1 - 2\delta + s(1-\delta) < 0.$$

Indeed, if $l = 2$, then $s = 0$, which is obvious. If $l \geqslant 3$, then $(l + 1)/(l + 3) \leqslant (l - 1)/l$. But $\delta \geqslant (l - 1)/l$ so that $\delta \geqslant (l + 1)/(l + 3)$. On the other hand, $s \leqslant (l - 1)/2$ and from the inequality $1 - 2\delta + s(1 - \delta) > 0$ it follows that $\delta < (l + 1)/(l + 3)$. The contradiction implies that the last sum on the right-hand side of (3.7) tends to 0 when $\lambda \to \infty$. We shall now prove that this is true also for the remaining terms under the norm sign on the right-hand side of this inequality, except $\partial_1\psi(z)$ and $(z^1)^s \partial_2 \psi(z)$.

Before we prove that $N_{\alpha,\beta} < 0$, note that the coefficient of α_2 in this expression for $N_{\alpha,\beta}$ equals $s(1 - \delta) - \delta$ and this value $< 1/2 - \delta \leqslant 0$ so that $s < 1/2$ and $l(1 - \delta) \leqslant 1$.

Now it is not difficult to verify that $N_{\alpha,\beta} < 0$ in the following cases:

1) $|\alpha'' + \beta''| \geqslant 2$;

2) $\alpha_1 \geqslant 1$, $|\alpha'| + |\beta| \neq 0$;

3) $\alpha_1 = 0$, $|\alpha'' + \beta''| = 1$, $\alpha_2 \geqslant 1$;

4) $\alpha_1 = 0$, $|\alpha'' + \beta''| = 1$, $\alpha_2 = 0$, $[\beta_1 + \beta_2(s + 1)](1 - \delta) > \delta - 1/2$;

5) $\alpha_1 = 0$, $\alpha'' = 0$, $\beta'' = 0$, $\alpha_2 \geqslant 2$;

6) $\alpha_1 = 0$, $\alpha'' = 0$, $\beta'' = 0$, $\alpha_2 = 1$, $\beta_2 \geqslant 1$;

7) $\alpha_1 = 0$, $\alpha'' = 0$, $\beta'' = 0$, $\alpha_2 = 1$, $\beta_2 = 0$, $\beta_1 > s$;

8) $\alpha_1 = 0$, $\alpha'' = 0$, $\beta'' = 0$, $\alpha_2 = 0$, $\beta_1 + \beta_2(s + 1) > \delta/(1 - \delta)$.

For $2 \leqslant |\alpha + \beta| \leqslant l - 1$, there are still combinations (α, β) for which

1) $\alpha_1 = 0$, $\alpha_2 = 0$, $|\alpha'' + \beta''| = 1$, $[\beta_1 + \beta_2(s + 1)](1 - \delta) \leqslant \delta - 1/2$;

2) $\alpha_1 = 0$, $\alpha_2 = 1$, $\alpha'' = 0$, $\beta' = 0$, $\beta_1 \leqslant s$;

3) $\alpha = 0$, $\beta'' = 0$, $[\beta_1 + \beta_2(s + 1)](1 - \delta \leqslant \delta$.

However $\delta/(1 - \delta) \leqslant l$ and $(\delta - 1/2)/(1 - \delta) \leqslant (l - 1)/2$. By (3.5) and (3.6) the derivatives $a_{1(\beta)}^{(\alpha)}(x_0, \xi^0)$ are equal to 0, except $\partial_1^s \partial_{\xi_2} a_1(x_0, \xi^0)$ for which $N_{\alpha,\beta} = 0$. Thus, inequality (3.7) changes to

$$\|\psi(z)\|_0 \leqslant C\|\partial_1\psi(z) + A(z^1)^s\partial_2\psi(z)\|_0, \text{ as } \lambda \to \infty, \tag{3.8}$$

where $A = i\partial_1^s\partial_{\xi_2}a_1(x_0, \xi^0)/s!$. This inequality must be fulfilled with the same constant for all functions $\psi \in C_0^\infty(\mathbb{R}^n)$. However, this is impossible. Indeed, let ψ_0 be an arbitrary function $\in C_0^\infty(\mathbb{R}^n)$ which is not equal to 0 identically. Let

$$\psi_m(z) = \psi_0(z^1/m, z^2/m^{s+1}, z^3, \ldots, z^n).$$

Obviously

$$\|\psi_m\|_0 = m^{(s+2)/2}\|\psi_0\|_0, \quad \|\partial_1\psi_m(z) + A(z^1)^s\partial_2\psi_m(z)\|_0 = m^{(s+1)/2}\|\partial_1\psi_0 + A(z^1)^s\partial_2\psi_0\|_0,$$

Substituting $\psi = \psi_m$ we arrive at a contradiction when $m \to \infty$. Theorem 3.1 is proved.

3.2. We shall now prove the necessity of condition (Ψ'). Suppose that this condition is violated at some point (x_0, ξ^0). By Theorem 4.1 of Chapter VI there exists a real number such that the numbers $k_1(x_0, \xi^0)$ and $k_0(x_0, \xi^0)$ are equal for operator $P_1 = (1 + i\rho)P$. If (3.1) is fulfilled for the operator P, then it will hold also for P_1. On the other hand, by the invariance of condition (Ψ') under constant multiplication, this condition will be violated also for the operator P_1 at (x_0, ξ^0). The latter implies that, if $p_1^0(x, \xi) = a_1(x, \xi) + ia_2(x, \xi)$ is the principal symbol of P_1 and grad $a_2(x_0, \xi^0) \neq 0$, then a_2 changes its sign from minus to plus on moving in the positive direction along the null bicharacteristic Γ of a_2 passing through the point (x_0, ξ^0). It is easily seen that since the number $k_1(x_0, \xi^0)$ is finite by condition $(\mathscr{B})$ it must be odd and $H_2^k a_1(x_0, \xi^0) > 0$. By Theorem 3.1 of Chapter VI, inequality (3.1) is not fulfilled in this case for any real δ. Thus, the necessity of the conditions of Theorem 1.2 has been proved.

§4. Estimates for First-Order Differential Operators

4.1. In this section a theorem analogous to Theorem 1.2 is proved in the model situation by examining first-order differential operators with polynomial coefficients instead of pseudo-differential operators. Use of these results has been made in proving the sufficiency of the conditions of Theorem 1.2.

Theorem 4.1. The estimate of the type

$$\lambda\|u\|_0 \leqslant C\|u'(t) + \lambda^{k+1}P(t)u\|_0 \tag{4.1}$$

for all $\lambda > 0$, $u \in C_0^\infty(\mathbb{R})$, where P(t) is a real polynomial of order k whose leading coefficient equals 1, holds if and only if the polynomial P does not change sign from minus to plus as t increases.

Theorem 4.2. If the conditions of Theorem 4.1 are fulfilled, then the estimate

$$\lambda^{k+1}\|P(t)u\|_0 + \|u'(t)\|_0 + \lambda\|u\|_0 \leqslant C_1\|u'(t) + \lambda^{k+1}P(t)u\|_0 \tag{4.2}$$

is valid for all $\lambda \geqslant 0$, $u \in C_0^\infty(\mathbb{R}^1)$.

Theorem 4.3. Let $Q = \{(t, x) \in \mathbb{R}^2, \ |t| \leqslant 1, \ |x| \leqslant 1\}$. The estimate of the type

$$\lambda \|u\|_0 \leqslant C \|u'(t) + A(t) D_x u + \lambda^{k+1} B(t, x) u\|_0 \tag{4.3}$$

for all $\lambda > 0$, $u \in C_0^\infty(Q)$, where $A(t)$ is a real polynomial of degree $l < k/2$ whose leading coefficient equals 1, and $B(t, x)$ is a real polynomial of degree k, holds if and only if

1) $A(t)$ does not change sign;

2) the function $C(t, x) = B(t, x)/A(t)$ is smooth and $\operatorname{sgn} A(t) \cdot \partial_t C(t, x) \leqslant 0$;

3) there exists a constant $c_0 > 0$ such that

$$\sum_{i+j \leqslant k-1} |D_t^i [A(t)^j D_x^j D_t B(t, x)]| \geqslant c_0.$$

Theorem 4.4. If the conditions of Theorem 4.3 are fulfilled, then the estimate

$$\lambda \|u\|_0 + \|u'(t)\|_0 + \|A(t) D_x u + \lambda^{k+1} B(t, x) u\|_0 \leqslant$$

$$c_1 \|u'(t) + A(t) D_x u + \lambda^{k+1} B(t, x) u\|_0 \tag{4.4}$$

holds for all $\lambda \geqslant 0$, $u \in C_0^\infty(Q)$.

4.2. Proof of Theorem 4.1.

Necessity. Let $Q(t)$ be a polynomial such that $Q'(t) = -P(t)$. After we have replaced $u = v \exp \lambda^{k+1} Q(t)$ inequality (4.1) changes to

$$\lambda \|v(t) e^{\lambda^{k+1} Q(t)}\|_0 \leqslant C \|v'(t) e^{\lambda^{k+1} Q(t)}\|_0, \qquad v \in C_0^\infty(\mathbb{R}).$$

If the polynomial $P(t)$ changes sign from minus to plus as t increases at a point, taken as origin, then the function $Q(t)$ has a local maximum at this point.

Therefore one can find m, ε_1, and ε_2 such that $\varepsilon_1 > 0$, $\varepsilon_2 > 0$ and

$$Q(t) \geqslant m \quad \text{when} \quad -\varepsilon_1 \leqslant t \leqslant \varepsilon_2,$$

$$Q(t) \leqslant m \ \text{when} \ -2\varepsilon_1 \leqslant t \leqslant -\varepsilon_1 \ \text{and when} \ \varepsilon_2 \leqslant t \leqslant 2\varepsilon_2.$$

If the function $v(t)$ is such that $v(t) = 1$ for $-\varepsilon_1 \leqslant t \leqslant \varepsilon_2$ and $v(t) = 0$ for $t \leqslant -2\varepsilon_1$ and for $t \geqslant 2\varepsilon_2$, then

$$\|v(t) e^{\lambda^{k+1} Q(t)}\|_0^2 \geqslant \int_{-\varepsilon_1}^{\varepsilon_2} e^{2\lambda^{k+1} Q(t)}\, dt \geqslant e^{2m\lambda^{k+1}} (\varepsilon_1 + \varepsilon_2),$$

$$\|v'(t) e^{\lambda^{k+1} Q(t)}\|_0^2 \leqslant \int_{-2\varepsilon_1}^{-\varepsilon_1} |v'(t)|^2 e^{2\lambda^{k+1} Q(t)}\, dt +$$

$$\int_{\varepsilon_2}^{2\varepsilon_2} |v'(t)|^2 e^{2\lambda^{k+1} Q(t)}\, dt \leqslant C_1 e^{2m\lambda^{k+1}} (\varepsilon_1 + \varepsilon_2).$$

Thus we arrive at the inequality

$$\lambda \leqslant CC_1$$

which, obviously, does not hold for large λ.

Sufficiency. The conditions of Theorem 4.1 imply that the polynomial $P(t)$ either does not change its sign at all or changes once from plus to minus as t increases. In the last case we shall denote by α the point where the change of sign takes place.

Let first $P(t) \geqslant 0$ for all t. Note that

$$\operatorname{Re}\int_{-\infty}^{t}[u'+\lambda^{k+1}P(t)u]\bar{u}\,dt=\frac{1}{2}|u(t)|^2+\lambda^{k+1}\int_{-\infty}^{t}P(t)|u(t)|^2\,dt \tag{4.5}$$

and therefore

$$2\int_{-\infty}^{\infty}|(u'+\lambda^{k+1}P(t)u)\bar{u}|\,dt\geqslant\frac{1}{2}\max_t|u(t)|^2+\lambda^{k+1}\int_{-\infty}^{\infty}|P(t)||u(t)|^2dt. \tag{4.6}$$

When $P(t) \leqslant 0$ everywhere, the inequality (4.6) is obtained, too, provided the integral is evaluated not from $-\infty$, but from $+\infty$

$$\operatorname{Re}\int_{+\infty}^{t}[u'+\lambda^{k+1}P(t)u]\bar{u}\,dt=\frac{1}{2}|u(t)|^2+\lambda^{k+1}\int_{+\infty}^{t}|P(t)||u(t)|^2\,dt. \tag{4.7}$$

Finally, if $P(t)(t-\alpha)$ we make use of equality (4.5) for $t < \alpha$ and of (4.7) for $t > \alpha$, and again obtain the estimate (4.6).

We shall now show how (4.1) follows from (4.6). Let $\mu_1, \ldots, \mu_k$ be the roots of the polynomial P(t) and let M be a set of points t on $\mathbb{R}$ for which $|t-\mu_j| \geqslant \lambda^{-1}$ for all j. It is easily seen that

$$\frac{1}{2}\max_t|u(t)|^2\geqslant(4k)^{-1}\lambda\int_{\mathbb{R}\setminus M}|u(t)|^2\,dt.$$

On the other hand, for $t \in M$ the inequality $|P(t)| \geqslant \lambda^{-k}$ is fulfilled, and therefore

$$\lambda^{k+1}\int|P(t)||u(t)|^2\,dt\geqslant\lambda\int_M|u(t)|^2\,dt.$$

Thus, from (4.6) it follows that

$$\int_{-\infty}^{\infty}|(u'+\lambda^{k+1}P(t)u)\bar{u}|\,dt\geqslant(8k)^{-1}\lambda\int_{-\infty}^{\infty}|u(t)|^2\,dt.$$

According to Cauchy's inequality

$$\int|(u'+\lambda^{k+1}P(t)u)\bar{u}|\,dt\leqslant(16k)^{-1}\lambda\int|u(t)|^2\,dt+4k\lambda^{-1}\int|u'+\lambda^{k+1}P(t)u|^2\,dt,$$

and we finally obtain

$$(16k)^{-1}\lambda\int|u(t)|^2\,dt\leqslant4k\lambda^{-1}\int|u'+\lambda^{k+1}P(t)u|^2\,dt,$$

that is, inequality (4.1) in which $C = 8k$. This completes the proof.

Remark 4.1. Note that if the leading coefficient of t^k in the polynomial P equals $b\,(b > 0)$, then (4.1) is fulfilled with constant

$$C=8kb^{-1/(k+1)}.$$

It immediately follows from the previous discussion if we replace λ by λb^{-1}.

4.3. In order to prove Theorem 4.2 we shall first prove some auxiliary assertions. Let us define the function

$$M(t)=\max_{0\leqslant j\leqslant k}|\delta_j\lambda^{k+1}P^{(j)}(t)/j!|^{1/(j+1)},$$

where $\delta_j > 0$ are sufficiently small constants which will be chosen later. Note that $M(t) \geqslant (\delta_k/k!)^{1/(k+1)}\lambda$.

Lemma 4.1. If δ_j are such that

$$\sum_{j=i+1}^{k} \frac{j!}{i!\,(j-i)!}\,\delta_j^{-1} \leqslant \frac{1}{4}\,\delta_i^{-1}, \qquad i=0,\ 1,\ \ldots,\ k-1, \tag{4.8}$$

then the inequalities

$$M\,(t_0)/2 \leqslant M\,(t) \leqslant 4M\,(t_0)/3$$

are fulfilled for $|t - t_0| \leqslant M^{-1}(t_0)$.

Proof. Let

$$M\,(t_0) = |\,\delta_i\lambda^{k+1}P^{(i)}\,(t_0)/i!\,|^{1/(i+1)},$$

where i is some integer, $0 \leqslant i \leqslant k$. Denote $M(t_0)$ by M. When $|t - t_0| \leqslant M^{-1}$ we have

$$|\,P^{(i)}\,(t) - P^{(i)}\,(t_0)\,| \leqslant \sum_{j=i+1}^{k} |\,P^{(j)}\,(t_0)\,|/(j-i)!\,M^{i-j}.$$

But from the definition of M it follows that

$$|\,P^{(j)}\,(t_0)\,| \leqslant j!\ (\delta_j\lambda^{k+1})^{-1}\,M^{j+1}.$$

Therefore

$$|\,P^{(i)}\,(t) - P^{(i)}\,(t_0)\,| \leqslant \sum_{j=i+1}^{k} \frac{j!}{(j-i)!}\,(\delta_j\lambda^{k+1})^{-1}\,M^{i+1},$$

and by (4.8) we have

$$|\,P^{(i)}\,(t)\,| \geqslant |\,P^{(i)}\,(t_0)\,| - \sum_{j=i+1}^{k} \frac{j!}{(j-i)!}\,(\delta_j\lambda^{k+1})^{-1}\,M^{i+1} =$$

$$\lambda^{-k-1}M^{i+1}\left(\delta_i^{-1}i! - \sum_{j=i+1}^{k} \frac{j!}{(j-i)!}\,\delta_j^{-1}\right) > \frac{3}{4}\,\lambda^{-k-1}M^{i+1}i!\,\delta_i^{-1}.$$

Hence

$$M\,(t)^{i+1} \geqslant \delta_i\lambda^{k+1}\,|\,P^{(i)}\,(t)\,|/i! \geqslant \frac{3}{4}\,M^{i+1},$$

and therefore

$$(3/4)^{1/(i+1)}\,M\,(t_0) \leqslant M\,(t).$$

On the other hand, from the same estimates it is seen that for all i

$$|\,P^{(i)}\,(t)\,| \leqslant \frac{5}{4}\,\lambda^{-k-1}M^{i+1}i!\,\delta_i^{-1},$$

so that $M\,(t) \leqslant \max\limits_i \left(\frac{5}{4}\right)^{1/(i+1)} M \leqslant \frac{5}{4}\,M$, which proves the lemma.

Lemma 4.2. Let for some s, $0 \leqslant s \leqslant k$,

$$M\,(t_0) = |\,\delta_s\lambda^{k+1}P^{(s)}\,(t_0)/s!\,|^{1/(s+1)}$$

and

$$P_0\,(t) = \sum_{j=0}^{s} P^{(j)}\,(t_0)\,(t-t_0)^j/j!$$

Then there exists a constant C_1 depending only on k such that

$$M\,(t_0)^2\,\delta_s^{-2/(s+1)}\,\|\,u\,\|^2 + M\,(t_0)\,\delta_s^{-1/(s+1)}\max|\,u\,(t)\,|^2 \leqslant C_1^2\,\|\,u'\,(t) + \lambda^{k+1}P_0\,(t)\,u\,(t)\,\|_0^2 \tag{4.9}$$

for all functions $u \in C_0^\infty(\mathbb{R})$.

Proof. Note that the leading coefficient of the polynomial $P_0(t)$ equals

$$b = M\,(t_0)^{s+1}\lambda^{-k-1}\delta_s^{-1}.$$

By Remark 4.1, the inequality

$$M(t_0)^2 \delta_s^{-2/(s+1)} \|u\|_0^2 \leqslant (8k)^2 \| u'(t) + \lambda^{k+1} P_0(t) u(t) \|_0^2$$

is valid. On the other hand, from (4.6) it follows that

$$\frac{1}{2} \max_t |u(t)|^2 \leqslant \int_{-\infty}^{\infty} |u' + \lambda^{k+1} P_0(t) u(t)| \, |\bar{u}(t)| \, dt,$$

and therefore

$$\max |u(t)|^2 \leqslant C \int_{-\infty}^{\infty} |u' + \lambda^{k+1} P_0(t) u|^2 dt + C^{-1} \int_{-\infty}^{\infty} |u(t)|^2 dt \leqslant$$

$$[C + C^{-1} (8k)^2 \delta_s^{2/(s+1)} M(t_0)^{-2}] \| u' + \lambda^{k+1} P_0(t) u \|_0^2.$$

Assuming $C = 8k\delta_s^{1/(s+1)} M(t_0)^{-1}$ we obtain inequality (4.9) with constant $C_1 = 64k^2 + 16k$. The lemma is proved.

Lemma 4.3. Suppose that the conditions of Lemma 4.2 are fulfilled and

$$\sum_{j=i+1}^{k} \delta_j^{-1} \leqslant \delta_i^{-1/(i+1)} (2C_1)^{-1} \tag{4.10}$$

(for all $i = 0, 1, \ldots, k-1$). Then $\lambda^{k+1} |P(t) - P_0(t)| \leqslant M(t_0) \delta_s^{-1/(s+1)} (2C_1)^{-1}$ for $|t - t_0| \leqslant M^{-1}(t_0)$.

Proof. We have

$$\lambda^{k+1} |P(t) - P_0(t)| = \lambda^{k+1} \left| \sum_{j=s+1}^{k} P^{(j)}(t_0)(t - t_0)^j / j! \right| \leqslant$$

$$\sum_{j=s+1}^{k} \delta_j^{-1} M^{j+1} |t - t_0|^j \leqslant M(t_0) \sum_{j=s+1}^{k} \delta_j^{-1}.$$

From (4.10) it then follows that

$$\lambda^{k+1} |P(t) - P_0(t)| \leqslant M(t_0) \delta_s^{-1/(s+1)} (2C_1)^{-1}$$

which proves the lemma.

Lemma 4.4. Let $\omega = \{t \in \mathbb{R}, \ |t - t_0| < M^{-1}(t_0)\}$ and let $\alpha \in \omega$ and $\beta \in \omega$. Then

$$\lambda^{k+1} |P(\beta) - P(\alpha)| \leqslant 2M(t_0) \sum_{j=1}^{k} \delta_j^{-1}.$$

Proof. We have

$$\lambda^{k+1} |P(\beta) - P(\alpha)| \leqslant \lambda^{k+1} \sum_{j=1}^{k} |P^{(j)}(\alpha)| \, |\beta - \alpha|^j / j! \leqslant M(\alpha) \sum_{j=1}^{k} \delta_j^{-1}.$$

By Lemma 4.1, $M(\alpha) \leqslant 2M(t_0)$. Therefore

$$\lambda^{k+1} |P(\beta) - P(\alpha)| \leqslant 2M(t_0) \sum_{j=1}^{k} \delta_j^{-1},$$

which proves the lemma.

4.4. Proof of Theorem 4.2. The interval set $\omega(t_0) = \{t; \ |t - t_0| < \frac{1}{2} M^{-1}(t_0)\}$ covers a straight line $\mathbb{R}$. From this we choose a subcovering by intervals $\omega_1, \ldots, \omega_l$ such that every point of $\mathbb{R}$ is covered by no more than two intervals. Let $h \in C_0^\infty(\mathbb{R})$, where $h(t) = 1$ when $|t| \leqslant 1/2$, $h(t) = 0$ when $|t| \geqslant 1$ and $|h'(t)| \leqslant 3$. Suppose that

$$\varphi_j(t) = h((t - t_j) M_j) \left(\sum h^2 ((t - t_r) M_r) \right)^{-1/2},$$

where t_j is the center of interval ω_j and $M_j = M(t_j)$. Then $\varphi_j \in C_0^\infty(\mathbb{R})$ and $\sum \varphi_j^2(t) = 1$, where $\varphi_j(t) = 0$ when $|t - t_j| > M^{-1}(t_j)$, so that the assertions of Lemmas 4.1-4.4 are fulfilled in the support of function φ_j. Also note that $|\varphi_j'(t)| \leqslant 4M_j$ by Lemma 4.1. Suppose that $u_j = \varphi_j u$. From Lemmas 4.2 and 4.3 it follows that

$$\| u' + \lambda^{k+1} Pu \|_0^2 = \sum \| \varphi_j (u' + \lambda^{k+1} Pu) \|_0^2 \geqslant \frac{1}{2} \sum \| u_j' + \lambda^{k+1} Pu_j \|_0^2 - \sum \| \varphi_j' u \|_0^2 \geqslant$$

$$\frac{1}{2} \sum \| u_j' + \lambda^{k+1} Pu_j \|_0^2 - 16 \sum M_j^2 \| u_j \|_0^2 \geqslant \frac{1}{4} \sum \| u_j' + \lambda^{k+1} Pu_j \|_0^2,$$

where the last inequality follows from Lemmas 4.2, 4.3 and the inequality

$$16 C_1 \delta_s^{1/(s+1)} < 1. \tag{4.11}$$

Indeed, let $M(t_j) = | \delta_s \lambda^{k+1} P^{(s)}(t_j)/s! |^{1/(s+1)}$. Then

$$\| u_j' + \lambda^{k+1} Pu_j \|_0^2 \geqslant \frac{1}{2} \| u_j' + \lambda^{k+1} P_0 u_j \|_0^2 - \lambda^{2(k+1)} \| (P - P_0) u_j \|_0^2 \geqslant$$

$$\frac{1}{2} C_1^{-2} \delta_s^{-2/(s+1)} M_j^2 \| u_j \|_0^2 - \frac{1}{4} C_1^{-2} \delta_s^{-2/(s+1)} M_j^2 \| u_j \|_0^2 = \frac{1}{4} C_1^{-2} \delta_s^{-2/(s+1)} M_j^2 \| u_j \|_0^2,$$

and therefore

$$\frac{1}{4} \| u_j' + \lambda^{k+1} Pu_j \|_0^2 \geqslant \frac{1}{16} C_1^{-2} \delta_s^{-2(s+1)} M_j^2 \| u_j \|_0^2 \geqslant 16 M_j^2 \| u_j \|_0^2.$$

It now remains to choose parameters δ_j so that the inequalities (4.8), (4.10), and (4.11) are fulfilled, which, after we have assumed $\delta^{-1} = \gamma_j$, can be written as

$$\sum_{j=i+1}^{k} \frac{j!}{i!\,(j-i)!} \gamma_j \leqslant \frac{1}{4} \gamma_i, \quad 2C_1 \sum_{j=i+1}^{k} \gamma_j \leqslant \gamma_i^{1/(i+1)},$$

$$\gamma_i^{1/(i+1)} > 16 C_1, \quad i = 0, 1, \ldots, k.$$

Clearly we can assume $\gamma_k = (16C_1)^{k+1} + 1$ and then define subsequently $\gamma_{k-1}, \ldots, \gamma_0$.

Since

$$\| u' + \lambda^{k+1} P(t) u \|_0^2 = \| u' \|_0^2 + \lambda^{2(k+1)} \| P(t) u \|_0^2 - \lambda^{k+1} \int P'(t) | u(t) |^2 dt,$$

inequality (4.2) follows from the inequality

$$\lambda^{k+1} \int P'(t) | u_j(t) |^2 dt \leqslant C_2 \| u_j' + \lambda^{k+1} P(t) u \|_0^2,$$

if C_2 is independent of j and λ. It is clear that the subset $\{t \in \mathbb{R}, P'(t) > 0\}$ intersects the support of function u_j in no more than k intervals. On each of these intervals

$$\lambda^{k+1} \int_{\alpha}^{\beta} P'(t) | u_j(t) |^2 dt \leqslant \max | u_j'(t) |^2 \lambda^{k+1} [P(\beta) - P(\alpha)].$$

From Lemmas 4.2 and 4.4 it follows that

$$\max | u_j(t) |^2 \lambda^{k+1} [P(\beta) - P(\alpha)] \leqslant$$

$$2 C_1^2 \delta_s^{1/(s+1)} \delta_0^{-1} \| u_j' + \lambda P(t) u_j \|_0^2 \leqslant C_3 \| u_j' + \lambda P(t) u_j \|_0^2,$$

where the constant C_3 depends only on k.

Thus, Theorem 4.2 is proved.

4.5. We shall present the proof of Theorem 4.3 in the form of lemmas.

Lemma 4.5. If estimate (4.3) is fulfilled, then for all $\lambda > 0$, x, and ξ the function $\xi A(t) + \lambda^{k+1} B(t, x)$ does not change its sign from minus to plus as t increases.

Proof. Suppose that (4.3) is valid and the function $\xi_0 A(t) + \lambda_0^{k+1} B(t, x_0)$ changes its sign from minus to plus at $t = 0$.

I. First we shall consider the case when

$$a = \xi_0 A'(0) + \lambda_0^{k+1} B_t'(0, x_0) > 0.$$

To simplify notations we assume that $x_0 = 0$.

Let $w(t, x)$ be a polynomial for which

$$iw_t + A(t) w_x + \lambda_0^{k+1} B(t, x) = 0, \tag{4.12}$$

$$w = x\xi_0 + ibx^2 \quad \text{when} \quad t = 0, \tag{4.13}$$

where $b > 0$ is a sufficiently large number.

Such a polynomial can be readily constructed as $\sum_{j=0}^{l} w_j(t, x)$, where $iw_{0t} + \lambda_0^{k+1} B(t, x) = 0$, $w_0 = x\xi_0 + ibx^2$ when $t = 0$ and

$$w_{jt} = i A(t)(w_{j-1})_x, \quad w_j = 0 \text{ when } t = 0$$

for $j = 1, \dots, l$. Clearly the degrees of polynomials w_j in x decrease so that l is equal to the degree of the polynomial B in x variables.

We shall now prove that $\operatorname{Im} w(t, x) \geqslant c_0(t^2 + x^2)$ in a small neighborhood of the origin of coordinates, where $c_0 = \text{const} > 0$. Indeed,

$$\frac{\partial w}{\partial x}(0, 0) = \xi_0, \quad \frac{\partial w}{\partial t}(0, 0) = 0, \quad \frac{\partial^2 w}{\partial t^2}(0, 0) = ia,$$

$$\frac{\partial^2 w}{\partial x^2}(0, 0) = 2ib, \quad \frac{\partial^2 w}{\partial x\,\partial t}(0, 0) = -2bA(0) + i\lambda_0^{k+1} B_x(0, 0).$$

Therefore, if $B_x^2(0, 0)\lambda_0^{2(k+1)} < 2ab$ then the form

$$bx^2 + \lambda_0^{k+1} B_x(0, 0)\, xt + \frac{a}{2} t^2$$

is positive definite and

$$\operatorname{Im} w(t, x) \geqslant c_0(t^2 + x^2)$$

in a sufficiently small neighborhood ω of the origin. Let now $\varphi \in C_0^\infty(\omega)$ and $\varphi(0, 0) = 1$. Suppose that

$$u(t, x) = \varphi(t, x) e^{i(\lambda/\lambda_0)^{k+1} w(t, x)}.$$

Let $\operatorname{Im} w(t, x) \leqslant C(t^2 + x^2)$ in ω. Clearly

$$\|u\|_0^2 = \iint |\varphi(t, x)|^2 e^{-2(\lambda/\lambda_0)^{k+1} \operatorname{Im} w(t, x)}\, dx\, dt \geqslant$$

$$\iint |\varphi(t, x)|^2 e^{-2(\lambda/\lambda_0)^{k+1} C(t^2 + x^2)}\, dx\, dt \geqslant C_1 \lambda^{-k-1}, \quad C_1 > 0.$$

On the other hand

$$\|u_t + A(t) D_x u + \lambda^{k+1} B(t, x) u\|_0^2 = \left\| \left(\varphi_t + \frac{1}{i} A(t)\varphi_x\right) e^{i(\lambda/\lambda_0)^{k+1} w} \right\|_0^2 \leqslant$$

$$C_2 \iint e^{-c_0(\lambda/\lambda_0)^{k+1}(t^2 + x^2)}\, dx\, dt \leqslant C_3 \lambda^{-k-1}.$$

Therefore, from (4.3) it follows that

$$C_1 \lambda^{-k} \leqslant C_3 C \lambda^{-k-1}.$$

Obviously, this estimate is not fulfilled for large λ.

2. Thus, we can assume that if the function $\xi A(t) + \lambda^{k+1} B(t, x)$ does not change its sign from minus to plus, then its first derivative with respect to t vanishes at points where the sign changes.

Suppose that a change in sign again takes place at $t = 0$ for the values $\xi = \xi_0 \neq 0$, $\lambda = \lambda_0$, $x = 0$, and the function $\xi_0 A(t) + \lambda_0^{k+1} B(t, 0) = at^l[1 + O(1)]$ when $t \to 0$. Since $a > 0$ and k is odd, the following conditions are necessarily fulfilled in view of Lemma 3.1 of Chapter VI:

$$A^{(j)}(0) = 0, \quad \frac{\partial^j B(0, 0)}{\partial t^j} = 0, \quad \frac{\partial^{j+1} B(0, 0)}{\partial x \partial t^j} = 0 \quad \text{for } j = 0, 1, \ldots, (l-1)/2. \tag{4.14}$$

Let w be again a solution of equation (4.12) satisfying (4.13) for b = 1. We shall now prove that

$$\operatorname{Im} w(t, x) \geqslant c_0 (x^2 + t^{1+l}), \quad c_0 = \text{const} > 0 \tag{4.15}$$

in some neighborhood ω of the origin. From (4.12) and (4.13) it follows that

$$\frac{\partial w}{\partial t}(0, 0) = 0, \quad \frac{\partial w}{\partial x}(0, 0) = \xi_0, \quad \frac{\partial^2 w}{\partial x^2}(0, 0) = 2i, \quad \frac{\partial^{j+1} w}{\partial x \partial t^j}(0, 0) = 0 \quad \text{for } j \geqslant \frac{l+1}{2},$$

$$\frac{\partial^i w}{\partial t^i}(0, 0) = 0 \quad \text{for } i \leqslant l, \quad \frac{\partial^{l+1} w}{\partial t^{l+1}}(0, 0) = ia.$$

Thus, $w(t, x) = x\xi_0 + i(x^2 + at^{l+1}/(l+1)!)(1 + o(1))$. Let C be a constant such that

$$\operatorname{Im} w(t, x) \leqslant C(x^2 + t^{l+1}) \text{ in } \omega.$$

Let again $\varphi \in C_0^\infty(\omega)$ and $\varphi(0, 0) = 1$. Suppose that

$$u(t, x) = \varphi(t, x) e^{i(\lambda/\lambda_0)^{k+1} w(t, x)}.$$

Then

$$\|u\|_0^2 = \iint |\varphi(t, x)|^2 e^{-2(\lambda/\lambda_0)^{k+1} \operatorname{Im} w} dx\, dt \geqslant$$

$$\iint |\varphi(t, x)|^2 e^{-2(\lambda/\lambda_0)^{k+1} C(t^{l+1} + x^2)} dx\, dt \geqslant C_1 \lambda^\varkappa,$$

where $\varkappa = -\left(\frac{1}{2} + \frac{1}{l+1}\right)(k+1)$, $C > 0$.

On the other hand,

$$\|u_t + A(t) D_x u + \lambda^{k+1} B(t, x) u\|_0^2 = \|[\varphi_t + A(t) D_x \varphi] e^{i(\lambda/\lambda_0)^{k+1} w}\|_0^2 =$$

$$C_2 \iint e^{-c_0 (\lambda/\lambda_0)^{k+1} (t^{l+1} + x^2)} dx\, dt \leqslant C_3 \lambda^\varkappa.$$

Therefore, from (4.3) it follows that

$$C_1 \lambda^{\varkappa+1} \leqslant C C_3 \lambda^\varkappa.$$

This estimate does not hold if λ is large enough.

Thus, Lemma 4.5 is proved.

Lemma 4.6. If inequality (4.3) is fulfilled, then the function A(t) does not change its sign.

Proof. Suppose A does not change its sign at $t = t_0$ so that

$$A(t) = a(t - t_0)^l + O(|t - t_0|^{l+1}), \quad t \to t_0,$$

where $a \neq 0$ and l is odd. Let $t_1 < t_0 < t_2$ and

$$A(t_1) A(t_2) < 0, \ |A(t_1)| = \delta, \ |A(t_2)| = \delta, \ \delta = \text{const} > 0.$$

Let $\lambda_0^{k+1}|B(t, 0)| \leqslant M$ for $t_1 \leqslant t \leqslant t_2$. Choose ξ so that $\operatorname{sgn} A(t_1) \cdot \xi\delta = 2M$. Then at $t = t_1$ the function $\xi A(t) + \lambda_0^{k+1} B(t, 0)$ has a value less than $-M$ and at $t = t_2$, a value more than M, that is, the function changes sign from minus to plus as t increases. Thus, by Lemma 4.5, inequality (4.3) is not fulfilled. This proves Lemma 4.6.

Lemma 4.7. If inequality (4.3) is fulfilled, then the function $C(t, x) = B(t, x)/A(t)$ is smooth and $\partial_t C(t, x)\cdot A(t) \leqslant 0$.

Proof. Let for definiteness $A(t) \geqslant 0$. Then the sign of the function $\xi A(t) + \lambda_0^{k+1} B(t, x)$ agrees with that of the function $\xi + \lambda_0^{k+1} C(t, x)$, where $C(t, x) = B(t, x)/A(t)$. By Lemma 4.5, this sign cannot change from minus to plus as t increases. Therefore, the function $C(t, x)$ monotonically decreases in t and takes finite values, in particular. Thus, it may be seen that if $A(t_0) = 0$, then the function $B(t, x)$ at $t = t_0$ has a zero at least of the same order for every x. It means that the function $C(t, x)$ will be a polynomial in x, with coefficients smoothly depending on t and $\partial_t C(t, x) \leqslant 0$. The lemma is proved.

Lemma 4.8. If all of the derivatives $D_t^i A(t)^j D_x^j D_t B(t, x)$ when $i + j \leqslant k - 1$ vanish at $t = 0$ and $x = 0$, then (4.3) is impossible.

Proof. Let $A(t) = at^s[1 + o(1)]$ as $t \to 0$. By Lemma 4.6, s must be even. Note that the possibility of replacing u by $ue^{i\Phi(x)}$ with the real-valued function Φ enables us to assume $B(0, x) = 0$.

Let us change the variables

$$t = t'\lambda^{-1}, \quad x = x'\lambda^{-s-1}.$$

Inequality (4.3) in new variables is written as:

$$\| u \|_0 \leqslant c \| u_{t'} + at'^s [1 + O(\lambda^{-1})] D_{x'} u + O(\lambda^{-1}) u \|_0,$$

since the derivatives $D_t^i D_x^j D_t B(0, 0)$ are equal to 0, for $i + j(s + 1) \leqslant k - 1$, by virtue of the conditions of the lemma. Thus, from (4.3) it follows that

$$\| u \|_0 \leqslant C \| u_t + at^s D_x u \|_0, \quad u \in C_0^\infty(\mathbb{R}^2)$$

as $\lambda \to \infty$. However, the last estimate cannot be fulfilled, for instance, for a sequence of functions

$$u_m = \varphi(t/m, x/m^{s+1}),$$

where $\varphi \in C_0^\infty(\mathbb{R}^2)$, $\varphi \not\equiv 0$, since

$$\| u_m \|_0 = c_0 m^{1+s/2}, \quad \| \partial_t u_m + at^s D_x u_m \|_0 \leqslant C_1 m^{s/2}, \quad c_0 > 0.$$

Lemma 4.8 is proved.

4.6. Lemmas 4.6-4.8 reveal the necessity of the conditions of Theorem 4.3. Now we shall prove the sufficiency of these conditions. This proof is also based on a number of lemmas.

As above, we define the function

$$M(t, x) = \max_{\substack{|\alpha| = i-1 \\ |\beta| = j,\ i+j \leqslant k}} | \delta_{ij} \lambda^{k+1} (D_t)^\alpha (A(t) D_x)^\beta D_t B(t, x) |^{1/(i+j+1)}, \tag{4.16}$$

where $\alpha = (\alpha_1, \ldots, \alpha_r)$, $\beta = (\beta_1, \ldots, \beta_r)$ and $(L_1)^\alpha (L_2)^\beta = L_1^{\alpha_1} L_2^{\beta_1} \ldots L_1^{\alpha_r} L_2^{\beta_r}$, where δ_{ij} are some small positive numbers which will be indicated later. In addition, we single out the most important coefficient a_s of the polynomial $A(t) = \sum_{j=0}^{[k/2]} a_j (t - t_0)^j$ after having assumed that

$$| a_s | M^{-s} \gamma_s = \max_j | a_j | M^{-j} \gamma_j, \quad M = M(t_0, x_0),$$

where γ_j are positive numbers which along with δ_{ij} will be given later.

Clearly s and $a = |a_s|$ depend on t_0. Note that $|a_j| \leqslant \gamma_s \gamma_j^{-1} a M^{j-s}$ when $j = 0, 1, \ldots, k/2$, while $M(t, x) \geqslant c_1 \lambda$, where $c_1 = \text{const} > 0$ and $M(t, x) \leqslant C_2 \lambda^{(k+1)/2}$. Depending on the value of a we shall consider the two possible cases:

I. If $a \leqslant M^{s+1} \lambda^{-(k+1)/2}$, we suppose that

$$\Omega(t_0, x_0) = \{(t, x) \in \mathbb{R}^2;\ |t - t_0| < M^{-1},\ |x - x_0| < \lambda^{-(k+1)/2}\},$$

where $M = M(t_0, x_0)$, $a = a(t_0)$.

4.7. Let us verify that a and M vary slightly in $\Omega(t_0, x_0)$ and in the considered case

$$M(t, x) = \max_{0 \leqslant i \leqslant k} |\delta_{i0} \lambda^{k+1} \partial_t^i B(t, x)|^{1/(i+1)}, \tag{4.17}$$

that is, to obtain M it is sufficient to differentiate function B only with respect to t.

Lemma 4.9. If $a \leqslant M^{s+1} \lambda^{-(k+1)/2}$, then equality (4.17) is valid.

Proof. From the analysis carried out in Section 4 of Chapter VI it is shown that use has to be made of the monomials τ, $a t^s \xi$, and $\lambda^{k+1} b t^i x^j$ for M to be obtained in a different manner. If in this case $j \geqslant 2$, then

$$|\lambda^{k+1} b a^j| \leqslant C \lambda^{k+1} M^{j(s+1)} \lambda^{-k-1} \ll M^{i+j(s+1)+1}$$

and we arrive at a contradiction. If $j = 1$, then to obtain M by differentiating with respect to x use is necessarily made of the coefficients containing a_j at least twice, since $B(t, x) = A(t)C(t, x)$. The above estimate indicates that this is impossible. Lemma 4.9 is proved.

Lemma 4.10. If $a \leqslant M^{s+1} \lambda^{-(k+1)/2}$, then the inequality

$$|D_t^j A(t)| \leqslant C \lambda^{-(k+1)/2} M^{j+1}, \quad j = 0, 1, \ldots, [k/2],$$

is fulfilled for $|t - t_0| \leqslant 2M^{-1}$, in which the constant C is independent of λ and $M = M(t_0, x_0)$.

Proof. By hypothesis

$$|D_t^j A(t_0)| \leqslant \gamma_s \gamma_j^{-1} a M^{j-s} j!$$

It is seen that when $0 \leqslant i \leqslant [k/2]$, $|t - t_0| \leqslant 2M^{-1}$, $\lambda \geqslant \lambda_0$,

$$|D_t^i A(t)| \leqslant \left| \sum_{j=i}^{[k/2]} \frac{1}{(j-i)!} D_t^j A(t_0)(t - t_0)^{j-i} + O(|t - t_0|^{[k/2]-i+1}) \right| \leqslant C M^{i+1} \lambda^{-(k+1)/2},$$

which proves the lemma.

Lemma 4.11. If $a \leqslant M^{s+1} \lambda^{-(k+1)/2}$, then the inequality

$$M(t_0, x_0) \leqslant \frac{3}{2} M(t, x) \leqslant \frac{9}{4} M(t_0, x_0)$$

is fulfilled everywhere in $\Omega(t_0, x_0)$.

Proof. That formula (4.17) is valid when $|t - t_0| \leqslant 2M^{-1}$ follows from Lemmas 4.9 and 4.10. It is sufficient to prove that $M(t, x) \leqslant \frac{3}{2} M(t_0, x_0)$ when $|t - t_0| \leqslant 2M(t_0, x_0)^{-1}$. Indeed, (t_0, x_0) then falls in an analogous neighborhood of any point (t, x) in $\Omega(t_0, x_0)$, and therefore $M(t_0\ x_0) \leqslant \frac{3}{2} M(t, x)$.

Note that $M(t, x) = |\delta_{i0} D_t^i B(t, x) \lambda^{k+1}|^{1/(i+1)}$ for some i, $0 \leqslant i \leqslant k$. We have

$$|D_t^i B(t, x) - D_t^i B(t_0, x_0)| \leqslant \sum_{j=i+1}^{k} |D_t^j B(t_0, x_0)| \frac{j!}{(j-i)!} (M/2)^{i-j} +$$

$$\sum_{j=i+1}^{k} |D_t^j D_x B(t_0, x_0)| \frac{j!}{(j-i)!} (M/2)^{i-j} \lambda^{-(k+1)/2} +$$

$$O(M^{i-k-1}) \leqslant \sum_{j=i+1}^{k} \delta_{j0}^{-1} \frac{j!}{(j-i)!} 2^{j-i} M^{i+1} \lambda^{-k-1} +$$

$$\sum_{j=i+1}^{[k/2]} \frac{\gamma_s}{\gamma_j} \frac{j!}{(j-i)!} 2^{j-i} M^{i+1} \lambda^{-k+1} + C\lambda^{-k-2} M^{i+1},$$

since $B(t, x) = A(t)C(t, x)$ and the derivatives $D_t^i C$ are bounded, and

$$|D_t^j A(t_0)| \leqslant \gamma_s \gamma_j^{-1} a M^{j-s}/j! \leqslant \gamma_s \gamma_j^{-1} M^{j+1} \lambda^{-(k+1)/2} j!$$

If the numbers δ_{j0}, γ_j are such that

$$\sum_{j=i+1}^{k} \delta_{j0}^{-1} \delta_{i0} \frac{j!}{(j-i)!} 2^{j-i} \leqslant \frac{1}{8}; \quad \delta_{i0} \sum \gamma_s \gamma^{-1} \frac{j!}{(j-i)!} 2^{j-i} \leqslant \frac{1}{8}; \quad C\delta_{i0} \leqslant \frac{1}{8} \lambda_0 \tag{4.18}$$

and $\lambda \geqslant \lambda_0$ with sufficiently large λ_0, then we have

$$|D_t^i B(t, x)| \leqslant M^{i+1} \delta_{i0}^{-1} \lambda^{-k-1} (1 + 3/8)$$

so that

$$|M(t, x)| = |\delta_{i0} D_t^i B(t, x) \lambda^{k+1}|^{1/(i+1)} \leqslant M \left(\frac{3}{2}\right)^{1/(i+1)} \leqslant \frac{3}{2} M$$

and the lemma is proved.

4.8. Let now $H(t, x) = h((t-t_0)M) h((x-x_0)\lambda^{(k+1)/2})$, $h \in C_0^\infty(\mathbb{R})$, $h \geqslant 0$, $h(s) = 0$ when $|s| \geqslant 1$, and $h(s) = 1$ when $s \leqslant 1/2$. If

$$Lu = u_t + A(t) D_x u + \lambda^{k+1} B(t, x) u,$$

then

$$\|[L, H]u\|_0 \leqslant M \|H_t' u\|_0 + CM \|H_x' u\|_0,$$

where the constant C depends only on γ_j.

We shall now prove

<u>Lemma 4.12.</u>

$$\|LHu\|_0^2 \geqslant c_0 M^2 \|Hu\|_0^2 + c_0 \|D_1 Hu\|_0^2, \tag{4.19}$$

where c_0 is a positive constant independent of λ and $c_0 > (C + 2)^2$.

<u>Proof.</u> Let $h_j \in C_0^\infty(\mathbb{R})$, $\sum h_j^2(t) = 1$, $h_j \geqslant 0$, where $|t - j| \leqslant 2$ for $t \in \operatorname{supp} h_j$, and $|D^l h_j| \leqslant C_l$ for all l.

Suppose that $H_j(D_x) = h_j(D_x \lambda^{-(k+1)/2})$. Note that

$$[H_j(D_x), \lambda^{k+1} B(t, x)] = \sum_{l=1}^{k} \frac{1}{l!} \lambda^{(k+1)(1-l/2)} D_x^l B(t, x) H_j^{(l)}(D_x).$$

If $(t, x) \in \Omega(t_0, x_0)$, then

$$|\lambda^{(k+1)(1-l/2)} D_x^l B(t, x)| \leqslant C_1 M.$$

Indeed, for $l \geqslant 2$ it is obvious, but for $l = 1$ this expression is estimated by $C_2 \lambda^{(k+1)/2} |A(t)|$ and $|A(t)| \leqslant C_3 M \lambda^{-(k+1)/2}$ in view of the assumption. Therefore

$$\sum_j \|[H_j(D_x),\ \lambda^{k+1}B(t,\ x)]\,H(t,\ x)\,u\|_0^2 \leqslant C_4M^2\|Hu\|_0^2, \tag{4.20}$$

where the constant C_4 is independent of λ.

Furthermore, it is clear that, if $\xi^j \in \operatorname{supp} H_j$, then

$$\sum_j \|A(t)(D_x-\xi^j)\,H_j(D_x)\,H(t,\ x)\,u\|_0^2 \leqslant C_5M^2\|Hu\|_0^2, \tag{4.21}$$

since $|A(t)| \leqslant C_3M\lambda^{-(k+1)/2}$ and $|\xi - \xi^j| \leqslant 2\lambda^{(k+1)/2}$, provided $\xi \in \operatorname{supp} H_j$.

Let us now consider the operator L_j for which

$$L_ju = u_t + \xi^jA(t)\,u + \lambda^{k+1}B(t,\ x)\,u.$$

It satisfies the conditions of Theorem 4.1 and the function M, defined for this operator in Section 4.3, agrees with the function considered by us. This function depends, of course, on x, but the dependence is unimportant by Lemma 4.11. Using Lemma 4.2 and Theorem 4.2 it may be asserted that

$$\delta_{k0}^{-2/(k+1)}M^2\|H_j(D_x)\,Hu\|_0^2+\|D_tH_j(D_x)\,Hu\|_0^2 \leqslant C_6\|L_jH_j(D_x)\,Hu\|_0^2.$$

From this inequality and estimates (4.20), (4.21) it follows that

$$\delta_{k0}^{-2/(k+1)}M^2\|Hu\|_0^2+\|D_tHu\|_0^2 \leqslant C_6\|LHu\|_0^2 - C_7M^2\|Hu\|_0^2,$$

and therefore inequality (4.19) holds, provided

$$\delta_{k0}^{-2/(k+1)} > (C+2)^2 + C_7.$$

Lemma 4.12 is proved.

We proceed to consider the second case.

II. $a > M^{s+1}\lambda^{-(k+1)/2}$.

In this case the problem becomes essentially two-dimensional and does not reduce to Theorem 4.1. Suppose that

$$\Omega(t_0,\ x_0) = \{(t,\ x) \in \mathbb{R}^2,\ |t-t_0| < M^{-1},\ |x-x_0| < aM^{-s-1}\}.$$

4.9. As in the first case, we shall first prove that a and M vary little in $\Omega(t_0,\ x_0)$.

<u>Lemma 4.13.</u> Let $B(t,\ x) = \sum b_{ij}(t-t_0)^i(x-x_0)^j$. Then

$$\lambda^{k+1}|b_{ij}| \leqslant Ca^{-j}M^{i+j(s+1)+1}, \tag{4.22}$$

where $C = \delta_{i+js,\ j}^{-1}$.

<u>Proof.</u> Having fixed i and j in such a manner that $i + j(s+1) \leqslant k - 1$, $i \geqslant 1$ (inequality (4.22) obviously holds for $i + j(s+1) \geqslant k$), we examine the expression

$$c_{ij} = \partial_t^{js}(A(t)\,\partial_x)^j\,\partial_t^iB(t,\ x) = \partial_t^{js}\sum_{p\geqslant i}\sum_{q\geqslant j}\frac{p!\,q!}{(p-i)!\,(q-j)!}\,b_{pq}(t-t_0)^{p-i}(x-x_0)^{q-j}A(t)^j =$$

$$\sum_{p\geqslant i}\sum_{q\geqslant j}\sum_{k_1,\ \dots,\ k_j}\frac{p!\,q!}{(p-i)!\,(q-j)!}\,b_{pq}a_{k_1}a_{k_2}\dots$$

$$a_{k_j}\partial_t^{js}[(t-t_0)^{p-i+k_1+\dots+k_j}](x-x_0)^{q-j}.$$

At $(t,\ x) = (t_0,\ x_0)$ this expression equals

$$c_{ij} = \sum_{k_1+\dots+k_j+p=i+js}\frac{(js)!\,p!\,q!}{(p-i)!}\,b_{pj}a_{k1}a_{k2}\dots a_{kj}.$$

If $i = k$, then $j = 0$ and $c_{k0} = b_{k0}$. Therefore,

$$|\delta_{k0} b_{k0} \lambda^{k+1}|^{1/(k+1)} \leqslant M \quad \text{and} \quad \lambda^{k+1} |b_{k0}| < \delta_{k0}^{-1} M^{k+1},$$

so that (4.22) is valid. Suppose it has been proved for the coefficients b_{pq} when $p > i$; we shall now prove it for $p = i$. Note that for $s = 0$ the proof is simplified, since $c_{ij} = i!j!b_{ij}a^j$. It may therefore be assumed that $s > 0$.

Suppose that the term for which $p = i$, $k_1 = \ldots = k_j = s$, dominates in the sum defining c_{ij}, that is, $(js)!j!i!a^j b_{ij}$.

By the definition of function M, we have

$$\lambda^{k+1} |c_{ij}| \leqslant \delta_{i+js,\,j}^{-1} M^{i+j(s+1)+1}.$$

On the other hand,

$$|c_{ij} - (js)!\, j!\, i!\, a^j b_{ij}| \leqslant \sum{}' i!\, j!\, (js)!\, |b_{ij}|\, a^j \frac{\gamma_s}{\gamma_{k_1}} \cdots \frac{\gamma_s}{\gamma_{k_j}} + \sum{}'' \frac{(js)!\, p!\, j!}{(p-i)!} \delta_{p+js,\,j}^{-1} \frac{\gamma_s}{\gamma_{k_1}} \cdots$$

$$\frac{\gamma_s}{\gamma_{k_j}} \lambda^{-k-1} M^{p+j(s+1)+1} M^{(k_1-s)+\cdots+(k_j-s)},$$

where here and below Σ' denotes summation over $k_1, \ldots, k_j$ for which $k_1 + \ldots + k_j = js$, $(k_1 - s)^2 + \ldots + (k_j - s)^2 \neq 0$ and Σ'' the sum over $p, k_1, \ldots, k_j$, for which $p > i$, $p + k_1 + \ldots + k_j = i + js$. Let

$$\sum{}' \frac{\gamma_s}{\gamma_{k_1}} \cdots \frac{\gamma_s}{\gamma_{k_j}} \leqslant \frac{1}{4}. \tag{4.23}$$

Then

$$\lambda^{k+1} |b_{ij}| \leqslant \frac{4}{3} a^{-j} \left(\lambda^{k+1} \frac{|c_{ij}|}{i!\, j!\, (js)!} + \sum{}'' \frac{p!}{i!\, (p-i)!} \delta_{p+js,\,j}^{-1} \frac{\gamma_s}{\gamma_{k_1}} \cdots \frac{\gamma_s}{\gamma_{k_j}} M^{i+j(s+1)+1} \right) \leqslant$$

$$\frac{4}{3} a^{-j} M^{i+j(s+1)+1} \left[\frac{1}{i!\, j!\, (js)!} \delta_{i+js,\,j}^{-1} + \sum{}'' \frac{p!}{i!\, (p-i)!} \delta_{p+js,\,j}^{-1} \frac{\gamma_s}{\gamma_{k_1}} \cdots \frac{\gamma_s}{\gamma_{k_j}} \right].$$

Inequality (4.22) follows immediately, if ($\Sigma''' = \sum_{p=i+1}^{k}$ subject to $k_1 + \ldots + k_j = i - p + js$)

$$\frac{1}{i!\, j!\, (js)!} \delta_{i+js,\,j}^{-1} + \sum{}''' \frac{p!}{i!\, (p-i)!} \delta_{p+js,\,j}^{-1} \frac{\gamma_s}{\gamma_{k_1}} \cdots \frac{\gamma_s}{\gamma_{k_j}} \leqslant \frac{3}{4} \delta_{i+js,\,j}^{-1},$$

or, since $s \geqslant 2$ and therefore $i!j!(js)! \geqslant 2$,

$$\sum{}''' \frac{p!}{i!\, (p-i)!} \delta_{p+js,\,j}^{-1} \frac{\gamma_s}{\gamma_{k_1}} \cdots \frac{\gamma_s}{\gamma_{k_j}} \leqslant \frac{1}{4} \delta_{i+js,\,j}^{-1}. \tag{4.24}$$

Thus, if the conditions (4.23) and (4.24) are fulfilled, then Lemma 4.13 is proved.

4.10. Let the condition

$$\sum_{j=s+1}^{[k/2]} \gamma_s \gamma_j^{-1} 2^{j-s} \leqslant c_0 \tag{4.25}$$

be fulfilled, where c_0 is a small number. Then on every interval $[t_1, t_2]$ of length $M^{-1}/4$ one may find an interval of length $2/kM$ on which $|A(t)| \geqslant c_1 a M^{-s}$, where $c_1 = 2^{-4s-2}(s + 1)^{-s}$.

Proof. Let for definiteness $A(t) \geqslant 0$. For $|t - t_0| \leqslant 2M^{-1}$ we have

$$\left| \sum_{j=s+1}^{[k/2]} a_j t - t_0^j \right| \leqslant \sum_{j=s+1}^{[k/2]} \gamma_s \gamma_j^{-1} a M^{-s} 2^{j-s} \leqslant c_0 a M^{-s}.$$

Let l be such that $1 \leqslant l \leqslant 2(s+1)$ and the polynomial $\frac{a}{2} t^s + \sum_{j=0}^{s-1} a_j t^j$ be nonvanishing for

$$t_1 + \frac{l-1}{8M(s+1)} \leqslant t \leqslant t_1 + \frac{l+1}{8M(s+1)}.$$

Then on some interval of length $[8M(s+1)]^{-1}$ this polynomial assumes values with absolute value $\geqslant \frac{a}{2}[16M(s+1)]^{-s} = a2^{-4s-1}M^{-s}(s+1)^{-s}$. If $c_0 < 2^{-4s-2}(s+1)^{-s}$, then we have

$$|A(t)| \geqslant 2^{-4s-2} a M^{-s} (s+1)^{-s}$$

on this interval. The lemma is proved.

Now we shall prove that M varies little in Ω.

Lemma 4.15. Let c_0 and c_1 be sufficiently small numbers which will be given later, and let the inequalities

$$\sum_{s+1}^{[k/2]} \gamma_s \gamma_l^{-1} \leqslant c_0, \tag{4.26}$$

$$\sum_{p+q\geqslant 1} \delta^{-1}_{\bar{i}+p+(j+q)(s+1),\, j+q} \leqslant c_1 \delta^{-1}_{\bar{i}+j(s+1),\, j} \tag{4.27}$$

be fulfilled. Then

$$\frac{2}{3} M(t_0, x_0) \leqslant M(t, x) \leqslant \frac{3}{2} M(t_0, x_0), \qquad \frac{2}{3} a \leqslant a_s(t) \leqslant \frac{3}{2} a,$$

provided $(t, x) \in \Omega(t_0, x_0) = \{(t, x) : |t - t_0| \leqslant M^{-1},\ |x - x_0| \leqslant aM^{-s-1}\}$, $a_s(t) = \partial_t^s A(t)/s!$

Proof.

a. Note that the estimates

$$|a_l(t) - a_l(t_0)| \leqslant \sum_{j=l+1}^{[k/2]} l!\, |a_j(t_0)(t-t_0)^{j-l}|/(j-l)! \leqslant$$

$$\sum_{j=l+1}^{[k/2]} l!\, \gamma_s \gamma_l^{-1} a M^{l-s} l/(j-l)! \leqslant [k/2]! c_0 \gamma_s \gamma_l^{-1} a M^{l-s}$$

are valid for $0 \leqslant l \leqslant [k/2]$.

In particular, we have

$$|a_s(t) - a| \leqslant c_0 a [k/2]! \text{ for } l = s.$$

b. First we shall prove that $M(t, x_0) \leqslant \frac{5}{4} M(t_0, x_0)$ for $|t - t_0| \leqslant 2M^{-1}$. Upon interchanging of the points t and t_0 it follows from this estimate that $M(t_0, x_0) \leqslant \frac{5}{4} M(t, x_0)$ for $|t - t_0| \leqslant M^{-1}$.

Let for some i and j, $i + j \leqslant k$,

$$M(t, x_0)^{i+j+1} = \delta_{ij} \lambda^{k+1} |\partial_t^{l_j} A(t) \partial_x \dots \partial_t^{l_1} A(t) \partial_x \partial_t^{l_0} B(t, x)|,$$

where $l_0 + \dots + l_j = i$, $l_0 \geqslant 1$. Then

$$M(t, x_0)^{i+j+1} \leqslant M(t_0, x_0)^{i+j+1} +$$

$$\delta_{ij}\lambda^{k+1}\sum_{q=1}^{k-i-j}\frac{1}{q!}\partial_t^{q+l_1}A(t)\partial_x\partial_t^{l_j}A(t)\partial_x\partial_t^{l_0}B(t,x)(t-t_0)^q|_{\substack{t=t_0\\x=x_0}}\leqslant$$

$$M(t_0, x_0)^{i+j+1}+\delta_{ij}\sum_{q=1}^{k-i-j}\frac{1}{q!}\delta_{c+q,j}^{-1}M(t_0, x_0)^{i+q+j+1}(2M^{-1})^q\leqslant\frac{5}{4}M(t_0, x_0)^{i+j+1},$$

provided

$$\sum_{q=1}^{k-i-j}\frac{2^q}{q!}\delta_{i+q,j}^{-1}\leqslant c_1\delta_{i,j}^{-1}$$

and c_1 is sufficiently small.

c. In order to show that M varies little as x is varied, we choose a point $t' \in [t_0 - M^{-1}, t_0 + M^{-1}]$ in accord with Lemma 4.14 so that $A(t') \geqslant caM^{-s}$. Note that $|A^{(l)}(t')| \leqslant c(\gamma)aM^{l-s}$, where $C(\gamma)$ represents the constants depending on $\gamma_0, \ldots, \gamma_{[k/2]}$ and k, but independent of δ_{ij}. Furthermore, $|\partial_t^l A(t')^{-q}| \leqslant c(\gamma)a^{-q}M^{l+sq}$, where by a and s we mean their values at (t', x_0).

Let x be any value for which $|x - x_0| \leqslant c_1(\gamma)aM^{-s-1}$ and

$$M(t', x)^{i+j+1}=\delta_{ij}\lambda^{k+1}|F_{l_0,\ldots,l_j}(t', x)|,$$

where $F_{l_0,\ldots,l_j}(t, x)=\partial_t^{l_j}A(t)\partial_x\ldots\partial_t^{l_1}A(t)\partial_x\partial_t^{l_0}B(t, x)$, $l_0\geqslant 1$, $l_0+\ldots+l_j=i$. Then $|\partial_t^q F_{l_0,\ldots,l_j}(t', x)|\leqslant\delta_{i+q,j}^{-1}\lambda^{-k-1}M(t', x)^{i+j+q+1}$ and

$$|\partial_x F_{l_0,\ldots,l_j}(t', x)|\leqslant C^{-1}a^{-1}M^s|A(t)\partial_x F_{l_0,\ldots,l_j}(t', x)|=$$

$$C^{-1}a^{-1}M^s|F_{l_0,\ldots,l_j;0}(t', x)|\leqslant C^{-1}\delta_{i,j+1}^{-1}a^{-1}\lambda^{-k-1}M^{i+j+2+s}.$$

Similarly,

$$|\partial_x^p F_{l_0,\ldots,l_j}(t', x)|\leqslant(CaM^{-s})^{-p}|A(t)^p\partial_x^p F_{l_0,\ldots,l_j}(t', x)|=$$

$$(CaM^{-s})^{-p}|F_{l_0,\ldots,l_j,0,\ldots,0}(t', x)|\leqslant\delta_{i,j+p}^{-1}C^{-p}a^{-p}\lambda^{-k-1}M^{i+j+1+p+ps}.$$

Therefore

$$M(t', x)^{i+j+1}\leqslant M(t', x_0)^{i+j+1}+\delta_{ij}\lambda^{k+1}\sum_{p=1}^{k-i-j}\frac{1}{p!}|\partial_x^p F_{l_0,\ldots,l_j}(t', x_0)(x-x_0)^p|\leqslant$$

$$M(t', x_0)^{i+j+1}+\delta_{ij}\sum_{p=1}^{k-i-j}\frac{1}{p!}a^{-p}M^{ps+i+j+p+1}\delta_{i,j+p}^{-1}a^pM^{-p(s+1)}c_p(\gamma)\leqslant$$

$$M(t', x_0)^{i+j+1}\left[1+\delta_{ij}\sum_{p=1}^{k-i-j}\delta_{i,j+p}^{-1}c_p(\gamma)\right]\leqslant\frac{5}{4}M(t', x_0)^{i+j+1},$$

if the constant c_1 in (4.27) is small enough.

Lemma 4.15 is proved.

Lemma 4.16. The estimates

$$|c_{ij}|\leqslant C(\delta)a^{-j-1}M^{i+(j+1)(s+1)}\lambda^{-k-1},\quad |C(t, x)|\leqslant C(\delta)a^{-1}M^{s+1} \tag{4.28}$$

are valid; here $c_{ij}=D_t^iD_x^jC(t_0, x_0)$, $C(t, x)=B(t, x)A^{-1}(t)$, $(t, x)\in\Omega(t_0, x_0)$, and the constant $C(\delta)$ depends on the choice of constants δ_{ij}.

Proof. It is sufficient to verify these inequalities for $i = (j + 1)(s + 1) \leqslant k$, as for other values of i and j they are automatically fulfilled since $M \geqslant \lambda$.

First suppose that $A(t_0) \geqslant c_0 a M^{-s}$, where $c_0 = (s+1)^{-s}/2$. Then $C_{0j} = A(t_0)^{-1} b_{0j}$ and therefore $|C_{0j}| \leqslant C a^{-j-1} M^{(j+1)(s+1)} \lambda^{-k-1}$ by Lemma 4.13. Applying the formula $b_{ij} = \sum a_l C_{lj}$, where the sum is over l from 0 to i, one can easily verify by induction on i that $|C_{ij}| \leqslant C(\delta) a^{-j-1} \lambda^{-k-1} M^{i+(j+1)(s+1)}$.

If the inequality $A(t_0) \geqslant c_0 a M^{-s}$ is not fulfilled, then there exists a point t' such that $A(t') \geqslant c_0 a M^{-s}$ and $|t_0 - t'| \leqslant M^{-1}$. Therefore, the estimates (4.28) hold at (t', x_0). But a shift by M^{-1} along the t-axis does not vary the estimates, only the constant $C(\delta)$ changes insignificantly.

The lemma is proved.

4.11. Let now

$$H(t, x, D_x) = h_1(t) h((x - x_0) a^{-1} M^{s+1}) h(\delta_0 a M^{-s-1}(D_x - \xi_0)),$$

where h is the function defined in Section 4.8 and $\delta > 0$ is a small number which will be chosen later, and $h_1 \in C_0^\infty(\mathbb{R}^1)$ is a function with values in [0, 1] such that $h_1(t) = 1$ when $|t - t_0| \leqslant 3/4M$, $h_1(t) = 0$ when $|t - t_0| \geqslant M^{-1}$, and the inequality $A(t) \geqslant c_1 a M^{-s}$ ($c_1 = \text{const} > 0$) is fulfilled in the support of function $h_1(t)$. The latter is possible by Lemma 4.14. We have

$$[H, L]u = MH'_t u + a^{-1} M^{s+1} A(t) H'_x u +$$

$$\lambda^{k+1} A(t) \sum_{p=1}^{k} \frac{(i\delta_0)^p}{p!} (aM^{-s-1})^p \partial_x^p C(t, x) (\partial_{D_x}^p H) u. \tag{4.29}$$

Now we prove

Lemma 4.17.

$$c_0 M \| Hu \|_0 + c_1 \| D_1 Hu \|_0 \leqslant$$

$$\| HLu \|_0 + M \| H'_t u \|_0 + M \| H'_x u \|_0 + M \sum_{p=1}^{k} \| (\partial_{D_x}^p H) u \|_0, \tag{4.30}$$

where c_0 and c_1 are positive constants, $c_0 = \delta_{i,j}^{-1/(i+j+1)} \gg 1$.

Proof. A. If s = 0, then $Ca \geqslant A(t) \geqslant C^{-1} a$ in Ω so that from the equality

$$\iint A^{-1} | Lv |^2 \, dx \, dt = \iint A^{-1} | \partial_t v |^2 \, dx \, dt + \iint A | lv |^2 \, dx \, dt - \lambda^{k+1} \iint (\partial_t C) | v |^2 \, dx \, dt,$$

where $l = D_x - \xi_0 + \lambda^{k+1} C(t, x)$, $v = Hu$, it follows that

$$\| \partial_t v \|_0^2 + \| Alv \|_0^2 \leqslant C^2 \| Lv \|_0^2,$$

since $\partial_t C \geqslant 0$. In Section 4.12 we shall prove that

$$\delta_{i,j}^{-2/(i+j+1)} \| Mv \|_0^2 \leqslant C_0 (\| D_t v \|_0^2 + \| Alv \|_0^2), \tag{4.31}$$

where i, j are some integers, $i + j \leqslant k - 1$, so that the estimate

$$\| D_t v \|_0^2 + \| Alv \|_0^2 + \delta_{i,j}^{-2/(i+j+1)} M^2 \| v \|_0^2 \leqslant C^2 (1 + c_0) \| Lv \|_0^2 \tag{4.32}$$

holds for s = 0, $\lambda \geqslant \lambda_0$.

From equality (4.29) and estimate (4.28) follows inequality (4.30), provided

$$\delta_{i,j}^{-2/(i+j+1)} \geqslant c_1, \quad \delta_0 C(\delta) \leqslant 1, \tag{4.33}$$

where c_1 is a constant independent of λ and δ_{ij}.

B. Let now $s \neq 0$ (because $s \geqslant 2$). Denote by G_1 the set of those points $(t, x) \in \Omega(t_0, x_0)$ for which $A(t) \geqslant \varepsilon a M^{-s}$, where ε is some positive number which will be indicated later, and $G_0 = \Omega(t_0, x_0) \setminus G_1$. (For definiteness we assume that $A(t) \geqslant 0$ in $\Omega(t_0, x_0)$.)

By (4.29) we have

$$\iint_{G_1} A^{-1}(t) |\partial_t v|^2 dx\,dt + \iint_{G_1} A(t) |lv|^2 dx\,dt + 2\,\mathrm{Re} \iint_{G_1} \partial_t v \overline{lv}\, dx\,dt \leqslant C_1 \Bigg[\iint_{G_1} A^{-1}(t) |HLu|^2 dx\,dt + M^2 \iint_{G_1} A^{-1}(t) |(\partial_t H) u|^2 dx\,dt + a^{-2} M^{2(s+1)} \iint_{G_1} A(t) |H'_x u|^2 dx\,dt + \sum_{p=1}^{k} \delta_0^{2p} C(\delta) M^2 \iint_{G_1} A(t) |(\partial^p_{D_x} H) u|^2 dx\,dt\Bigg], \tag{4.34}$$

where the constant C_1 depends only on k.

Note that

$$2\,\mathrm{Re} \iint_{G_1} \partial_t v \overline{lv}\, dx\,dt = -2\,\mathrm{Re} \iint_{G_0} \partial_t v \overline{lv}\, dx\,dt - \lambda^{k+1} \iint \partial_t C(t, x) |v|^2 dx\,dt. \tag{4.35}$$

Let $A_0 = aM^{-s}$. Since in G_1 we have $\varepsilon A_0 \leqslant A(t) \leqslant CA_0$, from (4.34) and (4.35) it follows that

$$\frac{1}{C} \iint_{G_1} (|\partial_t v|^2 + |Alv|^2)\, dx\,dt \leqslant C_1 \Bigg[\frac{1}{\varepsilon} \iint_{G_1} |HLu|^2 dx\,dt + 2\,\mathrm{Re}\, A_0 \iint_{G_0} \partial_t v \cdot \overline{lv}\, dx\,dt + \frac{1}{c_0^2} M^2 \iint_{G_1} |(\partial_t H) u|^2 dx\,dt + CM^2 \iint_{G_1} |H'_x u|^2 dx\,dt + CM^2 \delta_0^2 C(\delta) \sum_{p=1}^{k} \iint_{G_1} |(\partial^p_{D_x} H) u|^2 dx\,dt\Bigg], \tag{4.36}$$

since $A(t) \geqslant c_0 A_0$, by construction, in the support of function $(\partial_t H)$. Furthermore, from (4.29) it is seen that

$$\frac{1}{C} \iint_{G_0} (|\partial_t v|^2 + |Alv|^2)\, dx\,dt \leqslant C_1 \Bigg[\iint_{G_0} |HLu|^2 dx\,dt + \varepsilon^2 A_0^2 \iint_{G_0} |lv|^2 dx\,dt + \varepsilon^2 M^2 \iint_{G_0} |H'_x u|^2 dx\,dt + \delta_0^2 \varepsilon^2 C(\delta) M^2 \iint_{G_0} \sum_{p=1}^{k} |(\partial^p_{D_x} H) u|^2 dx\,dt\Bigg].$$

Note that

$$2\,\mathrm{Re}\, A_0 \iint_{G_0} \partial_t v \overline{lv}\, dx\,dt = 2\,\mathrm{Re}\, A_0 \iint_{G_0} Lv \overline{lv}\, dx\,dt - 2\,\mathrm{Re}\, A_0 \iint_{G_0} A(t) |lv|^2 dx\,dt \leqslant$$

$$2\,\mathrm{Re}\, A_0 \iint_{G_0} Lv \overline{lv}\, dx\,dt \leqslant \frac{1}{\varepsilon} \iint_{G_0} |Lv|^2 dx\,dt + \varepsilon A_0^2 \iint_{G_0} |lv|^2 dx\,dt.$$

It is not hard to see that

$$\iint_{G_0} |lv|^2 dx\,dt \leqslant [C\delta_0^{-2} + C(\delta)]\, A_0^{-2} M^2 \iint_{G_0} |v|^2 dx\,dt + A_0^{-2} M^2 \iint_{G_0} |H_x u|^2 dx\,dt,$$

$$\iint_{G_0} |Lv|^2 dx\,dt \leqslant \iint_{G_0} |HLu|^2 dx\,dt + C\varepsilon^2 M^2 \iint_{G_0} |H'_x u|^2 dx\,dt +$$

$$\delta_0^2 \varepsilon^2 C(\delta) M^2 \iint_{G_0} \sum_{p=1}^{k} |(\partial^p_{D_x} H) u|^2 \, dx\, dt.$$

Thus

$$2 \operatorname{Re} A_0 \iint_{G_0} \partial_t v \bar{l v} \, dx\, dt \leq \frac{1}{\varepsilon} \iint_{G_0} |HLu|^2 \, dx\, dt +$$

$$\varepsilon [C\delta_0^{-2} + C(\delta)] M^2 \iint_{G_0} |v|^2 \, dx\, dt + \varepsilon M^2 (1 + C\varepsilon) \iint_{G_0} |H'_x u|^2 \, dx\, dt +$$

$$C(\delta)\, \delta_0^2 \varepsilon^2 M^2 \iint_{G_0} \sum_{p=1}^{k} |(\partial^p_{D_x} H) u|^2 \, dx\, dt.$$

Combining the inequalities obtained, we see that

$$\frac{1}{C} (\| \partial_t v \|^2 + \| Alv \|^2) \leq C_3 \Bigg[\frac{1}{\varepsilon} \| HLu \|^2 + \varepsilon [C\delta_0^{-2} + C(\delta)] M^2 \iint_{G_0} |v|^2 \, dx\, dt +$$

$$M^2 \iint_{G_1} |(\partial_t H) u|^2 \, dx\, dt + (1 + \varepsilon + \varepsilon^2) M^2 \iint |H'_x u|^2 \, dx\, dt +$$

$$\delta_0 C(\delta) (1 + \varepsilon + \varepsilon^2) M^2 \iint \sum_{p=1}^{k} |(\partial^p_{D_x} H) u|^2 \, dx\, dt \Bigg].$$

Hence the desired inequality (4.30) follows by (4.31), provided another inequality,

$$\varepsilon [C\delta_0^{-2} + C(\delta)] \leq 1, \tag{4.37}$$

is fulfilled besides (4.33).

Thus, Lemma 4.17 is proved.

4.12. Proof of inequality (4.31).

Let $L_1 = \partial_t$, $L_2 = A(t)\partial x + i\lambda^{k+1}B(t, x)$. By $e^{\tau L_j} v(t, x)$ we denote the solution of the Cauchy problem:

$$\partial_\tau z = L_j z, \quad z_j = v \text{ when } \tau = 0.$$

Clearly $\| e^{\tau L_j} \| = 1$.

Differentiating $r_j(\tau) = \| e^{\tau L_j} v - v \|^2$ with respect to τ, when $j = 1, 2$, we arrive at the inequality $r'_j(\tau) \leq 2\sqrt{r_j(\tau)} \| L_j v \|$ from which it follows that $\| e^{\tau L_j} v - v \| \leq |\tau| \| L_j v \|$. In the formula defining M suppose that a maximum is attained when we take an operator of the form

$$L_1^{l_\beta} L_2 L_1^{l_{\beta-1}} L_2 \ldots L_1^{l_1} L_2 L_1^{l_0} \lambda^{k+1} B(t, x),$$

where $l_0 + l_1 + \ldots + l_\beta = \alpha$, $l_0 \geq 1$. Lemma 4.15 implies that the inequalities

$$\frac{1}{2} M(t_0, x_0)^{\alpha+\beta+1} \leq c_{\alpha\beta}(t, x) \leq \frac{3}{2} M(t_0, x_0)^{\alpha+\beta+1},$$

where $c_{\alpha\beta}(t, x) = \delta_{\alpha\beta} \lambda^{k+1} L_1^{l_\beta} L_2 \ldots L_1^{l_1} L_2 L_1^{l_0} B(t, x)$, are fulfilled in $\Omega(t_0, x_0)$.

The Campbell-Hausdorff formula makes it possible to represent the operator of multiplication by $\exp(i\tau^{\alpha+\beta+1} c_{\alpha\beta} \delta_{\alpha\beta}^{-1})$ as the product of a large number of factors each equal to $e^{\pm\tau L_j}$ ($j = 1, 2$) (see L. Hörmander[8]). Assuming $\tau = M^{-1}(t_0, x_0)(\pi\delta_{\alpha\beta})^{1/(\alpha+\beta+1)}$ we obtain

$$| e^{i\tau^{\alpha+\beta+1} c_{\alpha\beta} \delta_{\alpha\beta}^{-1}} - 1 | \geq \sqrt{2}.$$

On the other hand, $\exp(i\tau^{\alpha+\beta+1} c_{\alpha\beta} \delta_{\alpha\beta}^{-1}) = (\Pi \exp(\pm \tau L_j))\, v$. Hence it follows that

$$| e^{i\tau^{\alpha+\beta+1} c_{\alpha\beta} \delta_{\alpha\beta}^{-1}} - 1 | \| v \| \leq \Sigma \| e^{\pm\tau L_j} v - v \| \leq C_1 \tau (\| L_1 v \| + \| L_2 v \|)$$

and therefore

$$M\delta_{\alpha\beta}^{-1/(\alpha+\beta+1)}\sqrt{2}\,\|v\|\leqslant C_1(\|L_1v\|+\|L_2v\|).$$

4.13. Selection of Parameters.

We write out all of the inequalities that are to be satisfied with parameters γ_j, δ_{ij}, ε, δ_0, λ_0 (here Σ' denotes summation subject to $k_1 + \ldots + k_j \geqslant js$, $(k_1 - s)^2 + \ldots + (k_j - s)^2 \neq 0$ and Σ'' summation from $p = i + 1$ to k over the cases $k_1 + \ldots + k_j = i - p + js$):

$$\sum_{j=i+1}^{k}\delta_{j0}^{-1}\frac{j!}{(j-i)!}2^{j-i}\leqslant\frac{\delta_{i0}^{-1}}{8},\quad \delta\delta_{i0}\sum\frac{\gamma_s}{\gamma_j}\leqslant\frac{1}{8},\quad C\delta_{i0}\leqslant\frac{1}{8}\lambda_0; \tag{4.18}$$

$$\sum{}'\frac{\gamma_s}{\gamma_{k_1}}\cdots\frac{\gamma_s}{\gamma_{k_j}}\frac{(k_1+\ldots+k_j)!}{(k_1+\ldots+k_j-js)!}\leqslant c_0, \tag{4.23}$$

$$\sum{}''\frac{p!}{i!\,(p-i)!}\frac{\gamma_s}{\gamma_{k_1}}\cdots\frac{\gamma_s}{\gamma_{k_j}}\delta_{p+js,\,j}^{-1}\leqslant\frac{1}{4}\delta_{i+js,\,j}^{-1}, \tag{4.24}$$

$$\sum_{j=1}^{[k/2]-i}2^j\gamma_{i+j}^{-1}\leqslant c_0\gamma_i^{-1}, \qquad (4.25),\ (4.26)$$

$$\sum_{p+q\geqslant1}\delta_{i+p+(j+q)(s+1),\,j+q}^{-1}\leqslant c_1\delta_{i+j(s+1),\,j}^{-1}, \tag{4.27}$$

$$\delta_{i,j}^{-2/(i+j+1)}\geqslant c_1,\quad \delta_0C(\delta)\leqslant1, \tag{4.33}$$

$$\varepsilon\,[C\delta_0^{-2}+C(\delta)]\leqslant1. \tag{4.37}$$

First we pick γ_j so that inequalities (4.23), (4.25), and (4.26) are fulfilled. For this to happen we assume that $\gamma_j' = \gamma_j^{-1}$. Conditions (4.23), (4.25), and (4.26) are written as

$$\sum\gamma_{k_1}'\cdots\gamma_{k_j}'\frac{(k_1+\ldots+k_j)!}{(k_1+\ldots+k_j-js)!}\leqslant c_0\gamma_s'^{\,j},\qquad \sum_{j=1}^{[k/2]-i}\gamma_{i+j}'2^j\leqslant c_0\gamma_i'.$$

Let $\gamma_0' = 1$ and $0 < \gamma_i' \leqslant 1$. Then, if among $k_1, \ldots, k_j$ there is even one index $>s$, these inequalities will be fulfilled, provided $\gamma_{s+1}' \leqslant c_0N^{-1}\gamma_s'^{\,k}$ for all $s = 0, 1, \ldots, [k/2]$, where N stands for the maximum possible number of terms in this sum. It may be assumed, for example, that $\gamma_j'=\rho^{(k+1)^j-1}$, where $\rho > 0$ is a sufficiently small number, $\rho < c_0/N$.

Having fixed γ_j we find δ_{ij} from (4.18) and (4.27). To do so we assume that $\delta_{ij}'=\delta_{i+js,\,j}^{-1}$. Inequalities (4.18) and (4.27) follow from the inequality

$$\sum_{\substack{p\geqslant i,\,q\geqslant j\\ p+q>i+j}}\delta_{p,\,q}'\leqslant c_1\delta_{i,\,j}' \tag{4.38}$$

with a sufficiently small constant c_1. For (4.38) to be fulfilled it is sufficient to assume that $\delta_{ij}'=C\rho^{(k+1)^{i+j}-1}$, where ρ is a sufficiently small number. We choose the constant C so large that

$$\delta_{ij}^{-2/(i+j+1)}\geqslant C_1$$

for all i, j, $i \geqslant 1$, $i + j \leqslant k - 1$. Then we pick δ_0 so small that (4.33) is fulfilled. Thereafter, ε is selected so that (4.37) is satisfied. Finally we fix the value of λ_0.

4.13. Proof of Theorems 4.3 and 4.4.

On the basis of the above results we can now prove inequalities (4.3) and (4.4).

Let (t_0, x_0) be an arbitrary point on the plane. Determine $M = M(t_0, x_0)$ by formula (4.16), and $a(t_0)$ as $|a_s(t_0)|$ for which

$$\gamma_j |a_j| M^{-j} \leqslant \gamma_s |a_s| M^{-s} \text{ when } j=0, 1, \ldots, k/2.$$

If $a \leqslant \lambda^{-(k+1)/2} M^{s+1}$, then with the point (t_0, x_0) we associate the neighborhood

$$\Omega(t_0, x_0) = \{(t, x) \in \mathbb{R}^2;\ |t-t_0| < M^{-1},\ |x-x_0| < \lambda^{-(k+1)/2}\}$$

and the function

$$H(t, x) = h((t-t_0)M)\, h((x-x_0)\lambda^{(k+1)/2}),$$

where $h \in C_0^\infty(\mathbb{R}^1)$ is a function with values in [0, 1], equal to 1 when $|t| \leqslant 1$ and 0 when $|t| \geqslant 3/2$.

If $a > \lambda^{-(k+1)/2} M^{s+1}$, then to (t_0, x_0) we associate a neighborhood

$$\Omega(t_0, x_0) = \{(t, x) \in \mathbb{R}^2;\ |t-t_0| < M^{-1},\ |x-x_0| < aM^{-s-1}\}$$

and a function

$$H(t, x) = h_1(t)\, h((x-x_0)a^{-1}M^{s+1}),$$

where h_1 is the function defined in Section 4.11.

The domain Q is covered with neighborhoods concentric with $\Omega(t_0, x_0)$ and having half the dimensions. Choose a covering of Q by neighborhoods such that every point of Q is covered by no less than four neighborhoods. Let $\Omega_1, \Omega_2, \ldots$ be domains forming this covering, and let $H_1, H_2, \ldots$ be functions corresponding to the covering. Then $1 \leqslant \sum H_j^2(t, x) \leqslant 4$.

Obviously, it will suffice to prove the inequality

$$\sum_j (c_0 M_j^2 \| H_j u \|_0^2 + \| D_t H_j u \|_0^2) \leqslant C \sum \| H_j L u \|_0^2, \tag{4.39}$$

where the constants c_0 and C are independent of j and λ, c_0 being $\gg C$.

Lemmas 4.12 and 4.16 reveal that

$$c_0 M_j^2 \| H_j u \|_0^2 + \| D_t H_j u \|_0^2 \leqslant C \| H_j L u \|_0^2 + C_1 M_j^2 \| (\partial_t H_j) u \|_0^2 + C_2 M_j^2 \| (\partial_x H_j) u \|_0^2 + C_3 M_j^2 \| H_j u \|_0^2$$

if the constants γ_j, δ_{ij} satisfy the above conditions. Thus, it is essential that $c_0 \gg C_1 + C_2 + C_3 + 1$. Therefore, summing these inequalities over j we arrive at (4.39).

Theorems 4.3 and 4.4 are completely proved.

§5. Sufficient Conditions for Subellipticity

5.1. In this section we shall prove that inequality (2.2) is fulfilled for conditions (Ψ') and ($\mathscr{B}$). By Theorem 2.2, estimate (2.3) follows from this inequality; that is, Theorems 1.2 and 1.3 are valid.

If the point (x_0, ξ_0) is such that $|\xi^0| = 1$, $p^0(x_0, \xi^0) \neq 0$, then there exist constants $c_0 > 0$ and $c > 0$ such that

$$\| p^0(x, \xi)\psi(y)\lambda^k \|_0 \geqslant c_0 \lambda^k \| \psi(y) \|_0,$$

$$\left\| \sum_{1 \leqslant |\alpha+\beta| \leqslant k} p^{0(\alpha)}_{(\beta)}(x, \xi) \frac{1}{\alpha!\,\beta!} y^\beta D^\alpha \psi(y) \lambda^{k-|\alpha|-|\beta|} \right\|_0 \leqslant C \sum_{1 \leqslant |\gamma+\delta| \leqslant k} \| y^\delta D^\gamma \psi \|_0 \lambda^{k-|\gamma|k-|\delta|}$$

for all (x, ξ) belonging to some neighborhood of the point (x_0, ξ^0) in $K \times S^{n-1}$. In this case, (2.2) follows from an inequality of the form

$$\| y^\delta D^\gamma \psi \|_0 \lambda^{k-|\gamma|k-|\delta|} \leqslant C_1 \lambda^{k-\varepsilon_0} \| \psi \|_0 + \lambda^{-\varepsilon_0} \sum_{|\alpha+\beta| \leqslant N} \| y^\beta D^\alpha \psi \|_0 \lambda^{-|\alpha|(k-1)} \tag{5.1}$$

for $1 \leqslant |\gamma + \delta| \leqslant k$, $\lambda \geqslant \lambda_0$.

Proposition 5.1. There exists a number C_1 such that (5.1) is fulfilled for all $\lambda \geqslant \lambda_0$ and all $\psi \in C_0^\infty(\mathbb{R}^n)$, provided $N > k + N\varepsilon_0$.

Proof. Note that

$$|y|^\delta \lambda^{k-|\gamma|k-|\delta|} \leqslant \lambda^l + |y|^{N-|\gamma|}\lambda^{-\varepsilon_0-|\gamma|(k-1)},$$

where $l = k - |\gamma|k - |\delta| + |\delta|(k - |\gamma| - |\delta| + \varepsilon_0)(N - |\gamma| - |\delta|)^{-1}$, since for $|y|^{N-|\gamma|-|\delta|} > \lambda^{k-|\delta|-|\gamma|+\varepsilon_0}$ this inequality is valid without the first term on the right-hand side, and in the complementary domain it holds without the second term. Hence it follows that

$$\| y^\delta D^\gamma \psi \|_0 \lambda^{k-|\gamma|k-|\delta|} \leqslant \lambda^l \| D^\gamma \psi \|_0 + \lambda^{-\varepsilon_0-|\gamma|(k-1)} \| |y|^{N-|\gamma|} D^\gamma \psi \|_0.$$

Similarly, $\lambda^l |\xi|^\gamma \leqslant \lambda^{k-\varepsilon_0} + \lambda^m |\xi|^N$, where $m = -\varepsilon_0 - N(k - 1)$ if $N - k > N\varepsilon_0$, and therefore $\lambda^l \| D^\gamma \psi \|_0 \leqslant \lambda^{k-\varepsilon_0} \| \psi \|_0 + \lambda^{-\varepsilon_0 - N(k-1)} \| |D|^N \psi \|_0$. From this follows our assertion.

5.2. Let now $p^0(x_0, \xi^0) = 0$. If we prove that (2.2) is fulfilled in some neighborhood ω of such a point (x_0, ξ^0) in $K \times S^{n-1}$, then by choosing a finite covering of compact $K \times S^{n-1}$ with such neighborhoods we find that this inequality with common constant C will be valid everywhere in $K \times S^{n-1}$.

Propositions 1.4, 1.8 and Theorem 2.1 of Chapter VI permit us to assume that $m = 1$ and $\operatorname{Im} p^0 = \xi_1$ in a neighborhood of the point (x_0, ξ^0) so that $p^0 = i\xi_1 + q(x, \xi)$. (Suppose that $\xi_1^0 = 0$.) In this case, (2.2) follows from the following estimate

$$\| \psi \|_0 + \| \partial_1 \psi \|_0 \leqslant c \Big[\| \partial_1 \psi + Q_\lambda \psi \|_0 + \lambda^{-\varepsilon_0} \sum_{|\alpha+\beta| \leqslant N} \lambda^{-|\alpha|(k-1)} (\| y^\beta D^\alpha \psi \|_0 + \| y^\beta D^\alpha \partial_1 \psi \|_0) \Big], \tag{5.2}$$

where

$$Q_\lambda \psi = \sum_{|\alpha+\beta| \leqslant k} \frac{1}{\alpha!\,\beta!} q^{(\alpha)}_{(\beta)}(x, \xi) y^\beta D^\alpha \psi \lambda^{k-|\alpha|k-|\beta|}.$$

Note that if $|\alpha + \beta| \geqslant 2$, $\alpha_1 \geqslant 1$, then

$$\| y^\beta D^\alpha \psi \|_0 \lambda^{k-|\alpha|k-|\beta|} \leqslant \lambda^{-1} \sum_{|\gamma+\delta| \leqslant k-1} \| y^\delta D^\gamma \partial_1 \psi \|_0 \lambda^{-|\gamma|(k-1)},$$

so that the terms of this type, which appear in Q_λ, could be eliminated, since they are estimated by the last term on the right-hand side of (5.2). Thereafter, we assume that

$$P_\lambda = (1 + i\rho)\partial_1 + \sum a_{\alpha\beta} y^\beta D^\alpha \lambda^{k-|\alpha|k-|\beta|},$$

where $|\alpha + \beta| \leqslant k$ and $\alpha_1 = 0$. Having divided P_λ by $1 + i\rho$ and having performed the canonical transformation for which the function $\xi_1 + \rho(1 + \rho^2)^{-1} \sum a_{\alpha\beta} x^\beta \xi^\alpha$ changes to ξ_1, we obtain an operator

$$P'_\lambda \psi = \partial_1 \psi + \sum b_{\alpha\beta} y^\beta D^\alpha \psi \lambda^{k-|\alpha|k-|\beta|},$$

where $|\alpha + \beta| \leqslant k$, $\alpha_1 = 0$, and $b_{\alpha\beta}$ are real, estimate (5.2) being equivalent to a similar estimate for the operator P'_λ.

To simplify notation, we replace y by $y\lambda$, and the number of variables n by n + 1 after t and the vector $(y^1, \ldots, y^n)$ have been substituted, respectively, for y^1 and $(y^2, \ldots, y^n)$. Then (5.2) takes the form

$$\lambda \| \psi \|_0 + \| \partial_t \psi \|_0 \leqslant$$

$$C\left[\| P_\lambda\psi\|_0+\lambda^{-\varepsilon_0}\sum_{\beta_0+|\alpha+\beta|\leqslant N}\lambda^{-|\alpha|k+|\beta|+\beta_0}(\| t^{\beta_0}y^\beta D^\alpha\psi\|_0+\| t^{\beta_0}y^\beta D^\alpha\partial_t\psi\|_0)\right], \qquad (5.3)$$

where

$$P_\lambda\psi=\partial_t\psi+Q_\lambda(t,\ y,\ D)\psi,\quad \alpha=(\alpha_1,\ \ldots,\ \alpha_n),\quad \beta=(\beta_1,\ \ldots,\ \beta_n),$$

$$Q_\lambda(t,\ y,\ D)=\sum_{\beta_0+|\alpha+\beta|\leqslant k}\frac{1}{\beta_0!\,\alpha!\,\beta!}\partial_t^{\beta_0}q_{(\beta)}^{(\alpha)}(t,\ x,\ \xi)\,t^{\beta_0}y^\beta D^\alpha\lambda^{(k+1)(1-|\alpha|)}$$

and $q(t, x, \xi)$ is a real-valued function. Inequality (5.3) is much stronger than (5.2), since the right-hand side has no derivatives with respect to t of order higher than 1.

5.3. Let h be a function $\in C_0^\infty(\mathbb{R})$ which assumes non-negative values equal to 1 for $|t| \leqslant 3/4$ and 0 for $|t| \geqslant 1$. We shall prove that (5.3) holds simultaneously with an analogous inequality for $H(t, y)\psi(y)$, where $H(t,\ y)=h(t\lambda^{1-\varepsilon_0})\prod_{i=1}^{n}h(y^i\lambda^{1-\varepsilon_0})$ (possibly with another constant). It is not hard to see that the norms in $L_2(\mathbb{R}^{n+1})$ of the commutators of H with P_λ, ∂_t, and $\lambda^{1-\varepsilon_0-|\alpha|k+|\beta|}y^\beta D^\alpha$, when $|\alpha + \beta| \leqslant N$, are estimated by

$$\sum_{|\alpha+\beta|\leqslant N}\| y^\beta D^\alpha\psi\|_0\,\lambda^{1-\varepsilon_0-|\alpha|k},$$

and the norm of the commutator with operator $\lambda^{1-\varepsilon_0-|\alpha|k+|\beta|}y^\beta D^\alpha\partial_t$ is estimated by

$$\sum_{|\alpha+\beta|\leqslant N}\lambda^{-\varepsilon_0-|\alpha|(k-1)}(\| y^\beta D^\alpha\partial_t\psi\|_0+\lambda\| y^\beta D^\alpha\psi\|_0).$$

On the other hand, (5.3) is always fulfilled for the function $(1 - H)\psi$, since in the support of this function $|y|+|t|\geqslant 3\lambda^{\varepsilon_0-1}/4$. Therefore

$$\lambda^{-\varepsilon_0+1}\left(\sum_{j=1}^{n}(\| y^j\psi\|_0\,\lambda+\| y^j\partial_t\psi\|_0)+\lambda\| t\psi\|_0+\| t\partial_t\psi\|_0\right)\geqslant\frac{3}{4}(\lambda\|\psi\|_0+\|\partial_t\psi\|_0).$$

Hence it is sufficient to verify (5.3) for the function $H\psi$. Arguing in the same fashion we see that it suffices to verify this inequality for the function

$$\prod_{i=1}^{n}h(D_i\lambda^{-\varepsilon_0-k})\,H(t,\ y)\,\psi(t,\ y).$$

Thus, in proving (5.3) we find of significance only those y where $|y| \leqslant \lambda^{\varepsilon_0-1}$, and only those points η of the support $\tilde{\psi}$ for which $|\eta|\leqslant\lambda^{\varepsilon_0+k}$.

5.4. The essence of the proof given below consists in using a sufficiently small partition of unity in the space of variables (t, y, η) and in a partial freezing of arguments (y, D) in the operator $P_\lambda(t, y, D)$ which transforms it into a first-order operator. We shall now define a function $M(t, y, \eta)$ for which the estimate

$$M(t,\ y,\ \eta)\|\psi\|\leqslant\| L_{(t,\ y,\ \eta)}\psi\|,$$

where $L_{(t,y,\eta)}$ stands for the first-order operator, is valid. This function is defined by the coefficients of the operator $P_\lambda(t, y, D)$.

Let i, j be non-negative integers and $i + j \leqslant k$. Let $a_1 = \tau$, $a_2 = Q_\lambda(t, y, \eta)$. Let $(\mathrm{ad}\ a_l)b = \{a_l, b\}$, where $\{a_l, b\}$ represents the Poisson bracket of functions a_l and b for $l = 1, 2$. Now examine the functions

$$c_{ij}(t, y, \eta) = \max_{\substack{i_1+\dots+i_k=i \\ j_1+\dots+j_k=j}} |(\mathrm{ad}\, a_2)^{j_k}(\mathrm{ad}\, a_1)^{i_k}\dots(\mathrm{ad}\, a_2)^{j_1}(\mathrm{ad}\, a_1)^{i_1} a_2|,$$

$$M(t, y, \eta) = \max_{i,j}[\delta_{ij}c_{ij}(t, y, \eta)]^{1/(i+j+1)}, \tag{5.4}$$

where δ_{ij} are positive constants defined in Section 4.13. For $i = j = 0$ we assume that $c_{00}(t, y, \eta) = |a_2(t, y, \eta)|$.

Note that $M \leqslant C\lambda^{k+1}$, where constant C depends only on $\max|D^\alpha q(t, x, \xi)|$ when $|\alpha| \leqslant k$. On the other hand, by condition $(\mathscr{B})$, the values $M(t, y, \eta)$ are bounded from below by $c^0\lambda$ with the constant $c^0 > 0$).

Until the end of this section we shall fix the point at which inequality (5.3) is proved. To simplify notations we shall assume that $t_0 = 0$, $x = 0$, $\xi = 0$ although formally this contradicts the condition $|\xi| = 1$.

Recall that $Q_\lambda(t, y, \eta) = \lambda^{k+1}T_k^{(t_0, x, \xi)}q(t_0+t, x+y, \xi+\eta\lambda^{-k-1})$.

5.5. As in Section 4, depending on the value of grad Q_λ we shall consider two cases:

$$\text{I. } |\lambda^{-k-1}\,\mathrm{grad}_y\, \partial_t^j Q_\lambda(t_0, y_0, \eta^0)| + |\mathrm{grad}_\eta\, \partial_t^j Q_\lambda(t_0, y_0, \eta^0)| \leqslant$$

$$\lambda^{-(k+1)/2}M^{j+1}(t_0, y_0, \eta^0); \qquad j = 0, 1, \dots, [k/2]; \tag{5.5}$$

$$\text{II. } \exists s,\ 0 \leqslant s \leqslant [k/2],\ \lambda^{-k-1}|\mathrm{grad}_y\, \partial_t^s Q_\lambda(t_0, y_0, \eta^0)| +$$

$$|\mathrm{grad}_\eta\, \partial_t^s Q_\lambda(t_0, y_0, \eta^0)| > \lambda^{-(k+1)/2}M^{s+1}(t_0, y_0, \eta^0). \tag{5.6}$$

<u>Case I.</u> Let

$$\Omega(t_0, y_0, \eta^0) = \Big\{(t, y, \eta) \in \mathbb{R}^{2n+1},\ |t - t_0| < M^{-1},\ |y - y_0| < \lambda^{-\frac{k+1}{2}},$$

$$|\eta - \eta^0| < \delta_0^{-1}\lambda^{\frac{k+1}{2}}\Big\}.$$

As in Section 4.7, we shall prove a number of lemmas on the structure of the symbol of operator P_λ in a neighborhood Ω.

<u>Lemma 5.1.</u> From (5.5) it follows that

$$M(t_0, y_0, \eta^0) = |\delta_{j0}\partial_t^j Q_\lambda(t_0, y_0, \eta^0)|^{1/(j+1)} \tag{5.7}$$

for some j, $0 \leqslant j \leqslant k$.

<u>Proof.</u> The results of Section 4 (Chapter V) indicate that M can be obtained in a different manner also, that is, by using only the monomials τ, $at^s\eta_1$, and $\lambda^{k+1}At^\alpha(y^1)^\beta$ for $s < k/2$, or a longer chain of monomials of this type.

Initially we shall consider the possibility when $M^{\alpha+\beta(s+1)+1} = \delta_{\alpha+\beta s,\beta} \times |Aa^\beta|\lambda^{k+1}$. From (5.5) it follows that $|a| \leqslant \lambda^{-(k+1)/2}M^{s+1}$ Therefore, for $\beta \geqslant 2$

$$|Aa^\beta|\lambda^{k+1} \leqslant C\lambda^{-(k+1)}M^{\beta(s+1)}\lambda^{k+1} = CM^{\beta(s+1)} \ll M^{\alpha+\beta(s+1)+1},$$

and we arrive at a contradiction. If $\beta = 1$, then, by virtue of (5.5), $|A| \leqslant \lambda^{-(k+1)/2}M^{\alpha+1}$ (for $\alpha \geqslant k/2$ this inequality also holds, since $M \geqslant c^0\lambda$) so that

$$|Aa|\lambda^{k+1} \leqslant M^{\alpha+1+s+1} < \delta^{1}_{\alpha+\beta s,\beta}M^{\alpha+s+2},$$

provided $\delta_{\alpha+\beta s,\beta}$ is small enough.

Similarly it can be verified that in the given case M cannot be obtained by using a longer chain: every time at least two coefficients satisfying (5.5) participate in the chain. Thus, Lemma 5.1 is proved.

Lemma 5.2. If (5.5) is fulfilled, then there exists a constant C independent of λ such that

$$\lambda^{-k-1}|\operatorname{grad}_y \partial_t^j Q_\lambda(t, y, \eta)| + |\operatorname{grad}_\eta \partial_t^j Q_\lambda(t, y, \eta)| \leqslant$$
$$C\lambda^{-(k+1)/2} M^{j+1}(t_0, y_0, \eta^0), \quad j = 0, 1, \ldots, [k/2], \tag{5.8}$$

when $(t, y, \eta) \in \Omega(t_0, y_0, \eta^0)$.

Proof. We have

$$(\lambda^{-k-1}\operatorname{grad}_y \partial_t^j Q_\lambda(t, y, \eta), \operatorname{grad}_\eta \partial_t^j Q_\lambda(t, y, \eta)) = \operatorname{grad}_{y,\eta} \partial_t^j q(t, y, \eta\lambda^{-k-1})$$

and

$$|\operatorname{grad}_{y,\eta} \partial_t^j q(t, y, \eta\lambda^{-k-1})| \leqslant \sum_{i=0}^{[k/2]-j} |\operatorname{grad}_{y,\eta} \partial_t^{i+j} q(t_0, y_0, \eta_0\lambda^{-k-1})| \frac{|t-t_0|^i}{i!} +$$
$$O(|t-t_0|^{[k/2]-j+1}) + O(|y-y_0| + |\eta-\eta^0|\lambda^{-k-1}) \leqslant$$
$$C_1\lambda^{-(k+1)/2}M^{j+1} + C_2\lambda^{-2-[k/2]}M^{j+1} + C_3\lambda^{-(k+1)/2} + C_4\lambda^{-(k+1)/2} \leqslant C\lambda^{-(k+1)/2}M^{j+1}$$

if $\lambda \geqslant \lambda_0$, λ_0 being large enough. The lemma is proved.

From Lemmas 5.1 and 5.2 it follows that (5.7) remains valid after the replacement of (t_0, y_0, η^0) by $(t, y, \eta) \in \Omega(t_0, y_0, \eta^0)$. Now we shall show that the values $M(t, y, \eta)$ in this domain hardly differ from $M(t_0, y_0, \eta^0)$.

Lemma 5.3. If $(t, y, \eta) \in \Omega(t_0, y_0, \eta^0)$, then

$$\frac{2}{3} M(t_0, y_0, \eta^0) \leqslant M(t, y, \eta) \leqslant \frac{3}{2} M(t_0, y_0, \eta^0).$$

Proof. It is sufficient to prove that $M(t, y, \eta) \leqslant \frac{3}{2} M$ for $|t-t_0| < 2M^{-1}$, $|y-y_0| < \lambda^{-(k+1)/2}$, and $|\eta-\eta^0| < \lambda^{(k+1)/2}$, since the second inequality is obtained by interchanging the positions of the points (t_0, y_0, η^0) and (t, y, η). Note that inequalities (5.8) remain valid also for the points (t, y, η), provided λ_0 is sufficiently large. Let

$$M(t, y, \eta) = |\delta_{i0} q_i \lambda^{k+1}|^{1/(i+1)},$$

where i is some integer, $0 \leqslant i \leqslant k$, $q_i = \partial_t^i q(t, y, \eta\lambda^{-k-1})/i!$ Then

$$M^{i+1}(t, y, \eta) = \delta_{i0}|q_i(t, y, \eta)|\lambda^{k+1} \leqslant$$
$$\delta_{i0}\lambda^{k+1}\Bigg|\sum_{j=0}^{k-i} \frac{1}{j!}\partial_t^{i+j} q(t_0, y_0, \eta^0\lambda^{-k-1})(t-t_0)^j +$$
$$\sum_{l=1}^{n} \partial_l \partial_t^i q(t_0, y_0, \eta^0)(y^l - y_0^l) +$$
$$\sum_{l=1}^{n} \partial_{\eta_l}\partial_t^i q(t_0, y_0, \eta^0\lambda^{-k-1})(\eta_l - \eta_l^0)\lambda^{-k-1} + O(\delta_0^{-2}\lambda^{-k-1})\Bigg| \leqslant$$
$$M^{i+1}\left(\sum_{j=0}^{k-i} 2^j\delta_{i0}\delta_{i+j,0}^{-1}/j! + C_1\delta_{i0} + C_2\delta_{i0}\delta_0^{-1} + C_3\lambda^{-1}\delta_0^{-2}\right),$$

since $M \geqslant c^0\lambda$. If

$$\sum_{j=1}^{k-i} 2^j \delta_{i,0} \delta_{i+j,0}^{-1} / j! \leqslant \frac{1}{4}, \qquad \delta_{i0} \leqslant c_0 \delta_0,$$

and $\lambda \geqslant \lambda_0$, λ_0 being sufficiently large, then it follows that $M^{i+1}(t, y, \eta) \leqslant \frac{3}{2} M^{i+1}$ and the lemma is proved.

5.6. Assuming

$$H(t, y, D) = h((t-t_0)M) \prod_{l=1}^{n} h((y^l - y_0^l)\lambda^{(k+1)/2}) h((D_l - \eta_l^0)\delta_0 \lambda^{-(k+1)/2}),$$

we find that the commutator $[H, P_\lambda]$ cannot be estimated and the function $P_\lambda H\psi$ is equivalent to the function

$$LH\psi = \partial_t H\psi + \lambda^{k+1} \sum_{j=0}^{k} \frac{1}{j!} \partial_t^j Q(t_0, x, \xi)(t-t_0)^j H\psi.$$

This enables us to make use of Theorems 4.1 and 4.2. Later we shall discuss this in more detail. Now we proceed to case II.

The proof above would have been sufficient if the estimates (5.5) had been fulfilled at all of the points (t, y, η). For example, this happens when $\mathrm{grad}_{x,\xi}\, q(t, x, \xi) = 0$ at every characteristic point.

5.7. Let now condition II be fulfilled and

$$|\mathrm{grad}_{y,\eta} \partial_t^s q(t_0, y_0, \eta^0)| > M^{s+1}\lambda^{-(k+1)/2}, \tag{5.9}$$

where s is an integer, $0 \leqslant s < k/2$. We may assume that in this case

$$\gamma_s M^{-s} |\mathrm{grad}_{y,\eta} \partial_t^s q(t_0, y_0, \eta^0)| = \max_j \gamma_j M^{-j} |\mathrm{grad}_{y,\eta} \partial_t^j q(t_0, y_0, \eta^0)|. \tag{5.10}$$

Suppose that $a = \mathrm{grad}_{y,\eta} |\partial_t^s q(t_0, y_0, \eta^0)|$.

Using the simplest canonical transformations: rotation and, perhaps, a Fourier transformation, one can obtain that

$$\mathrm{grad}_{y,\eta'} \partial_t^s q(t_0, y_0, \eta^0) = 0.$$

Now we define a canonical transformation of variables $(y, \eta) \mapsto (z, \zeta)$ after which

$$\mathrm{grad}_{z,\zeta'} \partial_1^j \partial_t^s q(t_0, z_0, \zeta^0) = 0, \qquad j = 0, 1, \ldots, l_0, \tag{5.11}$$

where $l_0 = [(k/2 - s)/(s+1)]$. To do so, we shift the origin to (t_0, y_0, η^0) and then define the canonical transformation using the generating function $y\zeta + S(y, \zeta)$ so that

$$z = y + S_\zeta(x, \zeta), \quad \eta = \zeta + S_y(y, \zeta).$$

Conditions (5.11) are equivalent to the equality

$$a\partial_1 S + \sum_{l=1}^{l_0} \sum_{j=2}^{n} [a_{jl}(y^1)^l y^s + b_l^j (y^1)^l [\zeta_j + \partial_j S(y, \zeta)]] + \sum_{l=1}^{l_0+1} c_l (y^1)^l = 0,$$

where

$$a_{jl} = \frac{1}{l!} \partial_j \partial_1^l \partial_t^s q(t_0, y_0, \eta^0), \quad b_l^j = \frac{1}{l!} \partial_{\eta_j} \partial_1^l \partial_t^s q(t_0, y_0, \eta^0), \quad c_l = \frac{1}{l!} \partial_1^l \partial_t^s q(t_0, y_0, \eta^0).$$

Here it may be assumed that $S = 0$ for $y^1 = 0$. Then

$$S(y, \zeta)=C(y^1)+\sum_{j=2}^{n} A_j(y^1) y^j+\sum_{j=2}^{n} B^j(y^1) \zeta_j, \tag{5.12}$$

where A_j, B^j, and C are polynomials in y^1, and

$$A_j(y^1)=-\frac{1}{a} \sum_{l=1}^{l_0} \frac{1}{l+1} a_{jl}(y^1)^{l+1}, \quad B^j(y^1)=-\frac{1}{a} \sum_{l=1}^{l_0} \frac{1}{l+1} b_l^j(y^1)^{l+1},$$

$$C(y^1)=-\frac{1}{a} \sum_{l=1}^{l_0+1} c_l(y^1)^{l+1} \frac{1}{l+1}-\frac{1}{a} \int_0^{y^1} \sum_{l=1}^{l_0} \sum_{j=2}^{n} b_l^j \tau^l A_j(\tau) d\tau .$$

Here a_{jl}, b_l^j, c_l are the values of the derivatives of q of order $\leqslant k$ and therefore are uniformly bounded. Hence for $|y^1| \leqslant aM^{-s-1}$ the estimates

$$|A_j^{(l)}(y^1)|+|B^{j(l)}(y^1)|+|C^{(l)}(y^1)| \leqslant Ca^{1-l} M^{(s+1)(l-2)}$$

are valid when $l = 0, 1$ and

$$|A_j^{(l)}(y^1)|+|B^{j(l)}(y^1)|+|C^{(l)}(y^1)| \leqslant Ca^{-1}$$

when $l \geqslant 2$. The desired transformation takes the form

$$z^1=y^1, \quad z^j=y^j+B^j(y^1), \quad \eta_j=\zeta_j+A_j(y^1), \quad j=2, \ldots, n,$$

$$\eta_1=\zeta_1+C'(y^1)+\sum_{l=2}^{n} A_l'(y^1) y^l+\sum_{l=2}^{n} B^{l\prime}(y^1) \zeta_l . \tag{5.13}$$

Therefore the function $q(t, y, \eta)$ maps to a function

$$q_0(t, z, \zeta)=q\Big(t, z^1, z'-B(z^1), \zeta_1+\sum_{l=2}^{n} B^{l\prime}(z^1) \zeta_l+$$

$$\sum_{l=2}^{n} A_l'(z^1) z^l+C_0(z^1), \zeta'+A(z^1)\Big).$$

in which $C_0(z^1)=C'(y^1)-\sum_{l=2}^{n} A_l'(y^1) B^l(y^1)$ and the inequalities

$$|\partial_\zeta^\alpha \partial_z^\beta q_0(t, z, \zeta)| \leqslant C_{\alpha, \beta} a^{1-\beta_1}$$

are valid for

$$|z^1| \leqslant CaM^{-s-1}, \quad |z^j| \leqslant \lambda^{-(k+1)/2}, \quad |\zeta_j| \leqslant \lambda^{(k+2)/2-1/3}.$$

5.8. Now we shall prove that

$$a^j \lambda^{(k+1)/2} |\operatorname{grad}_{z', \zeta'} \partial_1^j \partial_t^i q_0(t_0, z_0, \zeta^0)| \leqslant KM^{i+j(s+1)+\frac{1}{2}} \tag{5.14}$$

when $1 + j(s + 1) \leqslant k/2$.

Lemma 5.4. The inequalities

$$|\partial_1^j \partial_t^i q_0(t_0, z_0, \zeta^0)| \leqslant \delta_{i+js, j}^{-1} a^{-j} M^{i+j(s+1)+1} \lambda^{-k-1} \tag{5.15}$$

are valid for $i + j(s + 1) \leqslant k$.

The proof of this lemma is analogous to that of Lemma 4.13.

Lemma 5.5. There exists a constant K independent of λ such that inequalities (5.14) are fulfilled.

Proof. For $i + j(s + 1) \geqslant k/2$ these inequalities are always valid. Suppose that there are i_0, j_0 such that $0 \leqslant i_0 + j_0(s + 1) \leqslant k/2$ and

$$|\partial_{\zeta_2}\partial_1^j\partial_t^i q_0(t_0, z_0, \zeta^0)| > C_0 M^{i_0+j_0(s+1)+\frac{1}{2}} a^{-j_0}\lambda^{-(k+1)/2},$$

where C_0 is some number which will be indicated later. Let

$$\varphi(t, z^1, \zeta_1, \zeta_2) = q_0(t, z, \zeta) \text{ when } z' = z_0', \; \zeta'' = \zeta^{0''}.$$

Then

$$\varphi(t, z^1, \zeta_1, \zeta_2) = Q(t, z^1) + A(t)(\zeta_1 - \zeta_1^0) + B(t, z^1)(\zeta_2 - \zeta_2^0) + R,$$

where

$$Q(t, z^1) = \sum_{i+j(s+1)\leqslant k} q_{ij}(t-t_0)^i (z^1 - z_0^1)^j, \quad A(t) = \sum_{j=0}^{k} a_j (t-t_0)^j,$$

$$B(t, z^1) = \sum_{i+j(s+1)\leqslant k} b_{ij}(t-t_0)^i (z^1 - z_0^1)^j$$

are polynomials of degree k, and

$$|R(t, z^1, \zeta_1, \zeta_2)| \leqslant c[|\zeta_1 - \zeta_1^0|^2 + |\zeta_2 - \zeta_2^0|^2 + |z^1 - z_0^1||\zeta_1 - \zeta_1^0| +$$

$$\sum_{k+1\leqslant i+j(s+1)\leqslant k+s+1} |t-t_0|^i |z^1 - z_0^1|^j a^{-j}(1 + |\zeta_1 - \zeta_1^0| + |\zeta_2 - \zeta_2^0|)].$$

By Lemma 5.4 we find that for $|t - t_0| \leqslant \lambda^{-1}$, $|z^1 - z_0^1| \leqslant aM^{-s-1}$ the inequality

$$|Q(t, z^1)| \leqslant C_1 M\lambda^{-k-1}$$

is valid. Without loss of generality, it may be assumed that

$$|a_l| \leqslant CaM^{l-s}, \quad |b_{ij}| \leqslant CM^{i-i_0+(j-j_0)(s+1)} a^{j_0-j} b,$$

where $0 \leqslant l \leqslant k/2$, $0 \leqslant i + j(s+1) \leqslant k/2$, $b = |b_{i_0 j_0}|$, $0 \leqslant i_0 + j_0(s+1) \leqslant k/2$, and

$$a \geqslant M^{s+1}\lambda^{-(k+1)/2}, \quad b \geqslant KM^{i_0+j_0(s+1)+1/2} a^{-j_0}\lambda^{-(k+1)/2}.$$

Let us consider the auxiliary polynomials

$$a(t) = a^{-1}M^s A(t_0 + tM^{-1}), \qquad b(t, x^1) = b^{-1}M^{i_0+j_0(s+1)} a^{-j_0} B(t_0 + tM^{-1}, z_1^0 + ax^1 M^{-s-1}).$$

In the polynomial a the coefficient of t^s is equal to 1 and the remaining coefficients do not exceed C in absolute value. In the polynomial b the $t^s(x^1)^j$ coefficients are equal to 0, provided $s + j(s + 1) \leqslant k/2$; the $t^{i_0}(x^1)^{j_0}$ coefficient is equal to 1 and the remaining coefficients do not exceed C in absolute value. By $\mathfrak{M}$ we denote the collection of all pairs of the polynomials a and b, of degree $\leqslant$k, satisfying these conditions. Making use of the compactness of this set it is easy to show that the value

$$c_0 = \inf_{a,\, b\in\mathfrak{M}} \; \sup_{|t|\leqslant 1,\; |t'|\leqslant 1,\; |x^1|\leqslant 1} \left|\det\begin{pmatrix} a(t) & b(t, x^1) \\ a(t') & b(t', x^1)\end{pmatrix}\right|$$

is positive and depends only on k, s, i_0, j_0, and C. For our polynomials $a(t)$, $b(t, x^1)$ we can find points t, t', x^1 such that

$$\left|\det\begin{pmatrix} a(t) & b(t, x^1) \\ a(t') & b(t', x^1)\end{pmatrix}\right| \geqslant c_0/2.$$

Let for definiteness $t < t'$. We can now find ζ_1, ζ_2 such that

$$A(t_0 + tM^{-1})(\zeta_1 - \zeta_1^0) + B(t_0 + tM^{-1}, z_0^1 + ax^1 M^{-s-1})(\zeta_2 - \zeta_2^0) = -(1 + C_1 + 2C)M\lambda^{-k-1},$$

$$A(t_0+t'M^{-1})(\zeta_1-\zeta_1^0)+B(t_0+t'M^{-1}, z_0^1+ax^1M^{-s-1})(\zeta_2-\zeta_2^0)=(1+C_1+2C)M\lambda^{-k-1}.$$

It is easily seen that

$$|\zeta_1-\zeta_1^0|\leqslant C_2a^{-1}M^{s+1}\lambda^{-k-1}, \quad |\zeta_2-\zeta_2^0|\leqslant C_2b^{-1}M^{i_0+j_0(s+1)+1}a^{-j_0}\lambda^{-k-1},$$

where C_2 is a constant independent of K, λ, and M.

For the chosen values of variables z^1, ζ_1, ζ_2, condition (Ψ') is violated if

$$|R(t, z^1, \zeta_1, \zeta_2)|\leqslant 2CM\lambda^{-k-1}, \tag{5.16}$$

since the values of function Q then vary between $\leqslant -M\lambda^{-k-1}$ and $\geqslant M\lambda^{-k-1}$ as t increases from $t_0 + tM^{-1}$ to $t_0 + t'M^{-1}$, and the values of Q differ from those of the function q by $O(\lambda^{-k-1})$.

To verify (5.16) we note that

$$|\zeta_1-\zeta_1^0|\leqslant C_2\lambda^{-(k+1)/2}, \quad |\zeta_2-\zeta_2^0|\leqslant C_2K^{-1}M^{1/2}\lambda^{-(k+1)/2}$$

by (5.9). Therefore, for $|t - t_0| \leqslant M^{-1}$ and $|z^1 - z_0^1| \leqslant aM^{-s-1}$

$$|R(t, z^1, \zeta_1, \zeta_2)|\leqslant C(C_2^2\lambda^{-k-1}+C_2^2K^{-2}M\lambda^{-k-1}+C_2\lambda^{-k-1}+C_3\lambda^{-k-1}).$$

Hence inequality (5.16) follows, since $M \geqslant c_0\lambda$ and $\lambda \geqslant \lambda_0$, λ_0 being large enough.

Lemma 5.5 is proved.

5.9. We shall now show that the values of a and M vary little in $\Omega(t_0, z_0, \zeta^0)$, where

$$\Omega(t_0, z_0, \zeta^0)=\{(t, z, \zeta);\ |t-t_0|<M^{-1},\ |z^1-z_0^1|<aM^{-s-1},$$

$$|z'-z_0'|<\lambda^{-(k+1)/2},\ |\zeta_1-\zeta_1^0|<\delta_0^{-1}a^{-1}M^{s+1},\ |\zeta'-\zeta^{0\prime}|<\delta_0^{-1}\lambda^{(k+1)/2}\}.$$

<u>Lemma 5.6.</u> If conditions (4.26) and (4.27) are fulfilled, then

$$\frac{2}{3}M\leqslant M(t, z, \zeta)\leqslant\frac{3}{2}M, \quad \frac{2}{3}a\leqslant|a_s(t, z, \zeta)|\leqslant\frac{3}{2}a,$$

where

$$(t, z, \zeta)\in\Omega(t_0, z_0, \zeta^0), \quad a_s(t, z, \zeta)=\partial_{\zeta_1}\partial_t^sQ_\lambda(t, z, \zeta)/s!,$$

$$M=M(t_0, z_0, \zeta^0), \quad a=|a_s(t_0, z_0, \zeta^0)|.$$

<u>Proof.</u> A. Note that the estimates

$$|a_s(t, z, \zeta)-a_s(t_0, z_0, \zeta^0)|\leqslant\left|\sum_{j=1}^{[k/2]-s}\frac{1}{j!}\partial_t^{j+s}\partial_{\zeta_1}Q_\lambda(t_0, z_0, \zeta^0)(t-t_0)^j\right|+$$

$$C(|t-t_0|^{[k/2]-s+1}+|z-z_0|+\lambda^{-k-1}|\zeta-\zeta^0|)\leqslant$$

$$\sum_{j=1}^{[k/2]-s}\frac{1}{j!}\gamma_{j+s}^{-1}\gamma_s aM^{j-s}M^{-j}+C(M^{s-1-[k/2]}+\lambda^{-(k+1)/2}+\delta_0^{-1}\lambda^{-(k+1)/2})$$

are valid for $0 \leqslant l \leqslant [k/2]$. Since $a > M^{s+1}\lambda^{-(k+1)/2}$ we see that

$$M^{s-1-[k/2]}\leqslant M^{s+1}\lambda^{-(k+3)/2}\leqslant\lambda^{-1}a$$

and, for $\lambda \geqslant \lambda_0$, from (4.26) it follows that

$$|a_s(t, z, \zeta)-a|\leqslant a/3,$$

which proves the lemma.

B. Let us verify that, by (5.14), M can be obtained only with the use of the monomials τ, $at^s\xi_1$, and $At^{\alpha}(x^1)^{\beta}$ (perhaps, for some possible values of s, α, and β).

This proof is analogous to the proof of Lemma 5.1. As we have already seen, if use is made of monomials of the type $a\tau^s\xi_1$, for $a > M^{s+1}\times \lambda^{-(k+1)/2}$, then M can be obtained only as $[(\mathrm{ad}\,\tau)^i q\lambda^{k+1}]^{1/(i+1)}$.

We shall prove that M cannot be obtained using a chain of monomials of the type τ, $at^s\xi_1$, $Bt^{\alpha}(x^1)^{\beta}\xi_2$, and $Ct^{\sigma}(x^1)^{\tau}(x^2)^{\varkappa}\lambda^{k+1}$. Indeed, if this had been possible, then we could have

$$|B^{\varkappa}Ca^{\tau+\beta\varkappa}|\lambda^{k+1} = c_0M^{\sigma+\varkappa(\alpha+1)+(\tau+\beta\varkappa)(s+1)+1}.$$

But from (5.14) it follows that

$$|B| \leqslant KM^{\alpha+\beta(s+1)+1/2}\lambda^{-(k+1)/2}a^{-\beta},$$

and if $\varkappa > 1$, then

$$|B^{\varkappa}Ca^{\tau+\beta\varkappa}|\lambda^{k+1} \leqslant C_1M^{\varkappa(\alpha+1)+\beta\varkappa(s+1)-\varkappa/2},$$

and we arrive at a contradiction. If $\varkappa = 1$, then by (5.14)

$$|C| \leqslant KM^{\sigma+\tau(s+1)+1/2}\lambda^{-(k+1)/2}a^{-\tau},$$

so that

$$|BCa^{\tau+\beta}|\lambda^{k+1} \leqslant K^2M^{\sigma+(\alpha+1)+(\beta+\tau)(s+1)+1}\lambda^{-1},$$

and we again reach a contradiction.

Similarly it is verified that M cannot be obtained without the use of a longer chain of monomials: every time in the chain there will be no less than two monomials for the coefficients of which the inequalities (5.14) are fulfilled.

C. We proceed to show that inequalities (5.14) are fulfilled everywhere in $\Omega(t_0, x_0, \zeta^0)$, that is,

$$|\mathrm{grad}_{z',\zeta'}\partial_1^j\partial_t^i q_0(t, z, \zeta)| \leqslant K_1M^{i+j(s+1)+1/2}\lambda^{-(k+1)/2}a^{-j} \qquad (5.17)$$

for $(t, z, \zeta) \in \Omega$. Clearly these inequalities are valid for $i + j(s+1) > k/2$. Furthermore, since $|z' - z_0'| < \lambda^{-(k+1)/2}$ and $|\zeta - \zeta^0| < \delta_0^{-1}\times \lambda^{-(k+1)/2}$ it is sufficient to verify that the inequalities (5.17) are retained on varying t and z^1, that is, for $z' = z_0'$ and $\zeta = \zeta^0$. We have

$$\partial_1^j\partial_t^i \mathrm{grad}_{z',\zeta'} q_0(t, z, \zeta) =$$

$$\sum_{i+l+(j+m)(s+1)\leqslant k/2} \frac{1}{m!\,l!}\partial_1^{j+m}\partial_t^{i+l}\mathrm{grad}_{z',\zeta'} q_0(t_0, z_0, \zeta^0)(t-t_0)^l(z^1-z_0^1)^m +$$

$$O\Big(\sum_{k/2+1\leqslant i+l+(j+m)(s+1)\leqslant k/2+s+1} a^{-m}|t-t_0|^l|z^1-z_0^1|^m\Big).$$

Therefore, for $|t - t_0| < M^{-1}$, $|z^1 - z_0^1| < aM^{-s-1}$ we have

$$|\partial_1^j\partial_t^i \mathrm{grad}_{z',\zeta'} q_0(t, z, \zeta)| \leqslant CM^{i+j(s+1)+1/2}\lambda^{-[k/2]-1/2}(a^{-j}+\lambda^{-1}).$$

D. We now prove that

$$|B(t, z, \zeta) - B(t_0, z_0, \zeta^0)| < \frac{1}{4}|B(t_0, z_0, \zeta^0)|,$$

where $B(t, z, \zeta) = \partial_t^{\alpha}\partial_1^{\beta}q_0(t, z, \zeta)$ and α and β are such that

$$M^{\alpha+\beta(s+1)+1}=\delta_{\alpha+\beta s,\,\beta}\,|B|\,a^{\beta}\lambda^{k+1}, \tag{5.18}$$

that is, M is obtained through the use of the monomials τ, $at^s\xi_1$, and $Bt^{\alpha}(x^1)^{\beta}\lambda^{k+1}$.

It is clear that

$$B(t,\ z,\ \zeta)-E(t_0,\ z_0,\ \zeta^0)=\sum_{s_{ij}=1}^{k}\partial_t^i\partial_1^jB(t_0,\ z_0,\ \zeta^0)\frac{1}{i!\,j!}(t-t_0)^i(z^1-z_0^1)^j+$$
$$\sum_{i=0}^{[k/2]}\frac{1}{i!}\left[\sum_{l=2}^{n}\partial_l\partial_t^iB(t_0,\ z_0,\ \zeta^0)(z^l-z_0^l)+\right.$$
$$\left.\sum_{l=1}^{n}\partial_{\xi_l}\partial_t^iB(t_0,\ z_0,\ \zeta^0)(\zeta_l-\zeta_l^0)\right](t-t_0)^i+$$
$$O\left(\sum_{s_{ij}=k+1}^{k+s+1}a^{-j}|t-t_0|^i|z^1-z_0^1|^j+|z^1-z_0^1|^2+\right.$$
$$\left.|\zeta-\zeta^0|^2+|t-t_0|^{k/2+1}(|\zeta-\zeta^0|+|z'-z_0'|)\right),$$

where $s_{ij}=i+\alpha+(j+\beta)(s+1)$. The remainder term is estimated by

$$\sum_{s_{ij}=k+1}^{k+s+1}\left(M^{-s_{ij}}+\lambda^{-k-1}\delta_0^{-2}\right)\leqslant C_1\lambda^{-k-1}\delta_0^{-2}\leqslant C_2\lambda^{-1}\delta_0^{-2}|B|,$$

since it is seen from (5.18) that $|B|\geqslant C\lambda^{-k}$.

It follows from (5.15) that (provided the condition (4.27) holds)

$$\left|\sum_{s_{ij}=1}^{k}\partial_t^i\partial_1^jB(t_0,\ z_0,\ \zeta^0)\frac{1}{i!\,j!}(t-t_0)^i(z^1-z_0^1)^j\right|\leqslant$$
$$\sum_{s_{ij}=1}^{k}\delta_{i+\alpha+(j+\beta)s,\ j+\beta}^{-1}a^{-\beta}M^{\alpha+\beta(s+1)+1}\frac{1}{i!\,j!}\lambda^{-k-1}\leqslant$$
$$|B|\,\delta_{\alpha+\beta s,\,\beta}\sum_{s_{ij}=1}^{k}\delta_{i+\alpha+(j+\beta)s,\ j+\beta}^{-1}\leqslant\frac{1}{16}|B|.$$

It is seen from (5.14) that

$$\left|\sum_{i=0}^{[k/2]}\sum_{l=2}^{n}\frac{1}{i!}\left[\partial_l\partial_t^iB(t_0,\ z_0,\ \zeta^0)(z^l-z_0^l)+\right.\right.$$
$$\left.\left.\partial_{\xi_l}\partial_t^iB(t_0,\ z_0,\ \zeta^0)(\zeta_l-\zeta_l^0)\right](t-t_0)^i\right|\leqslant$$
$$K\delta_0^{-1}M^{\alpha+\beta(s+1)+1/2}a^{-\beta}\lambda^{-k-1}\cdot 2(n-1)\leqslant C\lambda_0^{-1/2}\delta_0^{-1}|B|.$$

Note that if $\beta\geqslant 1$, then

$$\lambda^{-k-1}|\zeta_1-\zeta_1^0|\leqslant\delta_0^{-1}a^{-1}M^{s+1}\lambda^{-k-1}\leqslant C|B|\lambda_0^{-1}.$$

Also for $\beta=0$ and $\alpha\geqslant k/2$, we have

$$\lambda^{-k-1}|\zeta_1-\zeta_1^0|\leqslant\delta_0^{-1}a^{-1}M^{s+1}\lambda^{-k+1}\leqslant\delta_0^{-1}\lambda^{-(k+1)/2}\leqslant C\delta_0^{-1}\lambda_0^{-1}|B|.$$

Finally, for $\beta=0$, $\alpha\leqslant k/2$, we have

$$\left|\sum_{i=0}^{[k/2]}\frac{1}{i!}\partial_t^i\partial_{\zeta_1}B(t_0,\ z_0,\ \zeta^0)(\zeta_1-\zeta_1^0)(t-t_0)^i\right|\leqslant$$

$$\sum_{i=0}^{[k/2]-\alpha} \frac{1}{i!} \gamma_s \gamma_{i+\alpha}^{-1} M^{i+\alpha-s} \delta_0^{-1} M^{s+1-i} + C\,|B|\,\lambda^{-1} \leqslant$$

$$|B| \left(\sum_{i=0}^{[k/2]-\alpha} \delta_0^{-1} \delta_{\alpha 0} \frac{1}{i!} \gamma_s \gamma_{i+\alpha}^{-1} + C\lambda^{-1} \right).$$

If

$$\delta_0^{-1} \delta_{\alpha 0} \sum_{i=0}^{[k/2]-\alpha} \frac{1}{i!} \gamma_s \gamma_{i+\alpha}^{-1} \leqslant \frac{1}{16}, \qquad (5.19)$$

we find that

$$|B(t, z, \zeta) - B(t_0, z_0, \zeta^0)| \leqslant \frac{1}{4} |B(t_0, z_0, \zeta^0)|,$$

provided $\lambda \geqslant \lambda_0$ and condition (5.19) is fulfilled.

It follows from (5.17) that if M were not obtained at $(t, z, \zeta) \in \Omega$ in the same manner as at (t_0, z_0, ζ^0) then in accordance with the arguments of B the result would be such that $M(t, z, \zeta) \leqslant CM\lambda^{-1/(2k+2)}$. But it can be seen from D that for $\bar{M}(t, s, \zeta)$ a value only slightly different from $M(t_0, z_0, \zeta^0)$ can be obtained by using the same monomials. To complete the proof it is sufficient to repeat the proof of Lemma 4.15.

Thus, Lemma 5.6 is proved.

5.10. Let inequality (5.9) be fulfilled at (t_0, z_0, ζ^0) and let $q(t, z, \zeta)$ be the function into which the function $q(t, x, \xi)$ has been transformed after the canonical transformations of Section 5.7.

Let $A(t) = \partial_{\zeta_1} q(t, z, \zeta)$ and $B(t) = q(t, z, \zeta)$ for $z = z_0$, $\zeta = \zeta^0$.

Lemma 5.7. Function A does not change its sign when $|t - t_0| \leqslant 2M^{-1}$.

Proof. Suppose that function A changes its sign when $t = t'$ so that $A(t') = 0, \ldots, A^{(l-1)}(t') = 0$, $A^{(l)}(t') \neq 0$ and l is odd. By Lemma 5.6 we have $|A^{(s)}(t)| \geqslant as!/2$ for $|t - t_0| < 2M^{-1}$ and therefore $l \leqslant s$. Shift the origin to $t - t_0$ so that $A(t) = \alpha t^l + O(|t|^{l+1})$, $\alpha \neq 0$, l is odd and $l \leqslant s$.

Let $B(t) = \sum_{i=0}^{k} b_i t^i$. By (5.15) we have $|b_i| \leqslant CM^{i+1}\lambda^{-k-1}$. It follows that the inequality

$$|B(t)| \leqslant C_1 M \lambda^{-k-1}$$

is fulfilled for $|t| \leqslant 2M^{-1}$. Suppose that $|\zeta'| = 4C_1\alpha^{-1}\rho^{-l}M^{l+1}\lambda^{-k-1}$, where ρ is a sufficiently small number so that $|A(t)| \geqslant \alpha|t|^l/2$ when $|t| \leqslant \rho M^{-1}$; choose the sign of ζ_1 so that function $\zeta_1 A(t)$ changes sign at $t = 0$ from minus to plus as t increases. For $z = z_0$, $\zeta' = \zeta'^0$, $t = \pm\rho M^{-1}$, and the chosen value of ζ_1 we have

$$|q(t, z, \zeta)| = |A(t)\zeta_1 + B(t) + O(\zeta_1^2)| \geqslant$$

$$2C_1 M \lambda^{-k-1} - C_1 M \lambda^{-k-1} - C_2 M^{2(l+1)} \lambda^{-2k-2}.$$

Note that from (5.9) it follows that $M^{s+1} \leqslant C_3 \lambda^{(k+1)/2}$ so that

$$M^{2(l+1)} \lambda^{-2k-2} \leqslant M^{2(s+1)} \lambda^{-2k-2} \leqslant \lambda^{-k-1} \leqslant C_4 M \lambda^{-k-1} \lambda^{-1}.$$

Therefore $|q(t, z, \zeta)| > C_1 M \lambda^{-k-1}/2$ for $\lambda \geqslant \lambda_0$. Consequently, the sign of function q changes from minus to plus as t increases. This contradicts condition (Ψ').

Lemma 5.7 is proved.

5.11. Let now for $z = z^0$, $\zeta' = \zeta^{0'}$, $\zeta_1^0 = 0$

$$q(t, z, \zeta) = B(t, z^1), \quad \partial_{\zeta_1} q(t, z, \zeta) = A(t, z^1),$$

so that

$$q(t, z, \zeta) = B(t, z^1) + \zeta_1 A(t, z^1) + O(\zeta_1^2),$$

and let $A|t, z_0^1| \geqslant 0$ in accordance with Lemma 5.7.

<u>Lemma 5.8.</u> Suppose that condition (5.9) is fulfilled. Then there exist polynomials $A_k(t)$, of degree $k/2$, and $C_k(t, z^1)$, of degree k, such that the following inequalities are fulfilled for $|t - t_0| \leqslant M^{-1}$, $|z^1 - z_0^1| \leqslant aM^{-s-1}$:

1°. $A_k(t) \geqslant -C_1 aM^{-s-1}$, $|A(t, z^1) - A_k(t)| \leqslant CaM^{-s-1}$.

2°. $|A_k(t) C_k(t, z^1) - B(t, z^1)| \leqslant C_2 \lambda^{-k-1}$.

3°. $|C_k(t, z^1)| \leqslant C(\delta) a^{-1} M^{s+1} \lambda^{-k-1}$.

4°. The inequality $\partial_t C_k(t, z^1) \leqslant C(\varepsilon) a^{-1} M^{s+1} \lambda^{-k-1}$ is fulfilled at those points where $A_k(t) \geqslant \varepsilon aM^{-s}$ for $\varepsilon > 0$.

<u>Proof.</u> To simplify notations we assume that $t_0 = 0$ and $z^1 = 0$. Suppose that

$$A_k(t) = \sum_{j=0}^{[k/2]} \partial_t^j A(0, 0) t^j / j!$$

Since $|z^1| \leqslant aM^{-s-1}$ and $a \geqslant M^{s+1} \lambda^{-(k+1)/2}$, it immediately follows that

$$|A(t, z^1) - A_k(t)| \leqslant CaM^{-s-1},$$

and therefore inequalities 1° are valid.

Note that

$$A(t) = \sum_{j=0}^{s-1} a_j t^j + at^s + \sum_{j=s+1}^{[k/2]} a_j t^j + O(|t|^{[k/2]+1}).$$

For $|t| \leqslant M^{-1}$ we have

$$\left| \sum_{j=s+1}^{[k/2]} a_j t^j + O(|t|^{[k/2]+1}) \right| \leqslant \sum_{j=s+1}^{[k/2]} \gamma_s \gamma_j^{-1} aM^{-s} + C\lambda^{-[k/2]-1} \leqslant c_0 aM^{-s}$$

if γ_j are such that

$$\sum_{j=s+1}^{[k/2]} \gamma_s \gamma_j^{-1} \leqslant c_0/2 \tag{5.20}$$

and λ_0 is large enough. Let l be an integer such that $0 \leqslant l \leqslant 2(s + 1)$ and the polynomial $at^s + \sum_{j=0}^{s-1} a_j t^j$ does not vanish for $\left(1 + \frac{l-1}{2(s+1)}\right) M^{-1} \leqslant |t| \leqslant \left(1 + \frac{l+1}{2(s+1)}\right) M^{-1}$. Then, for $|t| = (1 + l/2(s + 1))M^{-1}$, this polynomial assumes values which in modulus are not less than $a[2(s + 1)M]^{-s}$. If $c_0 < [2(s + 1)]^{-s}/2$, then at these points

$$|A_k(t)| \geqslant \frac{a}{2} [2(s+1) M]^{-s}.$$

Shift the origin to such a point. Let

$$B(t, z^1) = \sum b_{ij} t^i (z^1)^j, \quad C_k(t, z^1) = \sum c_{ij} t^i (z^1)^j.$$

Now define c_{ij} so that the k-th order expansions of functions $A_k C_k$ and B by Taylor series expansion would agree.

Note that $c_{0j} = b_{0j}/a_0$. It follows from this, by Lemma 5.4, that $|c_{0j}| \leqslant Ca^{-j-1}M^{(j+1)(s+1)}\lambda^{-k-1}$, where the constant C depends only on δ_{ij}. Using the formula

$$b_{ij} = \sum_{l=0}^{i} a_{i-l}c_{lj}$$

and the inequality $|a_j| \leqslant CaM^{j-s}$, it can be readily verified by induction that

$$|c_{ij}| \leqslant Ca^{-j-1}M^{i+(j+1)(s+1)}\lambda^{-k-1}, \quad i+j(s+1) \leqslant k-1. \tag{5.21}$$

Inequalities 2° and 3° follow from this.

Now we shall prove inequality 4°. Note that

$$q(t, z, \zeta) = A_k(t)[\zeta_1 + C_k(t, z^1)] + O(\zeta_1^2 + \lambda^{-k-1}).$$

Let $A_k(t_0) \geqslant \varepsilon a M^{-s}$ and $\varepsilon > 0$. Assume that

$$\partial_t C_k(t_0, z_0^1) > Ka^{-1}M^{s+1}\lambda^{-k-1}.$$

If $\rho > 0$ is a sufficiently small number, then

$$A_k(t) \geqslant \frac{\varepsilon}{2} aM^{-s}, \quad \partial_t C_k(t, z_0^1) \geqslant \frac{K}{2} a^{-1}M^{s+1}\lambda^{-k-1}$$

for $|t - t_0| \leqslant \rho M^{-1}$. In this case ρ is independent of λ. Suppose that $\zeta_1^0 = -C_k(t_0, z_0^1)$. Then by 3°, we have

$$q(t, z_0, \zeta_1^0) = \pm A_k(t_0)\,\partial_t C_k(t_0, z_0^1)\,\rho M^{-1} + O(a^{-2}M^{2(s+1)}\lambda^{-2(k+1)} + (\rho^2 + \lambda^{-1})M\lambda^{-k-1})$$

at some point of this interval, since $|\partial_t^2 C_k| \leqslant Ca^{-1}M^{s+3}\lambda^{-k-1}$ by virtue of (5.21). But $a \geqslant M^{s+1}\lambda^{-(k+1)/2}$ and therefore

$$a^{-2}M^{2(s+1)}\lambda^{-2(k+1)} \leqslant \lambda^{-k-1} \leqslant \rho^2 M\lambda^{-k-1}.$$

From this it is seen that

$$q(t, z, \zeta) > \frac{\varepsilon K}{4}\rho M\lambda^{-k-1} - C_1\rho M\lambda^{-k-1} \quad \text{for} \quad t = t_0 + \rho M^{-1},$$

$$q(t, z, \zeta) < -\frac{\varepsilon K}{4}\rho M\lambda^{-k-1} + C_1\rho M\lambda^{-k-1} \quad \text{for} \quad t = t_0 - \rho M^{-1},$$

and if $\varepsilon K > 4C_1$, then the sign of q changes from minus to plus on passing from the point $t = t_0 - \rho M^{-1}$ to $t = t_0 + \rho M^{-1}$. Since this contradicts condition (Ψ'), Proposition 4° is proved.

5.12. Let us examine the point (t_0, y_0, η^0), where (5.9) is fulfilled and which lies on the surface $\partial_t^s q(t, y, \eta) = 0$. By condition (5.9), this surface is smooth in a neighborhood of the point considered.

Make the canonical transformation described in Section 5.7. Consider the neighborhood

$$\Omega_0(t_0, z_0, \zeta^0) = \{(t, z, \zeta) \in \mathbb{R}^{2n+1};\ |t - t_0| < M^{-1},$$

$$|z^1 - z_0^1| < aM^{-s-1},\ |z' - z_0'| < \lambda^{-(k+1)/2},\ |\zeta - \zeta^0| < \delta^{-1}\lambda^{(k+1)/2}\}.$$

This neighborhood is different from that Ω defined in Section 5.9 only in dimension in the direction of the ζ_1 axis.

Inequalities (5.14) and (5.15) enable us to conclude that

$$Q(t, z, \zeta) = \lambda^{k+1}B(t, z^1) + A(t, z^1)(\zeta_1 - \zeta_1^0) + O(\lambda^{k+1}|z' - z_0'|^2 + \lambda^{-k-1}|\zeta - \zeta^0|^2 +$$

$$\sum_{i+j(s+1)\geqslant k+1} a^{-j}\lambda^{k+1}|t - t_0|^i|z^1 - z_0^1|^j + \lambda^{k+1}\sum_{i,j}\sum_{l=2}^{n}\left(|\partial_l\partial_1^j\partial_t^i q_0(t_0, z_0, \zeta^0)||z^l - z_0^l| +\right.$$

$$\left.|\partial_{\zeta_l}\partial_1^j\partial_t^i q_0(t_0, z_0, \zeta^0)||\zeta_l - \zeta_l^0|\right)|t - t_0|^i|z^1 - z_0^1|^j\frac{1}{i!\,j!} +$$

$$\sum_{i+j(s+1)\geqslant k/2} a^{-j}|t - t_0|^i|z^1 - z_0^1|^j|\zeta_1 - \zeta_1^0|) =$$

$$A(t, z^1)(\zeta_1 - \zeta_1^0) + \lambda^{k+1}B(t, z^1) + O(1 + M^{1/2} + \delta^{-2}),$$

where $M = M(t_0, z_0, \zeta^0)$. We could replace $A(t, z^1)$ by $A(t)$ in $\Omega(t_0, z_0, \zeta^0)$, but not in $\Omega_0(t_0, z_0, \zeta^0)$ if

$$\lambda^{(k+1)/2} \gg \delta^{-1}a^{-1}M^{s+1}.$$

By Lemma 5.6, the function $M(t, z, \zeta)$ varies slightly in $\Omega(t_0, z_0, \zeta^0)$. At (t_0, z_0, ζ^0)

$$M^{i+j+1} = \delta_{ij}\lambda^{k+1}\left|\partial_t^{l_j}A(t, z^1)\partial_1\ldots\partial_t^{l_1}A(t, z^1)\partial_1\partial_t^{l_0}B(t, z^1)\right|,$$

where $l_0 \geqslant 1$, $l_0 + l_1 + \ldots + l_j = i$. The right-hand side of this equality is independent of ζ_1.

Therefore $M(t_0, z_0, \zeta^1, \zeta^{0\prime}) \geqslant M(t_0, z_0, \zeta^0)$ because with the increase of $|\zeta_1 - \zeta_1^0|$ other methods of finding M appear and, for $|\zeta_1 - \zeta_1^0| > C\delta_0^{-1}a^{-1}M^{s+1}$, the inequality becomes strict, since

$$M(t, z, \zeta)^{s+1} \geqslant \delta_{s0}|\partial_t^s Q_\lambda(t_0, z_0, \zeta^1, \zeta^{0\prime})| \geqslant$$

$$\delta_{s0}s!\,\frac{a}{2}C\delta_0^{-1}a^{-1}M^{s+1} = \delta_{s0}s!\,\frac{C}{2}\delta_0^{-1}M^{s+1} > M^{s+1}$$

if $C\delta_{s0}s! > 2\delta_0$, that is, $|\zeta_1 - \zeta_1^0| > 2\delta_{s0}^{-1}a^{-1}M^{s+1}/s!$.

By Lemma 5.6, we see that $M(t, z, \zeta) \geqslant c_0M(t_0, z_0, \zeta^0)$, where the constant c_0 is independent of M and λ, and $(t, z, \zeta) \in \Omega(t_0, z_0, \zeta^0)$. Also note that inequalities (5.14) hold everywhere in Ω_0 with $M = M(t_0, z_0, \zeta^0)$, as is seen from the proof of Lemma 5.6, part C. Therefore, if Ω_0 is split into domains similar to Ω, by planes ζ_1 = const at intervals of length $\delta_0^{-1}aM^{-s-1}$, then M will change little in such strips despite the fact that $M(t_0, z_0, \zeta^0)$ may be large and that we have taken domains with similar dimensions in the t and z^1 directions.

Let us consider a covering of the set of points

$$\omega = \{(t, y, \eta) \in \mathbb{R}^{2n+1};\ |t| + |y| \leqslant \lambda^{\varepsilon_0 - 1},\ |\eta| \leqslant \lambda^{\varepsilon_0 + k}\}$$

with neighborhoods of the type Ω_0.

If inequalities (5.5) are fulfilled at (t_0, y_0, η^0), then

$$\Omega_0(t_0, y_0, \eta^0) = \{(t, y, \eta) \in \mathbb{R}^{2n+1};\ |t - t_0| < M^{-1},$$

$$|y - y_0| < \lambda^{-\frac{k+1}{2}},\ |\eta - \eta^0| < \delta^{-1}\lambda^{\frac{k+1}{2}}\}.$$

Let now (5.9) be fulfilled at (t_0, y_0, η^0). Note that this point is then $\leqslant \delta_{s0}^{-1}\lambda^{(k+1)/2}$ distant from the surface $\partial_t^s q(t, y, \eta) = 0$. Indeed, $Q_\lambda(t, z, \zeta) = \lambda^{k+1}B(t, z^1) + A(t, z^1)(\zeta_1 - \zeta_1^0) + O(M^{1/2})$ when $\lambda \geqslant \lambda_0$. Therefore, at points lying at a distance $> \delta_{s0}^{-1}\lambda^{(k+1)/2}$ from this surface, the inequality

$$M^{s+1} \geqslant \delta_{s0}(\delta_{s0}^{-1}\lambda^{(k+1)/2})\,a = a\lambda^{(k+1)/2},$$

that is, inequality (5.5) is fulfilled. Since $\delta \ll \delta_{s0}$, for every point (t_0, y_0, η^0) at which condition (5.9) is fulfilled we may take as the domain covering it the domain Ω_0 corresponding to the point (t', y', η') lying on the surface $\partial_t^s q(t, y, \eta) = 0$ and nearest to it.

Also note that the domains concentric to Ω_0 and having half the dimensions cover the set ω as well. From this, choose such a covering for which every point is covered by no more than N neighborhoods; here N depends only on n, but not on λ.

We number these neighborhoods $\Omega_0^1\Omega_0^2, \ldots$ and associate to each of them the function

$$H_j(t, z, \zeta) = h_1(t)\, h((z^1 - z_0^1)\, a^{-1}M^{s+1})\, h((\zeta_1 - \zeta_1^0)\, \lambda^{-(k+1)/2}\delta) \times$$

$$\prod_{l=2}^{n} h\left((z^l - z_0^l)\, \lambda^{\frac{k+1}{2}}\right) h\left((\zeta_l - \zeta_l^0)\, \lambda^{-\frac{k+1}{2}}\delta\right),$$

where (t_0, y_0, η^0) is the center of neighborhood Ω_0^j and M stands for its value at (t_0, y_0, η^0), h is a standard function in $C_0^\infty(\mathbb{R}^1)$, $0 \leqslant h(t) \leqslant 1$, $h(t) = 1$ for $|t| \leqslant 1/2$, $h(t) = 0$ for $|t| \geqslant 1$. For a neighborhood of the first kind h_1 may be taken in the form $h_1(t) = h((t - t_0)M)$. For a neighborhood of the second kind, the function h_1 is constructed in accordance with the rule of Section 4.11 on the basis of the function $A(t) = \partial_{\xi_1} q(t, z, \zeta)$.

5.13. Let now $G_j(t, y, \eta) = H_j(t, z, \zeta)$, where the variables (y, η) and (z, ζ) are related by the canonical transformations of Section 5.7. Let us write out these transformations. The change of coordinates (y, η) to (z, ζ) is a product of the following transformations:

(1) Transformation $(y, \eta) \longmapsto (\eta\lambda^{-k-1}, y\lambda^{k+1})$ or, perhaps, of a similar transformation for some of the variables y.

(2) Rotations

$$y^j = \sum_{l=1}^{n} \alpha_l^j t^l, \quad \eta_j = \sum \alpha_j^l \tau_l + \sum_{l=1}^{n} \beta_{jl} t^l \lambda^{k+1}, \qquad j = 1, \ldots, n.$$

(3) Transformations of the type

$$t^1 = z^1, \quad t' = z' - B(z^1), \quad \tau_1 = \zeta_1 + \lambda^{k+1}(A_1'(z^1), z' - B(z^1)) -$$

$$(B'(z^1), \zeta^1) - \lambda^{k+1}C(z^1), \quad \tau' = \zeta' + \lambda^{k+1}A(z^1).$$

Let $G(t, y, \eta) = \sum_j G_j^2(t, y, \eta)$. By construction,

$$1 \leqslant G(t, y, \eta) \leqslant C(n).$$

In Section 6 we shall verify that

$$|D_t^l D_\eta^\alpha D_x^\beta G(t, y, \eta)| \leqslant C_{l,\alpha,\beta} M^l \lambda^{(k+1)(|\beta| - |\alpha|)/2} \delta^{|\alpha|}. \tag{5.22}$$

Based on the results of Section 7, Chapter II, we can conclude that

$$\frac{1}{2} I \leqslant \sum_j G_j^* G_j(t, y, D_y) \leqslant 2C(n)\, I \tag{5.23}$$

if δ is sufficiently small.

5.14. Let $P_\lambda\psi = f$. By (5.23) it is clear that

$$\|f\|^2 \geqslant c_0 \sum \|G_j(t, y, D_y) f\|^2.$$

The operator $P_\lambda(t, y, D_t, D_y) = iD_t - Q_\lambda(t, y, D_y)$ is a polynomial in y and D_y. That is why the commutator $[G_j, P_\lambda]$ always has a finite number of terms.

It may be seen from (5.22) that in these commutators the terms containing derivatives of Q_λ with respect to y and D_y, of orders $\geqslant 2$, are bounded operators and may be neglected. As regards the terms containing first derivatives, they have as their symbol the Poisson bracket $\{G, Q_\lambda\}$ and can therefore be computed in the (z, ζ) coordinates.

5.15. If a neighborhood Ω_j with its center at (t_0, y_0, η^0) belongs to neighborhoods of the first kind, that is, inequalities (5.5) are fulfilled in supp G_j, then

$$\|[G_j, Q_\lambda]\psi\|^2 \leqslant M_j^2 \|(\partial_t G_j)\psi\|^2 + R_j(\psi), \tag{5.24}$$

where $\sum \|R_j(\psi)\| \leqslant C_1 \sum M_j^2 \|G_j\psi\|^2$. Since

$$\|Q_\lambda^{(l)}(t, y_0, \eta^0)\lambda^{(k+1)/2} G_{j(l)}\psi\|^2 \leqslant CM_j^2 \|G_{j(l)}\psi\|^2,$$

$$\|Q_{\lambda(l)}(t, y_0, \eta^0)\lambda^{-(k+1)/2} G_j^{(l)}\psi\|^2 \leqslant CM_j^2 \|G_j^{(l)}\psi\|^2.$$

It follows from Theorems 4.1 and 4.2 that in this case

$$\delta_{\alpha 0}^{-2/(\alpha+1)} M_j^2 \|G_j\psi\|^2 + \|\partial_t G_j\psi\|^2 \leqslant C\|L_j G_j\psi\|^2,$$

where $L_j = iD_t + Q_\lambda(t, y_0, \eta^0)$. Therefore, from (5.24) it follows that

$$\delta_{\alpha 0}^{-2/(\alpha+1)} M_j^2 \|G_j\psi\|^2 + \|\partial_t G_j\psi\|^2 \leqslant C\|G_j P_\lambda\psi\|^2 + M_j^2 \|(\partial_t G_j)\psi\|^2 + R_j(\psi)). \tag{5.25}$$

5.16. Let now Ω_j be a neighborhood of the second kind, so that inequality (5.6) is fulfilled. Function A(t) is defined in the original coordinates, since the variable t does not change under the canonical transformations.

Note that $[G_j, \partial_t] = -\partial_t G_j$ and

$$[G_j, Q_\lambda] = [G_j, A(t, z^1)(D_1 - \zeta_1^0) + \lambda^{k+1} B(t, z^1)] + R'_j =$$

$$A(t)(D_1 G_j) + \lambda D_1 B(t, z^1)\,\partial_{D_1} G_j + R''_j,$$

where $\sum \|R'_j\psi\|_0^2 \leqslant C \sum M_j \|G_j\psi\|_0^2$ and $\sum \|R''_j\psi\|_0^2 \leqslant C \sum M_j^2 \|G_j\psi\|_0^2$. Indeed $A(t, z^1)D_1G_j$ can be replaced by $A(t)D_1G_j$, since $|z_1 - z_1^0| \leqslant aM^{-s-1}$, $|\zeta_1 - \zeta_1^0| \leqslant \delta^{-1}\lambda^{(k+1)/2}$ in supp G_j. On the other hand, the operator $(\partial_1 B)\delta_0\lambda^{k+1}(\partial_{D_1} G_j)$ is also bounded, since $\delta C(\delta) a^{-1} M^{s+2} \lambda^{(k+1)/2} \leqslant \delta C(\delta) M$ and $\delta C(\delta) < 1$.

So, we see that

$$G_j P_\lambda = L_j G_j - \partial_t G_j - A(t)(D_1 G_j) - \lambda D_1 B(t, z^1)\,\partial_{D_1} G_j + R_j,$$

where $\sum \|R_j\psi\|_0^2 \leqslant C \sum M_j^2 \|G_j\psi\|_0^2$ and the operator L_j is equivalent to the operator

$$L_j^0 = \Phi_j^* L_j \Phi_j = \partial_t + A(t, z^1)(D_1 - \zeta_1^0) + \lambda^{k+1} B(t, z^1).$$

Thus, we find that

$$\|f\|_0^2 \geqslant c_0 \sum \|G_j f\|_0^2 \geqslant c_1 \sum \|L_j G_j\psi - \partial_t G_j\psi -$$

$$A(t)(\partial_1 G_j)\psi - \lambda^{k+1} D_1 B(t, z^1)(\partial_{D_1} G_j)\psi\|_0^2 - C \sum M_j^2 \|G_j\psi\|_0^2 =$$

$$c_1 \sum \|L_j^0 v_j - \Phi_j^*(\partial_t G_j)\psi - A(t)\Phi_j^*(D_1 G_j)\psi -$$

$$\lambda^{k+1}\Phi_j^* D_1 B(t, z^1)(\partial_{D_1} G_j)\psi\|_0^2 - C \sum M_j^2 \|G_j\psi\|_0^2,$$

where $v_j = \Phi_j^* G_j \psi$.

When $a^{-1}M^{s+1} \leqslant \lambda^{(k+1)/2}$ we make use of an additional partition of unity in the ζ_1-variable:

$$\sum g_m'^2(\delta_0 a M^{-s-1}\zeta_1) = 1,$$

where each of the functions g'_m is smooth, assumes values in [0, 1], and has support on a section of unit length, g'_m being equal to 1 on a section of length 1/2 and $|D^\alpha g'_m| \leqslant C_\alpha$ for all α. Suppose that $g_m(D_1) = g'_m(\delta_0 aM^{-s-1}D_1)$ and $\psi_{jm} = g_m v_j$. Let

$$\Omega_j^{(m)} = \{(t, z, \zeta) \in \Omega_j, \ \zeta_1 \in \operatorname{supp} g'_m\}.$$

For $(t, z, \zeta) \in \Omega_j^{(m)}$ we have

$$Q_\lambda(t, z, \zeta) = A(t, z^1)(\zeta_1 - \zeta_1^m) + \lambda^{k+1}B_m(t, z^1) + Q_m,$$

where $B_m = B + A(\zeta_1^m - \zeta_1^0)\lambda^{-k-1}$ and $\|Q_m\psi_{jm}\| \leqslant C\|\psi_{jm}\|_0$. Furthermore, by Lemma 5.8, there exist $A_k = A_{kjm}(t)$ and $C_k = C_{kjm}(t, z^1)$ for which

$$\begin{gathered}\|[A(t, z^1) - A_k(t)](D_1 - \zeta_1^m)\psi_{jm}\|_0 \leqslant C_1\delta_0^{-1}\|\psi_{jm}\|_0, \\ \lambda^{k+1}\|[B_m(t, z^1) - A_k(t)C_k(t, z^1)]\psi_{jm}\|_0 \leqslant C_2\|\psi_{jm}\|_0.\end{gathered} \tag{5.26}$$

Note that

$$\sum_{m,j}\|g_m L_j^0 v_j - L_{jm}^0\psi_{jm}\|_0^2 \leqslant \sum_{m,j}\|[g_m, A(t, z^1)](D_1 - \zeta_1^m)v_j +$$

$$[g_m, \lambda^{k+1}B_m(t, z^1)]v_j + (L_j^0 - L_{jm}^0)\psi_{jm}\|_0^2 \leqslant$$

$$C\sum[1 + C(\delta)\delta_0]^2 M_{jm}^2\|\psi_{jm}\|_0^2 + C\delta_0^{-2}\|\psi\|_0^2 + C\|\psi\|_0^2,$$

where

$$B_m(t, z^1) = B(t, z^1) + (\zeta_1^m - \zeta_1^0)A_k(t)\lambda^{-k-1} = A_k(t)C_{km}(t, z^1),$$

$$L_{jm}^0 = A_k(t)(D_1 - \zeta_1^m) + \lambda^{k+1}A_k(t)C_{km}(t, z^1).$$

Therefore

$$\|f\|_0^2 \geqslant \sum_{j,m}\|L_{jm}^0\psi_{jm} - A_k(t)g_{jm}\|_0^2 - C_1\delta_0^{-2}\|\psi\|_0^2 - C_2\sum M_{jm}^2\|\psi_{jm}\|_0^2, \tag{5.27}$$

where

$$g_{jm} = A_k^{-1}(t)g_m\Phi_j^*\partial_t G_j\psi - g_m\Phi_j^*(D_1G_j)\psi - \lambda^{k+1}g_m D_1C_k(t, z^1)\Phi_j^*(\partial_{D_1}G_j)\psi.$$

Recall that $A_k(t) \geqslant c_0 aM^{-s}$ in supp $\partial_t G_j$. Therefore,

$$\sum_{j,m} M_{jm}^{-2(s+1)}a_j^2\|g_{jm}\|^2 \leqslant C\|\psi\|^2$$

and

$$\sum_{m,j}\|A_k g_{jm}\|_0^2 \leqslant C\sum_{m,j} M_{jm}^2\|\psi_{jm}\|_0^2.$$

5.17. The desired estimate

$$\begin{gathered}M_{jm}^2\delta_{\alpha\beta}^{-2/(\alpha+\beta+1)}\|\psi_{jm}\|_0^2 + \|\partial_t\psi_{jm}\|_0^2 \leqslant \\ C\left(\|L_{jm}^0\psi_{jm} - A_k(t)g_{jm}\|_0^2 + a^{-2}M_{jm}^{2(s+1)}\|g_{jm}\|_0^2\right)\end{gathered} \tag{5.28}$$

is proved virtually in the same manner as in Lemma 4.16 with the difference that here we use conditions 1° and 4° of Lemma 5.8 instead of conditions $A(t) \geqslant 0$ and $\partial_t C(t, z^1) \leqslant 0$. Clearly the desired inequality (5.3) follows from this estimate and from (5.23) and (5.25). In so doing we first sum up inequalities (5.28) over m and only then, having used that $\sum G_m^* G_m = I$, add the inequalities obtained and (5.25) over j, which by virtue of (5.23) yields the desired result.

The parameters ε, γ_j, δ_{ij}, and δ_0 were chosen earlier. A small difference is introduced by the only new parameter δ which appeared in Section 5.12. Its final value will be specified in Section 5.19.

5.18. Proof of Estimate (5.28).

A. First consider the case when $s = 0$. Then $Ca \geqslant A_k(t) \geqslant C^{-1}a$ in Ω_{jm} so that from the equality

$$\iint A_k^{-1} |L_{jm}v|^2 \, dz\, dt = \iint A_k^{-1}(t) |\partial_t v|^2 \, dz\, dt + \iint A_k |lv|^2 \, dz\, dt + \lambda^{k+1} \iint \partial_t C |v|^2 \, dz\, dt,$$

where $l = D_1 - \zeta_1^m + \lambda^{k+1} C_k(t, z^1)$, $v = \psi_{jm}$, it follows that

$$\|\partial_t v\|_0^2 + \|Alv\|_0^2 \leqslant C \|L_{jm}v\|_0^2 + CM_{jm}^2 \|v\|_0^2.$$

From this we obtain (5.28) from (4.31), since

$$\sum_{j,m} (\|(\partial_t G_j) g_m \psi\|_0^2 + \|a^{-1} M A_k(t) (\partial_1 G_j) g_m \psi\|_0^2 + \|\lambda^{k+1} [G_j g_m, C_k] A_k(t) \psi\|_0^2 \leqslant C \sum_{j,m} M_{jm}^2 \|\psi_{jm}\|_0^2.$$

B. Let now $s \neq 0$ (so that $s \geqslant 2$). Let G_1 be a set of points (t, z), where $A_k(t) \geqslant \varepsilon a M^{-s}$. Here ε is some positive number and G_0 is a set of points which are not in G_1 and belong to the projection Ω_{jm} on the space (t, z). Let $A_0 = aM^{-s}$, $M = M_{jm}$. By (5.27), we have

$$\iint_{G_1} A_k^{-1}(t) |\partial_t v|^2 \, dz\, dt + \iint_{G_1} A_k(t) |lv|^2 \, dz\, dt + 2\,\mathrm{Re} \iint_{G_1} \partial_t v \overline{lv} \, dz\, dt \leqslant 2\left(\iint_{G_1} A_k^{-1}(t) |f_{jm}|^2 \, dz\, dt + \iint_{G_1} A_k(t) |g_{jm}|^2 \, dz\, dt\right), \tag{5.29}$$

where $f_{jm} = L_{jm}^0 \psi_{jm} - A_k(t) g_{jm}$, $v = \psi_{jm}$.

Note that

$$2\,\mathrm{Re} \iint_{G_1} \partial_t v \overline{lv} \, dz\, dt = -2\,\mathrm{Re} \iint_{G_0} \partial_t v \overline{lv} \, dz\, dt - \lambda^{k+1} \iint_{G_0 + G_1} \partial_t C(t, z) |v|^2 \, dz\, dt. \tag{5.30}$$

Since the inequalities $\varepsilon A_0 \leqslant A_k(t) \leqslant \varepsilon A_0$ are fulfilled in G_1, it follows from (5.29) and (5.30), with regard to Lemma 5.8, that

$$\frac{1}{C} \iint_{G_1} (|\partial_t v|^2 + |A_k lv|^2) \, dz\, dt \leqslant \frac{2}{\varepsilon} \iint_{G_1} |f_{jm}|^2 \, dz\, dt + 2\,\mathrm{Re}\, A_0 \iint_{G_0} \partial_t v \cdot \overline{lv} \, dz\, dt + 2CA_0^2 \iint_{G_1} |g_{jm}|^2 \, dz\, dt + C(\delta) M^2 \iint_{G_0} |v|^2 \, dz\, dt + C(\varepsilon) M \iint_{G_1} |v|^2 \, dz\, dt. \tag{5.31}$$

Furthermore, from the equality

$$L_{jm}^0 \psi_{jm} = f_{jm} + A_k(t) g_{jm},$$

it follows that

$$\frac{1}{C} \iint_{G_0} (|\partial_t v|^2 + |A_k lv|^2) \, dz\, dt \leqslant C_1 \left(\iint_{G_0} |f_{jm}|^2 \, dz\, dt + \varepsilon^2 A_0^2 \iint_{G_0} |g_{jm}|^2 \, dz\, dt + \varepsilon^2 A_0^2 \iint_{G_0} |lv|^2 \, dz\, dt\right).$$

Note that $\mathrm{mes}\{t, A_k(t) \leqslant \varepsilon a M^{-s}\} \leqslant c_0 \varepsilon^{2/k} M^{-1}$. Therefore,

$$M^2 \iint_{G_0} |v|^2 \, dz \, dt \leqslant c_0 \varepsilon^{2/k} M \max_t \int v^2 \, dz \leqslant c_0 \varepsilon^{2/k} \iint |\partial_t v|^2 \, dz \, dt. \qquad (5.32)$$

Furthermore, we have

$$2 \operatorname{Re} A_0 \iint_{G_0} \partial_t v \overline{lv} \, dz \, dt = 2 \operatorname{Re} A_0 \iint_G L_{jm} v \cdot \overline{lv} \, dz \, dt - 2 \operatorname{Re} A_0 \iint_{G_0} A_k(t) \, |lv|^2 \, dz \, dt \leqslant$$

$$2 \operatorname{Re} A_0 \iint_{G_0} L_{jm} v \cdot \overline{lv} \, dz \, dt + C_2 A_0^2 \varepsilon \iint_{G_0} |lv|^2 \, dz \, dt \leqslant$$

$$\frac{1}{\varepsilon} \iint_{G_0} |L_{jm} v|^2 \, dz \, dt + (C_2 + 1) \, \varepsilon A_0^2 \iint_{G_0} |lv|^2 \, dz \, dt.$$

It is not hard to see that

$$\iint_{G_0} |lv|^2 \, dz \, dt \leqslant [C_1 \delta_0^{-2} + C(\delta)] \, M^2 A_0^{-2} \iint_{G_0} |v|^2 dz \, dt +$$

$$C_3 M^2 A_0^{-2} \iint_{G_0} \sum_{i=1}^{k} |g_m^{(i)}(D) \, v_j|^2 \, dz \, dt,$$

$$\iint_{G_0} |L_{jm} v|^2 \, dz \, dt \leqslant 2 \iint_{G_0} |f_{jm}|^2 \, dz \, dt + 2\varepsilon^2 A_0^2 \iint_{G_0} |g_{jm}|^2 \, dz \, dt.$$

Thus,

$$2 \operatorname{Re} A_0 \iint_{G_0} \partial_t v \overline{lv} \, dz \, dt \leqslant \frac{2}{\varepsilon} \iint_{G_0} |f_{jm}|^2 \, dz \, dt + 2\varepsilon A_0^2 \iint_{G_0} |g_{jm}|^2 \, dz \, dt +$$

$$(C_2 + 1) \, \varepsilon \, [C_1 \delta_0^{-2} + C(\delta)] \, M^2 \iint_{G_0} |v|^2 dz \, dt +$$

$$(C_1 + 1) \, C_3 \varepsilon M^2 \iint \sum_{i=1}^{k} |g_m^{(i)}(D) \, v_j|^2 \, dz \, dt.$$

Combining the inequalities obtained, we arrive at the inequality

$$\frac{1}{C} (\|\partial_t v\|_0^2 + \|A_k lv\|_0^2) \leqslant C_4 \Big(\frac{1}{\varepsilon} \|f_{jm}\|_0^2 + \varepsilon [\delta_0^{-2} + C(\delta)] M^2 \|v\|_0^2 + C A_0^2 \|g_{jm}\|_0^2 +$$

$$C(\delta) \, c_0 \varepsilon^{2/k} \|\partial_t v\|_0^2 + C(\varepsilon) \, M \|v\|_0^2 + \varepsilon M^2 \sum_{i=1}^{k} \|g_m^{(i)}(D) \, v_j\|_0^2 \Big).$$

If

$$C_4 C(\delta) \, C_0 \varepsilon^{\frac{2}{k}} < \frac{1}{4C}; \quad \lambda_0 > \frac{1}{4C} \delta_{\alpha\beta}^{-\frac{2}{\alpha+\beta+1}} C_4 C(\varepsilon); \quad \varepsilon [\delta_0^{-2} + C(\delta)] < 1, \qquad (5.33)$$

then, by virtue of (4.31), inequality (5.28) follows.

5.19. Let us now go back to the question of parameters. Besides the conditions considered in Section 4.13, account must be taken of (5.19) and (5.33).

That the constant $C(\delta)$ depends only on δ_{ij}, for $j \geqslant 1$, in conditions (4.33) and (5.33) is significant. This is the reason why we first choose γ_j, then δ_{ij} for $j \geqslant 1$; then we choose δ_0 from (4.33), and only after this we pick $\delta_{\alpha 0}$ for $0 \leqslant \alpha \leqslant k$. Afterwards, we select ε from (5.33) and then λ_0 and δ.

§6. Canonical Transformations and Partition of Unity

6.1. In this section the inequalities (5.22) used earlier in proving the theorem on sufficient conditions for subellipticity are proved.

First we construct integral operators corresponding to the canonical transformations of Section 5.7. Fix a second-kind neighborhood Ω_j and for the moment shift the origin to its center. Let $a_j = \partial_j \partial_i^s q(0)$, $b^j = \partial_{\eta_j} \partial_i^s q(0)$ $(j = 1, \ldots, n)$ and let for definiteness $\sum (b^j)^2 \geqslant \sum a_j^2$ (otherwise, the variables y and η may be interchanged after replacing ψ in (5.33) by its Fourier transform). Suppose that $\Pi = \| \pi_i^j \|$, $\mathrm{P} = \| \rho_{ij} \|$ are square matrices of order n such that

$$\sum_{i=1}^{n} \pi_i^j b^i = 0, \quad j = 2, \ldots, n, \quad \sum_{i=1}^{n} \pi_i^1 b^i = |b| = \sqrt{\sum (b^i)^2},$$

$$\sum_{i=1}^{n} \rho_{1j} b^i = -a_j, \qquad j = 1, \ldots, n.$$

It is not hard to see that such matrices always exist and it may be assumed that $\Pi^* \Pi = I$, $\mathrm{P}^* = \mathrm{P}$. The canonical transformation $(y, \eta) \longmapsto (\tilde{y}, \tilde{\eta})$ with generating function

$$S_\pi (y, \tilde{\eta}) = (\Pi y, \tilde{\eta}) + \frac{1}{2} (\mathrm{P} y, y)$$

transforms the function $\sum a_j y^j + \sum b^j \eta_j$ into $|b| \tilde{\eta}_1$. In accordance with Section 4 of Chapter II, we suppose that

$$\Phi_\pi u (y) = (2\pi)^{-n} \int e^{i\lambda^{k+1} S (y, \eta\lambda^{-k-1})} \tilde{u} (\eta)\, d\eta,$$

where $\tilde{u}(\eta) = \int u(x) e^{-ix\eta}\, dx$. Φ_π is the transformation described in the beginning of Section 5.7.

6.2. The integral operator Φ_1 corresponding to the generating function $S(y, \zeta)$, defined by (5.12), is very simple. Let

$$\Phi_1 u (y) = (2\pi)^{-n} \int e^{iy\xi + i\lambda^{k+1} S (y, \xi\lambda^{-k-1})} \tilde{u} (\xi)\, d\xi.$$

Since $S(y, \zeta) = (A(y^1), y') + (B(y^1), \zeta') + C(y^1)$ we obtain that

$$\Phi_1 u (y) = e^{i\lambda^{k+1} [(A (y^1),\, y') + C (y^1)]} u (y^1,\ y' + B (y^1)).$$

The variable y^1 plays the role of a parameter which does not change under the canonical transformations.

6.3. Now find the operator $\Phi = \Phi_\pi \Phi_1$. We have

$$\Phi u (x) = (2\pi)^{-2n} \iiint \tilde{u} (\xi)\, e^{iy\xi + i\lambda^{k+1} S (y, \xi\lambda^{-k-1})}\, e^{-iy\eta + i\lambda^{k+1} S_\pi (x, \eta\lambda^{-k-1})}\, d\xi\, dy\, d\eta.$$

Note that

$$(2\pi)^{-n} \int e^{-iy\eta + i\lambda^{k+1} S_\pi (x, \eta\lambda^{-k-1})}\, d\eta = e^{i\lambda^{k+1} (\mathrm{P}x,\, x)/2} \delta (y - \Pi x),$$

and therefore

$$\Phi u (x) = (2\pi)^{-n} \int \tilde{u} (\xi)\, e^{i\lambda^{k+1} S_0 (x, \xi\lambda^{-k-1})}\, d\xi,$$

in which

$$S_0 (x, \xi) = (\Pi x, \xi) + (A_1 (x), \Pi x) + (B_1 (x), \xi) + C_1 (x) + \frac{1}{2} (\mathrm{P} x, x),$$

where $A_1(x) = (0, A(\sum \pi_i^1 x^i))$, $B_1(x) = (0, B(\sum \pi_i^1 x^i))$, $C_1(x) = C(\sum \pi_i^1 x^i)$. It is significant that function S_0 linearly depends on ξ so that the transformation Φ reduces to a change of variables:

$$\Phi u(x) = e^{i\lambda^{k+1}T(x)} u(\Pi x + B_1(x)), \tag{6.1}$$

where $T(x) = (A_1(x), \Pi x) + C_1(x) + \frac{1}{2}(Px, x)$.

6.4. Let L be an operator for which

$$L(x, D)u(x) = (2\pi)^{-n} \int L(x, \xi)\tilde{u}(\xi) e^{ix\xi} d\xi.$$

By (6.1) we have

$$(\Phi^* L \Phi u, v) = (L\Phi u, \Phi v) =$$

$$(2\pi)^{-n} \iiint L(x, \xi) e^{i(x-y, \xi)} e^{i\lambda^{k+1}[T(y) - T(x)]}\, u(\Pi y + B_1(y))\, \overline{v(\Pi x + B_1(x))}\, dy\, d\xi\, dx.$$

Replacing $z = \Pi y + B_1(y)$, $t = \Pi x + B_1(x)$ yields

$$(\Phi^* L \Phi u, v) = (2\pi)^{-n} \iiint L_1(t, \xi) e^{i(t-z, \Pi\xi)} e^{i\lambda^{k+1}[T_1(z) - T_1(t)]} u(z)\overline{v(t)}\, dz\, d\xi\, dt,$$

where

$$L_1(t, \xi) = L(x, \xi), \quad T_1(z) = T(y), \quad T_1(t) = T(x).$$

Note that

$$T_1(z) = (A(z^1), z^1) + C(z^1) - (A(z^1), B(z^1)) + \frac{1}{2}(P\Pi^*(z - B(z^1)), \Pi^*(z - B(z^1))). \tag{6.2}$$

Let $T_1(z) - T_1(t) = R(z, t)(z - t)$ and $\eta = \Pi\xi - \lambda^{k+1}R(z, t)$. Then

$$\Phi^* L \Phi u(t) = L_1 u(t) = (2\pi)^{-n} \iint L_2(t, z, \eta) e^{i(t-z, \eta)} u(z)\, dz\, d\eta, \tag{6.3}$$

where

$$L_2(t, z, \eta) = L_1(t, \Pi^*(\eta + \lambda^{k+1}R(z, t))) = L(\Pi^*(t - B(t^1)), \Pi^*(\eta + \lambda^{k+1}R(z, t))).$$

Note immediately that if $L(x, \xi) \equiv 1$, that is, $L(x, D) = I$, then from (6.3) it follows that $\Phi^*\Phi = I$, that is, Φ is a unitary operator.

6.5. Let now $L(x, \xi)$ be a polynomial of x, ξ. Since S_0 is a polynomial, the functions T_1, R, and L_2 will also be polynomials. Expand the function $L_2(t, z, \eta)$ in a Taylor series in powers of $z - t$:

$$L_2(t, z, \eta) = \sum_{\alpha} \frac{1}{\alpha!} \partial_z^{\alpha} L_2(t, t, \eta)(z - t)^{\alpha}.$$

Represent $(z - t)^{\alpha} e^{i(t-z, \eta)}$ in the form $(-D_{\eta})^{\alpha} e^{i(t-z, \eta)}$ and integrate by parts in (6.3) with respect to η:

$$L_1 u(t) = (2\pi)^{-n} \sum_{\alpha} \frac{i^{|\alpha|}}{\alpha!} \int D_{\eta}^{\alpha} D_z^{\alpha} L_2(t, t, \eta)\tilde{u}(\eta) e^{it\eta} d\eta.$$

It is seen that the operator L_1 will be differential and its principal symbol equals

$$L_0(z, \zeta) = L(\Pi^*(z - B(z^1)), \Pi^*(\zeta\lambda^{-k-1} + R(z, z))),$$

where $R_i(z, z) = \partial_i T_1(z)$ and T_1 is defined by formula (6.2).

The formula obtained serves as the basis for choosing the transformation Φ. From this we see what the functions A, B, C and matrices Π, P should be for condition (5.11) to be fulfilled.

6.6. We now compare the operators Φ and Φ_1 which correspond to two adjacent neighborhoods Ω and Ω_1.

Let first the number s be common for Ω and Ω_1. The matrices Π and P are defined by the conditions

$$\Pi\nabla_\xi\partial_t^s q(0)=(|\nabla_\xi\partial_t^s q(0)|,\ 0,\ \ldots,\ 0),\qquad P\nabla_\xi\partial_t^s q(0)=-\nabla_x\partial_t^s q(0).$$

These conditions are, of course, insufficient for determining the matrices P and Π. In the orthogonal matrix Π only the first row is uniquely defined. If we fix Π, then also only the first row of the matrix $P\Pi^*$ will be uniquely defined. In the general case, we cannot choose matrices P and Π everywhere in the domain where conditions (5.9) are fulfilled in such a way that their values are close in adjacent neighborhoods. However, this domain can be divided into a finite number of cells $\Omega_1, \ldots, \Omega_N$, where the matrices P and Π can be chosen to be smooth, N being not greater than some constant depending only on k and n.

Now we shall see how close the matrices Π and Π_1, P and P_1 will be for the "best" choice of Π_1 and P_1 (assuming that Π and P are fixed). This closeness is defined by the extent to which vector $\nabla_{x,\xi}\partial_t^s q(x, \xi)$ varies in going from the center of neighborhood Ω to the center of neighborhood Ω_1 (to be more exact, it is not the vector, but its image in the canonical transformation corresponding to neighborhood Ω). The canonical coordinates of this vector at the center of Ω are equal to

$$(0,\ \ldots,\ 0;\ a,\ 0,\ \ldots,\ 0).$$

Let (t_0, x_0, ξ^0) be the coordinates of the center of Ω. Then for $|t-t_0|\leqslant M^{-1}$, $|x^1-x_0^1|\leqslant aM^{-s-1}$, $|x'-x_0'|\leqslant\lambda^{-(k+1)/2}$, $|\xi-\xi^0|\leqslant\delta^{-1}\lambda^{(k+1)/2}$, the coordinates of this vector vary so that

$$|\partial_1\partial_t^s q(t,\ x,\ \xi)|\leqslant C_1a^{-1}M^{s+j(s+1)+1}\lambda^{-k-1},\qquad |\nabla_{x',\xi'}\partial_t^s q(t,\ x,\ \xi)|\leqslant C_1M^{s+\frac{1}{2}}\lambda^{-(k+1)/2},$$

since, by Lemmas 5.4 and 5.5,

$$\begin{gathered}\nabla_{x',\xi'}\partial_t^s\partial_1^j q(t_0,\ x_0,\ \xi^0)=0\quad\text{for}\quad s+j(s+1)\leqslant k/2,\\ |\nabla_{x',\xi'}\partial_t^i\partial_1^j q(t_0,\ x_0,\ \xi^0)|\leqslant C_2a^{-j}M^{i+j(s+1)+\frac{1}{2}}\lambda^{-(k+1)/2},\\ |\partial_t^i\partial_1^j q(t_0,\ x_0,\ \xi^0)|\leqslant C_2a^{-j}M^{i+j(s+1)+1}\lambda^{-k-1}.\end{gathered}\tag{6.4}$$

In Lemma 5.6 it was proved that

$$|\partial_{\xi_1}\partial_t^s q(t,\ x,\ \xi)-a|\leqslant a/2.$$

Therefore it can be asserted that the estimates

$$1/2\leqslant\alpha_{0j}^j\leqslant 2,\quad |\alpha_{0j}^1|+|\beta_{1j}^0|\leqslant CM^{-1/2},\qquad j=2,\ \ldots,\ n$$

are fulfilled for the coefficients α_l^j, β_{jl} of the transformation

$$x^j=\sum\alpha_{0l}^j t^l,\quad \xi_j=\sum\alpha_{0j}^l\tau_l+\sum\beta_{jl}^0 t^l,\qquad j=1,\ \ldots,\ n.$$

By the orthogonality of matrix A, it follows that $|\alpha_{01}^1-1|\leqslant CM^{-1/2}$, and by the orthogonality of the entire transformation considered, we find that $|\beta_{11}^0|\leqslant CM^{-1/2}$. For transformation in the space of variables (y, η), where $y = x$, $\eta = \xi\lambda^{k+1}$, it follows that

$$\begin{gathered}1/2\leqslant\alpha_1^1\leqslant 2,\quad |\alpha_{j1}|\leqslant CM^{-1/2},\quad |\beta_{11}|\leqslant C\lambda^{k+1}M^{-1/2},\\ |\beta_{1j}|\leqslant C\lambda^{k+1}M^{-1/2},\qquad j=2,\ \ldots,\ n.\end{gathered}\tag{6.5}$$

6.7. Now we shall estimate the values of the polynomials $A_0 = A - A_1$, $B_0 = B - B_1$, $C_0 = C - C_1$ in $\Omega\cup\Omega_1$. One may see from the definition of these polynomials that

$$A_0(y^1) = -\frac{1}{a}\sum_{j=1}^{l_0}\frac{1}{(j+1)!}\nabla_{y'}\partial_1^j\partial_0^s q(P)(y^1)^{j+1},$$

$$B_0(y^1) = -\frac{1}{a}\sum_{j=1}^{l_0}\frac{1}{(j+1)!}\nabla_{\xi'}\partial_0^s q(P)(y^1)^{j+1},$$

$$C_0(y^1) = -\frac{1}{a}\sum_{j=1}^{l_0+1}\frac{1}{(j+1)!}\partial_1^j\partial_0^s q(P)(y^1)^{j+1} - \int_0^1 (A_0(\tau),\ B_0'(\tau))\,d\tau,$$

where P denotes the center of Ω, the derivatives $\partial_0^s q$ being computed after the canonical transformation corresponding to Ω. By (6.4), for $|y^1| \leqslant CaM^{-s-1}$

$$|\partial_1^j A_0(y^1)| \leqslant C\lambda^{-(k+1)/2}a^{-j}M^{j(s+1)-1/2}, \qquad |\partial_1^j B_0(y^1)| \leqslant C\lambda^{-(k+1)/2}a^{-j}M^{j(s+1)-1/2} \tag{6.6}$$

and by (6.4)

$$|\partial_1^j C_0(y^1)| \leqslant C\lambda^{-k-1}a^{-j}M^{j(s+1)}. \tag{6.7}$$

6.8. We shall now prove that

$$|\partial_t^l\partial_\eta^\alpha\partial_y^\beta H(t,\ z,\ \zeta)| \leqslant C_{l,\alpha,\beta}M^l\delta^{|\alpha|}\lambda^{(k+1)(|\beta|-|\alpha|)/2}, \tag{6.8}$$

where H is any of the functions H_j. By Theorem 7.1 of Chapter II and the inequality

$$1 \leqslant \sum H_j^2 \leqslant C(n)$$

it follows that

$$\frac{1}{2}\|u\|^2 \leqslant \sum \|H_j u\|^2 \leqslant 2C(n)\|u\|^2,$$

provided δ_0 is sufficiently small.

The transformation of coordinates from $(y,\ \eta)$ to $(z,\ \zeta)$ is the product of two transformations:

(1) $z^j = \sum_{l=1}^n \alpha_l^j t^l,\ \zeta_j = \sum_{l=1}^n \alpha_j^l \tau_l + \sum_{l=1}^n \beta_{jl}t^l,\ j = 1,\ \dots,\ n;$

(2) $t^1 = y^1,\quad t' = y' - B(y^1),\quad \tau' = \eta' + \lambda^{k+1}A(y^1),\quad \tau_1 = \eta_1 + \lambda^{k+1}(A'(y^1),\ y' - B(y^1)) - (B'(y^1),\ \eta') - \lambda^{k+1}C'(y^1)$. Compute the derivatives of H_j with respect to $(y,\ \eta)$. We have

$$\frac{\partial H}{\partial y^1} = \sum_{j=1}^n \frac{\partial H}{\partial z^j}\left(\alpha_1^j - \sum_{l=2}^n \alpha_l^j B^{l'}\right) + \sum_{j=1}^n \frac{\partial H}{\partial \zeta_j}\left(\beta_{j1} - \sum_{l=2}^n \beta_{jl}B^{l'} + \alpha_j^1 D + \sum \alpha_j^l A_l'\lambda^{k+1}\right),$$

$$\frac{\partial H}{\partial y^l} = \sum_{j=1}^n \frac{\partial H}{\partial z^j}\alpha_l^j + \sum_{j=1}^n \frac{\partial H}{\partial \zeta_j}(\alpha_j^1 A_l'\lambda^{k+1} + \beta_{jl}), \qquad l = 2,\ \dots,\ n,$$

$$\frac{\partial H}{\partial \eta_1} = \sum_{j=1}^n \frac{\partial H}{\partial \zeta_j}\alpha_j^1, \qquad \frac{\partial H}{\partial \eta_l} = \sum_{j=1}^n \frac{\partial H}{\partial \zeta_j}(\alpha_j^l - \alpha_j^1 B^{l'}), \qquad l = 2,\ \dots,\ n,$$

where

$$D(y,\ \eta) = (A''(y^1),\ y'\lambda^{k+1} - B\lambda^{k+1}) - (A',\ B')\lambda^{k+1} - (B'',\ \zeta') - C''\lambda^{k+1}.$$

Using inequalities (6.5), (6.6), and (6.7), we obtain

$$\left|\frac{\partial H}{\partial y^1}\right| \leqslant Ca^{-1}M^{s+1}, \qquad \left|\frac{\partial H}{\partial y^l}\right| \leqslant C\lambda^{(k+1)/2}, \qquad l = 2,\ \dots,\ n;$$

$$\left|\frac{\partial H}{\partial \eta_j}\right| \leqslant C\delta\lambda^{-(k+1)/2}, \qquad j=1, \ldots, n.$$

Since the coefficients in the above formulas for first derivatives are such that their differentiation with respect to y^1 yields a factor $a^{-1}M^{s+1}$, differentiation with respect to y' gives $\lambda^{(k+1)/2}$, and with respect to η — a factor $\delta_0\lambda^{-(k+1)/2}$, we obtain the required estimate (6.8).

6.9. Let us now consider the case when s varies in going from Ω to Ω_1 by a value $s_1 \neq s$. In this case

$$a = c_0 b M^{s - s_1},$$

where c_0 is independent of λ and M, $b = |\operatorname{grad}_{z,\xi}\partial_1^{s_1} q_0(P)|$. The transition from matrices Π, P to Π_1, P_1 will depend on the angle between the vectors $\operatorname{grad}_{x,\xi}\partial_t^s q(0)$ and $\operatorname{grad}_{x,\xi}\partial_t^{s_1} q(P)$. By (6.5) it is sufficient to consider the transformation corresponding to the change from $\operatorname{grad}_{x,\xi}\partial_t^s q(0) = (0, \ldots, 0; a, 0, \ldots, 0)$ to $\operatorname{grad}_{x,\xi}\partial_t^{s_1} q(0)$. Since

$$|\operatorname{grad}_{x,\xi}\partial_t^{s_1} q(0)| \leqslant CM^{s_1+1/2}\lambda^{-(k+1)/2}, \qquad |\partial_1\partial_t^{s_1} q(0)| \leqslant Ca^{-1}M^{s_1+s+1}\lambda^{-k-1},$$

(6.5) are fulfilled for the transformation of the rotation which is accomplished in going from $\operatorname{grad}_{x,\xi}\partial_t^s q(0)$ to $\operatorname{grad}_{x,\xi}\partial_t^{s_1} q(0)$. These estimates hold for inverse transformations. Therefore, the product of the two indicated transformations from $\operatorname{grad}_{x,\xi}\partial_t^s q(0)$ to $\operatorname{grad}_{x,\xi}\partial_t^{s_1} q(0)$ and then to $\operatorname{grad}_{x,\xi}\partial_t^{s_1} q(P)$ again satisfies conditions (6.5).

Now consider a canonical transformation with the generating function $x\xi + (A(x^1), x') + (B(x^1), \xi') + C(x^1)$, where

$$A(x^1) = -\frac{1}{b}\sum_{j=0}^{m_0} \partial_t^{s_1}\partial_1^j\partial_{x'} q(P)(x^1)^{j+1}/(j+1)!,$$

$$B(x^1) = -\frac{1}{b}\sum_{j=0}^{m_0} \bar{\partial}_t^{s_1}\partial_1^j\partial_{\xi'} q(P)(x^1)^{j+1}/(j+1)!,$$

$$C(x^1) = -\frac{1}{b}\sum_{j=0}^{m_0+1} \partial_t^{s_1}\partial_1^j q(P)(x^1)^{j+1}/(j+1)! - \int_0^{x^1} (B'(\tau), A(\tau))\,d\tau,$$

and $m_0 = [(k/2 - s_1)/(s_1 + 1)]$. As above, one can verify that (6.6) and (6.7) are fulfilled for the polynomials A, B, and C. Thus, all of the arguments given in estimating the derivatives of the function H remain valid. Hence, inequalities (6.8) are fulfilled also when s varies.

§7. Applications to the Oblique Derivative Problem

7.1. In this section we shall show how the results above can be utilized in studying the oblique derivative problem considered in Chapter IV.

First we shall consider a model problem. Let $\mathbb{R}_+^n$ be the part of the n-dimensional space $\mathbb{R}^n$ for the points $x = (x^1, \ldots, x^n) = (x', x^n)$ of which the inequality $x^n \geqslant 0$ is fulfilled. Suppose that $n \geqslant 3$. Consider the equation

$$\Delta u = 0 \quad \text{in} \quad \mathbb{R}_+^n \tag{7.1}$$

with boundary condition

$$(x^1)^k \frac{\partial u}{\partial x^n} + a\frac{\partial u}{\partial x^1} = g(x) \quad \text{for} \quad x^n = 0, \tag{7.2}$$

where $a \neq 0$ is a real constant. Note that for this problem the Lopatinski condition is violated only at $x^1 = 0$. Let $\tilde{u}(\xi', x^n) = \int u(x)e^{-ix'\xi'}dx'$ be the Fourier transform of function u in x'. By (7.1), $\tilde{u}(\xi', x^n)$ satisfies the equation

$$\frac{\partial^2 \tilde{u}}{(\partial x^n)^2} - |\xi'|^2 \tilde{u} = 0,$$

the general solution of which is written as: $\tilde{u}(\xi', x^n) = v(\xi')e^{-x^n|\xi'|} + w(\xi')\, e^{-x^n|\xi'|}$. Since we are interested in solutions which for $x^n > 0$ are Fourier transforms of distributions, we should assume $w = 0$. Inserting the result in (7.2) we obtain that for $x^n = 0$

$$\int [-(x^1)^k |\xi| + ia\xi_1]\, v(\xi')\, e^{ix'\xi'}\, d\xi' = (2\pi)^{n-1} g(x').$$

This is a pseudo-differential equation with symbol $p(x', \xi') = -(x^1)^k|\xi| + ia\xi_1$. The bicharacteristics of the function Im p coincide with lines parallel to the x^1 axis. From Theorem 2.1 it follows that this operator will be subelliptic when k is even; in the case of odd k, the operator is subelliptic if and only if $a > 0$. In these cases, Theorem 1.2 yields an *a priori* estimate of the solution:

$$\|u\|_s \leqslant C_s (\|g\|_{s-3/2+k/(k+1)} + \|u\|_{s-1}),$$

where the constant C_s is independent of u, and a corresponding theorem on smoothness (see Appendix 1.2). In Chapter IV we obtained a very rough estimate that does not take into account how many times the field touches the boundary, but on the other hand, relies on a far more elementary proof.

7.2. We shall show that in the general case the oblique derivative problem

$$Lu = f \quad \text{in} \quad \Omega \quad (\Sigma a_{ij}\xi_i\xi_j > 0,\ \xi \in R_n\setminus 0), \tag{7.3}$$

$$\frac{\partial u}{\partial l} = g \quad \text{on} \quad \Gamma \tag{7.4}$$

can be studied with the use of Theorem 1.2. Let $L_\lambda u = Lu - \lambda u$ when $\lambda \geqslant \lambda_0 > 0$. Let $\frac{d}{dN} = \sum a_{ij} \cos(\nu, x^i)\, \partial_j$, where ν is the outward normal to Γ, and $\cos(\nu, x^i)$ is the cosine of the angle between ν and the x^1 axis. Define the operator A, assuming that $Af = \frac{d}{dN} F\Big|_\Gamma$, where $f \in C^\infty(\Gamma)$ and F is the solution to the Dirichlet problem $L_\lambda F = 0$ in Ω, $F = f$ on Γ. The operator A is defined correctly if λ_0 is large enough. As seen in Chapter III, A is a first order pseudo-differential operator. Besides, it is elliptic and

$$\|f\|_s \leqslant C\|Af\|_{s-1} \leqslant C_1\|f\|_s, \qquad f \in C^\infty(\Gamma).$$

It follows that the operator allows closure and may be extended on $\cup H_s(\Gamma)$.

<u>Lemma 7.1.</u> Operator $A - A^*$ is of zero order and, therefore, the principal symbol $a^0(x, \xi)$ of operator A is a real-valued symbol.

<u>Proof.</u> Let $A_1 f = \frac{d}{dN} F_1\Big|_\Gamma$, where $(L_\lambda + l_1)F_1 = 0$ in Ω, $F_1 = f$ on Γ and l_1 is a first-order operator such that $L_\lambda + l_1$ is a symmetric operator. From Green's formula

$$\left(\frac{\partial u}{\partial N}, v\right)_\Gamma - \left(u, \frac{\partial v}{\partial N}\right)_\Gamma = \int [(L_\lambda + l_1)u \cdot \bar{v} - \overline{(L_\lambda + l_1)v} \cdot u]\, dx$$

it is seen that A_1 is a symmetric operator.

Note that $F - F_1 = 0$ on Γ, $L_\lambda(F - F_1) = l_1F_1 = T_{1/2}f$, where $T_{1/2}$ is an operator of order 1/2. Hence it follows that $F - F_1 = T_{-3/2}f$ and therefore

$$Af - A_1 f = \frac{d}{dN}(F - F_1)|_\Gamma = T_0 f,$$

where $T_{-3/2}$ and T_0 are operators or order (—3/2) and 0, respectively. But then $A^*f - A_1f = T_0^*f$ and therefore

$$A^*f - Af = (A^* - A_1^*)f + (A_1^* - A_1)f + (A_1 - A)f = T_0'f,$$

where T_0' is a zero-order operator. This proves the lemma.

Suppose that in a neighborhood ω of point P_0, lying on Γ, we have chosen a system of coordinates such that the directions $x^1, \ldots, x^{n-1}$ are tangent to Γ_n, and the x^n axis is directed along the normal to the boundary and $\frac{\partial}{\partial l(x)} = \sum_{j=1} \nu_j(x) \frac{\partial}{\partial x^j}$. Thus, the field is tangent to the boundary Γ only at those points of ω at which $\nu_n(x) = 0$. Then if $f \in C^\infty(\Gamma)$ and λ is large, we can solve the Dirichlet problem: $L_\lambda w = 0$ in Ω, $w = f$ on Γ and can assume that $Pf = \frac{\partial \omega}{\partial l}\Big|_\Gamma = \nu_n(x)Af + Bf$, where B is a first-order differential operator. The principal part of the symbol of operator P, for $x \in \omega$, equals

$$P^0(x, \zeta) = \nu_n(x)a(x, \zeta) + ib(x, \zeta), \tag{7.5}$$

where a and b are real-valued functions, $a(x, \xi) \geqslant C|\xi|$ for $x \in \omega$, $b(x, \zeta) = \sum_{j=1}^{n-1} b^j(x)\zeta_j, \quad C > 0.$

As shown earlier, the solution of problem (7.3) and (7.4) is reduced to the solution of the pseudo-differential equation

$$Pf = g \text{ in } \Gamma \tag{7.6}$$

and to the subsequent solution of the Dirichlet problem

$$Lu = 0 \text{ in } \Omega \quad u = f \text{ on } \Gamma.$$

Since the last problem has been well studied, it remains to consider equation (7.6).

Here we shall not write out the formal conditions for the construction of field $\partial/\partial l$, for which Theorem 1.2 is applicable. Only note that, unlike Chapter IV, we shall not make any assumptions on the structure of the set on which ν_n has a root of even order. In addition, we can consider the case when $b^j(x)$ and $c(x)$ are complex-valued functions and $\frac{\partial}{\partial l} = \frac{\partial}{\partial l_1} + i\frac{\partial}{\partial l_2}$. The symbol of operator P in local coordinates will in this case have the same form as in (7.5), but with complex-valued functions ν_n and b.

§8. Application to the $\bar{\partial}$-Neumann Problem

8.1. Let Ω be a relatively compact domain in the space $\mathbb{C}^2$ of complex variables z^1, z^2 with smooth boundary Γ.

Suppose that a smooth real-valued function r is defined in a neighborhood of the boundary Γ such that $r < 0$ in Ω, $t > 0$ outside Ω, and $dr \neq 0$.

Domain Ω is called *pseudoconvex* if the inequality

$$r_{z^1\bar{z}^1}(P)|t^1|^2 + r_{z^1\bar{z}^2}(P)t^1\bar{t}^2 + r_{\bar{z}^1z^2}(P)\bar{t}^1t^2 + r_{z^2\bar{z}^2}(P)|t^2|^2 \geqslant 0 \tag{8.1}$$

is fulfilled at every point $P \in \Gamma$ and for arbitrary vector $(t^1, t^2) \in \mathbb{C}^2$ satisfying the condition $r_{z^1}(P)t^1 + r_{z^2}(P)t^2 = 0$. Here we have used standard notation:

$$\partial_{\bar{z}^j}=\frac{1}{2}(\partial_{2j-1}+i\partial_{2j}),\quad \partial_{z^j}=\frac{1}{2}(\partial_{2j-1}-i\partial_{2j}),\qquad z^j=x^{2j-1}+ix^{2j};\quad j=1,\ 2.$$

Consider the system of equations

$$\frac{\partial u}{\partial \bar{z}^1}=f_1,\quad \frac{\partial u}{\partial \bar{z}^2}=f_2 \text{ in } \bar{\Omega} \tag{8.2}$$

assuming that Ω is a pseudoconvex domain and the compatibility condition

$$\frac{\partial f_1}{\partial \bar{z}^2}=\frac{\partial f_2}{\partial \bar{z}^1} \tag{8.3}$$

is fulfilled. In this case, we will be interested in a regular solution, that is a solution such that

$$\operatorname{sing\,supp} u=\operatorname{sing\,supp} f,\quad f=(f_1,\ f_2). \tag{8.4}$$

Since (8.2) is elliptic in Ω, property (8.4) for interior points follows from the general properties of elliptic operators. Only the points lying at the boundary Γ are studied with difficulty.

The solution to this problem is closely associated with the solution of E. Levi's problem on holomorphy domains and with other important problems of complex analysis (for details see B. V. Shabat,[1] R. C. Gunning and H. Rossi,[1] G. B. Folland and J. J. Kohn[1]).

8.2. The vector field L is called *holomorphic* in the open set $U\subset\mathbb{C}^2$, if it can be written as

$$L=a^1\frac{\partial}{\partial z^1}+a^2\frac{\partial}{\partial z^2},\quad a^i\in C^\infty(U).$$

The vector field L is said to be *tangent* if $L(r) = 0$ for $r = 0$. As usual, we assume

$$\bar{L}=\bar{a}^1\frac{\partial}{\partial \bar{z}^1}+\bar{a}^2\frac{\partial}{\partial \bar{z}^2}.$$

If T_1, T_2 are two vector fields, then we define their Lie bracket assuming

$$[T_1,\ T_2]=T_1T_2-T_2T_1.$$

The Lie algebra generated by the fields T_1 and T_2 over the space of smooth functions is the least module over this space which is closed in relation to the operation, and is denoted by $\mathscr{L}(T_1,\ T_2)$. A filtration is defined in this algebra, that is,

$$\mathscr{L}(T_1,\ T_2)=\bigcup_{k=0}^{\infty}\mathscr{L}_k(T_1,\ T_2),$$

where $\mathscr{L}_0(T_1,\ T_2)$ is the module generated by T_1, T_2, and $\mathscr{L}_{k+1}(T_1,\ T_2)$ is the module is obtained by adding to the elements of $\mathscr{L}_k(T_1,\ T_2)$ the elements $[T,\ T_1]$, $[T,\ T_2]$ for $T\in\mathscr{L}_k(T_1,\ T_2)$.

Let L be a holomorphic tangent field in some neighborhood of a point $P\in\Gamma$ which is different from zero at point P. Suppose $\mathscr{L}=\mathscr{L}(L,\ \bar{L})$, $\mathscr{L}_k=\mathscr{L}_k(L,\ \bar{L})$.

Let

$$\partial u=\frac{\partial u}{\partial z^1}dz^1+\frac{\partial u}{\partial z^2}dz^2,\quad \bar{\partial}u=\frac{\partial u}{\partial \bar{z}^1}d\bar{z}^1+\frac{\partial u}{\partial \bar{z}^2}d\bar{z}^2,$$

so that (8.2) can be written as

$$\bar{\partial}u=f,\quad f=f_1\,d\bar{z}^1+f_2\,d\bar{z}^2$$

and (8.3) as $\bar{\partial}f = 0$.

<u>Definition 8.1.</u> The point $P\in\Gamma$ is called a point of *finite type* if there exists an element $F\in\mathscr{L}$ such that

$$\langle\partial r(P),\ F(P)\rangle\neq 0.$$

If in this case $F \in \mathcal{L}_k$ and k is a minimum, then the point is called a *point of type* k.

8.3. Let P be a point on Γ and let ω^1, ω^2 be an orthonormal basis for (0, 1) forms in a neighborhood U of the point P assuming $\omega^2 = f\partial r$ and $f = 1$ on Γ. We denote by L_1, L_2 the corresponding basis in the dual space of vector fields. Clearly L_1 is a tangent holomorphic vector field, since

$$fL_1(r) = f\langle\partial r,\ L_1\rangle = \langle\omega^2,\ L_1\rangle = 0.$$

Using the equality

$$(\bar\partial u,\ \alpha) = (u,\ \bar\partial^*\alpha),$$

where α is a (0, 1) form and $u \in C_0^\infty(\bar\Omega)$, we define operator $\bar\partial^*$ which is formally adjoint to operator $\bar\partial$. It follows that, if α is in the domain of definition of the operator $\bar\partial^*$, then α satisfies a natural boundary condition: $\langle\alpha,\ \bar\partial r\rangle = 0$ on Γ, that is, in the indicated local basis $\alpha_2 = 0$ on Γ. We denote by $\mathscr{B}$ the set of (0, 1) forms with compact support $U \cup \bar\Omega$ which satisfy this boundary condition.

Now we shall prove the following important result.

Theorem 8.1. (J. J. Kohn). If Ω is a pseudoconvex domain, then there exists a constant C such that

$$\|L_1\varphi_1\| + \|\bar L_1\varphi_1\| + \|\bar L_2\varphi_1\| + \|\varphi_2\|_1 \leqslant C(\|\bar\partial\varphi\| + \|\bar\partial^*\varphi\| + \|\varphi\|) \tag{8.5}$$

for all $\varphi \in \mathscr{B}$. Here $\|\cdot\|$ and $\|\cdot\|_1$ are the norms in $L_2(\Omega)$ and $H^1(\Omega)$, respectively.

Proof. If $\varphi \in \mathscr{B}$, then

$$\varphi = \varphi_1\overline{\omega^1} + \varphi_2\overline{\omega^2}, \qquad \bar\partial\varphi = (\bar L_1\varphi_2 - \bar L_2\varphi_1)\,\overline{\omega^1} \wedge \overline{\omega^2} + \dots, \tag{8.6}$$

where the dots denote a form whose coefficients are combinations of functions φ_1 and φ_2. Furthermore,

$$\partial^*\varphi = -L_1\varphi_1 - L_2\varphi_2 +, \dots, \tag{8.7}$$

in which again the dots stand for a linear combination of functions φ_1 and φ_2. It follows from (8.6) that

$$\|\bar\partial\varphi\|^2 = \|\varphi\|_{\bar z}^2 - \sum(\bar L_i\varphi_j,\ \bar L_j\varphi_i) + O(\|\varphi\|_{\bar z}\|\varphi\| + \|\varphi\|^2), \tag{8.8}$$

in which $\|\varphi\|_{\bar z}^2 = \sum\|\bar L_i\varphi_j\|^2$. From (8.7) we obtain

$$\|\bar\partial^*\varphi\|^2 = \sum(L_i\varphi_i,\ L_j\varphi_j) + O(\|\varphi\|_{\bar z}\|\varphi\| + \|\varphi\|^2).$$

Here, boundary integrals do not appear when we integrate by parts in the expression $(L_i\varphi_i, \dots)$, because L_1 is a tangent field and $\varphi_2 = 0$ on Γ. Suppose that

$$[L_i,\ \bar L_j] = \sum c_{ij}^k L_k + \sum d_{ij}^k \bar L_k.$$

Integrating (8.8) by parts yields

$$-\sum(\bar L_i\varphi_j,\ \bar L_j\varphi_i) = \sum(L_j\bar L_i\varphi_j,\ \varphi_i) + O(\|\varphi\|_{\bar z}\|\varphi\| + \|\varphi\|^2) =$$

$$-\sum(L_j\varphi_j,\ L_i\varphi_i) + \sum(c_{ji}^2 L_2\varphi_j,\ \varphi_i) + O(\|\varphi\|_{\bar z}\|\varphi\| + \|\varphi\|^2) =$$

$$-\|\bar\partial^*\varphi\|^2 + \sum\int_\Gamma c_{ji}^2\varphi_j\bar\varphi_i\,dS + O(\|\varphi\|_{\bar z}\|\varphi\| + \|\varphi\|^2).$$

Since $\varphi_2 = 0$ on the boundary, the boundary integral equals $\int_\Gamma c_{11}^2|\varphi_1|^2\,dS$. It now follows that

$$\|\varphi\|_z^2 + \int_\Gamma c_{11}^2 |\varphi_1|^2 dS \leqslant C(\|\partial\varphi\|^2 + \|\partial^*\varphi\|^2 + \|\varphi\|^2). \tag{8.9}$$

Using the equality ($\varphi_2 = 0$ on Γ) again we obtain

$$\|L_i\varphi_2\|^2 = -(\bar{L}_i L_i \varphi_2, \varphi_2) + O(\|\varphi\|_{\bar{z}} \|\varphi\|) =$$

$$\|\bar{L}_i\varphi_2\|^2 + O(\|\varphi\|_{\bar{z}}\|\varphi\| + \|\varphi\|^2) \leqslant C(\|\varphi\|_z^2 + \|\varphi\|^2). \tag{8.10}$$

On the other hand,

$$\|L_1\varphi_1\|^2 = -(\bar{L}_1 L_1\varphi_1, \varphi_1) + O(\|\varphi\|_{\bar{z}}\|\varphi\|) = \|\bar{L}_1\varphi_1\|^2 + (c_{11}^2 L_2\varphi_1, \varphi_1) + O(\|\varphi\|_{\bar{z}}\|\varphi\|) \leqslant$$

$$\int_\Gamma c_{11}^2 |\varphi_1|^2 ds + O(\|\varphi\|_z^2 + \|\varphi\|^2). \tag{8.11}$$

Inequality (8.5) follows from (8.9), (8.10), and (8.11).

8.4. The earlier definition of pseudoconvexity is equivalent in the chosen local basis to the condition

$$\lambda \equiv \langle \bar{\partial} r, [\bar{L}, L_1] \rangle \geqslant 0 \text{ in } \Gamma \cap U. \tag{8.12}$$

Therefore, $c_{11}^2 \geqslant 0$ on Γ and both the terms on the left-hand side of (8.9) are non-negative.

We shall now consider the uniqueness of the solution of (8.2).

The equation $L_1(P, \xi) = 0$ defines a straight line in the space $T_p^*(\Gamma)$. Of the two directions defined by this straight line, one is elliptic for the set of operators $\bar{L}_1$, $\bar{L}_2$ (in accordance with the theory developed in Chapter III). Freezing the coefficients of the operators and performing a tangent-variable Fourier transformation, we obtain from operator L_2 a first-order operator in the normal direction ν which is of the form $\partial/\partial\nu + \tau$, where τ stands for the linear form of ξ. If the normal ν is directed inside Ω, then the values $\tau > 0$ ($\tau < 0$) correspond to solutions of an ordinary differential equation which decrease (increase) in the direction of ν. Call $\zeta(P)$ the cotangent vector defined by condition $L_1(P, \zeta(P)) = 0$ and such that $\langle \tau, \zeta(P) \rangle < 0$. Then

$$\mathrm{WF}(u) \setminus \mathrm{WF}(f) \subset \bigcup_{P \in \Gamma} (P, \zeta(P)). \tag{8.13}$$

The regularity conditions of the solution depend on the structure of the field L_1 in a neighborhood of the point in $T^*(\Gamma)$.

The pseudoconvexity conditions are insufficient for (8.4) to hold, as will be seen from the following

<u>Example</u>. (J. J. Kohn). Let $c_{11}^2 \equiv 0$ in a neighborhood of a point $P \in \Gamma$. Then one can choose local holomorphic coordinates z^1, z^2 with origin at P such that all points of the form $(z^1, 0)$ lie on Γ in some neighborhood of this point. Let $\rho \in C_0^\infty(U)$ and $\rho = 1$ in a neighborhood of the point P. Let $\alpha = \bar{\partial}(\rho/z^2) = \bar{\partial}\rho/z^2$. Then there exists a function u for which $\bar{\partial}u = \alpha$, provided the support of ρ is small enough. In this case, the function $h = u - \rho/z^2$ is holomorphic and equals $u - 1/z^2$ in the domain where $\rho = 1$. If sing supp u $\subset$ sing supp α, then u would have to be smooth in the domain ω in which $\rho = 0$ and would have singularities in the domain where $\rho = 1$ and which lies within ω. However, this is at variance with the well-known theorems of classical analysis.

8.5. Below we shall find conditions sufficient for regularity of the $\bar{\partial}$-problem. One such condition is based on subelliptic estimates of the form

$$\|\varphi\|_{1/(k+1)} \leqslant C(\|\bar{\partial}\varphi\| + \|\bar{\partial}^*\varphi\| + \|\varphi\|), \quad \varphi \in \mathscr{B}. \tag{8.14}$$

We shall prove the following theorem.

Theorem 8.2. Estimate (8.14) exists if and only if all the points belonging to $\Gamma \cap U$ are points of the type not higher than k.

We shall not dwell here on the corollaries of Theorem 8.2 but refer to J. J. Kohn,[2-4] G. B. Folland and J. J. Kohn,[1] and J. J. Kohn and L. Nirenberg.[1]

8.6. Let the boundary equation in local coordinates be written as $x^4 = \varphi(x^1, x^2, x^3)$. The change of variables $\tilde{x}^j = x^j$, $j = 1, 2, 3$, $\tilde{x}^4 = x^4 - \varphi$ straightens the boundary of the domain Ω so that Γ has an equation $x^4 = 0$ in new coordinates and the domain $\Omega \cap U$ is situated in $x^4 > 0$. Without any loss of generality, it may be assumed that

$$L_1 = D_1 + i(D_2 + aD_3), \quad L_2 = D_4 + ibD_3,$$

where $a, b \in C^\infty(U \cap \Omega)$, $b \geqslant b_0 > 0$. Also it is clear that

$$\lambda \equiv \partial_1 a \geqslant 0 \quad \text{for} \quad x^4 = 0. \tag{8.15}$$

It is not hard to see that P is a point of type k, provided

$$\sum_{j=1}^{k} |\partial_1^j a(x)| \geqslant c_0 > 0 \quad \text{for} \quad x \in \Gamma. \tag{8.16}$$

Suppose that U is a neighborhood so small that the set $\bigcup_{P \in \Gamma \cup U} \zeta(P)$ fills only a small portion of the unit sphere S^2 completely lying in the ball $|\xi - \xi_0| \leqslant 1/8$. Let the function $\psi \in C^\infty(S^2)$ be such that $\psi = 1$ in a neighborhood of this portion of the sphere and $\psi(\xi) = 0$ for $|\xi - \xi_0| \geqslant 1/4$. Extend this function to the space $\mathbb{R}^3$ so that $\psi(t\xi') = \psi(\xi')$ for $|\xi'| \geqslant 1$, $t \geqslant 1$ and $\psi \in C^\infty(\mathbb{R}^3)$.

As mentioned, the theory of elliptic boundary problems gives the estimate

$$\|(1 - \Psi(D))\varphi\|_1 \leqslant C(\|(1 - \Psi(D))\bar{\partial}\varphi\|_0 + \|(1 - \Psi(D))\bar{\partial}^*\varphi\|_0 + \|\varphi\|_0). \tag{8.17}$$

By H we shall denote the Sobolev space $H_s(U \cap \bar{\Omega})$ and by H_s' the space with the norm

$$\|u\|_s' = \left(\int\int (1 + |\zeta'|^2)^s |\tilde{u}(\zeta', x^4)|^2 d\zeta' dx^4\right)^{1/2}, \quad \zeta' = (\zeta_1, \zeta_2, \zeta_3).$$

Lemma 8.1. If $u \in H_s'$, $\bar{L}_2 u \in H_{s-1}'$, $s > 0$, then $u \in H_s$ and

$$\|u\|_s \leqslant C(\|u\|_s' + \|\bar{L}_2 u\|_{s-1}'), \quad u \in \mathring{H}_s(\bar{\Omega} \cap U).$$

Proof. It is clear that

$$\|u\|_s \leqslant \|u\|_s' + \|D_4 u\|_{s-1}' \leqslant \|u\|_s' + \|\bar{L}_2 u\|_{s-1}' + C_1\|D_3 u\|_{s-1}' \leqslant C_2(\|u\|_s' + \|\bar{L}_2 u\|_{s-1}').$$

If $u \in H_s'$, $\bar{L}_2 u \in H_{s-1}'$, then $D_4 u \in H_{s-1}'$ and hence $u \in H_s$.

8.7. Proof of Theorem 8.2.

Sufficiency. By (8.5), (8.17), and Lemma 8.1 it suffices to obtain

$$\|h\Psi\varphi_1\|_{1/(k+1)}' \leqslant C(\|h\Psi L_1\varphi_1\|_0 + \|h\Psi\bar{L}_1\varphi\|_0 + \|\varphi_1\|_0), \tag{8.18}$$

where $h \in C_0^\infty(\Gamma)$, $h = 1$ in a neighborhood of the set $U \cap \Gamma$. Clearly

$$L_1\Psi\varphi_1 = P\Psi\varphi_1, \quad \bar{L}_2\Psi\varphi_1 = \bar{P}\Psi\varphi, \tag{8.19}$$

where P is a pseudo-differential operator

$$P = D_1 + i(D_2 + aB(D'))$$

and B(D') is a first-order pseudo-differential elliptic operator in $\mathbb{R}^3$ whose synbol $B(\xi')$ coincides with ζ_3 in supp ψ and $|B(\zeta')| \geqslant C_0|\zeta'|$. The subellipticity of P follows from (8.15) and (8.16) by Theorem 1.2. By (8.19) this yields

$$\|h\Psi\varphi_1\|'_{1/(k+1)} \leqslant C(\|L_1(h\Psi\varphi_1)\|_0 + \|L_1(h\Psi\varphi_1)\|_0 + \|\varphi_1\|_0)$$

and therefore (8.18) is valid.

Necessity. Note that (8.18) follows from (8.14) and (8.17) and from the last inequality and from (8.19) it follows that

$$\|u\|'_{1/(k+1)} \leqslant C(\|Pu\|_0 + \|P^*u\|_0 + \|u\|_0),$$

at least for functions u of the type $u = h\Psi\varphi_1$.

That the estimate

$$\|P^*u\|_0^2 \leqslant \|Pu\|_0^2 + C_1\|u\|_0^2$$

exists for $u = h\Psi\varphi_1$ follows from the equality

$$\|Pu\|_0^2 = \|P^*u\|_0^2 + ((P^*P - PP^*)u, u)$$

by (8.15), and therefore

$$\|u\|'^2_{1/(k+1)} \leqslant C_2(\|Pu\|_0^2 + \|u\|_0^2).$$

By Theorem 2.1, this estimate will imply the local estimate (2.1) at the point $(P_0, \zeta(P_0))$ for $P_0 \in U$, from which, according to Theorem 3.1, the validity of inequalities (8.15) and (8.16) follows.

CHAPTER IX. THE CAUCHY PROBLEM

§1. Statement of the Problem

In this chapter we shall examine one of the basic problems for evolution equations considered in the general theory of linear partial differential equations, the Cauchy problem. In the most general form, the problem consists of finding a solution to the equation Pu = f in some neighborhood Ω which coincides with the given function g in a subdomain $\Omega' \subset \Omega$. Usually Ω is a neighborhood of the origin, and Ω' is that part of Ω in which $x^1 \leqslant 0$.

In this case it is quite natural to assume the plane $x^1 = 0$ to be noncharacteristic for the operator P, that is, $p^0(0, x^1, 1, 0) \neq 0$ when the point $(0, x') \in \Omega'$. If P is a differential operator, then for this condition the values of all the derivatives of the solution are uniquely defined on this plane.

The Cauchy problem is not well posed for all equations, but only for the so-called hyperbolic equations; this implies, in particular, that the roots ξ^1 of the equation $p^0(x, \xi^1, \xi') = 0$ are real for all $x \in \Omega$, $\xi' \in \mathbb{R}^{n-1}$. The Cauchy problem for hyperbolic equations was studied in the classic works of J. Hadamard,[1] I. G. Petrowsky,[1-2] J. Leray,[1] and L. Gårding.[1]

Uniqueness in the Cauchy problem can be proved for a much wider class of operators. The strongest result was obtained by A. P. Calderón,[1] which was further generalized in K. Smith,[1] F. Trèves,[9] B. Malgrange,[3] and L. Nirenberg.[2] We shall also present a generalization of their results. Our proof is based on the use of Carleman estimates; such estimates were first used by Carleman in studying the Cauchy problem for elliptic differential equations.

Examples of equations for which the Cauchy problem has a nonunique solution are given in A. Plis[1,2] and P. Cohen.[1] L. Hörmander[3] contains the equation

$$\frac{\partial u}{\partial t} = ia(t, x)\frac{\partial u}{\partial x},$$

for which P. Cohen has proved the existence of a nontrivial solution equal to 0 when $t \leqslant 0$. Here $a \in C^\infty$ is a real-valued function.

With the development of the general theory of differential equations, the S. V. Kovalevsky theorem — one of the earliest theorems pertaining to the theory of the Cauchy problem — is finding ever-increasing application. This theorem states that there exists a unique solution of the Cauchy problem for systems of ordinary differential equations of very general form in the class of analytic functions. The classical theorem of Kovalevsky was generalized by L. V. Ovsyannikov in his works, which have a very wide range of application. Based on Ovsyannikov's ideas, we present below a proof of the Kovalevsky theorem due to F. Trèves.[2,11]

§2. Kovalevsky Theorem

2.1. Consider a system of equations of the form

$$D_t^m u = \sum_{\substack{|\alpha|+j\leqslant m\\ j\leqslant m-1}} a_{\alpha,j}(t, x) D_x^\alpha D_t^j u + f(t, x). \tag{2.1}$$

Here $u = (u_1, \ldots, u_N)$, $f = (f_1, \ldots, f_N)$, $a_{\alpha,j}$ are matrices of order N. The system of equations of this form is called the Kovalevsky system. We shall find a solution of this system satisfying the initial conditions:

$$D_t^j u = \varphi_j, \quad j = 0, 1, \ldots, m-1 \quad \text{for} \quad t = 0. \tag{2.2}$$

The main result of this section is

Theorem 2.1 (S. V. Kovalevsky). If the functions $a_{\alpha,j}$, f are analytic in a neighborhood $\Omega = \{(t, x) \in \mathbb{R}^{n+1}, |x| < \delta, |t| < T\}$ of the origin and the functions φ_j are analytic in $\Omega \cap \{t = 0\}$, then there exists one and only one vector-valued function u(t, x) analytic in some neighborhood $\Omega' = \{(t, x) \in \mathbb{R}^{n+1}, |x| < \delta', |t| < T'\}$ of the origin and satisfying equation (2.1) and conditions (2.2) in Ω', $0 < \delta' \leqslant \delta$, $0 < T' \leqslant T$.

We shall also prove a similar theorem under somewhat broader conitions.

Theorem 2.2. Let $a_{\alpha,j}$, f be continuous functions of t, for $|t| \leqslant T$, with values in the space of vector-valued functions holomorphic in a neighborhood $\omega = \{x \in \mathbb{R}^n, |x| < \delta\}$ of the origin in $\mathbb{R}^n$, and let the functions φ_j be analytic in ω. Then there exists one and only one vector-valued function u(t, x) which is m times continuously differentiable with respect to t when $|t| \leqslant T'$ $(0 < T' \leqslant T)$ with values in the space of vector-valued functions holomorphic in a neighborhood $\omega' = \{x \in \mathbb{R}^n, |x| < \delta\}$ of the origin in $\mathbb{R}^n$, $0 < \delta' \leqslant \delta$ and satisfies equation (2.1) for $x \in \omega'$, $|t| < T'$, and condition (2.2) for $x \in \omega'$, $t = 0$.

2.2. First we shall make some simplifications.

Lemma 2.1. The Cauchy problem (2.1)-(2.2) is equivalent to the Cauchy problem for the system of first-order equations

$$\frac{\partial U}{\partial t} = \sum_{j=1}^{n} A^j(t, x) \frac{\partial U}{\partial x^j} + B(t, x) U + f(t, x), \tag{2.3}$$

$$U(0, x) = U_0(x), \tag{2.4}$$

where $U = (U_1, \ldots, U_M)$, $f = (f_1, \ldots, f_M)$, A^j and B are square matrices of order M, M being a large number. In this case, the properties of the coefficients and the solutions to be holomorphic or continuous, which we have mentioned in Theorems 2.1 and 2.2, are retained.

Proof. Let u be the solution of the problem (2.1)-(2.2). Suppose that

$$u_0 = u, \quad u_j = D_j u, \quad 1 \leqslant j \leqslant n, \quad u_{n+1} = D_t u, \tag{2.5}$$

and let $v = (u_0, u_1, \ldots, u_{n+1})$ be a vector in the (n + 2)N-dimensional space. It is clear that

$$D_t^{m-1} u_0 = D_t^{m-2} u_{n+1}, \quad D_t^{m-1} u_j = D_t^{m-2} D_j u_{n+1}, \quad j = 1, \ldots, n,$$

$$D_t^{m-1} u_{n+1} = \sum_{l=0}^{n+1} \sum_{j=0}^{m-2} \sum_{|\alpha|=0}^{m-1-j} b_{\alpha,j,l} D_x^\alpha D_t^j u_l + f(t, x), \tag{2.6}$$

where the last equation changes to (2.1), provided u_l is replaced by u as per formulas (2.5). These equations can be written by various methods. We choose one of them. Conditions (2.2) generate:

$$D_t^l u_0 = \varphi_l, \quad D_t^l u_j = D_j \varphi_l, \quad j = 1, \ldots, n,$$
$$D_t^l u_{n+1} = \varphi_{l+1}, \quad l = 0, 1, \ldots, m-2 \tag{2.7}$$

when $t = 0$, $l = 0, 1, \ldots, m-2$.

Equations (2.6) and conditions (2.7) can now be written as the system of equations

$$D_t^{m-1} v = \sum_{j=0}^{m-2} \sum_{|\alpha|=0}^{m-1-j} b_{\alpha,j}(t, x) D_x^\alpha D_t^j v + g(t, x), \tag{2.8}$$

$$D_t^j v = \psi_j, \quad j = 0, 1, \ldots, m-2, \quad t = 0. \tag{2.9}$$

Now we shall prove that problem (2.8)-(2.9) is equivalent to problem (2.1)-(2.2).

Indeed, if problem (2.8)-(2.9) has a solution for any g and ψ_j, then by solving it for

$$g = (0, \ldots, 0, f), \quad \psi_j = (\varphi_j, D_1\varphi_j, \ldots, D_n\varphi_j, \varphi_{j+1}), \quad 0 \leqslant j \leqslant m-2,$$

we obtain a solution of problem (2.1)-(2.2) if for u we take the vector defined by the first N components of the vector v.

If the homogeneous problem (2.1)-(2.2) has a nontrivial solution u, then the corresponding vector v is different from the null vector and is a solution of the homogeneous problem (2.8)-(2.9).

If the homogeneous system (2.8)-(2.9) has a nontrivial solution v, then the vector u formed by the first N components of the vector v will be a null vector. Indeed, let u = 0, that is, $u_0 = 0$ in (2.6). Then $D_t^{m-2} u_{n+1} = 0$, that is, u_{n+1} is a polynomial in t, of degree $\leqslant m-3$. From (2.7) it follows that $D_t^l u_{n+1} = 0$ for $t = 0$, $l = 0, 1, \ldots, m-2$. Therefore, $u_{n+1} = 0$. From (2.6) it then follows that $D_t^{m-1} u_j = 0$ for $j = 1, \ldots, n$. Since $D_t^l u_j = 0$ for $t = 0$, $j = 1, \ldots, n$, $l = 0, 1, \ldots, m-2$ we find that $u_j = 0$. Hence, v = 0 which contradicts the assumption.

We shall now show that the solvability of problem (2.8)-(2.9) for any g and φ_j follows from the solvability of problem (2.1)-(2.2) for any f and ψ_j. First note that if we replace v by $v - \sum_{j=0}^{m-2} \psi_j t^j / j!$ the problem (2.8)-(2.9) can always be reduced to a problem in which $\psi_j = 0$. If $g = (g_0, g_1, \ldots, g_{n+1})$, where every vector g_j is N-dimensional, then one can find a vector $(w_0, w_1, \ldots, w_n)$ for which

$$D_t^{m-1} w_0 = g_0, \quad D_t^{m-1} w_j = g_j, \quad j-1, \ldots, n,$$

$$D_t^l w_j = 0 \text{ for } t = 0, \ l = 0, 1, \ldots, m-2, \ j = 0, 1, \ldots, n.$$

Then the vector $\tilde{v} = v - (w_0, w_1, \ldots, w_n, 0)$ satisfies (2.8) with $\tilde{g} = (0, \ldots, 0, g_{n+1})$ and (2.9) with $\psi_j = 0$. But this problem is also equivalent to (2.1)-(2.2) when $f = \tilde{g}$ and $\varphi_j = 0$. Since, by hypothesis, this problem has a solution, problem (2.8)-(2.9) is also solvable.

That the coefficients, initial conditions, and the solution in the initial and new problem have the same regularity is obvious from the construction. Thus, the equivalence of these problems is proved.

Since the reduction to a system of equations of lower order can be continued, Lemma 2.1 is proved.

2.3. We obtain Theorems 2.1 and 2.2 as corollaries of the general abstract theorem of Cauchy and Kovalevsky.

Let E_s, for $0 \leqslant s \leqslant 1$, be a Banach space, and if $s' < s$, then $E_s \subset E_{s'}$, and

$$\|u\|_{s'} \leqslant \|u\|_s \quad \text{for} \quad u \in E_s. \tag{2.10}$$

Consider a set L_α of linear operators A from E_s into $E_{s'}$ for $0 \leqslant s' < s \leqslant 1$, and satisfying the condition

$$\|Au\|_{s'} \leqslant \frac{\alpha}{s-s'} \|u\|_s. \tag{2.11}$$

Theorem 2.3. Let $u_0 \in E_1$, $f \in C([-T, T]; E_1)$ and $A(t) \in C([-T, T]; L_\alpha)$. Then

I. There exists a function $u_0 \in C^1((-T', T'); E_0)$, u being in $C^1((-T_s, T_s); E_s)$, where $T' = 1/\alpha e$, $T_s = T'(1 - s)$ for which

$$\frac{\partial u}{\partial t} = A(t) u + f(t), \quad u(0) = u_0. \tag{2.12}$$

II. If for some T', $0 < T' \leqslant T$, and some s, $0 < s \leqslant 1$, there are two functions belonging to $C^1((-T', T'), E_s)$ and satisfying (2.12), then they coincide.

Proof of Theorem 2.3.

I. Let

$$v_0(t) = \int_0^t f(\tau)\, d\tau, \qquad v_{k+1}(t) = u_0 + \int_0^t f(\tau)\, d\tau + \int_0^t A(\tau) v_k(\tau)\, d\tau, \qquad k = 0, 1, \ldots.$$

Clearly $v_k \in C([-T, T]; E_s)$ for $0 \leqslant s < 1$. Let

$$w_0 = v_0, \quad w_{k+1} = v_{k+1} - v_k, \quad k = 0, 1, \ldots$$

Clearly

$$w_{k+1}(t) = \int_0^t A(\tau) w_k(\tau)\, d\tau.$$

We shall prove that

$$\|w_k(t)\|_s \leqslant M(t) \left(\frac{\alpha e |t|}{1-s}\right)^k, \quad |t| < T, \tag{2.13}$$

where $M(t) = \|u_0\|_1 + \left| \int_0^t \|f(\tau)\|_1 d\tau \right|$, $0 \leqslant s < 1$.

Inequality (2.13) is trivial for $k = 0$. Assume it to be valid for k. Verify it for $k + 1$. If $0 \leqslant s' < s < 1$, then

$$\|w_{k+1}(t)\|_{s'} \leqslant \frac{\alpha}{s-s'} \left| \int_0^t \|w_k(\tau)\|_s d\tau \right| \leqslant M(t) \frac{\alpha}{s-s'} \left(\frac{\alpha e}{1-s}\right)^k \frac{|t|^{k+1}}{k+1}.$$

Let

$$(s - s')(k + 1) = 1 - s.$$

Then

$$(1 - s)/(1 - s') = k/(k + 1)$$

and therefore

$$\|w_{k+1}(t)\|_{s'} \leqslant M(t) \left(\frac{\alpha e |t|}{1-s'}\right)^{k+1} \frac{1}{e} \left(\frac{k+1}{k}\right)^k \leqslant M(t) \left(\frac{\alpha e |t|}{1-s'}\right)^{k+1}.$$

Since $M(t) \leqslant M$, absolute convergence of the series $\sum_{k=0}^{\infty} w_k(t)$ to E_s follows from (2.13), the convergence being uniform in every closed segment inside the interval $|t| < 1 - s/\alpha e$. Its sum u(t) satisfies the conditions of the theorem and

$$u(t) = u_0 + \int_0^t f(\tau)\,d\tau + \int_0^t A(\tau)\,u(\tau)\,d\tau.$$

II. Let $v \in C^1((-T', T'), E_s)$ for some s, $0 < s \leqslant 1$, and

$$v_t = A(t)\,v, \quad v(0) = 0.$$

Show that $v = 0$ in $(-T', T')$. We have

$$v(t) = \int_0^t A(\tau)\,v(\tau)\,d\tau.$$

Let $s' < s$ and $M(t) = \sup\|v(\tau)\|_s$ for $0 \leqslant \tau \leqslant t$. Verify that

$$\|v(t)\|_{s'} \leqslant M(t)\left(\frac{\alpha e\,|t|}{s-s'}\right)^k, \qquad k = 0, 1, \ldots \tag{2.14}$$

For $k = 0$ it is trivial. If it is valid for some $k \geqslant 0$, then for $s'' < s'$

$$\|v(t)\|_{s''} \leqslant \frac{\alpha}{s'-s''}\left|\int_0^t \|v(\tau)\|_{s'}\,d\tau\right| \leqslant \frac{\alpha M(t)}{s'-s''}\left(\frac{\alpha e}{s-s'}\right)^k \frac{|t|^{k+1}}{k+1}.$$

Let

$$(s'-s'')(k+1) = s-s''.$$

Then

$$(s-s')/(s-s'') = k/(k+1)$$

and therefore

$$\|v(t)\|_{s''} \leqslant M(t)\left(\frac{\alpha\,|t|}{s-s''}\right)^{k+1} e^k \left(\frac{k+1}{k}\right)^k \leqslant M(t)\left(\frac{\alpha e\,|t|}{s-s''}\right)^{k+1}.$$

Thus, (2.14) is valid for all k. Hence it follows that $v(t) = 0$ for $|t| < s - s'/\alpha e$. By continuity, $v(t) = 0$ for $|t| \leqslant s - s'/\alpha e$. Repeating this argument, we can show that $v(t) = 0$ for $s - s'/\alpha e \leqslant t \leqslant 2(s - s')/\alpha e$ and so on. In a finite number of steps we can exhaust any interval $[-t_0, t_0]$ contained in $(-T', T')$. Hence it follows that $v(t) = 0$ in $(-T', T')$, and the theorem is proved.

2.4. Proof of Theorem 2.2. This theorem can be obtained from Theorem 2.3 if as E_s we take a space of vector-valued functions holomorphic in $\Omega_s = \{z \in \mathbb{C}^n, |z| < s\delta\}$ for $0 < s \leqslant 1$, with the norm

$$\|u\|_s = \sup_{\Omega_s} |u(z)|.$$

We shall now verify the validity of inequality (2.11).

If u is a complex-valued scalar function, then for $z_0 \in \Omega_{s'}$, $s' < s$ we have

$$u(z_0) = \frac{1}{2\pi i} \int\limits_{|\zeta^j - z_0^j| = (s-s')\delta} \frac{u(\zeta)}{\zeta^j - z_0^j}\,d\zeta^j,$$

where $z^l = z_0^l$ for $l \neq j$. It follows that

$$\frac{\partial u(z_0)}{\partial z_0^j} = \frac{1}{2\pi i} \int\limits_{|\zeta^j - z_0^j| = (s-s')\delta} \frac{u(\zeta)}{(\zeta^j - z_0^j)^2}\,d\zeta^j,$$

and therefore

$$\left|\frac{\partial u(z_0)}{\partial z_0^j}\right| \leqslant \frac{1}{\delta(s-s')}\|u\|_s.$$

Thus

$$\|D_j u\|_{s'} \leqslant \frac{1}{\delta(s-s')}\|u\|_s.$$

Obviously, the inequality

$$\|D_j u\|_{s'} \leqslant \frac{\sqrt{N}}{\delta(s-s')}\|u\|_s$$

is fulfilled for functions u with values in $\mathbb{C}^N$. Therefore,

$$\left\|\sum_{j=1}^{n} A^j(t, z)\frac{\partial u}{\partial z^j} + B(t, z)u\right\|_{s'} \leqslant \frac{\alpha}{s-s'}\|u\|_s$$

with some constant α depending on N and the maximum of the moduli of the elements of matrices A^j and B.

By Lemma 2.1, problem (2.1)-(2.2) is equivalent to problem (2.3)-(2.4). The existence and uniqueness of the solution of problem (2.3)-(2.4) follows from Theorem 2.3, the solution being continuously differentiable with respect to t. From the proof of Lemma 2.1 it is seen that in this case the problem (2.1)-(2.2) has a solution which is m times continuously differentiable with respect to y.

Theorem 2.2 is proved.

2.5. Proof of Theorem 2.1. First note that the solution u(t) of problem (2.12), constructed in Theorem 2.3, will be analytic in t, provided A(t) and f(t) are analytic functions of t. Indeed, for this condition, the functions v_k and w_k constructed in the course of the proof will be analytic. From (2.13) follows uniform convergence of the series $\sum w_k(t)$, for $|t| \leqslant q(1-s)/\alpha e$, where $q < 1$ is an arbitrary number. Since the sum of a uniformly convergent series of analytic functions is an analytic function, the solution u(t) of the problem will be an analytic function in t with values in E_s for $|t| < (1-s)/\alpha e$.

Hence it follows that problem (2.3)-(2.4) has an analytic solution U(t, x) in t and in x, provided A^j, B, and f are analytic functions in t and x. From this it follows, by Lemma 2.1, that Theorem 2.1 is valid.

Remark. Theorems 2.1, 2.2, and 2.3 are valid also for nonlinear equations. Besides, by choosing different spaces E_s, one can obtain from Theorem 2.3 assertions on solvability of the Cauchy problem in other Banach spaces, for instance, in the Gevrey spaces. Using Holmgren's scheme, a theorem on the uniqueness of solution of the Cauchy problem for nonanalytic functions can be obtained.

§3. Hyperbolic Equations

3.1. This section very briefly treats the theory of the Cauchy problem for hyperbolic equations, which serves as an illustration of the use of pseudo-differential operators and Fourier integral operators. To the reader interested in hyperbolic equations, we recommend L. Hörmander, *Linear Partial Differential Operators*,[3] J. Hadamard, *Le problême de Cauchy*,[1] S. Mizohata, *The Theory of Partial Differential Equations*,[1] and L. Gårding, *Linear Hyperbolic Partial Differential Equations with Constant Coefficients*.[1]

Lemma 2.1 makes it possible to restrict consideration to the Cauchy problem for the system of first-order equations

$$\frac{\partial u}{\partial t} = \sum_{j=1}^{n} A^j(t, x)\frac{\partial u}{\partial x^j} + B(t, x)u + f(t, x), \tag{3.1}$$

$$u(0, x) = \varphi(x). \tag{3.2}$$

Here $u = (u_1, \ldots, u_N)$, $f = (f_1, \ldots, f_N)$, $\varphi = (\varphi_1, \ldots, \varphi_N)$, A^j and B are square matrices of order N.

Definition 3.1. The system of equations (3.1) is called *strictly hyperbolic*, provided the equation with respect to variable λ

$$\det \left| \sum_{j=1}^{n} A^j(t, x)\xi_j - \lambda I \right| = 0$$

has N distinct real roots for all $(x, \xi) \in T^*\mathbb{R}^n \setminus 0$.

We shall prove that problem (3.1)-(3.2) is well posed for a strictly hyperbolic system of equations.

Since the same method enables us to consider a somewhat wider class of systems, we shall introduce another definition.

Definition 3.2. The system of equations (3.1) is called *symmetrizable* if there exists a matrix $R(t, x, \xi)$ of order $N \times N$ such that

1) $R(t, x, \xi) \in C^\infty([0, T] \times T^*(\mathbb{R}^n))$;
2) $R(t, x, \lambda\xi) = R(t, x, \xi)$ when $\lambda \geqslant 1$, $|\xi| \geqslant 1$;
3) $|\det R(t, x, \xi)| \geqslant C_0 > 0$ when $|\xi| \geqslant 1$;
4) $R\left(\sum_{j=1}^{n} A^j\xi_j\right)R^{-1}$ is Hermitian for $|\xi| \geqslant 1$.

3.2. Lemma 3.1. The strictly hyperbolic system (3.1) is symmetrizable.

Proof. Eigenvalues $\lambda_j(t, x, \xi)$ are uniquely defined and are smooth functions when $\xi \neq 0$, since they are real and do not coincide when $|\xi| = 1$.

Let $A(t, x, \xi) = \sum_{j=1}^{n} A^j(t, x)\xi_j$. By M_{ij} we shall denote the cofactor of the element of matrix $A - \lambda_1 I$ situated at the intersection of the i-th row and the j-th column. Then the vector

$$M_j = (M_{1j}, \ldots, M_{Nj}),\ j = 1, 2, \ldots, N,$$

will be an eigenvector of matrix A. At every point (t, x, ξ), for $|\xi| \geqslant 1$, there are two eigenvectors of unit length $e_j = M_j/|M_j|$ and $e_j' = M_j/|M_j|$ corresponding to the eigenvalue λ_j. Therefore, the problem is reduced to the selection of sign at every point so that smooth functions are obtained for $|\xi| = 1$. Since for $n \geqslant 3$ the fundamental group of the manifold $[0, T] \times S^*(\mathbb{R}^n)$ is trivial, by fixing the sign of every eigenvector at one point (t_0, x_0, ξ^0) we can uniquely determine it everywhere on this manifold.

If $n = 2$, then this manifold can be identified with $[0, T] \times \mathbb{R}^2 \times S^1$. It will suffice to construct a smooth field of eigenvectors on $(0, 0, 0) \times S^1$. Let θ be a coordinate on S^1, $0 \leqslant \theta \leqslant 2\pi$. Find vectors e_j for $j = 1, \ldots, N$, $\theta = 0$ and then move along the circle. If $e_j(2\pi) = e_j(0)$, then the vector $e_j(\theta)$ is constructed on the entire circle. If $e_j(2\pi) = -e_j(\theta)$, we replace $e_j(\theta)$ by $e_j(\theta)e^{i\theta/2}$. Thus, $e_j(\theta)$ will be defined on the entire circle. Having constructed the vectors e_j on $[0, T] \times T^*(\mathbb{R}^n)$, we extend them on $[0, T] \times S^*(\mathbb{R}^n)$ so that they are homogeneous in ξ, for $|\xi| \geqslant 1$, and smooth, for $|\xi| \leqslant 1$. Let now R be the matrix whose rows are formed by vectors e_j. The matrix RAR^{-1} is a diagonal matrix and therefore is Hermitian.

Lemma 3.1 is proved.

3.3. In that case when the roots of the characteristic equation are real, but some of them coincide, the system may not be symmetrizable although in some cases it is possible.

The study of the Cauchy problem for a symmetrizable system is based on the energy inequality.

Theorem 3.1. If system (3.1) is symmetrizable and u is a smooth solution of problem (3.1)-(3.2), then for every real number s there exists a constant C(s) such that

$$\|u(\cdot, t)\|_s \leqslant C(s)(\|\varphi\|_s + \int_0^t \|f(\cdot, \tau)\|_s d\tau). \tag{3.3}$$

Proof. For every t, $Q = R^*(t, x, D)R(t, x, D)$ is a pseudo-differential operator of zero order, $Q^* - Q$ being a smoothing operator of $-\infty$.

By condition, RAR^{-1} is a Hermitian matrix, that is,

$$RAR^{-1} = R^{*-1}A^*R^*.$$

It follows that

$$R^*RA = A^*R^*R,$$

that is, R^*RA is also a Hermitian matrix. Therefore, the operator

$$Q(t, x, D)A(t, x, D),$$

where $A(t, x, D) = \sum A^j(t, x)D_j$, is different from the operator of zero order adjoint to it.

Using the equalities $\frac{\partial u}{\partial t} = iA(t, x, D)u + B(t, x)u + f(t, x)$ and $(u, v)_s = (\Lambda^{2s}u, u)_0$, where Λ^{2s} is an operator with symbol $(1 + |\xi|^2)^s$ and $(u, v)_0 = \int u(x)\overline{v(x)}\,dx$, we obtain

$$\frac{\partial}{\partial t}(Qu, u)_s = \left(Q\frac{\partial u}{\partial t}, u\right)_s + \left(Qu, \frac{\partial u}{\partial t}\right)_s + (Q_t u, u)_s =$$

$$i(QAu, u)_s - i(Qu, Au)_s + (Q_t u, u)_s + (QBu, u)_s + (Qu, Bu)_s +$$

$$(Qf, u)_s + (Qu, f)_s = i(\Lambda^{2s}QAu, u)_0 - i(A^*\Lambda^{2s}Qu, u)_0 +$$

$$(\Lambda^{2s}Q_t u, u)_0 + (\Lambda^{2s}QBu, u)_0 + (\Lambda^{2s}Qu, Bu)_0 + (\Lambda^{2s}Qf, u)_0 + (\Lambda^{2s}Qu, f)_0.$$

Note that the operator

$$\Lambda^{2s}QA - A^*\Lambda^{2s}Q$$

is of order 2s by construction. Since the operator $\Lambda^{2s}Q_t + \Lambda^{2s}QB + B^* \times \Lambda^{2s}Q$ is also of order 2s and

$$|\operatorname{Im}(QAu, u)_s| + |\operatorname{Im}(Qu, Au)_s| \leqslant C_1\|u\|_s^2,$$

the inequality

$$\frac{\partial}{\partial t}\operatorname{Re}(Qu, u)_s \leqslant C_2(\|u\|_s^2 + \|f\|_s^2)$$

is valid, provided u is a smooth function of t, for $0 \leqslant t \leqslant T$, with values in $C_0^\infty(\mathbb{R}^n)$. Since $\Lambda^{2s}Q$ is an elliptic operator of order 2s, it may be assumed, by adding an operator of order -1 to Q, that the Gårding inequality

$$\operatorname{Re}(Qu, u) \geqslant c_0\|u\|_s^2,$$

is valid, where c_0 = const > 0. Thus

$$c_0\|u(t,\cdot)\|_s^2 \leqslant \operatorname{Re}(Q(t,x,D)u,u)_s \leqslant$$

$$\operatorname{Re}(Q(0,x,D)u,u)_s + C_2\int_0^t \|u(\tau,\cdot)\|_s^2\,d\tau + C_2\int_0^t \|f(\tau,\cdot)\|_s^2\,d\tau \leqslant$$

$$C_3\|\varphi\|_s^2 + C_2\int_0^t \|u(\tau,\cdot)\|_s^2\,d\tau + C_2\int_0^t \|f(\tau,\cdot)\|_s^2\,d\tau,$$

which by Gronwall's inequality readily yields (3.3). The theorem is proved.

3.4. Now we can prove the theorem on the existence and uniqueness of the solution of the Cauchy problem (3.1)-(3.2). Denote by $L_2(H_s)$ the space of functions square integrable with respect to t with values in the space H_s and with norm

$$\|u\|_{L_2(H_s)} = \left(\int_0^T \|u(\tau,\cdot)\|_s^2\,d\tau\right)^{1/2}.$$

Theorem 3.2. Let system (3.1) be symmetrizable, given $\varphi \in H_s(\mathbb{R}^n)$ and $f \in L_2(H_s)$, where s is an arbitrary real number. Then there is a solution u(t, x) to the problem (3.1)-(3.2) in the class $L_2(H_s)$; this solution is unique and satisfies inequality (3.3).

Proof. Suppose that $Pu = \frac{\partial u}{\partial t} - \sum_{j=1}^n A^j(t,x)\frac{\partial u}{\partial x^j} - B(t,x)u$. Note that the operator $P^* = -\frac{\partial}{\partial t} + \sum_{j=1}^n A^{j*}\frac{\partial}{\partial x^j} + \left(\sum_{j=1}^n \frac{\partial A^{j*}}{\partial x^j} - B^*\right)$ is also symmetrizable, the role of the matrix R, the matrix corresponding to the operator P, being played by R^{*-1}.

Here P^* stands for the operator adjoint to P with respect to the scalar product

$$\langle u, v\rangle = \int_0^T (u(t,x), v(t,x))_0\,dt.$$

It is clear that from (3.3) the uniqueness of the solution u in the class $L_2(H_s)$ follows.

On the other hand, if $v \in C^\infty([0,T]\times\mathbb{R}^n)$ and v(T, x) = 0, then

$$\|v(t,\cdot)\|_{-s} \leqslant C(s)\int_0^T \|P^*v(t,\cdot)\|_{-s}\,dt$$

by (3.3). Using Cauchy's inequality, we can readily find that

$$\int_0^T \|v(t,\cdot)\|_{-s}^2\,dt \leqslant C^2(s)T^2\int_0^T \|P^*v(t,\cdot)\|_{-s}^2\,dt,$$

$$\|v(0,\cdot)\|_{-s}^2 \leqslant C^2(s)T\int_0^T \|P^*v(t,\cdot)\|_{-s}^2\,dt.$$

We have

$$\left|\int_0^T (v(t,\cdot), v(t,\cdot))_0\,dt\right| + |(\varphi(\cdot), v(t,\cdot))_0| \leqslant$$

$$\int_0^T \|f(t,\cdot)\|_s\|v(t,\cdot)\|_{-s}\,dt + \|\varphi\|_s\|v(t,\cdot)\|_{-s} \leqslant$$

$$\|f\|_{L_2(H_s)} C(s) T \|P^*v\|_{L_2(H_{-s})} + \|\varphi\|_s C(s)\sqrt{T}\,\|P^*v\|_{L_2(H_{-s})}.$$

According to the Riesz Representation Theorem, there exists an element u $\in L_2(H_s)$ for which

$$\int_0^T (f(t,\cdot), v(t,\cdot))\,dt + (\varphi(\cdot), v(t,\cdot)) = \int_0^T (u(t,\cdot), P^*v(t,\cdot))\,dt.$$

This equality implies that u is a generalized solution (in $L_2(H_s)$) of the equation Pu = f satisfying the condition $u(0, x) = \varphi(x)$, that is, a solution of problem (3.1)-(3.2). In this case

$$\|u\|_{L_2(H_s)} \leqslant C(s)\left[T\|f\|_{L_2(H_s)} + \sqrt{T}\,\|\varphi\|_s\right],$$

that is, (3.3) is valid.

Theorem 3.2 is proved.

3.5. For the constructed solution we have a theorem on smoothness depending on the smoothness of functions φ and f.

<u>Theorem 3.3.</u> Let s be an arbitrary real number and let k be an integer, $k \geqslant 0$. If $f \in H_k(H_s)$, that is, the norm

$$\|f\|_{H_k(H_s)} = \left(\sum_{j=0}^{k}\int_0^T \|D_t^j f(t,\cdot)\|_s^2\,dt\right)^{1/2}$$

is finite, and $\varphi \in H_{s+k}$, then the solution u to problem (3.1)-(3.2) belongs to the class $H_k(H_s)$; if $f \in C^\infty([0, T]\times\mathbb{R}^n)$ and $\varphi \in C^\infty(\mathbb{R}^n)$, then $u \in C^\infty([0, T]\times\mathbb{R}^n)$.

<u>Proof.</u> Having differentiated (3.1) with respect to t we can obtain an equation for $\partial_t u$. From Theorem 3.2 it follows that $\partial_t u \in L_2(H_s)$ if $k \geqslant 1$. Repeating this argument k times, we obtain

$$u, \partial_t u, \ldots, \partial_t^k u \in L_2(H_s),$$

that is, $u \in H_k(H_s)$ and the inequality

$$\|u\|_{H_k(H_s)} \leqslant C\left(\|f\|_{H_k(H_s)} + \|\varphi\|_{k+s}\right)$$

is valid.

If f and φ are infinitely differentiable, then $u \in H_k(H_s)$ for all k and s, that is, $u \in C^\infty([0, T]\times\mathbb{R}^n)$, which proves the theorem.

3.6. In the case of the strictly hyperbolic system (3.1), the solution to the Cauchy problem can be constructed as a result of the application of Fourier integral operators to vectors f and φ.

Note first that it is sufficient to know how to solve the Cauchy problem with f = 0. If $v(\tau, t, x)$ is the solution of the problem

$$\frac{\partial v}{\partial t} = \sum_{j=1}^{n} A^j(t, x)\frac{\partial v}{\partial x^j} + B(t, x)v, \quad v(\tau, \tau, x) = f(\tau, x),$$

then the function

$$u(t, x) = \int_0^t v(\tau, t, x)\,d\tau$$

satisfies the condition $u(0, x) = 0$ and the equation

$$\frac{\partial u}{\partial t} = \sum_{j=1}^{n} A^j(t, x)\frac{\partial u}{\partial x^j} + B(t, x)u + f(t, x).$$

Using the construction of Lemma 3.1 and substituting $u = Rv$, we can reduce the problem to the following:

$$\frac{\partial v}{\partial t} = \Lambda(t, x, D)v + B(t, x, D)v, \tag{3.4}$$

$$v(0, x) = \varphi(x), \tag{3.5}$$

where $\Lambda(t, x, \xi)$ is a diagonal matrix in which the eigenvalues $\lambda_l(t, x, \xi)$ of the matrix $\sum_{j=1}^{n} A^j(t, x)\xi_j$ for $l = 1, \dots, N$, serve as elements, and $B(t, x, D)$ is a zero-order matrix operator.

Now assume that $w_1 = (I + K_1)v$, where K_1 is an operator of order -1 which is pseudo-differential in x at every t, $0 \leqslant t \leqslant T$. We have

$$\frac{\partial w_1}{\partial t} = (I + K_1)\Lambda(I + K_1)^{-1}w_1 + (I + K_1)B(I + K_1)^{-1}w_1 +$$

$$\frac{\partial K_1}{\partial t}(I + K_1)^{-1}w_1, \qquad w_1(0, x) = (I + K_1)\varphi.$$

Choose a matrix $K_1(t, x, \xi)$ so that

$$K_1\Lambda - \Lambda K_1 + B_0 \tag{3.6}$$

will be a diagonal matrix and the diagonal elements of the matrix K_1 will be equal to 0. We have

$$\frac{\partial w_1}{\partial t} = \Lambda w_1 + B_1 w_1 + \Lambda_1 w_1, \qquad w_1(0, x) = (I + K_1)\varphi(x),$$

where Λ_1 is a diagonal matrix in which zero-order operators serve as elements, and B_1 is an operator of order -1.

To prove the existence of matrix K_1 it is sufficient to prove that the solution of the homogeneous system

$$(K\Lambda - \Lambda K)_{ij} = 0, \qquad i \neq j, \quad i, j = 1, \dots, N \tag{3.7}$$

is trivial. Since by hypothesis, (3.6) is a diagonal matrix, it can be regarded as a system of $N^2 - N$ linear equations with respect to nondiagonal elements of the matrix K_1. However it follows from (3.7) that

$$\lambda_i k_{ij} = \lambda_j k_{ij}, \qquad i \neq j, \quad i, j = 1, \dots, N,$$

and since $\lambda_i \neq \lambda_j$, for $i \neq j$, it follows that $k_{ij} = 0$.

Furthermore, we can choose a matrix K_2 so that vector $w_2 = (I + K_2) \times w_1$ will satisfy the system of equations

$$\frac{\partial w_2}{\partial t} = \Lambda w_2 + B_2 w_2 + (\Lambda_1 + \Lambda_2)w_2,$$

where Λ_2 represents a diagonal matrix with operators of order -1 on the diagonal, and B_2 stands for an operator of order -2, and so on.

Let now K be an operator of order -1 such that

$$I + K = \dots(I + K_2)(I + K_1).$$

This operator can be constructed by the method described in Lemma 3.1 of Chapter II.

Then the vector $z = (I + K)v$ satisfies the system of equations

$$\frac{\partial z}{\partial t} = \Lambda z + Tz, \quad z = (I + K)\varphi \quad \text{for} \quad t = 0,$$

where T is a smoothing operator of order $-\infty$, and Λ is a diagonal operator with elements on the diagonal that are first-order operators.

3.7. Thus, our problem is reduced to the construction of a solution to the Cauchy problem of the scalar equation

$$Lu \equiv \partial_t u - \lambda(t, x, D_x) u = 0, \tag{3.8}$$

$$u(0, x) = \varphi(x), \tag{3.9}$$

where $\lambda(t, x, \xi)$ is a real-valued smooth function and $\lambda(t, x, D_x)$, for every $t \in [0, T]$, is a first-order pseudo-differential operator.

The solution of the problem (3.8)-(3.9) can be obtained by using the operator

$$U(t)\varphi = \int_{R^n} a(t, x, \xi)\tilde{\varphi}(\xi) e^{iS(t, x, \xi)} d\xi,$$

where $S(0, x, \xi) = x\cdot\xi$, $a(0, x, \xi) = 1$.

In proving Theorem 4.1 in Chapter II we obtained

$$\lambda(t, x, D_x) U(t)\varphi = (2\pi)^{-n} \int k(t, x, \xi)\tilde{\varphi}(\xi) e^{iS(t, x, \xi)} d\xi,$$

where

$$k(t, x, \xi) = \sum_\alpha \frac{i^{|\alpha|}}{\alpha!} D_y^\alpha [a(t, y, \xi) D_\eta^\alpha \lambda(t, x, A)]_{\substack{y=x \\ \eta=\xi}};$$

$$A(t, x, y, \xi, \eta) = \eta + \int_0^1 (1-\tau) \frac{\partial^2 S(t, x+\tau(y-x), \xi)}{\partial x^2} d\tau \cdot (y-x).$$

It is seen that the function S is defined by the conditions

$$\frac{\partial S}{\partial t} - \lambda_0\left(t, x, \frac{\partial S(t, x, \xi)}{\partial x}\right) = 0, \quad S(0, x, \xi) = x\cdot\xi.$$

The amplitude $a(t, x, \xi)$ is constructed as a sum of the asymptotic series $\sum_j a_j(t, x, \xi)$, where $a_j(t, x, \tau\xi) = \tau^{-j} a_j(t, x, \xi)$ for $\tau \geqslant 1$, $|\xi| \geqslant 1$ and

$$\frac{\partial a_0}{\partial t} - \sum_{j=1}^n \frac{\partial \lambda_0(t, x, \partial S(t, x, \xi)/\partial x)}{\partial \xi_j} \frac{\partial a_0}{\partial y^j} + i a_0 \lambda\left(t, x, \frac{\partial S(t, x, \xi)}{\partial x}\right) = 0, \quad a_0(0, x, \xi) = 1.$$

For functions a_j, when $j > 0$, differential equations with the same principal part as for a_0 are fulfilled, and the initial conditions are written as: $a_j(0, x, \xi) = 0$.

The operator U so constructed satisfies the equality

$$LU(t) = U(t) D_t + T, \quad U(0) = I + T_1,$$

where T and T_1 are operators of order $-\infty$. Actually, the construction agrees with that given in Section 4, Chapter II, and enables the operator L to be reduced to an equivalent operator D_t.

The solvability of the Cauchy problem (3.1)-(3.2) for a strictly hyperbolic system can be proved in the same manner. The advantages of this construction are based on the fact that one can follow the process of propagation of singularities, prove the theorem on finite domain dependence and describe it in terms of the behavior of bicharacteristics for eigenvalues λ_j.

§4. Carleman Estimates

4.1. In this section we shall examine Carleman estimates:

$$\sum_{|\alpha|\leqslant m-1} \tau^{2(m-\delta-|\alpha|)} \int |D^\alpha u|^2 \exp[2\tau\varphi(x)]\, dx \leqslant$$

$$C\int |P(x, D)u|^2 \exp[2\tau\varphi(x)]\, dx, \tag{4.1}$$

where $u \in C_0^\infty(\Omega)$, and $\tau > 0$. Here Ω is an open domain in $\mathbb{R}^n$, the constant C is independent of u and τ, P(x, D) is a pseudo-differential operator of order m of principal type, $\varphi \in C^\infty(\Omega)$ is a real-valued function, grad $\varphi(x) \neq 0$, $0 \leqslant \delta < 1$.

Assume the function $q_\tau(x, \xi) = p^0(x, \xi + i\tau\varphi'(x))$ to be defined for all real $\tau > 0$ as a smooth function of τ, x, ξ. This condition is fulfilled, in particular, if p^0 is a polynomial in ξ in the direction grad $\varphi(x)$. For example, if $\varphi(x) = t$, then the operator P, which is differential in t, satisfies this condition. Since we shall be interested in (4.1) mainly to prove the uniqueness of solutions to the Cauchy problem, such a restriction is natural.

Note that inequality (4.1) is equivalent to an analogous inequality with operator P^0 instead of P. Replacing $u = v(x, \tau)\exp(-\tau\varphi(x))$, (4.1) changes into

$$\sum_{|\alpha|\leqslant m-1} \tau^{2(m-\delta-|\alpha|)} \int |D_x^\alpha v|^2\, dx \leqslant c\int |P^0(x, D_x + i\tau\varphi'(x))v|^2\, dx. \tag{4.2}$$

4.2. Let us consider $P^0(x, D_x + i\tau\varphi'(x))$ as a pseudo-differential operator in spaces $H_{s,\tau}$ with the norms

$$\|v\|_{s,\tau} = \Big(\int_{R^n} (1+|\xi|^2+\tau^2)^s |\tilde{v}(\tau, \xi)|^2\, d\xi\Big)^{1/2}.$$

Then inequality (4.2) follows from

$$\|v\|_{m-\delta,\tau} \leqslant c\|P^0(x, D + i\tau \operatorname{grad}\varphi(x))v\|_{0,\tau}. \tag{4.3}$$

Note that for all α and β the inequalities

$$|D_x^\beta D_\xi^\alpha P^0(x, \xi + i\tau\operatorname{grad}\varphi(x))| \leqslant C_{\alpha,\beta,k}(1+|\xi + i\tau\operatorname{grad}\varphi(x)|)^{m-|\alpha|}$$

are fulfilled when $x \in K$, K being a compact subset in Ω. Since $|\operatorname{grad}\varphi(x)| \neq 0$, the inequalities

$$c_0(|\xi|+\tau) \leqslant |\xi + i\tau\operatorname{grad}\varphi(x)| \leqslant C_1(|\xi|+\tau)$$

are valid for $\tau > 0$, $c_0 = \text{const} > 0$.

4.3. For $v \in C_0^\infty(\Omega\times\mathbb{R}_+)$, inequality (4.3) is an estimate of the same type as those studied in Chapter VIII, condition (Ψ') being natural for the function $p^0(x, \xi + i\tau \operatorname{grad}\varphi(x))$. Note that for $\tau \leqslant \tau_0$ inequality (4.1) is equivalent to the usual estimate

$$\|u\|_{m-\delta} \leqslant C\|P(x, D)u\|_0$$

and therefore condition (Ψ') must be fulfilled for the function $p^0(x, \xi)$.

Furthermore, as shown in Chapter VIII, ($\mathscr{B}$) is a necessary condition for the estimate (4.3) to hold; in the given case it is written as:

($\mathscr{B}_\tau$) Let $a(x, \xi, \tau) = \operatorname{Re} p^0(x, \xi + i\tau\operatorname{grad}\varphi(x))$, $b(x, \xi, \tau) = \operatorname{Im} p^0(x, \xi + i\tau\operatorname{grad}\varphi(x))$. If $p^0(x, \xi + i\tau\operatorname{grad}\varphi(x)) = 0$, $x \in \Omega$, $\tau > 0$, $|\xi|^2+\tau^2 = 1$, then there exist integers r, α_s, β_s, s = 1,..., r, such that $\alpha_s \geqslant 0$, $\beta_s \geqslant 0$, $\sum(\alpha_s+\beta_s) \leqslant k$ and $|(ada)^{\alpha_1}(adb)^{\beta_1}\ldots(ada)^{\alpha_r}(adb)^{\beta_r}a(x, \xi, \tau)| \geqslant c_0$, where $c_0 = \text{const} > 0$,

$$(adf)g(x, \xi) = \sum_{j=1}^n \Big(\frac{\partial f}{\partial \xi_j}\frac{\partial g}{\partial x^j} - \frac{\partial f}{\partial x^j}\frac{\partial g}{\partial \xi_j}\Big)(x, \xi).$$

If conditions (Ψ') and ($\mathscr{B}_\tau$), are fulfilled, then the proof of Theorem 1.2, from Chapter VIII, can be carried over to the operator $Q_\tau(x, D) = P(x, D + i\tau \operatorname{grad} \varphi(x))$ in spaces $H_{s,\tau}$ and we have

Theorem 4.1. If

$$q_\tau(x, \xi) = p^0(x, \xi + i\tau \operatorname{grad} \varphi(x))$$

satisfies conditions (Ψ') and ($\mathscr{B}_\tau$) for all $\tau > 0$, $\xi \in \mathbb{R}^n$ and x from some domain $\Omega \subset \mathbb{R}^n$, then there exists a constant $C > 0$ such that

$$\sum_{|\alpha| \leqslant m-1} \tau^{2(m-|\alpha|-k/(k+1))} \int |D^\alpha u|^2 e^{2\tau\varphi(x)} dx \leqslant$$

$$C \int |P(x, D) u|^2 e^{2\tau\varphi(x)} dx \tag{4.4}$$

for all $u \in C_0^\infty(\Omega)$. The operator P on the right-hand side of the inequality can be replaced by the operator P^0.

4.4. Now we shall show how the uniqueness of the solution to the Cauchy problem is proved using Carleman estimates. For simplicity, we shall restrict our discussion to a differential operator.

Theorem 4.2. Let $P(x, D)$ be a differential operator of order $m \geqslant 1$ of principal type with smooth coefficients. Suppose that (Ψ') and ($\mathscr{B}_\tau$) are fulfilled for the function $q_\tau(x, \xi) = p^0(x, \xi + i\tau \operatorname{grad} \varphi(x))$ when x is in some $\Omega \in \mathbb{R}^n$, $\xi \in \mathbb{R}_n \setminus 0$, $\tau > 0$. If $u \in H_m^{loc}(\Omega)$, $P(x, D)u = 0$ in Ω and $u = 0$ in the domain $\{x;\ x \in \Omega,\ \psi(x) > \psi(x_0)\}$, where $\psi \in C^\infty(\Omega)$, $\psi(x_0) = \varphi(x_0)$ and $\psi(x) < \varphi(x)$ for $x \in \Omega \setminus x_0$, then $u = 0$ in some neighborhood of x_0.

Proof. According to Theorem 4.1, inequality (4.4) is fulfilled in a neighborhood of Ω. Let $h \in C_0^\infty(\Omega)$ and $h = 1$ in a neighborhood ω of x_0. Suppose that $v = hu$. Then $v \in H_m$ and $Pv = 0$ in ω. Clearly, (4.4) holds for functions $v \in H_m$, provided $\operatorname{supp} v \subset \Omega$. Therefore, function v can be inserted in this inequality.

By construction there is a number $\rho > 0$ such that $\varphi(x) \leqslant \varphi(x_0) - \rho$ in $\operatorname{supp} Pv \cap \Omega$. Let $\omega_1 = \{x;\ x \in \omega,\ \varphi(x) > \varphi(x_0) - \rho\}$. Then from (4.4) it follows that

$$\sum_{|\alpha| \leqslant m-1} \tau^{2(m-|\alpha|-\delta)} \int_{\omega_1} |D^\alpha v(x)|^2 e^{2\tau[\varphi(x_0)-\rho]} dx \leqslant$$

$$C \int |P(x, D) v(x)|^2 e^{2\tau[\varphi(x_0)-\rho]} dx,$$

and therefore

$$\tau^{2(m-\delta)} \int_{\omega_1} |v|^2 dx \leqslant C \int |P(x, D) v|^2 dx$$

for all $\tau > 0$. Hence it follows that $v = 0$ in ω_1 and therefore $u = 0$ in ω_1. Since ω_1 is a neighborhood of x_0, the theorem is proved.

§5. Calderón's Theorem

5.1. In this section a theorem generalizing the well-known Calderón theorem on the uniqueness of solutions of the Cauchy problem for a system of differential equations is proved.

Let us consider a solution to the system of equations

$$Lu \equiv D_t^m u - \sum_{j=0}^{m-1} a_j(t, x, D_x) D_t^j u = f(t, x), \tag{5.1}$$

where $u = (u_1, \ldots, u_N)$, $f = (f_1, \ldots, f_N)$, and a_j is a square matrix of order N whose elements are pseudo-differential operators in x of order m − j with conditions

$$D_t^j u = 0 \quad \text{for} \quad t = 0, \quad j = 0, 1, \ldots, m-1.$$

Since we shall make use of pseudo-differential operators in x variables, it is convenient to transfer the plane t = 0, on which the initial conditions are given, to the surface $t = \delta|x|^2$. To do so we shall introduce new variables (t', x') setting

$$x' = x, \quad t' = t + \delta|x|^2, \quad \delta > 0.$$

It is not hard to see that (5.1) in the new variables remains a system of the same type, that is, it is a Kovalevsky system. Here, the solution u of the Cauchy problem can be assumed to be equal to 0 for $t' \leqslant \delta|x|^2$ and therefore it has compact support in x' for every t', $0 < t' \leqslant T$, where T is a positive number.

Furthermore, we shall denote t' and x' again by t and x.

The main result of this section is

Theorem 5.1. Suppose that the roots $\lambda_1, \ldots, \lambda_{mN}$ of the characteristic equation

$$\det\left(\lambda^m I_N - \sum_{j=0}^{m} a_j(t, x, \xi)\lambda^j\right) = 0 \tag{5.2}$$

satisfy the conditions (I)-(IV) below. Then for sufficiently small T and $u \in [C_0^\infty(\omega)]^N$, where $\omega = \{(t, x) \in \mathbb{R}^{n+1}, T \geqslant t \geqslant \delta|x|^2\}$, the inequality

$$\int_0^T \sum_{|\alpha| \leqslant m-1} \|D^\alpha u(t, \cdot)\|^2 e^{2\tau\varphi(t)}\, dt \leqslant$$

$$C(T^2 + \tau^{-1}) \int_0^T \|Lu(t, \cdot)\|^2 e^{2\tau\varphi(t)}\, dt \tag{5.3}$$

is fulfilled. Here $\varphi(t) = (t - T)^2$, $\tau \geqslant \tau_0$, the constant C is independent of u, T, and τ, and

$$Lu = D_t^m u - \sum_{j=0}^{m} a_j(t, x, D) D_t^j u.$$

As in the previous section, the theorem on uniqueness of solutions to the Cauchy problem can be readily deduced from this theorem.

Theorem 5.2. Let Lu = 0 in the domain $\Omega = \{(t, x) \in \mathbb{R}^{n+1}, t < T\}$ and supp $u \subset \omega = \{(t, x) \in \mathbb{R}^{n+1}, t \geqslant \delta|x|^2\}$. If the roots of the characteristic equation (5.2) satisfy the conditions (I)-(IV) below and T is small enough, then u = 0 in Ω.

Proof of Theorem 5.2. Let $h \in C^\infty(\mathbb{R})$, $h \geqslant 0$, where h(t) = 0 for $t \geqslant T$, h(t) = 1 for $t \leqslant T(1 - \delta)$; here δ is any number, $0 < \delta < 1/2$. Then L(hu) = 0 for $t \leqslant T(1 - \delta)$.

In (5.3) we substitute the function hu instead of u. Then

$$\int_0^{T(1-2\delta)} e^{2\tau\varphi(t)} \|u(t, \cdot)\|^2 dt \leqslant \int_0^T e^{2\tau\varphi(t)} \|h(t)u(t, \cdot)\|^2 dt \leqslant$$

$$C(T^2 + \tau^{-1}) \int_0^T e^{2\tau\varphi(t)} \|L(hu)(t, \cdot)\|^2 dt = C(T^2 + \tau^{-1}) \int_{T(1-\delta)}^T e^{2\tau\varphi(t)} \|L(hu)(t, \cdot)\|^2 dt.$$

Since $\varphi'(t) < 0$ for $0 < t < T$, it is seen that

$$e^{2\tau\varphi(T(1-2\delta))} \int_0^{T(1-2\delta)} \|u(t, \cdot)\|^2 dt \leqslant C(T^2 + \tau^{-1}) e^{2\tau\varphi(T(1-\delta))} \int_{T(1-\delta)}^T \|L(hu)(t, \cdot)\|^2 dt.$$

Letting $\tau \to +\infty$ we find that $u(t, x) = 0$ for $t \leqslant T(1 - 2\delta)$. Since δ is any non-negative number, it follows that $u(t, x) = 0$ for $t \leqslant T$. The theorem is proved.

Remark 5.1. Instead of the surface $t = \delta|x|^2$ we can examine any smooth surface of the type $t = \psi(x)$, where $\psi(x) > \Psi(0)$ for $x \neq 0$ if the diameter of the intersection of domain $\{(t, x), t > \psi(x)\}$ with the plane $t = \varepsilon$ tends to zero when $\varepsilon \to 0$.

5.2. Let us replace the Cauchy problem for (5.1) by the equivalent problem for a system of first-order equations. However, the technique used in proving Lemma 2.1 will not suit us, since superfluous characteristics appear in this case. The Calderón method enables system (5.1) to be reduced to a first-order system of N_m equations, which has the same characteristic equation. Unlike Lemma 2.1, the system obtained will contain pseudo-differential equations in x variables even when the system (5.1) is a differential system.

Let Λ be an operator with symbol $(1 + |\xi|^2)^{1/2}$ so that $\Lambda^2 = 1 - \Delta$. Let

$$u_j = \Lambda^{m-j} D_t^{j-1} u, \quad j = 1, \ldots, m,$$

so that u_j is an N-dimensional vector. Suppose that U is a column vector $[u_1, u_2, \ldots, u_m]$. Then the first-order system for the vector U, equivalent to (5.1), is written as:

$$D_t u_j = \Lambda u_{j+1}, \quad j = 1, \ldots, m-1, \quad D_t u_m = \sum_{j=0}^{m-1} a_j(t, x, D_x) \Lambda^{j+1-m} u_{j+1} + f. \tag{5.4}$$

The operators on the right-hand side of this system are pseudo-differential operators of first order. At the same time, the characteristic equation takes the form

$$\left| \lambda I_{mN} - \begin{pmatrix} 0 & \\ \vdots & |\xi| I_{(m-1)N} \\ 0 & \\ a_0^0(t, x, \xi)|\xi|^{1-m}, & a_1^0(t, x, \xi)|\xi|^{2-m}, \ldots, a_{m-1}^0(t, x, \xi) \end{pmatrix} \right| = 0.$$

(Recall that a_j stands for square matrices of order N.) This characteristic equation is in accord with (5.2). Therefore the characteristic roots of the initial system (5.1) agree with those of the first-order system (5.4).

5.3. Now write (5.4) as

$$D_t U = P(t, x, D_x) U + F.$$

Here P is a matrix composed of first-order pseudo-differential operators acting on x-variables.

Below we shall prove that the inequality

$$\int_0^{T_0} e^{2\tau\varphi(t)} \| U(t, \cdot) \|_0^2 \, dt \leqslant C(T_0^2 + \tau^{-1}) \int_0^{T_0} e^{2\tau\varphi(t)} \| D_t U - PU \|_0^2 \, dt + C \int_0^{T_0} e^{2\tau\varphi(t)} \| U \|_{-1}^2 \, dt \tag{5.5}$$

is fulfilled for the conditions (I)-(IV), if T_0 is sufficiently small.

This suffices to deduce the estimate (5.3). Indeed, from the definition of vector U it follows that the estimates

$$C^{-1} \| U \|_0^2 \leqslant \sum_{|\alpha| \leqslant m-1} \| D_{t,x}^\alpha u \|^2 \leqslant C \| U \|_0^2, \qquad \| U \|_{-1}^2 = \sum_{j=1}^{m} \| \Lambda^{m-j-1} D_t^{j-1} u \|_0^2$$

with some constant C are valid. On the other hand,

$$\| D_t U - PU \|_0^2 = \| f \|_0^2 = \| Lu \|_0^2.$$

Therefore, from (5.5) it follows that

$$\int_0^{T_0} e^{2\tau\varphi(t)} \sum_{|\alpha| \leqslant m-1} \| D^\alpha u(t, \cdot) \|_0^2 \, dt \leqslant$$

$$C_1 (T_0^2 + \tau^{-1}) \int_0^{T_0} e^{2\tau\varphi(t)} \| Lu(t, \cdot) \|_0^2 \, dt + C_1 \sum_{j=1}^{m} \int_0^{T_0} e^{2\tau\varphi(t)} \| \Lambda^{m-j-1} D_t^{j-1} u(t, \cdot) \|_0^2 \, dt.$$

Note that the inequality diam $\omega \leqslant \sqrt{T}$ holds for $0 \leqslant t \leqslant T$, and therefore

$$\sum_{j=1}^{m} \| \Lambda^{m-j-1} D_t^{j-1} u \|_0^2 \leqslant C_2 T \sum_{j+|\alpha| \leqslant m} \| D_t^j D_x^\alpha u \|_0^2.$$

If T is so small that $C_1 C_2 T < 1/2$, we obtain the estimate (5.3). Thus, our task amounts to proving inequality (5.5).

5.4. We now assume that *condition* I is fulfilled

(I). The multiplicity of the roots λ of the characteristic equation (5.2) is constant.

This condition enables us locally, that is, in a neighborhood of the point $(0, \xi^0) \in \mathbb{R}^{n+1} \times S^{n-1}$, to reduce the matrix P_0 to Jordan normal form in which every diagonal block has constant dimensions. It may be asserted that in the given neighborhood there exists a nonsingular matrix $Q(t, x, \xi)$ with smooth elements such that the matrix

$$J = QP_0Q^{-1}$$

contains diagonal blocks of the form

$$\begin{pmatrix} \lambda(t, x, \xi) & 1 & 0 & \dots & 0 \\ 0 & \lambda(t, x, \xi) & 1 & \dots & 0 \\ \dots & \dots & \dots & \dots & \vdots \\ 0 & 0 & 0 & \dots & \lambda(t, x, \xi) \end{pmatrix}.$$

Having extended the matrix Q smoothly in such a manner that

$$Q(t, x, \tau\xi) = \tau Q(t, x, \xi) \quad \text{for} \quad |\xi| \geqslant 1, \ \tau \geqslant 1,$$

we can consider the pseudo-differential operator $Q(s, D)$. Since ξ^0 is an arbitrary point on the sphere S^{n-1}, we obtain a covering of the sphere with neighborhoods in which the matrix Q is defined. By the Heine-Borel theorem, from this covering we can select a finite subcovering with neighborhoods $\omega_1, \dots, \omega_p$ of the given type and then construct a partition of unity subject to the subcovering:

$$\sum_{j=1}^{p} \alpha_j^2(\xi) = 1, \quad \alpha_j \in C^\infty(S^{n-1}).$$

Extend a_j to smooth functions on the whole space, the functions being positively homogeneous of zero degree for $|\xi| \geqslant 1$. Let $A_j(D)$ be a pseudo-differential operator with symbol $\alpha_j(\xi)$ and $U_j = A_j(D)U$. Then $\sum_j \| U_j \|_0^2 = \| U \|_0^2$ and

$$\| D_t U - PU \|_0^2 = \sum_j \| A_j(D)(D_t U - PU) \|_0^2 = \sum_j \| D_t U_j - PU_j + [A_j, P] U \|_0^2.$$

The operator $[A_j, P]$ is of zero order and therefore

$$\left| \| D_t U - PU \|_0^2 - \sum_j \| D_t U_j - PU_j \|_0^2 \right| \leqslant C \| U \|_0^2.$$

Here $\|\cdot\|_0$ represents the norm in the space $L_2(\mathbb{R}^n)$ so that the considered functions and their norms depend on t as on a parameter.

Now we shall construct the symbol $P_j(t, x, \xi)$ agreeing with $P_0(t, x, \xi)$ in supp α_j and which is reducible to Jordan normal form simultaneously on the entire sphere S^{n-1}. For this it is convenient to construct a smooth mapping ψ_j: $S^{n-1} \to \omega_j$ identical in supp α_j. Furthermore, we shall assume that $P_j(t, x, \xi) = P_0(t, x, \psi_j(\xi))$ for $|\xi| = 1$ and similarly $Q_j(t, x, \xi) = Q(t, w, \psi_j(\xi))$, $J_j(t, x, \xi) = J(t, x, \psi_j(\xi))$, where Q and J are the matrices defined above for a neighborhood ω_j. Clearly $Q_jP_jQ_j^{-1} = J_j$ for j = 1, ..., p. We extend these matrix functions to smooth functions on the entire space of variables ξ, homogeneous for $|\xi| \geqslant 1$, so that the relation $Q_jP_jQ_j^{-1} = J_j$ is fulfilled for all ξ, $|\xi| \geqslant 1$. For this, the homogeneity order for Q_j equals 0 and for P_j and J_j it is equal to 1. Now let $Q_j(t, x, D)U_j = V_j$. Then $U_j = S_jV_j + T_jU_j$, where S_j is a parametrix for the elliptic operator Q_j, the orders of S_j and Q_j are equal to 0, and

$$\|U_j\|_0 \leqslant C\|V_j\|_0 + C_1\|U_j\|_{-1}, \quad \|V_j\|_0 \leqslant C_1\|U_j\|_0. \tag{5.6}$$

Thus,

$$Q_j(D_tU_j - PU_j) = D_tV_j - Q_jPS_jV_j + T_0U_j = D_tV_j - J_jV_j + T_0'U,$$

where T_0 and T_0' are of zero order. It follows that

$$\|D_tV_j - J_jV_j\|_0 \leqslant C_2\|D_tU_j - PU_j\|_0 + C_3\|U_j\|_0$$

and therefore

$$\sum_j \|D_tV_j - J_jV_j\|_0^2 \leqslant C_4\|D_tU - PU\|_0^2 + C_5\sum_j\|U_j\|_0^2. \tag{5.7}$$

Let us now verify that inequality (5.5) follows from the estimate

$$\int_0^T e^{2\tau\varphi(t)}\|v(t, \cdot)\|_0^2\,dt \leqslant C(T^2 + \tau^{-1})\int_0^T e^{2\tau\varphi(t)}\|(D_t - J_j)v(t, \cdot)\|_0^2\,dt \tag{5.8}$$

for j = 1,..., p, $v \in C_0^\infty(\Omega)$, where $\Omega = \{(t, x) \in \mathbb{R}^{n+1},\ 0 \leqslant t < T\}$.

To do so we substitute in (5.8) $v = V_j$ and sum up over j the inequalities obtained.

We have

$$\sum_j \|V_j\|_0^2 \geqslant C_6\sum\|U_j\|_0^2 - C_7\sum\|U_j\|_{-1}^2 = C_6\|U\|_0^2 - C_7\|U\|_{-1}^2,$$

where C_6 = const > 0. On the other hand, from (5.7) it follows that

$$\sum_j \|(D_t - J_j)V_j\|_0^2 \leqslant C_4\|D_tU - PU\|_0^2 + C_5\sum\|V_j\|_0^2 \leqslant C_4\|D_tU - PU\|_0^2 + C_8\|U\|_0^2.$$

If $T^2 + \tau^{-1} \leqslant (2CC_8)^{-1}C_6$, from (5.8) we obtain

$$\int_0^T e^{2\tau\varphi(t)}\|U(t, \cdot)\|_0^2\,dt \leqslant C_6(T^2 + \tau^{-1})\int_0^T e^{2\tau\varphi(t)}\|D_tU - PU\|_0^2\,dt + C_{10}\int_0^T e^{2\tau\varphi(t)}\|U\|_{-1}^2\,dt,$$

that is, (5.5) is valid.

5.5. Now we prove inequality (5.8).

To do so we assume that the *conditions* II, III, and IV hold:

(II). If the order of the Jordan block corresponding to the root $\lambda = a + ib$ is equal to 1, then the inequality

$$-b_t'\varphi' + \{a, b\}\varphi' + b\varphi'' \leqslant 0 \tag{5.9}$$

is fulfilled at every point (t, x, ξ), $|\xi| > 1$, where $b(t, x, \xi) > 0$. Besides, if the function $b(t, x, \xi)$ changes sign in (t_0, x_0, ξ^0), then at this point

$$\sum_{j=1}^{k} |\partial_t^j b(t_0, x_0, \xi^0)| \geqslant c_0 |\xi|, \tag{5.10}$$

where $k_1 \geqslant 1$, c_0 = const > 0,

$$(adf)g(x, \xi) = \sum_{j=1}^{n} \left(\frac{\partial f(x, \xi)}{\partial \xi_j} \frac{\partial g(x, \xi)}{\partial x_j} - \frac{\partial f(x, \xi)}{\partial x_j} \frac{\partial g(x, \xi)}{\partial \xi_j} \right).$$

(III). If the order of the Jordan block corresponding to the root $\lambda = a + ib$ is equal to 2, then

$$|b| \geqslant c_0 |\xi| \text{ for } |\xi| \geqslant 1, \text{ where } c_0 = \text{const} > 0.$$

(IV). If the order of the Jordan block corresponding to the root $\lambda = a + ib$ is more than or equal to 3, then

$$b(t, x, \xi) \geqslant c_0 |\xi| \text{ for } |\xi| \geqslant 1, \text{ where } c_0 = \text{const} > 0. \tag{5.11}$$

Obviously, it suffices to prove inequality (5.8) for different blocks in the Jordan normal form. We shall prove it in three stages conforming to the conditions (II)-(IV).

(II). Let $u \in C_0^\infty(\omega)$, $D_t u - A(t, x, D)u - iB(t, x, D)u = f$ and let the conditions (5.9) and (5.10) be fulfilled for the leading symbol $a + ib$ of the operator $A + iB$. Substitution $u = ve^{-\tau\varphi(t)}$ leads us to the equation

$$D_t v - A(t, x, D)v - iB(t, x, D)v + i\tau\varphi'(t)v = g,$$

where $g = fe^{\tau\varphi(t)}$. Our task is to prove the inequality

$$\int_0^T \|v\|_0^2 dt \leqslant CT^2 \int_0^T \|g\|_0^2 dt. \tag{5.12}$$

Here condition (5.9) means that condition (Ψ') is fulfilled for all $\tau > 0$. Indeed, condition (Ψ') implies that the function $b(t, x, \xi) - \tau\varphi'(t)$ does not change its sign from minus to plus along the bicharacteristic of the function $\sigma - a(t, x, \xi)$ (σ is the variable dual to t). Therefore, the function $b/\varphi' - \tau$ must not change its sign from plus to minus for all $\tau > 0$ as t increases. This, in turn, however, implies that the function b/φ' must be a monotonically increasing function in that domain where it has positive values, that is,

$$b(t, x, \xi) < 0 \Rightarrow \left\{\sigma - a, \frac{b}{\varphi'}\right\} \geqslant 0.$$

But this is condition (5.9), which thus suffices for condition (A) of Theorem 1.4 of Chapter VII to hold in a neighborhood of the point (t, x, ξ) at which $b(t, x, \xi) < 0$. Condition (B) of the same theorem is fulfilled in a neighborhood of (t, x, ξ) where function b assumes a positive value. Finally, (5.10) implies that in a neighborhood of every point (t, x, ξ) at which function b changes sign the condition of Theorem 1.2 of Chapter VIII is fulfilled. By the constructions made in Section 2.5 of Chapter VII, we can assert that inequality (5.12) holds.

(III). Let

$$\begin{aligned} D_t u_1 - A(t, x, D)u_1 - iB(t, x, D)u_1 - \Lambda u_2 &= f_1, \\ D_t u_2 - A(t, x, D)u_2 - iB(t, x, D)u_2 \qquad &= f_2 \end{aligned} \tag{5.13}$$

and let B be an elliptic operator of order 1 with real symbol so that

$$\|\Lambda v\| \leqslant C_1(\|Bv\| + \|v\|)$$

for $v \in C_0^\infty(\omega)$. Substitution $u_j = v_j e^{-\tau\varphi(t)}$ for j = 1, 2 transforms (5.13) into the system

$$\begin{aligned} &D_t v_1 - A(t, x, D) v_1 - iB(t, x, D) v_1 + i\tau\varphi'(t) v_1 - \Lambda v_2 = g_1, \\ &D_t v_2 - A(t, x, D) v_2 - iB(t, x, D) v_2 + i\tau\varphi'(t) v_2 = g_2, \end{aligned} \tag{5.14}$$

where $g_j = f_j e^{\tau\varphi(t)}$. From the second equation of this system it follows that

$$\|g_2\|^2 = \|D_t v_2 - A v_2\|^2 + \|(B - \tau\varphi') v_2\|^2 + 2\,\mathrm{Re}\, i\,(D_t v_2 - A v_2,\ (B - \tau\varphi') v_2).$$

As in case (II), the estimate

$$\int_0^T \|D_t v_2 - A v_2\|^2\, dt \geqslant c_0 T^{-2} \int_0^T \|v_2\|^2\, dt$$

holds. Furthermore,

$$2\,\mathrm{Re}\, i \int_0^T (D_t v_2 - A v_2,\ (B - \tau\varphi') v_2)\, dt =$$

$$\mathrm{Re} \int_0^T ((-B_t - i[B, A]) v_2,\ v_2)\, dt + \mathrm{Re} \int_0^T \left(\frac{\partial v_2}{\partial t},\ (B - B^*) v_2\right) dt +$$

$$\tau \int_0^T \varphi'' \|v_2\|^2\, dt - \mathrm{Re}\, i \int_0^T (((B^* - B) A - (A^* - A) B) v_2,\ v_2)\, dt +$$

$$\mathrm{Re}\, i\tau \int_0^T ((A^* - A) v_2,\ v_2)\, \varphi'(t)\, dt. \tag{5.15}$$

Since the operators $A^* - A$ and $B^* - B$ are of zero order, and the operators B_t and B, A are of order 1, it follows that

$$2\,\mathrm{Re}\, i \int_0^T (D_t v_2 - A v_2,\ (B - \tau\varphi') v_2)\, dt \geqslant \tau \int_0^T \varphi'' \|v_2\|^2\, dt -$$

$$C_2 \int_0^T \|\Lambda v_2\| \|v_2\|\, dt - C_2 \tau \int_0^T |\varphi'(t)| \|v_2\|^2\, dt - C_2 \int_0^T \|g_2\| \|v_2\|\, dt.$$

Since

$$\|(B - \tau\varphi') v_2\| \geqslant \|B v_2\| - \tau |\varphi'| \|v_2\| \geqslant c_3 \|\Lambda v_2\| - (C_4 + \tau |\varphi'|) \|v_2\|,$$

where $c_3 = \mathrm{const} > 0$, we have

$$\int_0^T \|\Lambda v_2\| \|v_2\|\, dt \leqslant C_5 \int_0^T \|(B - \tau\varphi') v_2\| \|v_2\|\, dt + C_6 \int_0^T \|v_2\|^2\, dt +$$

$$C_5 \tau \int_0^T |\varphi'| \|v_2\|^2\, dt \leqslant \frac{1}{2} \int_0^T \|(B - \tau\varphi') v_2\|^2\, dt + C_7 \int_0^T \|v_2\|^2\, dt +$$

$$C_8 \tau \int_0^T |\varphi'| \|v_2\|^2\, dt \leqslant \frac{1}{2} \int_0^T \|(B - \tau\varphi') v_2\|^2\, dt + \frac{\tau}{2} \int_0^T \|v_2\|^2\, dt,$$

if τ is sufficiently large and T is small enough.

Thus

$$\int_0^T \|g_2\|^2\, dt \geqslant \frac{\tau}{4} \int_0^T \|v_2\|^2\, dt + c_0 T^{-2} \int_0^T \|v_2\|^2\, dt + \frac{1}{2} \int_0^T \|(B - \tau\varphi') v_2\|^2\, dt. \tag{5.16}$$

From this one sees that

$$\int_0^T \|\Lambda v_2\|^2 dt \leqslant C_9 \int_0^T \|(B-\tau\varphi')v_2\|^2 dt + C_9 \int_0^T \|v_2\|^2 dt +$$

$$C_9\tau^2 \int_0^T \varphi'^2 \|v_2\|^2 dt \leqslant C_{10}(1+\tau T^2) \int_0^T \|g_2\|^2 dt. \qquad (5.17)$$

Using the same technique, from the inequality

$$\|D_t v_1 - Av_1 - iBv_1 + i\tau\varphi' v_1\| \leqslant \|g_1 + \Lambda v_2\|$$

we find

$$T^{-2}\int_0^T \|v_1\|^2 dt + \tau^{-1}\int_0^T \|\Lambda v_1\|^2 dt + \tau \int_0^T \|v_1\|^2 dt \leqslant C_{11}\int_0^T \|g_1 + \Lambda v_2\|^2 dt.$$

Hence

$$\int_0^T \|v_1\|^2 dt \leqslant C_{11}\tau^{-1}\int_0^T \|g_1\|^2 dt + C_{11}(\tau^{-1}+T^2)\int_0^T \|g_2\|^2 dt,$$

which together with (5.16) yields the original estimate.

(IV). Here we shall examine the system of equations

$$\begin{aligned} &D_t u_1 - Au_1 - iBu_1 - \Lambda u_2 = f_1,\\ &\dots\dots\dots\dots\dots\dots\\ &D_t u_{r-1} - Au_{r-1} - iBu_{r-1} - \Lambda u_r = f_{r-1},\\ &D_t u_r - Au_r - iBu_r = f_r, \end{aligned}$$

where r is some number, $3 \leqslant r \leqslant N$. After the substitution $u_j = v_j e^{-\tau\varphi(t)}$ this system changes into

$$\begin{aligned} &D_t v_1 - Av_1 - iBv_1 + i\tau\varphi' v_1 - \Lambda v_2 = g_1,\\ &\dots\dots\dots\dots\dots\dots\\ &D_t v_{r-1} - Av_{r-1} - iBv_{r-1} + i\tau\varphi' v_{r-1} - \Lambda v_r = g_{r-1},\\ &D_t v_r - Av_r - iBv_r + i\tau\varphi' v_r = g_r. \end{aligned}$$

We shall start estimating from the last equation of this system. As above, we have

$$\|D_t v_r - Av_r\|^2 + \|(B-\tau\varphi')v_r\|^2 + 2\,\mathrm{Re}\, i\,(D_t v_r - Av_r,\ Bv_r - \tau\varphi' v_r) = \|g_r\|^2.$$

Thus

$$2\,\mathrm{Re}\, i\int_0^T (D_t v_r - Av_r,\ Bv_r - \tau\varphi' v_r)\,dt =$$

$$\mathrm{Re}\int_0^T ((-B_t - i[B,A])v_r,\ v_r)\,dt + \tau\int_0^T \varphi''\|v_r\|^2 dt +$$

$$\mathrm{Re}\int_0^T ((B-B^*)v_r,\ \partial_t v_r)\,dt + \mathrm{Re}\, i\tau\int_0^T ((A^*-A)v_r,\ v_r)\,\varphi'(t)\,dt -$$

$$C_1\int_0^T \|\Lambda v_r\|\|v_r\|\,dt \geqslant \frac{\tau}{2}\int_0^T \|v_r\|^2 dt - C_2\int_0^T \|\Lambda v_r\|\|v_r\|\,dt - \frac{C_2}{\tau}\int_0^T \|g_r\|^2 dt.$$

Since by condition (IV), $b(t, x, \xi) \geqslant c_0|\xi|$ for $|\xi| \geqslant 1$ and $\varphi' < 0$, the inequality

$$|b(t, x, \xi) - \tau\varphi'| \geqslant c_0|\xi| + \tau|\varphi'|$$

holds, and therefore

$$\|(B-\tau\varphi')v\|^2 \geqslant c_0\|\Lambda v\|^2 + \tau^2\|\varphi' v\|^2 - C_2\|v\|^2$$

for $v \in C_0^\infty(\omega)$, where $c_0 = \mathrm{const} > 0$.

Then

$$\int_0^T \|g_r\|^2\,dt \geq \frac{c_0}{2}\int_0^T \|\Lambda v_r\|^2\,dt + c_1\tau\int_0^T \|v_r\|^2\,dt + c_2T^{-2}\int_0^T \|v_r\|^2\,dt,$$

where c_j = const > 0, j = 0, 1, 2.

Using the estimate for Λv_r we can further estimate the function v_{r-1} so that

$$\int_0^T (\|g_r\|^2 + \|g_{r-1}\|^2)\,dt \geq c_0'\int_0^T \|\Lambda v_{r-1}\|^2\,dt + c_1'\tau\int_0^T \|v_{r-1}\|^2\,dt + c_2'T^{-2}\int_0^T \|v_{r-1}\|^2\,dt.$$

Continuing this process we arrive at the estimate

$$\sum_{j=1}^{r}\int_0^T (\|\Lambda v_j\|^2 + \tau\|v_j\|^2 + T^{-2}\|v_j\|^2)\,dt \leq C\sum_{j=1}^{r}\int_0^T \|g_j\|^2\,dt.$$

Thus, inequality (5.8) is proved. This completes the proof of Theorem 5.1.

Remark 5.1. Condition (I) could be replaced by the more precise condition that the normal divisors have constant multiplicity, since the characteristic roots λ in different blocks of the matrix J can of course be equal.

COMMENTS

Chapter I

The development of distribution theory began with the publication of L. Schwartz's books,[1-2] though its underlying basis was determined by S. L. Sobolev's papers (see Ref. 1). A more detailed discussion of this theory is available in the books of V. S. Vladimirov,[1-2] L. Schwartz,[2] S. L. Sobolev[2], and I. M. Gelfand and G. E. Shilov.[1] Our presentation in most instances is close to that of L. Hörmander.[3] The construction of the fundamental solution has been taken from the work of I. M. Gelfand and G. E. Shilov, *Generalized Functions*.[1] (At most points our presentation is close to that of L. Hörmander.[3] The construction of the fundamental solution is due to I. M. Gelfand and G. E. Shilov.[2]) The theory of distributions on a manifold is discussed in greater detail in J. de Rham, *Variétiēs Différentiables*.[1]

Chapter II

The book by S. G. Mikhlin[1] and the papers of A. P. Calderón and A. B. Zygmund[1-2] are dedicated to the theory of singular integral operators. The algebra of pseudo-differential operators was first constructed in J. J. Kohn and L. Nirenberg, *Harmonic Integrals on Strongly Pseudoconvex Manifolds*;[1] it is described in the books of M. Taylor,[1] M. A. Shubin,[1] K. Friedrichs[1] and in some papers of L. Nirenberg[1-2] and L. Hörmander.[4]

In Section 4 we discuss mainly our own work.[10] Note that the results, close in form in Theorem 4.1, were earlier obtained by V. A. Fock[1] in application to quantum mechanic operators. Other statements of this theorem are given in the papers of M. V. Fedoryuk,[1] L. Nirenberg and F. Trèves,[2] and L. Hörmander.[13]

Gårding's inequality, of significance for these applications, was first obtained in a correct form by L. Hörmander.[6] In Section 5 we follow mainly the paper of P. Lax and L. Nirenberg.[1] This theorem is generalized in A. Melin, *Lower Bounds for Pseudo-Differential Operators*[1]; L. Hörmander, *The Weyl Calculus of Pseudo-Differential Operators*;[16] and C. Fefferman and D. H. Phong, *On Positivity of Pseudo-Differential Operators; The Uncertainty Principle and Sharp Gårding Inequalities*.[1-2]

For other generalizations of the theory of pseudo-differential operators see: L. Hörmander,[7,16] R. Beals,[1-2] H. Kumano-go.[2] The main result of Section 7 is due to A. P. Calderón and R. Vaillancourt.[1-2] Our discussion is close to that of L. Boutet de Monvel.[1]

Chapter III

Boundary problems for elliptic equations with constant coefficients were studied in Y. B. Lopatinski, *On a Method of Reducing Boundary Problems for a System of Differential Equations of Elliptic Type to Regular Equations*[1]; and Z. Ya. Shapiro, *On General Boundary Problems for Equations of Elliptic Type*[1]; the proposed construction of the parametrix is close to that given in F. E. Browder, *Estimates and Existence Theorems for Elliptic Boundary Value Problems*.[1] The method of reduction to an equation on the boundary of manifold is developed by R. T. Seeley[1] and L. Hörmander.[6]

For a discussion of the theory of elliptic boundary problems see the papers of S. Agmon, A. Douglis, L. Nirenberg,[1] the book of L. Hörmander[1], and the papers of M. Shechter[1] and J. Peetre.[1]

Chapter IV

Our presentation is close to that of Yu. V. Egorov and V. A. Kondratiev.[1] An example given in Section 1 is due to A. B. Bitsadze (see Ref. 1). The papers of M. B. Malyutov,[1] V. G. Mazya and V. P. Paneyakh,[1] V. G. Mazya,[1] K. Taira,[1-4] B. Winzell[1], and Sh. A. Alimov[1] are dedicated to oblique derivative problems.

Chapter V

The concept of a wave front set was introduced in L. Hörmander, *Fourier Integral Operators.*[13] Earlier a similar concept of a singular support of hyperfunction was introduced by M. Sato.[1] The method discussed in Section 3 was suggested in the paper of V. Guillemin and D. Schaeffer[1] (see also V. Guillemin and S. Sternberg[1]). The results presented in Section 6 were obtained in the papers of L. Hörmander,[13] G. Duistermaat and L. Hörmander.[1]

Chapter VI

Example 1.1 is due to V. V. Grushin[2] who revised the examples of P. Garabedyan.[1] The main theorem of Section 2 was proved in L. Hörmander, *Differential Equations Without Solutions.*[2] The finite-order zero theorem was studied in the papers of L. Hörmander and F. Trèves[2] and of the author.[11] The structure of the symbol was studied by the author in, *On the Structure of Subelliptic Operators*[13] (see also Ref. 14). The theorem of Section 5 was not published earlier; it generalizes the theorem from the paper of Yu. V. Egorov and P. R. Popivanov.[1]

Chapter VII

Here we follow mainly our work.[20] The proof of a more general theorem, announced in Egorov[16,19] is still not published.

Chapter VIII

Subellitpic operators with $\delta = 1/2$ were obtained for the first time in L. Hörmander, *Pseudo-Differential Operators and Nonelliptic Boundary Problems.*[6] Certain classes of elliptic operators are described in the papers of Yu. V. Egorov.[1,3,5,7,8] The estimate for first-order operators were studied in Yu. V. Egorov, *Estimates for First-Order Differential Operators.*[2] The theorem on necessary and sufficient conditions for subellipticity was first formulated in the author's paper.[4] Complete proofs of the theorem were published in the papers of Yu. V. Egorov,[14,15] (see also the correction in Ref. 18). An improvement for the final part of the proof is suggested in the paper of L. Hörmander,[18] which simplifies the construction and estimation with the partition of unity from the author's work.[15] Use of this idea has been made in Section 5.

Theorem 8.1 was obtained by J. J. Kohn in *Boundary Behavior of $\bar{\partial}$ on Weakly Pseudoconvex Manifolds of Dimension Two.*[3] The necessity of conditions of Theorem 8.2 is proved in the paper of P. Greiner.[1] The sufficiency (in a somewhat weaker form) is proved in the paper of J. J. Kohn.[3]

Theorem 8.3 was obtained in the author's paper.[17] In 1964, subelliptic operators were completely described by J. J. Kohn[2] for a strictly pseudoconvex domain in the space $\mathbb{C}^n$ for any n. Later J. J. Kohn[5] obtained final results on subelliptic estimates in pseudoconvex subdomains in the space $\mathbb{C}^n$.

Chapter IX

The Kovalevsky theorem has been described mainly following the book of F. Trèves.[11] The Cauchy problem for symmetrizable systems is studied in the book of S. Mizohata,[1] also see the paper of L. Nirenberg.[1] The process of diagonalization of a strictly hyperbolic system, given in Section 3, was suggested in M. E. Taylor, *Reflection of Singularities of Solutions to Systems of Differential Equations*.[2] Theorems 4.1-4.3 and 5.1-5.2 were proved in the author's paper.[22] Our discussion in Section 5 is close to that given in the paper of L. Nirenberg.[2]

REFERENCES

Agmon, S., Douglis, A., and Nirenberg, L.:

1. Estimates near boundary for solutions of elliptic differential equations satisfying general boundary conditions, Commun. Pure Appl. Math., 12, (1959), 623-727.

Agranovich, M. S.:

1. Elliptic singular integro-differential operators, Usp. Mat. Nauk, 20, No. 5 (1965), 3-120 (Russian).
2. Boundary problems for systems with a parameter, Mat. Sb., 84, No. 1 (1971), 27-65.

Alimov, Sh. A.:

1. On an oblique derivative problem, Differ. Uravn., 17, No. 10 (1981), 1738-1751 (Russian).

Andersson, K. G.:

1. Propagation of analyticity of solutions of pseudo-differential equations with constant coefficients, Ark. Mat., 81 (1970), 277-302.

Andersson, K. G. and Melrose, R.:

1. The propagation of singularities along gliding rays, Invent. Math., 41 (1977), 197-232.

Arnol'd, V. I.:

1. Lectures on Classical Mechanics, Moscow State University, Moscow (1968) (Russian).
2. Mathematical Methods of Classical Mechanics, Nauka, Moscow (1974) (Russian).

Atanasov, A. I.:

1. On a class of subelliptic systems of pseudo-differential operators, Byull. Mos. Gos. Univ., No. 2 (1974), 3-8 (Russian).

Beals, R.:

1. Spatially inhomogeneous pseudo-differential operators, II, Commun. Pure Appl. Math., 27 (1974), 161-205.
2. A general calculus of pseudo-differential operators, Duke Math. J., 42, No. 1 (1975), 1-42.
3. Characterization of pseudo-differential operators and applications, Duke Math. J., 44, No. 1 (1977), 45-57.

4. On the boundedness of pseudo-differential operators, Commun. Partial Differ. Eq., 2 (1977), 1063-1070.

Beals, R., and Fefferman, C.:

1. On local solvability of linear partial differential equations, Ann. Math., 97, No. 3 (1973), 482-498.
2. Spatially inhomogeneous pseudo-differential operators, I, Commun. Pure Appl. Math., 27 (1974), 1-24.

Bers, L., John, F., and Schechter, M.:

1. Partial Differential Equations, Wiley, New York (1964).

Berezin, F. A., and Shubin, M. A.:

1. Lectures on Quantum Mechanics, Moscow State University, Moscow (1972) (Russian).

Bitsadze, A. V.:

1. Boundary Problems for Second-Order Elliptic Equations, Nauka, Moscow (1966) (Russian).

Bony, J.-M.:

1. Principe du maximum, inégalité de Harnack et unicité du probleme de Cauchy pour les operateurs elliptiques degeneres, Ann. Inst. Fourier, 19, No. 1 (1969), 277-304.
2. Propagation des singularities differentiables pour une class d'operateurs differentiels a coefficients analytiques, Asterisque, 34-35 (1976), 43-91.

Borelli, R.:

1. The singular, second-order oblique derivative problem, J. Math. Mech., 1 (1966), 51-81.

Boutet de Monvel, L.:

1. Hypoelliptic operators with double characteristics and related pseudo-differential operators, Commun. Pure Appl. Math., 27 (1974), 585-639.

Boutet de Monvel, L., and Guillemin, V.:

1. The spectral theory of Toeplitz operators, Ann. Math. Stud., No. 99 (1981).

Browder, F. E.:

1. Estimates and existence theorems for elliptic boundary value problems, Proc. Natl. Acad. Sci. USA, 451 (1959), 365-372.
2. Functional analysis and partial differential equations, Math. Ann., 138 (1959), 55-79.

Calderón, A. P.:

1. Uniqueness in the Cauchy problem for partial differential equations, Am. J. Math., 80 (1958), 16-36.

2. Singular integrals, Bull. Am. Math. Assoc., 72 (1966), 426-465.

Calderón, A. P., and Vaillancourt, R.:

1. On the boundedness of pseudo-differential operators, J. Math. Soc. Jpn., 23 (1971), 374-378.
2. A class of bounded pseudo-differential operators, Proc. Natl. Acad. Sci. USA, 69 (1972), 436-447.

Calderón, A. P., and Zygmund, A.:

1. On the existence of certain singular integrals, Acta Math., 88 (1952), 85-139.
2. Singular integral operators and differential equations, Am. J. Math., 79 (1957), 901-921.

Caratheodory, C.:

1. Variationsrechnung, Berlin (1935).

Cardoso, F.:

1. On the existence of local solutions of pseudo-differential equations, Bol. Soc. Brazil Mat., 4, No. 2 (1973), 121-137.
2. Wave front sets, Fourier integrals, and propagation of singularities, Bol. Soc. Brazil Mat., 6 (1975), 39-52.

Cardoso, F., and Trèves F.:

1. On subelliptic pseudo-differential operators, Anal. Fonct. Appl., C. R. Colloq. Anal. (Rio de Janeiro, 1972), Hermann (1975), p. 61.

Carleman, T.:

1. Sur une probleme d'unicité pour les systemes d'equations aux d derivees partielles a deux variables independantes, Ark. Mat. Fys., 17 (1939), 1-9.

Cohen, P.:

1. The nonuniqueness of the Cauchy problem, Off. Nav. Res. Tech. Rep., 93, Stanford (1960).

Denker, N.:

1. On the propagation of singularities for pseudo-differential opertors of principal type, Thesis, Lund (1981).

De Rham, J.:

1. Varieties differentiables, Hermann, Paris (1955).

Duistermaat, J.:

1. Fourier integral operators, Courant Inst. Lect. Notes, New York (1971).

Duistermaat, J., and Hörmander, L.:

1. Fourier integral operators, II, Acta Math., 128 (1972), 183-269.

Duistermaat, J., and Sjostrand, J.:

1. A global construction for pseudo-differential operators with non-involute characteristics, Invent. Math., 20 (1973), 209-225.

Dynin, A. S.:

1. Singular integral operators of arbitrary order on manifolds, Dokl. Akad. Nauk SSSR, 141 (1961), 21-23 (Russian).

Egorov, Yu. V.:

1. Pseudo-differential operators of principal type, Mat. Sb., 73 (1967), 356-374 (Russian).
2. Estimates for first-order differential operators, Funkts. Anal., 3, No. 3 (1968), 59-77 (Russian).
3. On a class of pseudo-differential operators, Mat. Sb., 79 (1969), 59-77 (Russian).
4. On subelliptic pseudo-differential operators, Dokl. Akad. Nauk SSSR, 188, No. 1 (1969), 20-22 (Russian).
5. On nonsingular subelliptic operators, Dokl. Akad. Nauk SSSR, 186, No. 5 (1969), 1006-1007 (Russian).
6. On canonical transformations of pseudo-differential operators, Usp. Mat. Nauk, 24, No. 5 (1969), 235-236 (Russian).
7. On local properties of pseudo-differential operators of principal type, Thesis, Moscow State University, Moscow (1970) (Russian).
8. Nonsingular subelliptic pseudo-differential operators, Mat. Sb., 82 (1970), 323-342 (Russian).
9. On local solvability of pseudo-differential equations, Proc. International Congress of Mathematicians, Nice, Nauka, Moscow (1972) (Russian).
10. Canonical transformations and pseudo-differential operators, Tr. Mosk. Mat. Ova., 24 (1971), 3-28 (Russian).
11. On necessary conditions of solvability of pseudo-differential equations of principal type, Tr. Mosk. Mat. Ova., 24 (1971), 29-41 (Russian).
12. On the solvability of differential equations with simple characteristics, Usp. Mat. Nauk, 26, No. 2 (1971), 3-28 (Russian).
13. On the structure of subelliptic operators, Dokl. Akad. Nauk SSSR, 198, No. 6 (1971), 1259-1262 (Russian).
14. Subelliptic operators, Usp. Mat. Nauk, 30, No. 2 (1975), 57-114 (Russian).
15. On subelliptic operators, Usp. Mat. Nauk, 30, No. 3 (1975), 57-104 (Russian).
16. On the solvability conditions of differential equations with simple characteristics, Dokl. Akad. Nauk SSSR, 229, No. 6 (1976), 1310-1312 (Russian).
17. On the regularity of the solution of $\bar{\partial}$-Neumann problem, Byull. Mosk. Gos. Univ., Ser. Mat. Mekh., No. 5 (1977), 12-21 (Russian).
18. On an equality for first-order differential operators, Usp. Mat. Nauk, 33, No. 6 (1978), 203-204 (Russian).
19. On the solutions of equations of principal type, in: Partial Differential Equations, Nauka, Novosibirsk (1980), 116-120 (Russian).
20. On sufficient conditions of local solvability of pseudo-differential equations of principal type, Tr. Mosk. Mat. Ova., 31 (1974), 59-84.

21. Operateurs souselliptiques, Colloque Intern. C. N. R. S., Asterisque, Paris (1973), 152-159.
22. On the uniqueness of the solution of the Cauchy problem, Dokl. Akad. Nauk SSSR, 264, No. 4 (1982), 812-814 (Russian).

Egorov, Yu. V., and Kondratiev, V. A.:

1. On the oblique derivative problem, Mat. Sb., 78 (1969), 148-176 (Russian).

Egorov, Yu. V., and Popivanov, P. R.:

1. On equations of principal type without solutions, Usp. Mat. Nauk, 29, No. 2 (1974), 172-189 (Russian).

Ehrenpreis, L.:

1. Solutions of some problems of division, I, Am. J. Math., 76 (1954), 883-903.

Eskin, G. I.:

1. Cauchy's problem for hyperbolic equations in convolutes, Mat. Sb., 74, No. 2 (1967), 262-297 (Russian).
2. Systems of pseudo-differential equations with simple real characteristics, Mat. Sb., 77, No. 2 (1968), 174-200 (Russian).
3. Degenerate elliptic pseudo-differential operators of principal type, Mat. Sb., 82 (1970), 585-628 (Russian).
4. Elliptic pseudo-differential operators with degeneracy of first-order space variables, Tr. Mat. Ova., 25 (1971), 83-118 (Russian).
5. Boundary Problems for Elliptic Pseudo-Differential Equations, Nauka, Moscow (1973) (Russian).
6. A parametrix for interior mixed problems for hyperbolic equations, J. Anal. Math., 32 (1977), 17-62.

Farris, M.:

1. Egorov's theorem on manifold with diffractive boundary, Commun. Partial Differ. Eq., 6 (1981), 651-687.

Fedoryuk, M. V.:

1. The stationary phase method and pseudo-differential operators, Usp. Mat. Nauk, 26, No. 1 (1971), 67-212 (Russian).

Fefferman, C., and Phong, D. H.:

1. On positivity of pseudo-differential operators, Proc. Natl. Acad. Sci. USA, 75 (1978), 4673-4674.
2. The uncertainty principle and sharp Gårding inequalities, Commun. Pure Appl. Math., 34 (1981), 285-331.

Fok, V. A.:

1. On canonical transformations in classical and quantum mechanics, Byull. Leningr. Gos. Univ., 16 (1959), 67 (Russian).

Folland, G. B.:

1. The tangential Cauchy—Riemann complex on spheres, Trans. Am. Math. Soc., 171 (1972), 83-133.

Folland, G. B., and Kohn, J. J.:

1. The Neumann problem for the Cauchy—Riemann complex, Ann. Math. Stud., 75 (1972).

Folland, G. B., and Stein, E. M.:

1. Parametrices and estimates for the $\bar{\partial}_b$-complex on strongly pseudo-convex boundaries, Bull. Am. Math. Soc., 80, No. 2 (1974), 253-258.

Friedlander, F.:

1. The wavefront set of a simple initial boundary value problem with glancing rays, Math. Proc. Cambridge Philos. Soc., 79 (1976), 145-149.

Friedrichs, K.:

1. Pseudo-differential operators, An Introduction, Lecture Notes, Courant Inst. Math. Sci., New York (1970).

Friedrichs, K., and Lax, P. D.:

1. Boundary value problem for first-order operators, Commun. Pure Appl. Math., 18, No. 1-2 (1965), 355-388.

Gabor, A.:

1. Remarks on the wave front of a distribution, Trans. Am. Math. Soc., 170 (1972), 239-244.

Garabedian, P.:

1. Partial Differential Equations, Wiley, New York (1964).

Gårding, L.:

1. Linear hyperbolic partial differential equations with constant coefficients, Acta Math., 85 (1951), 1-62.
2. Cauchy's problem for hyperbolic equations, Lecture Notes, Chicago (1957).

Gelfand, I. M., and Shilov, G. E.:

1. Generalized Functions, Volumes 1-3, Fizmatgiz, Moscow (1959) (Russian).

Greiner, P.:

1. Subelliptic estimates for the $\bar{\partial}$-Neumann problem in C^2, J. Differ. Geomet., 9 (1974), 239-250.

Grossman, A., Loupias, G., and Stein, E. M.:

1. An algebra of pseudo-differential operators and quantum mechanics in phase space, Ann. Inst. Fourier, 18 (1969), 343-368.

Grushin, V. V.:

1. The extension of smoothness of solutions of differential equations of principal type, Dokl. Akad. Nauk SSSR, 148, No. 6 (1963), 1241-1244 (Russian); English translation in Sov. Math., 4 (1963), 248-252.
2. Pseudo-differential operators in R^n with bounded symbols, Funkts. Anal., 4, No. 3 (1970), 37-50 (Russian).
3. An example of a differential equation without solutions, Mat. Zametki, 10, No. 2 (1971), 125-128 (Russian).
4. On a class of elliptic pseudo-differential operators degenerate on submanifolds, Mat. Sb., 84 (1971), 163-195 (Russian).
5. Hypoelliptic differential equations, Mat. Sb., 88 (1972), 504-521 (Russian).

Guillemin, V., and Schaeffer, D.:

1. Fourier integral operators from the Radon transform point of view, Proc. Symp. Pure Math., 27 (1975), 297-300.

Gunning, R. C., and Rossi, H.:

1. Analytic Functions of Several Complex Variables, Prentice-Hall, Englewood Cliffs, New Jersey (1965).

Guillemin, V., and Sternberg, S.:

1. Geometric asymptotics, Math. Surv., 14 (1977).

Hadamard, J.:

1. Le probleme de Cauchy, Paris (1932).
2. On a certain class of Fuchsian pseudo-differential equations, Duke Math. J., 44, No. 1 (1977), 157-199.

Henkin, G. M.:

1. Integral representation of functions in strictly pseudo-convex domains, Mat. Sb., 78 (1969), 611-632 (Russian).

Hörmander, L.:

1. On the theory of general partial differential operators, Acta Math., 94 (1955), 161-248.
2. Differential equations without solutions, Math. Ann., 140 (1960), 169-173.
3. Linear Partial Differential Operators, Springer, Berlin—Heidelberg—New York (1963).
4. Pseudo-differential operators, Commun. Pure Appl. Math., 18 (1965), 501-517.
5. An Introduction to Complex analysis in Several Variables, Van Nostrand, Princeton, New Jersey (1966).
6. Pseudo-differential operators and nonelliptic boundary problems, Ann. Math., 83 (1966), 129-209.

7. Pseudo-differential operators and hypoelliptic equations, Proc. Symp. Pure Math., 10 (1966), 138-183.
8. Hypoelliptic second-order equations, Acta Math., 119 (1967), 147-171.
9. The spectral function of an elliptic operator, Acta Math., 121 (1968), 193-218.
10. Linear differential operators, Actes Congr. Int. Math., 1 (1970), 121-133.
11. On the singularities of solutions of partial differential equations, Commun. Pure Appl. Math., 23 (1970), 329-358.
12. On the existence and regularity of solutions of linear pseudo-differential equations, L'Enseignement Math., 17 (1971), 99-163.
13. Fourier integral operators, I, Acta Math., 127 (1971), 79-183.
14. Uniqueness theorems and wave front sets for solutions of linear differential equations with analytic coefficients, Commun. Pure Appl. Math., 24 (1971), 671-704.
15. Propagation of singularities and semiglobal existence theorems for (pseudo-) differential operators of principal type, Ann. Math., 108, No. 3 (1978), 569-609.
16. The Weyl calculus of pseudo-differential operators, Commun. Pure Appl. Math., 32 (1979), 1-25.
17. Spectral analysis of singularities. Seminar on singularities of solutions of linear partial differential equations, Ann. Math. Stud., No. 91 (1979), 3-49.
18. Subelliptic operators, Ann. Math. Stud., No. 91 (1979), 127-208.

Ivrii, V. Ya.:

1. On wave fronts of solutions of certain pseudo-differential equations, Funktsional. Analiz Ego Prilozhen., 10, No. 2 (1976), 71-72 (Russian).

Kannai, Y.:

1. An unsolvable hypoelliptic differential operator, Isr. Math. J., 9 (1979), 306-315.

Kashiwara, M., and Kawai, T.:

1. Microhyperbolic Pseudo-Differential Equations, Lecture Notes, No. 287, Springer (1973).

Kato, Y.:

1. On a class of nonelliptic boundary problems, Nagoya Math. J., 54 (1974), 7-20.

Kawai, T.:

1. Construction of a local elementary solution for linear partial differential operators, Proc. Jpn. Acad., 47 (1971), 19-23.

Kohn. J. J.:

1. Harmonic integrals on strongly pseudoconvex manifolds, I, Ann. Math., 78 (1963), 112-148; II, 79 (1964), 450-472.
2. Boundaries of complex manifolds, Proceedings of the Conference on Complex Analysis, Springer, Berlin (1965), 113-119.
3. Boundary behavior of $\bar{\partial}$ on weakly pseudoconvex manifolds of dimension 2, J. Differ. Geom., 6 (1972), 523-542.

4. Global regularity for $\bar{\partial}$ on weakly pseudoconvex manifolds, Trans. Am. Math. Soc., 181 (1973), 273-292.
5. Subellipticity of the $\bar{\partial}$-Neumann problem on pseudoconvex domains: sufficient conditions, Acta Math., 142 (1979), 79-122.

Kohn, J. J., and Nirenberg, L.:

1. An algebra of pseudo-differential operators, Commun. Pure Appl. Math., 18 (1965), 269-305.

Kolmogorov, A. N., and Fomin, C. V.:

1. Elements of the Theory of Functions and of Functional Analysis, Nauka, Moscow (1976) (Russian).

Kumano-go, H.:

1. Pseudo-differential operators and the uniqueness of the Cauchy problem, Commun. Pure Appl. Math., 22 (1969), 73-129.
2. Algebras of pseudo-differential operators, J. Fac. Sci. Univ. Tokyo, Sec. 1A, 17 (1970), 31-50.
3. Pseudo-differential operators of multiple symbol and the Calderón-Vaillancourt theorem, J. Math. Soc. Jpn., 27, No. 1 (1975), 113-119.

Lax, P. D.:

1. Asymptotic solutions of oscillatory initial value problems, Duke Math. J., 24 (1957), 627-646.

Lax, P. D., and Nirenberg, L.:

1. On stability of difference schemes; a sharp form of Gårding's inequality, Commun. Pure Appl. Math., 19, No. 4 (1966), 473-492.

Leray, J.:

1. Probleme de Cauchy, Bull. Soc. Math., I, 85 (1957), 389-430; II, 86 (1958), 75-96; III, 87 (1959), 81-179.
2. Solutions asymptotiques et groupe symplectique; Fourier integral operators and partial differential equations; Lect. Notes Math., 459 (1975), 473-497.

Lewy, H.:

1. An example of a smooth linear partial differential equation without solution, Ann. Math., 66 (1957), 155-158.

Lions, J., and Magenes, E.:

1. Problemes aux limites non homogenes et applications, Vol. 1, Dunod, Paris (1968).

Lopatinski, Ya. B.:

1. On a method of reducing boundary problems for a system of differential equations of elliptic type to regular equations, Ukr. Mat. Zh., 5, No. 2 (1953), 123-151.

Ludwig, D.:

1. Exact and asymptotic solutions of the Cauchy problem, Commun. Pure Appl. Math., 13 (1960), 473-508.

Malgrange, B.:

1. Equations aux derivees partielles a coefficients constants, I, Solution elementaire, C. R. Acad. Sci., 237, No. 25 (1953), 1620-1622.
2. Existence et approximation des solutions des equations aux derivees partielles et des equations de convolution, Ann. Inst. Fourier Grenoble, 6 (1956), 271-355.
3. Unicité du probleme de Cauchy, Seminaire Schwartz, No. 60 (1959-1960), Lectures 8-10.

Malyutov, M. B.:

1. Poincaré's boundary problem, Tr. Mosk. Mat. Ova., 20 (1969), 173-203 (Russian).

Markushevich, A. I.:

1. Theory of Analytic Functions, Vol. 2, Nauka, Moscow (1968) (Russian).

Maslov, V. P.:

1. Theory of Perturbations and Asymptotic Methods, Moscow State University, Moscow (1965) (Russian).
2. Operator Methods, Nauka, Moscow (1973) (Russian).

Maslov, V. P., and Fedoryuk, M. V.:

1. Semiclassical Approximation for Equations of Quantum Mechanics, Nauka, Moscow (1976) (Russian).

Mazya, V. G.:

1. On a degenerate oblique derivative problem, Mat. Sb., 87 (1972), 417-454 (Russian).

Mazya, V. G., and Paneyakh, V. P.:

1. Degenerate elliptic pseudo-differential operators and the oblique derivative problem, Tr. Mosk. Mat. Ova., 31 (1974), 237-255.
2. Coercive estimates and regularity of the solutions of degenerate elliptic equations, Funktsional. Analiz Ego Prilozhen., 4, No. 4 (1970), 41-56.

Melin, A.:

1. Lower bounds for pseudo-differential operators, Arkiv. Mat., 9, No. 1 (1971), 117-140.

Melin, A., and Sjostrand, J.:

1. Fourier integral operators with complex phase functions and parametrix for an interior boundary value problem, Commun. Partial Differ. Eq., 1 (1976), 313-400.

Melrose, R.:

1. Equivalence of glancing hypersurfaces, Invent. Math., 37 (1976), 165-191.
2. Differential boundary value problems of principal type: in Seminar on Singularities of Solutions of Linear Partial Differential Equations, L. Hörmander, ed., Princeton Univ. Press (1979), 81-112.
3. Transformation methods for boundary value problems, Singularities in Boundary Value Problems (1981), 131-168.

Melrose, R., and Sjostrand, J.:

1. Singularities of boundary value problems, I, Commun. Pure Appl. Math., 31 (1978), 593-617.

Mikhlin, S. G.:

1. Many-Dimensional Singular Integrals and Integral Equations, Fizmatgiz, Moscow (1962) (Russian).

Miranda, C.:

1. Equazioni alle Derivate Parziali di Tipo Ellittico, Springer, Berlin (1955).

Mishchenko, A. S., Sternin, B. Yu., and Shatalov, V. E.:

1. Lagrangian Manifolds and the Canonical Operator Method, Nauka, Moscow (1978) (Russian).

Mizohata, S.:

1. The Theory of Partial Differential Equations, Mir, Moscow (1977) (Russian translation).

Morrey, C., and Nirenberg, L.:

1. On the analyticity of the solutions of linear elliptic systems of partial differential equations, Commun. Pure Appl. Math., 10 (1957), 271-290.

Nazaikinski, V. E., Oshmyan, V. G., Sternin, B. Yu., and Shatalov, V. E.:

1. Fourier integral operators and the canonical operator, Usp. Mat. Nauk, 36, No. 2 (1981), 81-140 (Russian).

Neri, U.:

1. Singular integral operators on manifolds, Proc. Symp. Pure Math., 10 (1967), 232-242.

Nirenberg, L.:

1. Pseudo-differential operators, Proc. Symp. Pure Math., 16 (1970), 149-167.
2. Lectures on linear partial differential equations, Am. Math. Soc. Regional Conference Series, No. 17 (1973), 1-58.
3. On elliptic partial differential equations, Ann. Sc. Norm. Super. Pisa, 13 (1959), 115-162.

Nirenberg, L., and Trèves, F.:

1. Solvability of a first-order linear partial differential equation, Commun. Pure Appl. Math., 14 (1963), 331-351.
2. On local solvability of linear partial differential equations, Commun. Pure Appl. Math., 23 (1970), I, Necessary conditions, 1-38; II, Sufficient conditions, 499-510.
3. Remarks on the solvability of linear equations of evolution, Symp. Math., 7 (1971), 325-338.

Oleinik, O. A., and Radkevich, E. V.:

1. Second-Order Equations with Non-Negative Characteristic Form, Math. Anal. 1969, VINITI, Moscow (1971) (Russian).

Ovsyannikov, L. V.:

1. Singular operator in the scale of Banach spaces, Dokl. Akad. Nauk SSSR, 163, No. 4 (1965), 819-822 (Russian).

Palais, R. S.:

1. Seminar on the Atiyah-Singer Index Theorem, Princeton University Press (1965).

Palamodov, V. P:

1. Linear Differential Operators with Constant Coefficients, Fizmatgiz, Moscow (1967) (Russian).

Paneyakh, B. P.:

1. Constant force pseudo-differential operators in the large, Mat. Sb., 73, No. 2 (1967), 204-226 (Russian).

Peetre, J.:

1. Another approach to elliptic boundary problems, Commun. Pure Appl. Math., 14, No. 4 (1961), 711-731.

Petrowsky, I. G.:

1. Uber das Cauchysche Problem fur Systeme von partieller Differentialgleichungen, Mat. Sb., 2 (44) (1937), 815-870.
2. Uber das Cauchysche Probleme fur ein System linearer partieller Differentialgleichungen im Gebiete der nichtanalytischen Funktionen, Bull. Univ. Moscow, Ser. Int., 1, No. 7 (1938), 1-72.
3. Sur l'analyticite des solutions des systemes d'equations differentielles, Mat. Sb., 5 (47) (1939), 3-70.
4. On some problems of the theory of partial differential equations, Usp. Mat. Nauk, 1, No. 3-4 (1946), 44-70 (Russian).
5. Lectures on Partial Differential Equations, Fizmatgiz, Moscow (1961) (Russian).

Plis, A.:

1. A smooth linear elliptic differential equation without any solution in a sphere, Commun. Pure Appl. Math., 14 (1961), 599-617.
2. Unique continuation theorems for solutions of partial differential equations, Proc. Intern. Congr. Math., Stockholm (1962).

Popivanov, P. R.:

1. On local solvability of a class of pseudo-differential equations with double characteristics, Tr. Sem. im. I. G. Petrovskogo, 1 (1975), 237-278 (Russian).
2. Elements of the symplectic differential geometry and partial differential equations, Akad. Wiss. DDR, Berlin (1977).

Radkevich, E. V.:

1. On an estimate of L. Hörmander, Usp. Mat. Nauk, 24, No. 2 (1969), 233-234 (Russian).
2. Shauder type estimate for a certain class of pseudo-differential operators, Usp. Mat. Nauk, 24 (1969), 199-200 (Russian).

Riesz F., and Sz.-Nagy, B.:

1. Lécons d'Analyse Fonctionelle, Akad. Kiado, Budapest (1972).

Sato, M.:

1. Hyperfunctions and partial differential equations, Conf. on Funct. Analysis and Related Topics, Tokyo (1969), 91-94.
2. Regularity of hyperfunction solutions of partial differential equations, Actes Cong. Int. Math., Nice (1970), Vol. 2, pp. 785-794.

Schwartz, L.:

1. Theorie des Distributions, I-II, Hermann, Paris (1950).
2. Methodes Mathematiques pour les Sciences Physiques, Hermann, Paris (1961).

Seeley, R. T.:

1. The resolvent of an elliptic boundary problem, Am. J. Math., 91 (1969), 889-920.
2. Integrodifferential operators on vector bundles, Trans. Am. Math. Soc., 117 (1965), 167-204.
3. Analytic extension of the trace associated with elliptic boundary problems, Am. J. Math., 91 (1969), 969-983.
4. Refinement of the fundamental calculus of Calderón and Zygmund, Proc. Acad. Wet. Ned., Ser. A, 68 (1965), 521-531.

Shabat, B. V.:

1. An Introduction to Complex Analysis, Part II, Nauka, Moscow (1976) (Russian).

Shapira, P.:

1. Une equation aux derivees partielles sans solutions dans l'espace des hyperfunctions, C. R. Acad. Sci., 265 (1967), 665-667.

Shapiro, Z. Ya.:

1. On general boundary problems for equations of elliptic type, Izv. Akad. Nauk SSSR, Ser. Math., 17 (1953), 539-562 (Russian).

Shechter, M.:

1. Various types of boundary conditions for elliptic equations, Commun. Pure Appl. Math., 13 (1960), 407-425.

Shilov, G. E.:

1. Local properties of the solutions of partial differential equations with constant coefficients, Usp. Mat. Nauk, 14, No. 5 (1955), 3-44 (Russian).

Shubin, M. A.:

1. Pseudo-Differential Operators and Spectral Theory, Nauka, Moscow (1978) (Russian).

Sjostrand, J.:

1. Operators of principal type with interior boundary conditions, Univ. Lund (1972).

Slobodetsky, L. N.:

1. Sobolev's generalized spaces and their application to boundary problems for partial differential equations, Uch. Zap. Leningr. Ped. Inst., 197 (1958), 54-112 (Russian).

Smith, K. T.:

1. Some remarks on a paper of Calderón on existence and uniqueness theorems for systems of partial differential equations, Commun. Pure Appl. Math., 18 (1965), 415-441.

Sobolev, S. L.:

1. Some Applications of Functional Analysis in Mathematical Physics, Leningrad State University, Leningrad (1950) (Russian).
2. An Introduction to the Theory of Cubic Formulas, Nauka, Moscow (1974) (Russian).

Stein, E. M.:

1. Singular Integrals and Differentiability Properties of Functions, Princeton University Press (1970).

Sternin, B. Yu.:

1. On the microlocal structure of differential operators in a neighborhood of the stationary point, Usp. Mat. Nauk, 32, No. 6 (1977), 235-236 (Russian).

Strauss, M., and Trèves, F.:

1. First-order linear PDE's and uniqueness in the Cauchy problem, J. Differ. Eq., 15 (1974), 195-209.

Suzuki, H.:

1. Improving estimates for differential operators in two independent variables, Publ. RIMS, Kyoto Univ., 5 (1970), 287-299.

Sweeney, W.:

1. A condition for subellipticity in Spencer's Neumann problem, J. Diff. Eq., 21, No. 2 (1976), 316-362.

Taira, K.:

1. On some degenerate oblique derivative problems, J. Fac. Sci. Univ. Tokyo, Sec. IA, 23 (1976), 259-287.
2. On some noncoercive boundary value problems for the Laplacian, J. Fac. Sci. Univ. Tokyo, Sec. IA, 23 (1976), 343-367.
3. Sur le probleme de derivees oblique, I, J. Math. Pures Appl., 57 (1978), 379-395; II, Ark. Mat., 17, No. 2 (1979), 177-191.
4. Sur l'existence de process de diffusion, Ann. Inst. Fourier, 29, No. 4 (1979), 99-126.

Tartakoff, D. S.:

1. Gevrey hypoellipticity for subelliptic boundary value problems, Commun. Pure Appl. Math., 26 (1973), 251-312.
2. Local analytic hypoellipticity for $\Box_b$ on nondegenerate Cauchy-Riemann manifolds, Proc. Natl., Acad. Sci. USA, 75 (1978), 3027.

Taylor, M.:

1. Pseudo-Differential Operators, Lecture Notes in Math., 416, Springer, New York (1974).
2. Reflection of singularities of solutions to systems of differential equations, Commun. Pure Appl. Math., 28 (1975), 457-478.
3. Propagation, reflection, and diffraction of singularities of solutions to wave equation, Bull. Am. Math. Soc., 84 (1978), 589-611.
4. Fourier integral operators and harmonic analysis on compact manifold, Proc. Symp. Pure Math., 35 (1979), 115-136.
5. Pseudo-Differential Operators, Princeton Univ. Press (1981).

Trèves, F.:

1. Lectures on linear partial differential equations with constant coefficients, Rio de Janeiro (1961).
2. Ovsyannikov theorem and hyperdifferential operators, Notas de Matematica, No. 46, Rio de Janeiro (1968).
3. On the local solvability of linear partial differential equations in two independent variables, Am. J. Math., 52, No. 1 (1970), 174-204.
4. On local solvability of linear partial differential equations, Bull. Am. Math. Soc., 76, No. 3 (1970), 552-571.
5. Linear Partial Differential Equations, Gordon and Breach, New York (1970).
6. Hypoelliptic equations of principal type, with analytic coefficients, Commun. Pure Appl. Math., 23 (1970), 637-651.
7. A new method of proof of the subelliptic estimates, Commun. Pure Appl. Math., 24 (1971), 71-115.
8. Hypoelliptic partial differential equations of principal type, sufficient conditions and necessary conditions, Commun. Pure Appl. Math., 24 (1971), 631-670.
9. A link between solvability of pseudo-differential equations and uniqueness in the Cauchy problem, Am. J. Math., 94, No. 1 (1972), 267-288.

10. Existence de regularite des solutions des equations aux derivees partielles lineaires, quelques resultats et quelques problems ouverts, Sem. Goulaouic-Schwartz, Paris (1973), Exp. No. XXII.
11. Basic Linear Partial Differential Equations, Academic Press, New York (1975).

Unterberger, A., and Bokobza, J.:

1. Les operateurs de Calderón-Zygmund precises, C. R. Acad. Sci., 259 (1964), 1612-1614.

Vaillancourt, R.:

1. A simple proof of the Lax-Nirenberg theorem, Commun. Pure Appl. Math., 23 (1970), 151-163.

Vishik, M. I., and Grishin, V. V.:

1. On a class of degenerate elliptic equations of higher order, Mat. Sb., 79 (1969), 3-36 (Russian).
2. Degenerate elliptic differential and pseudo-differential operators, Usp. Mat. Nauk, 25, No. 4 (1970), 29-56 (Russian).

Vladimirov, V. S.:

1. Equations of Mathematical Physics, 3rd edn., Nauka, Moscow (1976) (Russian).
2. Generalized Functions in Mathematical Physics, Nauka, Moscow (1979) (Russian).

Volevich, L. R.:

1. Hypoelliptic equations in convolutes, Dokl. Akad. Nauk SSSR, 168, No. 6 (1966), 1232-1235 (Russian).
2. Pseudo-differential operators with holomorphic symbols and the Gevrey classes, Tr. Mosk. Mat. Ova., 24 (1971), 43-68 (Russian).

Volevich L. R., and Paneyakh, B. P.:

1. Certain spaces of generalized functions and of the imbedding theorem, Usp. Mat. Nauk, 20, No. 1 (1965), 3-74 (Russian).

Weinstein, A.:

1. Symplectic manifolds and their Lagrangian submanifolds, Adv. Math., 6 (1971), 329-346.

Weyl, H.:

1. Gruppentheorie and Quantenmechanik, Verlag von S. Hirzel, Leipzig (1928).

Winzell, B.:

1. Solutions of second-order elliptic pseudo-differential equations with prescribed directional derivative on the boundary, Dissertation, Linkoping (1975).

Yosida, K.:

1. Functional Analysis, Springer, Berlin (1965).

Zachmanoglou, E. C.:

1. Uniqueness of the Cauchy problem when the initial surface contains characteristic points, Arch. Rat. Mech. Math. Anal., 23 (1966), 317-326.
2. Nonuniqueness of the Cauchy problem for linear partial differential equations with variable coefficients, Arch. Rat. Mech. Math. Anal., 27 (1968), 373-384.

Zerner, M.:

1. Solutions singulieres d'equations aux derivees partielles, Bull. Soc. Math., 91 (1963), 203-226.